# Challenges and Advances in Computational Chemistry and Physics

Volume 23

**Series editor**

Jerzy Leszczynski
Department of Chemistry and Biochemistry
Jackson State University, Jackson, MS, USA

This book series provides reviews on the most recent developments in computational chemistry and physics. It covers both the method developments and their applications. Each volume consists of chapters devoted to the one research area. The series highlights the most notable advances in applications of the computational methods. The volumes include nanotechnology, material sciences, molecular biology, structures and bonding in molecular complexes, and atmospheric chemistry. The authors are recruited from among the most prominent researchers in their research areas. As computational chemistry and physics is one of the most rapidly advancing scientific areas such timely overviews are desired by chemists, physicists, molecular biologists and material scientists. The books are intended for graduate students and researchers.

More information about this series at http://www.springer.com/series/6918

Minh Tho Nguyen · Boggavarapu Kiran
Editors

# Clusters

## Structure, Bonding and Reactivity

*Editors*
Minh Tho Nguyen
Department of Chemistry
KU Leuven
Leuven
Belgium

Boggavarapu Kiran
Department of Chemistry and Physics
McNeese State University
Lake Charles, LA
USA

ISSN 2542-4491 ISSN 2542-4483 (electronic)
Challenges and Advances in Computational Chemistry and Physics
ISBN 978-3-319-48916-2 ISBN 978-3-319-48918-6 (eBook)
DOI 10.1007/978-3-319-48918-6

Library of Congress Control Number: 2016957864



Printed on acid-free paper

This Springer imprint is published by Springer Nature
The registered company is Springer International Publishing AG
The registered company address is: Gewerbestrasse 11, 6330 Cham, Switzerland

# Preface

Nanochemistry, a branch of nanoscience, is focused on building nanoscale systems (1–100 nm) mainly using atom or molecule as building block. Nanochemistry brings together multiple disciplines using physical methods and chemical theories and experiments to investigate properties of materials at the nanoscale. Of the latter, the nanoclusters continue to attract much interest due not only to their intriguing physicochemical features but also to a manifold of potential industrial applications.

What are atomic clusters? A cluster is first defined as a finite group of atoms that are held together mainly, or at least to a significant extent, by chemical interactions directly between the atoms. The concept of cluster can also be used for characterizing molecular complexes as they are formed in plasmas by clusterification processes.

Clusters $M_N$ are thus aggregates of a finite number N of atoms or molecules M, thus, bridging the gap between the isolated atom/molecule and the macroscopic solid state of matter. Depending on the methodology and experimental techniques, clusters and their aggregation can either be formed in gas or liquid phase or be deposited on solid surfaces. The continuous variations in properties of clusters in going from a small number of atoms to a larger scale is a consequence of the quantum size regime in which each additional atom makes a unique difference in the properties. In addition, the cluster growth mechanism allows the transition from atoms to clusters and to bulk materials to be understood (bottom-up approach) in a systematic way.

Cluster science has deep roots in both physics and chemistry, but also has unique features making it relevant to many fields including materials science, environment, biology, medicine, etc. Cluster science is thus regarded as a bridge across many scientific disciplines. The stabilities and properties of clusters depend on both the nature of constituent elements and their interactions and sizes. The well-known carbon buckyball $C_{60}$ is a cluster having high symmetry and thermodynamic stability whereas that of the element next to the carbon, the boron buckyball $B_{80}$, is not stable in the high symmetry form. In addition, combination of various types of elements results in a large number of mixed clusters. Binary clusters composed of two elements are of particular interest due to the high propensity for control, tuning properties and a wide range of potential applications. Depending on the atoms ratio in a binary cluster the elements can be classified as host (having larger percentage) or dopant (having smaller percentage). Certain clusters that are referred to as superatoms, have a *magic number*, both in terms of electronic structure and atomic number, exhibit high thermodynamic stability and can be used as potential building blocks for cluster-assembled materials. The latter are currently of great interest as they take advantage of novel properties of clusters and make them accessible in bulk materials. Cluster-based nanostructures are promising to become pivotal for a new generation of catalysts for chemical reactions, semiconducting materials and electronic devices with tailored properties.

In recent years, a large number of papers have been published on elemental clusters. Stability, structural and electronic characteristics of clusters generated from different experimental techniques have been interpreted with the help of theoretical models, or by quantum chemical computations. Bottlenecks still remain in the prediction of their geometric and electronic structures, especially for clusters with a large number of atoms. While concepts and rules for rationalizing and predicting structural motifs, properties and reactivities of different classes of organic and inorganic compounds have long reached maturity, relatively less has been formulated for atomic clusters. In this context, much effort has been devoted to the establishment of a set of general rules for nanoclusters using quantum chemical computations and theoretical models.

Although a number of excellent reviews and books on elemental clusters are available in the literature, it appears to us that there is a need for an account on recent developments in the field. This is the purpose of the present volume with ten chapters that have been written by experts and active researchers in the field.

This collection begins with Chapter "Global Optimisation Strategies for Nanoalloys" by Heard and Johnston describing the efforts involved in finding the global minima for nanoalloys. Identifying the most stable configuration state of a given system is vital for understanding its properties. The authors discuss the advantages and shortcomings of various methods of global optimizations. In addition, the authors draw our attention to more complex environments such as surface deposition and ligand passivation on which little attention has been paid.

The following two chapters are devoted to the characterization of atomic clusters using spectroscopic techniques. Li, Fielicke, Lievens and Janssens report in Chapter "Structural Identification of Doped Silicon Clusters" a critical review of the current state-of-the-art mass, infrared, photoelectron and X-ray and magnetic circular dichroism spectrometric methods in experimental characterizations of pure and doped silicon clusters. In Chapter "Structural Evolution, Vibrational Signatures and Energetics of Niobium Clusters from $Nb_2$ to $Nb_{20}$", Nhat, Majumdar, Leszczynski and Nguyen analyze the determination of structures using experimental vibrational spectra with the help of density functional theory computations for a series of small pure niobium clusters. The results point out how a combined experimental and theoretical investigation can allow the structural motifs of clusters to be determined with confidence.

Falvo analyzes in Chapter "Submersion Kinetics of Ionized Impurities into Helium Droplets by Ring-Polymer Molecular Dynamics Simulations" the characteristics of the clusters of the lightest rare gas, the helium droplets that provide a unique medium sensitive to the weakest interactions. The kinetics of the submersion of ionized alkali dopants into helium droplets are investigated using ring polymer molecular dynamics simulations. The interplay between experiment and theory on this challenging class of compounds is also discussed.

Next in Chapter "Structure, Stability and Electron Counting Rules in Transition Metal Encapsulated Silicon and Germanium Clusters", Sen discusses in depth the structures and stability of transition metal encapsulated silicon and germanium clusters. Arguably silicon is still the most important element in the semiconductor industry. It has been known for some time that pure silicon or germanium clusters are unstable and not suitable for rational design of materials and applications. However, TM-doped Si/Ge clusters showed enhanced stability over a certain size range. In this chapter a comprehensive review of this field is provided in order to answer some fundamental questions on the clusters' stability and the use of electron counting rules. This issue is pursued further in Chapter "Transition Metal Doped Boron Clusters: Structure and Bonding of $B_nM_2$ Cycles and Tubes" in which Pham and Nguyen attempt to rationalize the structural motifs and growth patterns of transition metal doped boron clusters. Orbital interactions and partitions of the total electron density into basins are extensively used to probe the chemical bonding phenomena. Establishment of simple electron count rules to understand and predict shapes of clusters remains a challenging subject for future research.

In Chapter "Silicate Nanoclusters: Understanding Their Cosmic Relevance from Bottom-Up Modelling", Bromley discusses in detail how the bottom-up computational modelling provides fundamental insights into one of the most abundant materials on earth and in cosmos as well. Despite the importance and long history of silicates, several important questions remain unanswered. How the silicon suboxide clusters transform to basic units seen in bulk, and what is the significance of the incorporation of metals such as Mg and Fe in silicate nucleation process and subsequently in the formation of silicate minerals, to name a few. This chapter highlights the importance of atomic-scale computational modelling in the understanding of the transition from sub-nano to nano to bulk.

In Chapter "Magnetic Anisotropy Energy of Transition Metal Alloy Clusters", Hoque, Baruah, Reveles and Zope examine the magnetic properties of transition metal alloy clusters such as the $As@Ni_{12}@As_{20}$ cluster. Transition metal clusters often show high spin moments but generally are also reactive with the environment. The doping effect leads to a passivation of the surface atoms stabilizing the doped clusters. In particular, the anisotropy energy is computed using density functional theory. As the growth patterns from atoms to bulk state remain a subject of current and intensive debate, the formation of baby-crystals from clusters is of great interest, and this issue is treated in Chapter "Growth Pattern and Size-Dependent Properties of Lead Chalcogenide Nanoclusters" by Gill, Sawyer, Salavitabar, Kiran and Kandalam. The structural evolution of lead chalcogenide $(PbX)_n$ ($X$= S, Se, and Te; $n = 1 - 32$) nanoclusters and how various properties of these clusters vary with increasing cluster size are investigated using different experimental techniques and computational (DFT-based) studies are also discussed in detail. The importance of the synergy between computation and experiment is again demonstrated.

Finally, one of the most promising applications of elemental clusters is the field of catalysis of chemical reactions. Lang and Bernhardt provide in Chapter "Chemical Reactivity and Catalytic Properties of Binary Gold Clusters: Atom by Atom Tuning in a Gas Phase Approach" insights into the reaction mechanisms, energetics, and kinetics of the catalytic processes of free clusters. Free clusters in the gas phase represent simplified but suitable model systems which allow insight into catalytic processes to be obtained on a rigorously molecular level. Experimental and theoretical studies are described on the reactivity and catalytic activity of free gold clusters and the change of their chemical properties induced by doping these clusters with transition metal atoms. Three catalytic reactions selected include the oxidation of carbon monoxide, the conversion of methane, and the coupling of methane and ammonia.

We hope that this monograph will prove to be a useful source of information for researchers on the current experimental techniques and theoretical methods and also results on different specific classes of elemental clusters.

We are grateful to the authors for their contributions and their patience. We sincerely thank the editorial staff of Springer, in particular Dr. Karin De Bie and Dr. Sathya Karuppaiya, for their efficient assistance. Finally, we would express our gratitude toward our spouses, Mai Phuong Le and Nallini Persaud, for their constant support.

Leuven, Belgium — Minh Tho Nguyen
Lake Charles, LA, USA — Boggavarapu Kiran

# Contents

# Contributors

**Tunna Baruah** Department of Physics, University of Texas El Paso, El Paso, TX, USA

**Thorsten M. Bernhardt** Institute of Surface Chemistry and Catalysis, Ulm University, Ulm, Germany

**Stefan T. Bromley** Departament de Ciencia de Materials i Química Física and Institut de Química Teòrica I Computacional (IQTCUB), Universitat de Barcelona, Barcelona, Spain; Institució Catalana de Recerca i Estudis Avançats (ICREA), Barcelona, Spain

**F. Calvo** University of Grenoble-Alpes and CNRS, LIPHy, Grenoble, France

**André Fielicke** Institut für Optik und Atomare Physik, Technische Universität Berlin, Berlin, Germany

**Ann F. Gill** Department of Physics, West Chester University, West Chester, PA, USA

**Christopher J. Heard** Department of Physics, Chalmers University of Technology, Gothenburg, Sweden

**Nabil M.R. Hoque** Department of Physics, University of Texas El Paso, El Paso, TX, USA

**Ewald Janssens** Laboratory of Solid State Physics and Magnetism, KU Leuven, Leuven, Belgium

**Roy L. Johnston** School of Chemistry, University of Birmingham, Edgbaston, Birmingham, UK

**Anil K. Kandalam** Department of Physics, West Chester University, West Chester, PA, USA

**Boggavarapu Kiran** Department of Chemistry, McNeese State University, Lake Charles, LA, USA

**Sandra M. Lang** Institute of Surface Chemistry and Catalysis, Ulm University, Ulm, Germany

**Jerzy Leszczynski** Interdisciplinary Center for Nanotoxicity, Department of Chemistry and Biochemistry, Jackson State University, Jackson, MS, USA

**Yejun Li** Laboratory of Solid State Physics and Magnetism, KU Leuven, Leuven, Belgium

**Peter Lievens** Laboratory of Solid State Physics and Magnetism, KU Leuven, Leuven, Belgium

**Devashis Majumdar** Interdisciplinary Center for Nanotoxicity, Department of Chemistry and Biochemistry, Jackson State University, Jackson, MS, USA

**Minh Tho Nguyen** Department of Chemistry, KU Leuven, Leuven, Belgium

**Pham Vu Nhat** Department of Chemistry, Can Tho University, Can Tho, Vietnam

**Hung Tan Pham** Institute for Computational Science and Technology (ICST), Ho Chi Minh City, Vietnam

**Kamron Salavitabar** Department of Physics, West Chester University, West Chester, PA, USA

**William H. Sawyer** Department of Physics, West Chester University, West Chester, PA, USA

**Prasenjit Sen** Harish-Chandra Research Institute, Jhunsi, Allahabad, India

**J. Ulises Reveles** Department of Physics, University of Texas El Paso, El Paso, TX, USA; Department of Physics, Virginia Commonwealth University, Richmond, VA, USA

**Rajendra R. Zope** Department of Physics, University of Texas El Paso, El Paso, TX, USA

# Global Optimisation Strategies for Nanoalloys

**Christopher J. Heard and Roy L. Johnston**

**Abstract** The computational prediction of thermodynamically stable metal cluster structures has developed into a sophisticated and successful field of research. To this end, research groups have developed, combined and improved algorithms for the location of energetically low-lying structures of unitary and alloy clusters containing several metallic species. In this chapter, we review the methods by which global optimisation is performed on metallic alloy clusters, with a focus on binary nanoalloys, over a broad range of cluster sizes. Case studies are presented, in particular for noble metal and coinage metal nanoalloys. The optimisation of chemical ordering patterns is discussed, including several novel strategies for locating low-energy permutational isomers of fixed cluster geometries. More advanced simulation scenarios, such as ligand-passivated, and surface-deposited clusters have been developed in recent years, in order to bridge the gap between isolated, bare clusters, and the situation observed under experimental conditions. We summarise these developments and consider the developments necessary to improve binary cluster global optimisation in the near future.

**Keywords** Nanoalloys · Global optimisation · Nature-inspired algorithms · Basin hopping · Coinage · Metals · Noble metals · Cluster deposition · Passivated clusters · Binary clusters

## 1 Introduction

In this chapter, we aim to cover the broad topic of computational global optimisation (GO) of clusters, particularly as it pertains to nanoalloys, which are clusters composed of two or more metallic elements, and vary in size from a few atoms to many

---

C.J. Heard (✉)
Department of Physics, Chalmers University of Technology, Gothenburg, Sweden
e-mail: heard@chalmers.se

R.L. Johnston
School of Chemistry, University of Birmingham, Edgbaston, Birmingham, UK



M.T. Nguyen and B. Kiran (eds.), *Clusters*, Challenges and Advances in Computational Chemistry and Physics 23,
DOI 10.1007/978-3-319-48918-6_1

millions [1]. The chapter is therefore organised into several sections which discuss the basics of GO, why it has become such an invaluable tool in chemistry and physics, and the various methods which have been applied, with a focus on metal particles. After introducing the computational techniques, we deal with the additional factors which must be considered to tackle the complexity of multicomponent clusters including the distinct optimisation of chemical ordering in nanoalloys. We follow this section with a discussion of combining and comparing GO methods, in essence, the attempt to optimise the optimisation process. The next section contains examples of studies which have binary metal clusters as their focus, where we wish to draw particular attention to investigations which consider trends over multiple compositions, or those which are explicitly concerned with chemical ordering. The final sections consider the extension of GO methods to more complex environments, including surface deposition, ligand passivation and ternary and higher order alloys.

The main focus of this chapter is direct, unbiased GO, which includes in its definition all methods which are automated, aim for the lowest energy cluster structure and require no human input once begun. However, multiple step optimisation, such as the subsequent reoptimisation of structures found at one level of theory at a higher level, must be given significant attention, as these methods make up the majority of computational structural studies at present, owing primarily to practical limitations of computing resources. We aim to make clear the strengths and weaknesses of such approaches, and will highlight which types of GO were used for the examples we discuss.

## *1.1 Global Optimisation*

Computational GO is so ubiquitous a concept in modern cluster physics and structural chemistry in general, that it scarcely needs promoting. This situation has been reached by a confluence of two major factors. First, the central importance of locating stable, low energy structures, as a basis for further examination. This allows the researcher to approach answers to the important question, “what configurational state is my system of interest most likely to be in”. Secondly, the profound success of computational GO methods in achieving this goal. Clusters, as a special case of nanoscale materials, constitute a bridge between the atomic and bulk regimes. In each situation in which they are found, the structure of the cluster is both a vital piece of information governing its properties, and a difficult one to ascertain with experimental methods. Until experiments can cheaply and reliably determine the geometry of a cluster in its native environment, with atomic resolution, and without affecting the structure, GO methods will play an important role in understanding cluster behaviour.

In addition to the structures of the $n$ most stable configurations a cluster may attain, a whole wealth of relevant information on rearrangement probabilities, electronic structure, dynamic and thermodynamic properties neatly follow from such a computational exploration of configurational space. However, we cannot hope to delve into these topics in any detail in one chapter, and so we restrict our discussion to exclude any structure sampling method which does not have as its end goal the determination of an optimal cluster structure. For in-depth studies on these topics, we direct the reader towards Wales' book [2], the review of Baletto and Ferrando [3] and the recent perspective by Calvo [4].

It is more than a matter of definition to divide a good GO strategy into its salient features, and some insight may be gained from defining these features. A good method, regardless of the subject under study, must balance the local and the global aspects of searching. Local searching is concerned with the sampling of individual "basins" which belong to a class of related structures, and the determination of the locally optimal structure within such a basin. Global searching, however, is tasked with exploring sufficiently diverse regions of configurational space, across multi-minimum regions, so-called superbasins (or alternatively, funnels [5, 6]), to capture all the energetically relevant basins. For the former, many highly efficient local optimisation routines have been developed [7, 8], which are robust and largely system-independent, and are described and compared elsewhere [9, 10]. The latter is the major concern of a GO strategy, particularly in the case of a cluster, for which there are often many superbasins, containing a vast number of minima, many of which are comparable in energy. For monometallic clusters, the omnidirectionally of bonding, and the high coordination of atoms promotes a complex energy landscape [2], while nanoalloys additionally introduce a combinatorial increase in the number of minima, according to the permutation of individual elements. It has been estimated that the number of minima upon the landscape grows exponentially with the number of atoms ($N$) [11, 12]. The exact value of the exponent varies with the system, but is often predicted to lie between $N$ and $2N$. Permutational isomers, or homotops, then add a binomial factor into the total number of energetically distinct minima.

$$n_{min} = \frac{N!}{N_A!(N - N_A!)} e^{\alpha N} \tag{1}$$

where $N_A$ is the number of atoms of type $A$ in the $N$ atom cluster, and $\alpha$ is an unknown exponent, generally between 1 and 2.

The problem of finding an appropriate GO for nanoalloys is thus reduced in practice to one which effectively searches a sufficient breadth of the landscape in a reasonable time, and can handle permutational isomers efficiently. The preferred chemical ordering of a binary cluster is primarily a balance between homo- and hetero-metallic cohesive energies, atomic radii and relative surface energies. As a result, a number of classes of common chemical ordering patterns are observed: the randomly mixed cluster, the core-shell segregated (and higher order derivatives, such as the onion layered core-shell) and the fully segregated, or "Janus particle" types, as depicted in Fig. 1. Algorithmic developments have been made to exploit these common motifs, by guiding the search along trajectories which vary the chemical ordering, which will be discussed in Sect. 5.

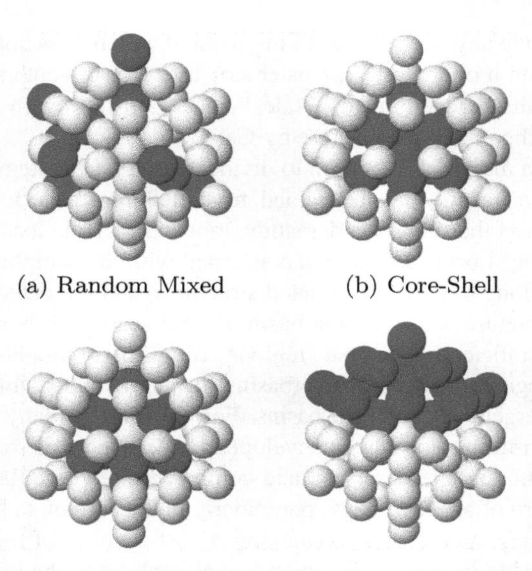

(a) Random Mixed (b) Core-Shell

(c) Onion-Layered, or Matryoshka (d) Phase segregated, or Janus

**Fig. 1** Common types of chemical ordering in binary nanoalloys

## *1.2 Choice of Energetic Model*

The energy landscape, and thus the relative stability of different structures, may be represented for metal clusters by a range of models, which can be considered to make up three main classes. The first is the (semi-)empirical potential (EP), which is usually a simple, closed function, mapping atomic coordinates to the energy, and parameterised either by fitting to higher levels of theory or experimental properties. Examples widely used for clusters and nanoalloys include the Sutton-Chen [13, 14], Gupta [15–19] and Glue embedded-atom-type [20, 21] potentials, and the Morse [22–24] and Lennard-Jones [25–29] pairwise potentials. These methods are primarily utilised for their lack of computational expense, good scaling with cluster size and ease of programming. As such, mixed cluster GO investigations have been performed in a select few cases for up to around two thousand atoms [30], with local optimisation coupled with permutational optimisation applied for several thousands of atoms [31]. For bimetallic clusters, the Gupta potential, which is a member of the class of potentials which utilise the second moment approximation to tight binding theory, is of particular note, and is defined in Eqs. 2–4.

$$U_{tot} = \sum_i \left[ U_i^{\mathrm{rep}} + U_i^{\mathrm{att}} \right] \tag{2}$$

which is equivalent to

$$U_{tot} = \sum_{i \neq j} \left[ \sum_j \theta(r_{ij}) - [\zeta^2 \sum_j \phi(r_{ij})]^{\frac{1}{2}} \right] \tag{3}$$

where

$$\theta(r_{ij}) = Ae^{-p(\frac{r_{ij}-r_0}{r_0})} \tag{4}$$

is the pairwise repulsive term between atoms i and j, and

$$\phi(r_{ij}) = e^{-2q(\frac{r_{ij}-r_0}{r_0})} \tag{5}$$

is the attractive term which aims to reproduce the manybody effect of embedding an atom into the density of the existing lattice. Parameters $A, q, p$ and $\zeta$ are fitted to empirically derived properties.

The second level of computational sophistication is the density functional theory (DFT) GO. This class of calculation technique is considered separately from other electronic structure methods due to its popularity, balance of accuracy and expense and its flexibility in treating many metal elements well. Size scaling of computational cost may be less than $N^3$ in current implementations of DFT [32–34], and many metal-specific formulations exist in order to capture features, such as scalar relativistic effects, or polarisation functions, which are absent from most empirical potentials. Rogan et al. recently suggested that a two phase procedure involving reoptimisation of EP minima with DFT is reasonable for small clusters [35], noting that well parameterised Gupta, Sutton-Chen and LJ potentials all give similar results when reoptimised with DFT, and thus that the choice of potential is not the most important factor. To our knowledge, no such examination has been performed for bimetallic clusters, however, and several instances where these simple EP are not appropriate will be covered in later sections, though it should be noted that EP can in principle be fitted very flexibly to capture a wide range of effects. To date, direct GO at the DFT level has been performed for monometallic clusters up to several tens of atoms, whilst such studies for nanoalloys are limited at present to around 10–20 atoms, if homotop optimisation is considered part of the process. For an in-depth discussion of the state of the art regarding DFT-GO for clusters, we refer the reader to the recent review by Heiles and Johnston [36], in which the various methods and their development are discussed.

For more accurate representations of the electronic structure of metal clusters, high level post-Hartree Fock methods such as the perturbative coupled cluster singles, doubles and triples (CCSD(T)) method may be employed [37], in addition to the popular Möller-Plesset perturbation theory techniques (MP2, MP4 etc.) [38]. These, and other highly sophisticated methods aim to accurately describe quantum mechanical contributions to the energy of a system, such as electron correlation and relativistic effects, but owing to their rapidly scaling computational cost, they are prohibitively expensive for a global search strategy for anything larger than a few atoms. Thus, they have currently found little use in the field of nanoalloy structure prediction, except in high level refinement of structures found with other methods [39–44].

For the GO strategy, in principle, varying the method by which the energy is calculated should not alter the algorithm, because the level of theory simply modifies the specific topology of the energy landscape, and in a scenario where computing power were infinite, this would indeed be the case. However, practical compromises must be made when changing from an empirical potential to the DFT level. The most obvious one is the smaller size of cluster which may be treated.

Figure 2 is a schematic representation of the approximate size ranges over which GO methods are currently applied for clusters. The very smallest range (not shown), contains the highly accurate, correlated wavefunction methods. Above this is the range of direct DFT-GO, which for small metallic and bimetallic clusters has become a standard technique in the last 10 years [36, 45–49]. Up to a few hundred atoms, the two phase GO involving DFT reoptimisation of promising candidate structures found with an EP is common [50–53]. EPs have been used for alloy clusters of up to 2000 atoms [30, 54]. From the thousand atom range upwards, GO methods tend not to be employed. High symmetry cluster structures, or those known from experimental observation are commonly investigated [55–57], commonly employing Monte Carlo (MC) methods [58, 59] and homotop optimisation [31, 60–62]. There is of course, no formal limit to the sizes which may be considered computationally, with Molecular Dynamics (MD) and MC methods popular for specific instances of cluster structural studies as the size becomes closer to the macroscopic [63–65].

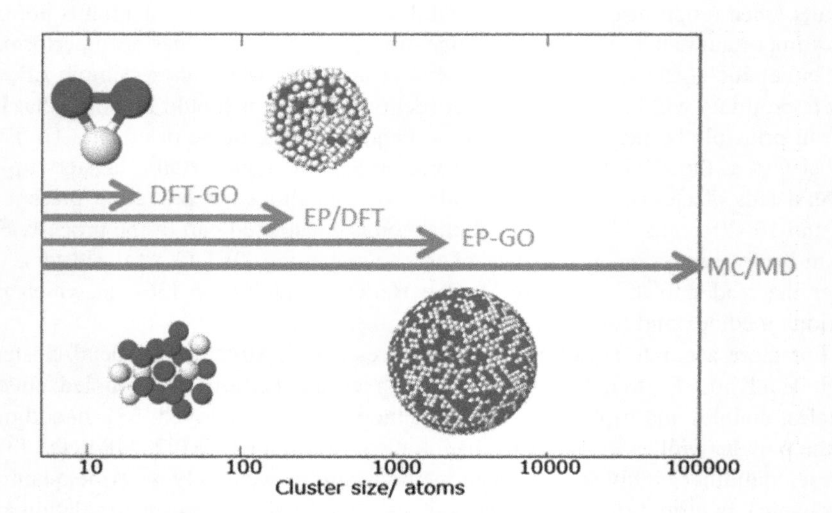


**Fig. 2** Schematic of the approximate size ranges available to various GO methods, with example cluster structures. The size range beyond GO includes Monte Carlo and Molecular Dynamics methods

Any attempt to separate theoretical methods into categories is an optimisation task in it's own right, insofar as seemingly distinct optimisation strategies may be closer in their mode of operation than is apparent, and separate methods convergently evolve the same solutions to problems. As a result, our definition, which broadly divides optimisation methods into nature-inspired, Monte Carlo-based and other classes, should be thought of as a permeable, flexible boundary at best. In reality, as methods establish themselves and develop, borrowing ideas from each-other and reimplementing them, it may be said that such boundaries are, in fact, artificial.

## 2 Nature-Inspired GO Methods

The natural world provides many enviable examples of the optimisation of complex strategies. Selection pressures, brought about by harsh and competitive environments, drive the need to efficiently complete tasks. As a result, organisms find methods which may be emulated computationally to solve the problem of structure prediction. Of course, the process of natural selection itself is a useful model for optimisation which may be included in the computational toolkit [66–69]. In the following sections, we focus on some of the nature-inspired methods which have shown success in the field of cluster and nanoalloy GO, in particular, evolutionary algorithms and swarm-based methods [70], although the list of nature-inspired GO strategies extends in other fields to neural networks, harmony searching, firefly algorithms, and many others [71].

### *2.1 Evolutionary Algorithms*

Evolutionary algorithms (EA) are a general class of computational methods which utilise the main principles which control evolution in organisms: random changes in genetic makeup, and natural selection under a selection pressure, in order to improve a cost function with respect to its cost. This definition leaves room to encode an almost unlimited number of problems, from optimisation of bus routes [72], to the shape of satellite antennae [73], to structure prediction in chemistry, each of which incorporate problem-specific implementations and features. The only required features in such an algorithm are:

(1) A means to determine the value of a cost function, and so rank the quality of a putative solution.
(2) A selection process which chooses, based on a fitness criterion, which individuals to mate.
(3) Crossover, or mating steps, in which individuals (solutions to the problem) may interact to product new individuals with features from both (Fig. 3).

(4) Mutation steps, which introduce new information to a selected individual, and increase diversity in the population.

The genetic algorithm (GA) is an example of an EA, in which evolutionary operators are employed to optimise the cost function, which is encoded as a gene. The operators act on the gene at the genetic level [74, 75], that is, the representation of the phenotype by the gene is important for the efficiency of the algorithm [76]. Many modern implementations of the EA do not in fact satisfy this criterion, utilising instead a direct, phenotypic action for the genetic operators [69, 77, 78]. However, the phrase genetic algorithm is commonly still used, and is evident in the naming conventions of several groups' codes, such as the Birmingham Cluster Genetic Algorithm (BCGA) [69] (a schematic of which is displayed in Fig. 3), or Sierka's Hybrid Ab Initio Genetic Algorithm (HAGA) [78]. We will use the more general EA in this chapter, which is guaranteed to be correct for all examples.

Most EA additionally employ, analogously to nature, the concept of populations, in which offspring created through mating steps define the next generation, upon which further genetic/phenotypic operators may subsequently operate. In fact, this is not necessary for the optimisation of a cost function, and the idea of pool GAs, in which parent and offspring are not distinguished from one another, and generations become redundant, has become popular. Hammer's EA code utilises such an implementation for cluster GO upon surfaces, wherein genetic operators act upon

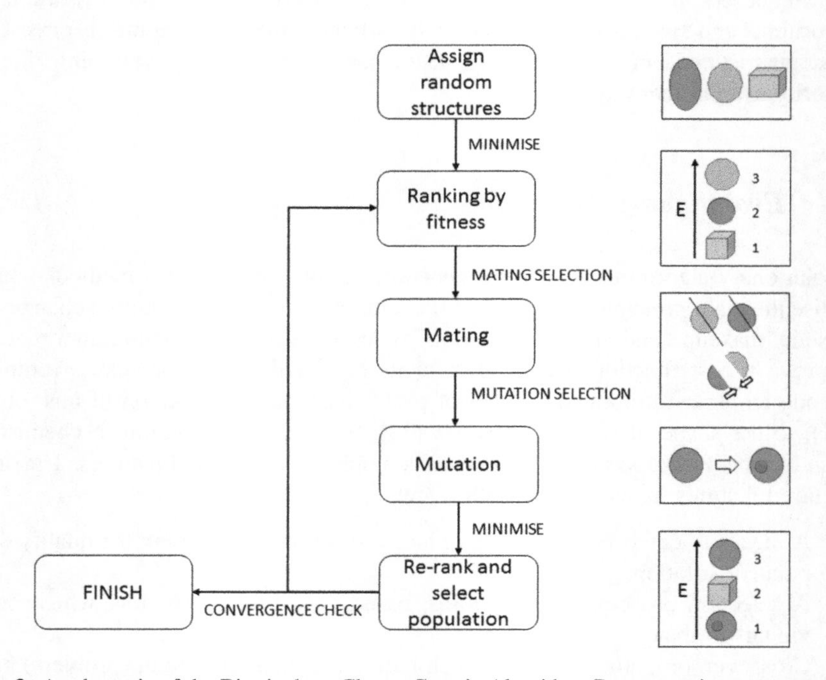


**Fig. 3** A schematic of the Birmingham Cluster Genetic Algorithm. Representative structures are shown as spherical, octahedral and cubic

all clusters generated thus far, and all solutions are retained and ranked by energy [48, 77, 79]. Shayeghi et al. have recently benchmarked a pool GA version of the BCGA [80], which is observed to improve the algorithm efficiency monotonically as the pools are decreased in size, and distributed over more subprocesses. This useful result is derived by removing the bottleneck inherent to any population-based method: that each generation is only processed as fast as its slowest member. Any calculation which proves difficult to converge holds up the entire algorithm until completion. The SAGA method employs sub-populations in a similar manner to the pool-GA, effectively parallelising, whilst allowing for semi-independent evolution of groups of individuals [81], which interact according to a defined probability.

The cost function in the case of nanoclusters is simply the energy function(al), such that the optimal solution is found once the energy is globally minimised. The structure which generates the lowest energy is thus the global minimum (GM). The mapping of structure to energy is dependent on the level of theory applied, and is commonly either DFT or an EP function as previously described. To encode the structure in a manner conducive to applying genetic operators is crucial, and was originally performed as a binary string, which represented, in the parlance of evolution, the gene (clusters in this way are organisms of just one gene, having no more information encoded genetically). A particular structure is thereby an allele of that gene, and in fact, the choice of mapping between structure and allele is important to the efficiency of the GO. This method, the true genetic algorithm, was originally applied in cluster science to the silicon tetramer by Hartke [76]. Mutation, then, was simply the flipping of binary "bits", which, when decoded, induced a change in cluster structure.

Major advancements to the efficiency of EA were established in the 1990s, which may be considered three separate features. The first was the change of representation from binary string to Cartesian coordinates of atom positions [27, 82]. The second, related step was the use of this representation to convert genetic operations to phenotypic operations, which essentially solved the representation problem. Deaven and Ho introduced a mating scheme in which the cut is made through the cluster itself, rather than a string of encoded bits [27]. This improvement meant cuts were more physical than previously, and with more of the actual structure of the cluster retained during crossover, the generation of good offspring candidates became significantly more likely. Similarly, mutation applied at the phenotypic level conveys an improved efficiency. The third improvement, again implemented first by Deaven and Ho, was the use of local optimisation to relax the cluster structure to its corresponding local minimum. In effect, the individual is converted to the lowest energy version of its structural class. This advancement is in fact general, and is applied in many GO methods. [6, 83–85], including basin hopping, for which it was explained to correspond to a particularly useful class of landscape deformation procedures [28, 86, 87].

In effect, the choice to locally relax structures once discovered, is equivalent to utilising a Lamarckian [88], rather than Darwinian [89] natural selection process. The latter, employed in nature, involves the selection of the individual based on her fitness to the environment, which is encoded genetically, and expressed phenotypically. Instead, the former allows for adaptations which occur during the life of an individual to be passed down to her descendants. This Lamarckian method is generally employed in modern EA, so that the selection or mating and mutation is made upon the locally best available structure. It is well established, and was systematically shown by Woodley and Catlow [90], in a study on condensed titania phases, that this procedure is more efficient than the Darwinian scheme, despite the additional local minimisations. This is because the Lamarckian scheme reduces the likelihood of multiple individuals occupying the same basin (assuming a filter to remove duplicates exists), by collapsing a basin down to a point at the minimum. Also it does not bias against basins which are narrow, or those which contain high energy regions (but low energy minima), which is an observed feature of the cluster energy landscape [23, 91].

Mutation and mating are the most important steps to control, and the easiest to modify for the particular problem. For clusters, the mutation step may be almost limitlessly varied, and as such, many varied examples have been developed [68, 69, 74, 77, 92, 93]. The most simple are the replacement of the cluster with a randomly generated new one [69], or application of a random displacement of one or many atoms to a new location. More advanced mutation schemes involve mirroring the cluster across a plane which passes through it [77], in order to enhance symmetry, or a SHAKE move [92, 94], in which all atoms are moved by a fixed amount. For mating, the cutting of the cluster into parts at the phenotypic level during crossover may take any form, and is often guided by chemical intuition. Multiple-cut crossover [69], or radially defined cuts [93], for clusters where core-shell structures are likely to dominate the low energy regions of the landscape are also common.

Applications of EA to binary clusters are manyfold, and many of the examples given in a later section are derived from EA GO, both at the EP and DFT levels of theory. Frequently the initial tests of an EA to multimetallic clusters are provided by binary Lennard-Jones (BLJ), or occasionally ternary Lennard-Jones (TLJ) clusters. For the former, a database is maintained by the Wales group. For example, Marques and Periera developed their EA specifically for binary cluster GO, employing it for BLJ clusters up to 50 atoms and $Ar_{(38-n)}Kr_n$ clusters [95], finding new GM structures lower than reported in the Cambridge database. Wu et al. employed an adaptive immune optimisation algorithm (AIOA) to model TLJ noble element clusters [96] as well as binary [97] and ternary coinage metal clusters [98]. Recently, Dieterich and Hartke employed their EA to study higher order mixed clusters, including binary, ternary and quinary alloys of rare gas elements [99].

## 2.2 Particle Swarms

Particle swarm optimisation (PSO) is a nature-inspired GO strategy, which follows the properties of animals which move in packs or swarms, with velocities which depend on eachother's position [70, 71, 100]. The method is a stochastic search of the energy landscape defined by a cost function, in which several walkers (putative structures) search the energy landscape in parallel. The position of an individual walker is varied in a stepwise fashion, by a function which includes both its personal best energy thus far, and the total best energy of the population. In this sense, there is global communication between walkers, which drives each to react dynamically to the position (structure) of the most promising locations visited thus far. One of the earliest uses of PSO for chemical structural investigations was for small LJ clusters, silicon hydride molecules and water clusters [101]. The authors included the advancements of minimum distances, to keep variation between individuals high. Small nickel and bimetallic Ni–Al clusters have since been optimised [102], with a dynamically varying inertia weight function, which reduces the amount of influence the best known structure has on the velocity for subsequent steps. This allows the search to gradually vary from global to local searching as the simulation progresses.

PSO has more recently been applied within the CALYPSO framework of Ma et al. [103, 104] to find GM for layered structures, crystals, and most recently, atomic clusters. In their formulation, local-PSO is applied, in which the global interaction between the best solution of the entire population, is replaced with one of a local group of individuals, which form a sub-set of the total population. This is choice designed to avoid trapping into a single funnel. This method was only applied to LJ clusters up to 150 atoms, and simple metal clusters, $Li_n{}^+$ [105]. The benchmark calculations however suggest promise for the more complex problem of bimetallics.

Similar to the PSO is the artificial bee colony (ABC) [106], which maps the salient features of a population of bees foraging for nectar and gathering information on good locations, onto the problem of continuous cost function optimisation. Bees are deemed to forage in a stochastic manner, weighted by visual information on the likely sources of nectar, to optimise the amount of nectar discovered, in a manner which couples the current path chosen, to paths already taken by other bees. In the algorithm, the amount of nectar is mapped to the fitness of the solution to the cost function (i.e. the energy of the cluster structure), and a number of parallel walkers explore configuration space in such a way that good paths to low energy structures are favoured. Three classes of walker are defined: one which moves randomly over configuration space, one which moves in the locality of good locations previously discovered, and one which move randomly, but only accepts "downhill steps" on the landscape. This division of labour thus simultaneously allows for efficient global and local searching. A version of this algorithm has been applied to LJ [107] and water clusters [108], but to our knowledge, no investigations for binary metallic systems have yet been made.

Ant colony optimisation is another popular algorithm for GO on continuous and discrete function spaces [109], and from a chemical perspective, has found utility for the structure prediction of peptides [110] and model clusters [111]. The algorithm uses the process by which efficient paths are reinforced over a sample of many trajectories by the laying down of a pheromone trail, which decays over time. The quality of the solution (energy of the structure) represents the quality of the path that an ant may take, and thus determines, through the amount of pheromone laid down, the likelihood of following this path. Thus walkers are biased towards directions which lead to low energy structures. The pheromone can be laid down either simultaneously by parallel walkers (ants), thus creating an ensemble of paths which gives rise to a statistical distribution of paths with varying quality, or the amount can be varied, based on the quality of the solution the path found [112].

# 3 Monte Carlo-Based Methods

Monte Carlo (MC) type landscape exploration is a stochastic search method involving sequential moves according to a well defined moveclass, followed by energy evaluation. Usually, this type of landscape search employs a Markov chain type walk, though several GO methods include memory of previously visited states in order to speed-up the search. Metropolis Monte Carlo is the variant of MC in which the Boltzmann distribution itself is used as the weighting function for importance sampling [113]. If only steps in configuration space which lead downhill in energy ($E_{new} < E_{old}$) are to be accepted, the Metropolis temperature (T*), as defined in Eq. 6 is chosen to be zero.

$$p_{acc} = e^{-(E_{\mathrm{new}} - E_{\mathrm{old}})/\mathrm{T}^*} \tag{6}$$

This type of exploration has been used since the earliest days of computational GO, but suffers from the drawback that the energy landscape that is searched still exhibit barriers which slow the exploration, both in downhill and uphill directions. There is a balance in efficiency to be made in the choice of T*, between allowing uphill steps to overcome energetic barriers, and biasing steps downhill towards local minima.

## *3.1 Basin Hopping*

Basin hopping (BHMC) is the result of combining MC steps with local geometry optimisation, such that every move is immediately followed by a downhill convergence to the local minimum of the basin to which the initial structure belongs. Doye and Wales remarked that by minimising after every step, the landscape is converted into a series of discrete plateaux, between which, barriers which lead downhill are removed [28]. Thus, a hypersurface deformation strategy which retains the connectivity of the landscape, along with the relative energies of all local minima is created, and is displayed schematically in Fig. 8. This locally minimised view of the energy landscape is in fact identical to the one produced in the Lamarckian GA.

Basin hopping was initially demonstrated upon a Lennard-Jones cluster [28], and has become a popular technique for multi-component cluster GO since its inception

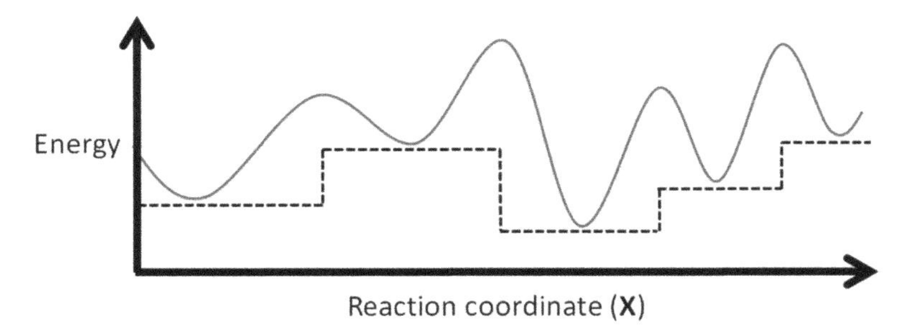


**Fig. 4** A schematic one-dimensional energy landscape, under the transformation of local minimisation. The true landscape (*blue curve*) is converted to the discrete set of plateaux (*black dashed line*)

[30, 45, 92, 114, 115]. A database of putative global minima for a wide range of empirical potentials, including Lennard-Jones, Glue-type, Murrell-Mottram and Sutton-Chen, is maintained on the Cambridge Cluster Database (CCD), and includes cluster sizes of well over 100 atoms for some unitary clusters [14, 116–118]. For BLJ clusters, Doye and Meyer produced putative GM to 100 atoms [114]. Successive authors have improved energies of their putative GM with newer algorithms [119, 120]. For DFT-BHMC GO, Bulusu et al. applied the method to $Au_n$ clusters, with n = 12–18, locating complex, capped prismatic cage structures [121, 122]. The same group applied DFT-BHMC GO to doped M-Au $^{\alpha}_{14}$, with (M,$\alpha$) = (Zr, 0), (Hf, 0), (Sc, −1) and (Y, −1) [45]. In 2011, Tao and colleagues implemented a variant of basin hopping which consisted of local optimisation plus three additional moveclasses, discovering nineteen new, lower energy GM than those reported in the CCD [123]. These operators involved moving higher energy atoms from the surface to the core (KNEAD), high energy core atoms to vacant surface sites (SMOOTH), and exchange of type A to type B atoms (FLIP). Each of these types of moves have been applied in previous GO algorithms, notably from Hartke [124] and Takeuchi [125] and appear to show great promise for binary cluster GO in general (Fig. 4).

## 3.2 *Advancements to MC*

Goedecker introduced the minima-hopping algorithm [83], applied later to DFT-GO of Mg clusters [126], and extended to binary systems [119]. This method involves MD, rather than MC steps to move in configuration space, with a dynamically controlled kinetic energy for the simulation, which determines the barrier over which the MD run can escape, and thus the search space generated by each walker. Once a new region is located, local minimisation finds the minimum of the basin. For systems in which the barriers between superbasins are high, minima-hopping is proposed to improve upon the performance of BHMC methods.

Rossi and Ferrando made further advancements to BHMC, by including an order parameter-based search coupled with a memory, which disfavours already-visited minima [30, 84, 90]. Their algorithm may be described as like a tabu-search [127] basin-hopping algorithm, but without the strict restriction that no minimum may be revisited. By defining good order parameters, including heterogeneous bonds and common neighbour analysis for binary clusters, they perform GO for standard test cases $LJ_{38}$ and $LJ_{75}$, along with $Ag_{32}Cu_6$ and $Au_{90}Cu_{90}$ bimetallic clusters. The authors describe the moves available to their BHMC-type algorithms, including the "BOND move" in which the lowest-coordinated atoms are moved preferentially, and the "SHAKE move", in which every atom is displaced in a random direction; claimed to be the most efficient Cartesian moveclass. Dynamic Cartesian steps have been shown to speed up the location of the GM in binary systems by Kim et al. for $Ag_{42}Pt_{13}$ [128], in which the moveclass is a simple Cartesian step of one atom, and the step size is proportional to the square of the distance from the centre of the cluster, in order to mimic the increased diffusion rates of atoms on the surface of a cluster. Figure 5 shows the progress of such a GO simulation, with the progress of the structure depicted.

Cheng et al. developed the "funnel-hopping" strategy for cluster GO [6], in which a second local geometry optimisation step is inserted into the GO process, to

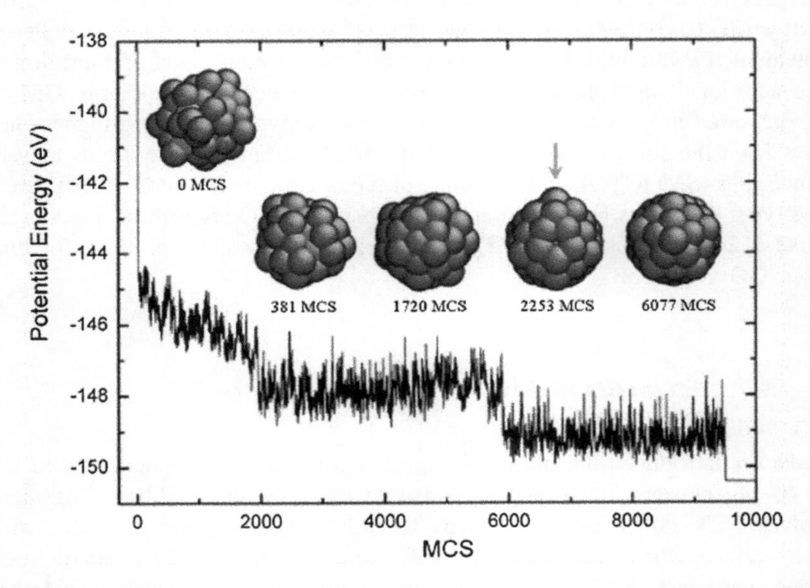


**Fig. 5** Progress of the GO for $Ag_{42}Pd_{13}$. The final surface defect (*highlighted by the arrow*) migrates in the final step, resulting in the perfect icosahedral GM. Reprinted with permission from Kim et al. [128] in the Journal of Chemical Physics, 128, 144702. Copyright 2008, AIP Publishing LLC

locate the lowest energy minimum of a superbasin of a cluster. By permuting through filled and vacant surface sites, the best minimum of the possible set defined by a rigid cluster core structure is located, i.e. the lowest minimum in the funnel.

Extending the algorithm to the parallel domain, replica exchange Monte Carlo methods have been developed, in which barriers between basins may be overcome by combining several parallel searches at different Metropolis temperatures. The creation of a superensemble which combines the low-energy searching required to local GM with the high energy search required to explore space has been applied for clusters [129, 130]. Calvo and Yurtsever [131] introduced a replica exchange MC procedure designed specifically for binary clusters. Each replica comprised a BHMC simulation at fixed composition, including particle exchange moves, while every replica was at a different composition. Exchange of configurations between replicas was found to speed up the landscape exploration for binary rare gas clusters of up to 38 atoms.

# 4 Other Methods

## *4.1 Simulated Annealing*

Simulated annealing (SA) was probably the first general GO approach for clusters which proved successful [132, 133], though it is now routinely outperformed by the more sophisticated BHMC and EA methods. The temperature, which defines the energy barrier over which a walker may traverse, is reduced according to a pre-defined schedule, trapping the walker in decreasing regions of configuration space, until a local minimum is obtained. The guiding principle behind this strategy is ergodicity, such that given sufficient time at each temperature, a structure search should find the optimal geometry, in the manner of annealing in experiment. The major flaw in this idea, as experienced in experiment for glass-forming materials, is that the basin which contains the GM does not necessarily have a large phase volume, and so slow annealing (and thus a slow algorithm) by no means guarantees the location of the GM. It should be noted that the SA concept applies traditionally to MD [134–136], though is extended equally to MC types of landscape search. For example, Tan et al. [137] used a DFT Monte Carlo SA simulation to optimise the homotop for compositions across the range of $Pd_{(55-n)}$–$Pt_n$, in order to determine the role of alloying on the hydrogen evolution reaction. The authors report a correlation between hydrogen adsorption energy and the fraction of Pt atoms in the (111) hollow site which may be of use for design of tunable binary catalysts.

The SA concept was adopted by Lee et al., and applied to atomic clusters, in the form of conformational space annealing (CSA) [85]. The distance between configurations in a population is calculated, and used to define a cutoff "distance" $r_{cut}$, which defines structural similarity between individuals. Steps are taken in configuration space to generate new individuals, replacing structures based on total energy. This is followed by a reduction in $r_{cut}$. This is the range over which a new cluster, which is always a modification of an existing one, may vary, and therefore decreases monotonically over time. The annealing of the cluster, by this stepwise reduction of the possible search manifold generates traps for configurations in a manner similar to SA.

## 4.2 Threshold Algorithm

The lid algorithm [138], or more importantly, its extension to continuous functions: the threshold algorithm [139], is another example of a local search GO method which has been applied to ionic [140, 141] and metallic clusters [142–144], among other chemical optimisation problems [145–147]. The threshold method relies primarily on the MC approach to explore the energy landscape, which is separated into disjoint sets of configuration space by the presence of a bounding energy threshold ($E_{lid}$), above which the walker may not proceed, as shown in Fig. 6. By repeating the stochastic search with varying threshold values, information regarding the activation barrier heights for intra-cluster rearrangements may be established. Additionally, by accumulating information regarding the energy and structure of visited configurations, and their frequencies, statistics regarding the density of configurational states, thermodynamics, and dynamical properties, such as escape rates may be determined [139]. However, within the scope of this section, our

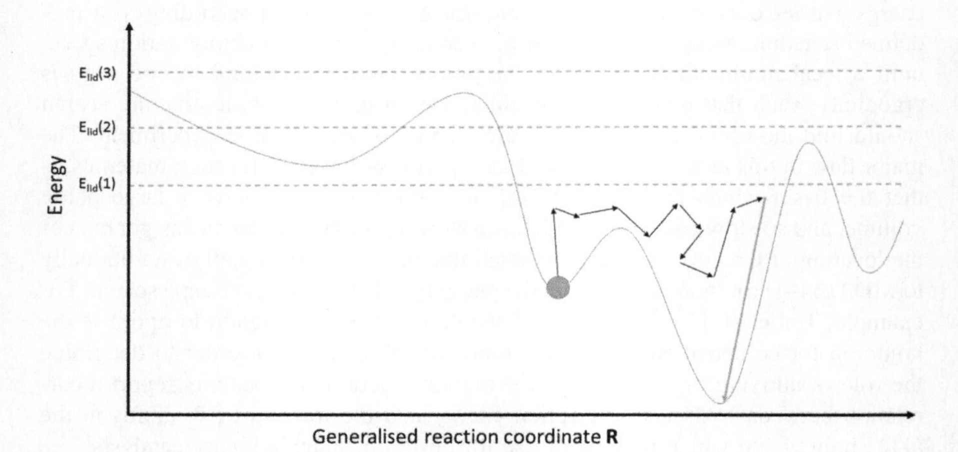


**Fig. 6** A schematic of the threshold algorithm. *Black arrows* correspond to the MC random walk below predetermined lid energy $E_{lid}(1)$ (*red dashed line*) As the lid energy increases, new basins become available. The *green arrow* is a steepest descent quench to the local minimum

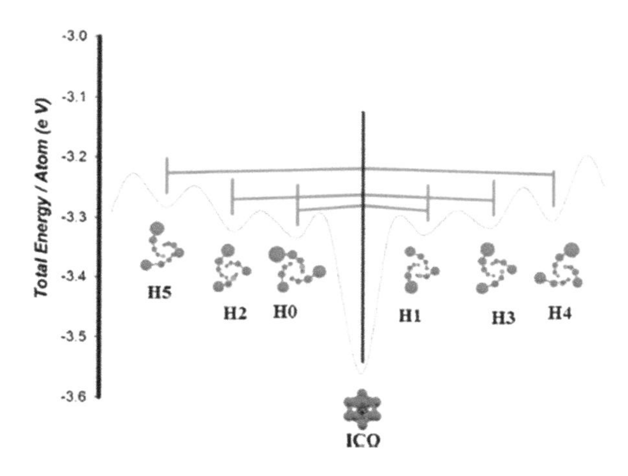


**Fig. 7** Schematic of the $Pt_{12}Pd$ energy landscape. Energetic barriers for conversion between tetrahelical homotops and the icosahedral GM are estimated (including enantiomers). Reprinted from Pacheco-Contreras et al. [142] in the Journal of Physical Chemistry A. **116** 2012, pp. 5235. Copyright 2012 American Chemical Society. Copyright 2012 American Chemical Society

interest lies in the method as a GO algorithm. If the threshold is systematically reduced, the search is equivalent to an SA run, insofar as the temperature defined in the SA schedule is related to the Metropolis temperature which defines the threshold value. The threshold algorithm has been incorporated into a broader, modular code by Schön, which includes a number of more efficient GO modes. Energy landscape transformation may be applied in order to recover basin hopping functionality, while collective cluster moves, reminiscent of EA crossover steps may be included if required. This method has been applied to bimetallic clusters [142–144]. Pacheco-Contreras et al. investigated ultrasmall $Pt_{12}Pt_1$ clusters with the Gupta potential [142], finding the GM, and additionally a range of interesting chiral, tetrahelical conformers, as shown in Fig. 7.

The prescribed path method is a further tool available for the energy landscape search of continuous systems [148], in which pathways which interconvert minima are found and mapped out by local searches along cartesian directions, starting from minima already located. Another method, which has proven very successful for the mapping of energy landscapes, for clusters [149–151] and other continuous systems [152–154], is the discrete path sampling approach of Wales et al. [151, 155] Utilising highly efficient transition state searches [156, 157], a library of the connectivity between minima, through the lowest energy saddle points which interconvert them, may be grown. As a result, this method has contributed significantly to the popularity of the tree, or disconnectivity graph as a means to quantitatively map out the salient thermodynamic features of energy landscapes [2].

## 4.3 *Discrete Lattice Search*

Discrete lattice search (DLS) is a method which may be described as a biased local search technique, in which the core of a structure is used as a template for the location of additional atoms upon sites around it [158]. The location of the potential sites is set up as a lattice, according to the symmetry of the core, and by sampling potential occupations, the best structure grown from the initial seed is developed. This idea has found many formulations, each implementing additional moveclasses, such as rotation [159], or two-phase local optimisation [160]. Very recently the binary-DLS (BDLS) was presented by Wu et al. [161]. These strategies have been applied variously to LJ [162] and metallic clusters [163], whilst the BDLS was used to optimise binary Gupta Pd–Au clusters with up to 79 atoms and BLJ up to 100 atoms [161]. In the case of the BDLS, the lattice is simply bifurcated into two overlapping sublattices for the two atomic types, while combination of this algorithm with the iterated local search (ILS) method of Lai et al. was shown to systematically improve performance for binary clusters [161, 164].

### 4.4 Metadynamics

Metadynamics is a landscape exploration method which aims to combine searching for low lying energetic regions according to an appropriate set of collective variables, while simultaneously gathering statistical information on the free energy of the system [165]. The metadynamics search is in essence a tabu MD walk in order parameter space, with the dynamical trajectories biased by a repulsive potential which acts against revisiting states previously found. As a means to enhance the sampling of rare events in MD, such as escaping deep potential energy basins, metadynamics has found use recently for metallic clusters [166–170]. Santarossa et al. [166] explored the PES and FES for the $Au_{12}$ cluster with a DFT metadynamics simulation, highlighting the differences between landscapes. On the FES, several local minima were found to merge, generating regions which are likely to intercovert freely, despite large interconversion barriers on the PES. Configurational entropy is thus found to play a significant role. Recently, Pavan et al. [170] studied the $Pt_{13}$ cluster with the Gupta potential, observing a broad landscape of basins. The authors suggest the use of metadynamics for efficient metal cluster GO. Extensions such as replica exchange metadynamics [171] look promising for the larger configurational space spanned by binary clusters (Fig. 8).

## 5 Optimisation of Chemical Ordering

Optimising the chemical ordering simultaneously with the structure may not be the most efficient procedure for any case in which there is a hierarchy of energy scales. If structural isomers differ in energy by a wider margin than homotops of a given structure, as is common for clusters, then optimising the homotop for a suboptimal cluster may waste computer time, finding the best instance of a poor structure. For this reason, several methods have been developed in order to efficiently locate low

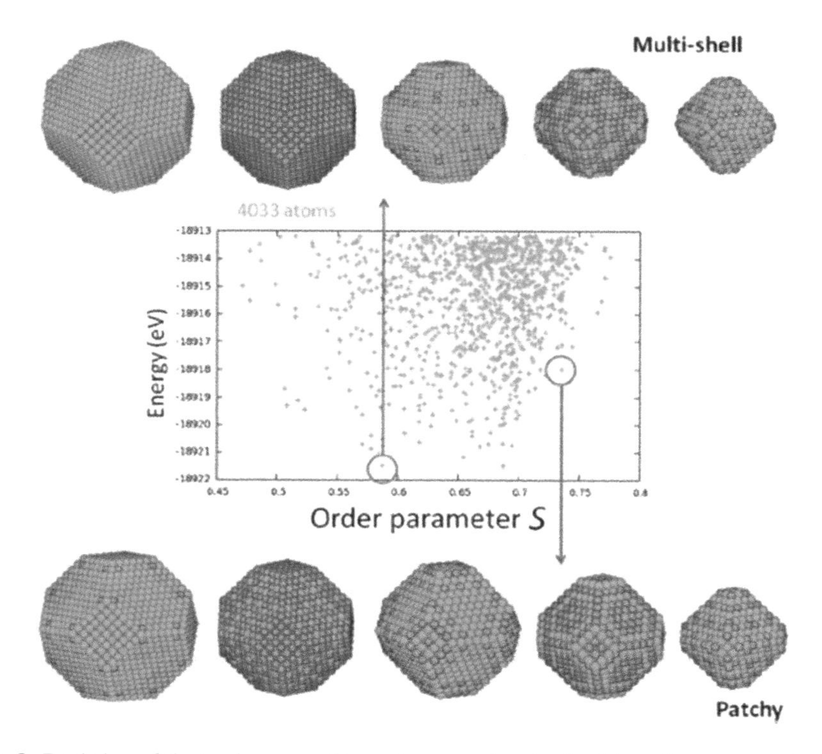


**Fig. 8** Depiction of the total energy of homotops sampled for the $Pd_{2433}$ $Pt_{1600}$ cluster located with the GGO method of Barcaro et al. Concentric shells are depicted for two instances of each cluster size. Reproduced from Ref [49] with permission of The Royal Society of Chemistry

energy homotops of a given, usually high symmetry structure. Indeed it is common that a GO process is applied to the homotop optimisation itself, just as the structural optimisation which precedes it.

For systems with a strong preference for a particular chemical ordering pattern, such as two metals with very weak heterometallic bonding (which promotes phase separation), the converse case is possible. A poor homotop of a good structure may be located, simply because the homotops span a broad energy range. In such an instance, care must be taken when decoupling structure and homotop searching, wherever the search is far from exhaustive. For subnanometre clusters, the effect of the homotop on the stability of the structure may be profound. The introduction of a silver or copper dopant into small gold clusters can drive a change in dimensionality which drastically alters the energetic balance [49, 172]. For this reason, many GO methods employ a simple exchange move step which is activated only rarely. An example of this is the work of Darby et al. [17], which employed a mutation which exchanges one third of atoms, and is activated for one in every ten individuals of the population, within the BCGA.

A recent modification to the EA for homotop optimisation was applied by Vegge and colleagues [93], based on chemical intuition regarding the chemical ordering of bimetallic clusters. New mutation operators were defined, which selectively exchange a pair of atoms of unlike type, based either on the local concentration of like atoms, or to replace core atoms with surface atoms. The former mutation serves to drive mixing on-the-fly, or segregation if run in the reverse mode. The latter modification is designed to find low energy configurations for clusters which exhibit core-shell structures, as many transition metal and noble metal bimetallic clusters are known to do. Calvo et al. applied a weighted exchange move homotop GO to Ag–M 3:1 clusters of up to 309 atoms [173], where M = Au, Ni and Pt. The authors note that it has been found that atoms of differing sizes particles have low acceptance rates in exchange moves, as the larger atom cannot easily occupy the small hole left by it's partner. Weighted schemes which utilise conformational bias have been previously applied to simulations of dilute solutions [174] and Lennard-Jones glasses [175]. In the article, a weighting function which takes into account the distance between pairs of exchanged particles is used. Very recently, Shao et al. employed a discrete PSO algorithm for homotop optimisation, for large Pt–Pd particles of up to 6323 atoms [176]. The authors point out that exchange moves between atomic sites in the fixed geometry are justified by the similar sizes of Pt and Pd atoms.

A significant reduction of the search manifold may be made for high symmetry clusters by mapping atom sites to the symmetry orbits of the point group to which the geometry belongs. By collecting all symmetry equivalent sites together, the number of possible exchange moves which may be made is diminished from $N!/(N_A!N_B!)$ to $2^S$, where $S$ is the number of orbits of the point group [177–179]. It should be noted, however, that this strategy limits the possible homotops to those in which only one atom may occupy the site of a particular orbit. Very recently, Barcaro et al. introduced a variant of the symmetry-based homotop optimisation strategy [31], in which the grouping of atoms into equivalence classes is relaxed to include translational symmetry, or a user-defined order parameter-based definition, and has been applied to Pt–Pd and Cu–Ag clusters. The variation of stability with order parameter for a large Pd–Pt cluster is depicted in Fig. 8. An advancement which utilises the principle of maximum symmetry was introduced by Oakley et al. [154], in which atomic moves which increase the symmetry of the cluster are preferentially made, by filling in missing sites in the orbits of the point group defined by a core of cluster atoms. This maximisation of approximate symmetry led to a reduction in mean first encounter time (MFE) for the various LJ clusters tested.

Another method utilised for fixed-symmetry clusters is the Iterative Lattice Search [164], which has been applied in conjunction with the BDLS for binary clusters of Pd $_{(79-5m)}$–Au $_{5m}$ [161]. The ILS was originally implemented for the homotop optimisation of Cu–Ag, Cu–Au and Au–Ag clusters up to 55 atoms, and is based on the generation of neighbourhood lists of homotop energies. The authors utilise the observation that low (high) energy homotops usually produce, on local minimisation, low (high) energy minima. By exchanging a subset of possible permutational isomers of rank $N_A \times N_B$ (which is vastly smaller than the complete set), and ranking these by energy before local minimisation, a further subset, comprising the neighbourhood are subsequently minimised. Acceptance of a new configuration is thus based on whether better candidates are found during this procedure. A related approach was introduced by Schebarchov and Wales [180], utilising graph theory partitioning methods, in which a BHMC search through permutation space is self-guided according to the approximate cost of a particle identity flip. Significant gains in optimisation efficiency are found in cases where the energy gain in flipping the particle identity is a good descriptor for the quality of the exchange move which could be made from a pair of such flips. The underlying principle for both Lai et al. and Schebarchov et al. is that guaranteeing the best swap move is prohibitively expensive for complex energy functions, and approximate methods can give good estimations of good swap choices without complete elucidation of the permutation space. Schebarchov and Wales very recently extended this concept to combined structure-homotop optimisation (i.e. complete GO), introducing the concept of biminima [181]. The biminimum is a local minimum on the combined manifold of configurational and permutational space, and their approach has been fruitfully applied to GO for the BLJ cluster, with extension to higher order nanoalloys possible.

Full simulations have been applied to the homotop optimisation problem, including semi-grand canonical MC simulations, to optimise the chemical ordering of bimetallic clusters along with composition, at fixed chemical potential difference ($\Delta\mu$) [182–184]. Cheng et al. found for the 55 atom Cu–Au cluster there exists a wide region of $\Delta\mu$ which generates $Cu_{core}Au_{shell}$ configurations [182], while three layer, onion-like configurations can be found where $\Delta\mu$ is suppressed. Rubinovich and Polak developed a sophisticated procedure to determine accurate segregation properties of binary clusters, named CBEV (coordination-dependent bond energy variation) in which the coordination dependence of bond energies is fitted from DFT surface energies [60]. By parameterising the energetic cost of occupying sites with varying degrees of under-coordination and mixing, the optimal chemical ordering has been calculated for several size-matched Pt–Pd [60] and mismatched Pt–Ir clusters, of up to 923 atoms [61, 185]. This method is promising as a highly transferable method which utilises DFT energetics to justify the surface behaviour. An alternative procedure has recently been developed by Kozlov et al. [186], using a topological energy formulation, wherein the energy of each coordination environment an atom can be found in, is parameterised against a set of DFT calculations. The stability of a given homotop is then determined by the sum of the contributions from homo- and heterometallic bonds and site-specific (vertex, edge and terrace) energies. For larger clusters still, a PSO has been applied to the Pt–Cu Pt–Pd and Cu–Pd binary clusters of up to 5.5 nm (approximately 15000 atoms) with a quantum Sutton-Chen potential [187]. In opposition to full simulations, Reyes-Nava et al. [136] reported a predictive tool for determining segregation properties for special case of binary alloy cluster without the need for atomistic optimisation. They claim that for clusters with both metals in the same period (for example Ni–Pt), or in adjacent positions in the periodic table (for example Pt–Au), the preference for core or shell sites is governed by the relative valence and core electron densities in the individual metals. In the former case, high core electron densities lead to core site occupation, while in the latter case, high valence electron densities lead to core occupation.

Before concluding this section, it is worth noting that while the optimisation of homotops from a thermodynamic perspective gives information on the likely chemical ordering pattern a cluster may adopt, it says nothing about the pathway by which such a homotop may be reached. In experiment, this consideration may be important, as biases inherent to the cluster production procedure will skew the product distribution. Homotops which are energetically stable once located, may well be kinetically unavailable in practice. An energy landscape approach, which involves mapping an interconversion pathway between homotops may be employed to probe this effect. Calvo et al. have performed such an analysis for $Ag_{27}Au_{28}$ and $Ag_{42}Ni_{13}$ clusters with a Gupta potential [149], finding long, multi-step pathways for the inversion of core-shell isomers, which significantly distort the cluster structure along the path. This approach therefore extends the analysis beyond merely the determination of optimal chemical ordering, towards the question of kinetic feasibility.

## 6 Comparing and Combining GO Methods

A combined approach, which utilises features of several algorithms, should in principle generate a more efficient GO strategy given sufficient flexibility. The individual drawbacks of a particular method can be overcome by mixing with another method whose weaknesses do not overlap. For example, while BHMC is extremely efficient for searching within a superbasin, the barriers to finding new funnels remain [6]. Large-scale steps in configuration space which are thus difficult to make, are trivial for an EA, which includes as standard, well-defined collective moves in crossover steps. Oakley et al. recently generated and benchmarked such a combined method, bringing together BHMC and the BCGA in a hierarchical manner, allowing BH local searching coupled to large scale phenotype moves [188]. The authors note that this combined method does not generally lead to improvements over both of the individual EA and BHMC methods, and this failure is attributed to the issue of algorithm optimisation. While both individual algorithms were previously optimised for their respective uses, when applied together, computational efficiencies are lost. It is predicted therefore, that if designed initially with a combined procedure in mind, such methods could become powerful tools for multicomponent cluster GO. One such possible approach was implemented by Hsu and Lai [189], called the "PTMBHGA" method. This algorithm combines features of parallel tempering, modified basin hopping and an evolutionary algorithm, for binary cluster GO, and was first applied to the 38 atom Cu–Au cluster. By performing BHMC steps on a population of individuals, which were then mated together, the authors reported lower energy isomers than previously found, although mean first encounter (MFE) times were not given.

Other distinct GO methods that have been combined in such a way previously, include (CSA) [85], which combines beneficial traits from EA, BHMC and SA. The local minimisation transformation in most efficient GO methods is utilised, while collective moveclasses based on configurations already generated and stored in a population, similar to mating in an EA are used. Where CSA is related most closely to SA, is in the scheduled decrease in the breadth of configuration space which may be sampled over time. Applied to LJ clusters of $< 201$ atoms, the algorithm was able to find all existing putative GM, and is an excellent example of the convergent evolution of GO methods for clusters.

While it is difficult to directly compare the performance of algorithms, owing to the diverse, home-made nature of most scientific GO codes, there is of course much value in objective comparisons of efficiency. Claims of generality, in which specific instances of an algorithm are taken to be representative of every implementation will naturally underestimate the performance of the "old method", when newer versions are proposed. The most reliable method of comparison is thus with success rates and MFE of the GM for specific test systems, within the literature. This ensures each developer compares with the best of her method's competitors.

Within BHMC-type algorithms, the tabu-like HISTO method is compared with unmodified BHMC and a parallel search method called parallel excitable walkers (PEW) [84]. The latter involves a BHMC search performed by several walkers simultaneously, which dynamically repel eachother in order parameter space (and thus indirectly, configuration space) according to a penalty term in the energy. Over a range of Metropolis temperatures, the number of times the GM is located is compared between the three methods [92]. This measure, while less useful than the MFE for determining the speed of the algorithm, demonstrates well the reliability of the HISTO and PEW methods, particularly at low temperatures, which exhibit around 300 % speed-up on unmodified BHMC. While HISTO improves the algorithm by reducing time wasted on repeated minimisations of already visited minima, the advancement which PEW represents is one of scope. By repelling walkers dynamically, the breadth of minima sampled per unit time is increased. These comparisons should be taken carefully, for the aformentioned reasons, though the claim that the average number of minima visited before finding the $LJ_{38}$ GM is over 100 fewer than for minima-hopping is encouraging.

## 7 Nanoalloy Case Studies

In this section we will describe a number of GO investigations for binary clusters, ranging from the large (several hundred atoms), where two phase GO is often required, by computational limitations, to the ultrasmall, in which direct DFT-GO methods become necessary. It is separated into parts on group 11 (coinage) metals, noble metals and main group metals. The noble metals section is further subdivided into pure group 10 (platinum group) clusters, and mixed-group clusters with a noble metal component.

## 7.1 Coinage Metal Clusters

For coinage metal nanoalloys, the majority of cluster GO falls into two categories: larger clusters, determined with potentials (and often subsequently reoptimised with DFT), and ultrasmall clusters, optimised directly with DFT-GO. For the former case, structures and chemical ordering are expected to be controlled by the balance of atomic radius, cohesive energies and surface energies, as given in Table 1 for a wide range of metals. For smaller clusters, relativistic effects are known to play a significant role [190–192]. The electronic structure of the coinage metals diverges from alkali metals, due to the energetic closeness between (n)d and (n + 1)s electrons, which induces orbital hybridisation. For gold, this effect is particularly acute, and leads to planar GM to unusually large sizes. $Au_n$: (up to 12 atpms). Because the electronic structure of these clusters is subtle, and not captured well by most of the commonly employed empirical potentials, necessitates the use of methods of at least the quality of DFT in GO for the smallest clusters.

Bonačić-Koutecký and colleagues determined general properties of ultrasmall bimetallic gold-silver clusters, including a propensity to form heterometallic bonds which maximised silver to gold charge transfer [197]. This work was extended by Weis et al. [198] and others [199, 200] later, using more sophisticated range-corrected exchange-correlation functionals, by Rohrdanz et al. [201] and Shayeghi et al. [202, 203], who combined EA-GO with simulated optical spectra to compare directly with experiment.

**Table 1** Physical properties of selected late group transition metals. Covalent radius $r_{cov}$, cohesive energy $E_{coh}$ and surface energy $E_S$ for low index planes

| Element | $r_{cov}$/pm [193] | $E_{coh}$ eV/atom [194] | $E_S$ J/m$^2$ [195, 196] | Plane |
|---|---|---|---|---|
| Fe | 140 | 4.28 | 2.417, 2.475 | (110) |
| Co | 135 | 4.39 | 2.522, 2.550 | (0001) |
| Ni | 135 | 4.44 | 2.380, 2.450 | (111) |
| Cu | 135 | 3.49 | 1.790, 1.825 | (111) |
| Ru | 130 | 6.74 | 3.043, 3.050 | (111) |
| Rh | 135 | 5.75 | 2.659, 2.700 | (111) |
| Pd | 140 | 3.89 | 2.003, 2.050 | (111) |
| Ag | 160 | 2.95 | 1.246, 1.250 | (111) |
| Os | 130 | 8.17 | 3.439, 3.450 | (0001) |
| Ir | 135 | 6.94 | 3.048, 3.000 | (111) |
| Pt | 135 | 5.84 | 2.489, 2.475 | (111) |
| Au | 135 | 3.81 | 1.506, 1.500 | (111) |

The next interesting feature of subnanometre doped coinage metal clusters is the 2D–3D transition. Direct DFT GO has had a notable impact on this problem. The BCGA was employed by Heiles et al. [49], and later, Heard and Johnston [204] to probe the 2D–3D transition in neutral $Au_{(8-n)}$–$X_n$ (X = Ag, Cu) directly at the DFT level. The structural and energetic progress of the EA simulation for $Ag_8$ and $Au_4$–$Ag_4$ is depicted in Fig. 9, from Heiles et al. [49]. Both studies show the conversion to 3D structures to have fully occurred by $n = 3$, with copper instigating the transition by $n = 2$, which suggests a greater propensity to form compact clusters than silver. This result is drawn in parallel with the decrease in d electron contribution to the highest occupied orbitals as a function of dopant level. As copper and silver have a smaller degree of s-d hybridisation than gold, the d/s ratio in these orbitals drops precipitously with increased doping, in line with the sharp change in structural preference. Wang et al. noted a reduction in the position of 2D–3D transition in M–$Au_n^-$ clusters where M = Cu, Ag, and n = 8–11 as compared to $Au_n^-$[205]. Wang, Kuang and Li investigated the effect of doping a single copper atom into $Au_n$ $n = 3$–9, with charge states of +1, 0 and −1 [172]. They find a crossover from 2D to 3D GM at $Au_6Cu^+$, but no three dimensional isomers for up to nine atoms for the neutral and anionic clusters. It should be noted however that this work was not a true GO investigation, but DFT reoptimisation of preconstructed isomers. More recently, Hong et al. applied an EA-DFT GO to the full range of compositions of $Au_n$–$Ag_n$ ($4 < n + m < 13$) clusters [200]. In agreement with previous calculations, they find that the 2D–3D transition, which only occurs for $n + m > 6$, is always observed before gold becomes the majority component. This implies the role of the silver s valence orbitals, which drive 3D motifs, is stronger than the gold s-d valence orbitals, which oppose this trend.

The preference of low coordinate gold was observed by Lang et al. [206], who combined optical spectra from photodissociation spectroscopy with DFT

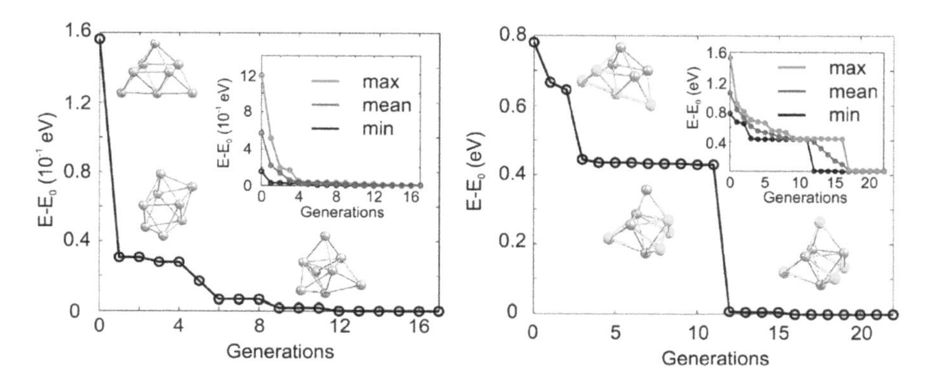


**Fig. 9** Progress of the BCGA for the DFT GO of $Ag_8$ (*left*) and $Ag_4$–$Au_4$ (*right*), from Heiles et al. The inserts display the maximum, average and minimum energy of the population during the simulation. Reproduced from Ref [50] with permission of The Royal Society of Chemistry

calculations for Ar-tagged Cu–Au tetramers. This result is in agreement with the results of Heard and Johnston, and Heiles et al., but disagrees with those of Hong et al.

In the intermediate size range between planar and compact structures, Pal et al. applied a BHMC DFT GO approach to locate the GM of $Au_n^-$, for n = 12–15, with 0 or 1 dopant atoms of Ag or Cu [47]. Comparing the resulting calculated photoelectron spectra allowed for unambiguous structure determination, including the discovery of a new structure for $Au_{13}^-$. The dopant atoms were found to have little structural effect, implying higher dopant loadings are required to distort the gold frame.

For larger clusters, Darby et al. [17] showed that the Gupta potential GM of a medium sized (55 atom) gold cluster was significantly perturbed by the presence of even a single copper dopant atom, funnelling the structure into the icosahedral superbasin to which the monometallic copper cluster belongs. Toai et al. [54] subsequently extended the size range to much larger clusters of selected sizes, 100, 160 and 200 atoms, with Cu:Au ratios of 3:1 1:1 and 1:3, each of which are known stable phases in the bulk. Applying the PEW method at the Gupta level of theory, they observed that icosahedral-based motifs dominate the low-lying regions of configuration space. The 160 and 200 atom clusters consist of a 147 atom icosahedron, covered by an incomplete shell of anti-Mackay and Mackay icosahedral atoms, respectively, while the 100 atom cluster varies between an icosahedron ($Cu_{25}Au_{75}$), double icosahedron ($Cu_{50}Au_{50}$) and defect Marks decahedron ($Cu_{75}Au_{25}$). Rodríguez-López et al. [135] extended the size range to 561 atoms with SA and a Sutton-Chen potential, finding icosahedral motifs dominant for the 75 and 10 % copper compositions considered. FCC-cuboctahedra were predominant in the 50 and 25 % copper regime.

Molayem et al. used a BHMC/embedded atom model GO for the full range of $Cu_m$–$Ag_n$ clusters up to 60 atoms [207], finding a preponderance of polyicosahedral (pIh) and 5-fold symmetry clusters, in agreement with previous work [53, 208]. Additionally, energetic analysis, most notably the mixing, or excess energy showed silver-rich compositions to be more stable, as displayed in Fig. 10. This result may be explained by the preference of Cu–Ag to form $Cu_{core}$ $Ag_{shell}$ structures, in which the shell atoms make up the major part of the cluster. Recently, Atanasov et al. reported BHMC GO for 1000 and 2000 atom clusters of $Cu_m$–$Ag_n$, where the ratio $m$:$n$ was 0.15, 0.25 and 0.50 [30]. In this work, the authors find that icosahedral motifs are the most stable, though fcc crystalline geometries are competitive across the composition range investigated. Copper and silver are essentially immiscible in the bulk, forming phase-separated structures. However, cluster alloying is affected by the large lattice mismatch of around 13 %, which generates a uneven distribution of strain in finite clusters, and the significant difference in surface energies, which drives core-shell chemical ordering. The work of Atanasov highlights the intermediate size range which exhibits off-centre copper cores, representing the middle ground between core-shell and Janus clusters.

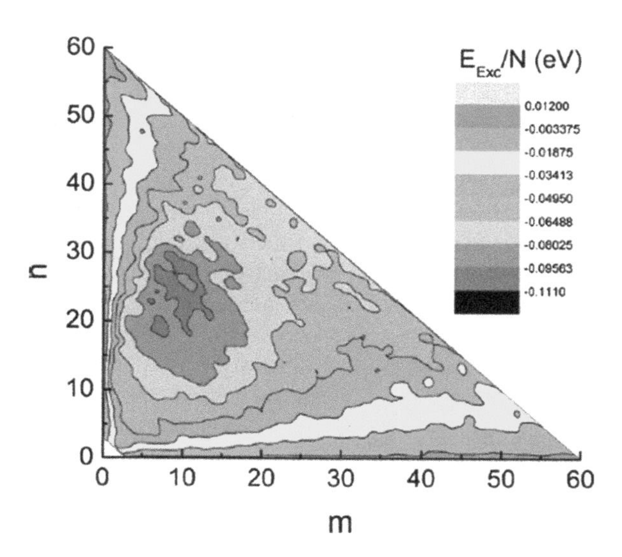


**Fig. 10** Heat map of the excess energy per atom for $Cu_mAg_n$ clusters, showing a preference for slightly silver-rich compositions and a very low propensity for heterometallic mixing. Reprinted from Molayem et al. [207], in the Journal of Physical Chemistry C. **115** 2011, p. 22148. Copyright 2011 American Chemical Society

Overall, a picture for coinage metal GO emerges, with a preference for planar structures which are controlled by s/d mixing and the symmetries of the valence orbitals for ultrasmall clusters. The size of the transition to 3D geometries is thus governed by the relative abundances of "s-like" metals Cu and Ag or "sd-like" Au. The exact position is then subject to additional factors, such as charge state and chemical ordering, the latter of which is difficult to predict in gold-containing clusters, with conflicting results regarding the preference for high and low coordinate sites for Au. Once the 3D cluster is established, there is a size range up to 50 atoms at which fivefold symmetric, pIh and pseudo-planar "pancake structures" are important, before the complete icosahedron at 55 atoms marks the dominance of that structural class. With exceptions near the magic sizes of decahedral, truncated octahedral, and Leary tetrahedral clusters, the Ih superbasin appears to control GM geometries into the thousand atom regime. The inclusion of a second element induces strain through lattice mismatch, and introduces the symmetry breaking effect of differing homo- and heterometallic bond strengths between metals. This feature makes specific predictions about chemical ordering, and thus reactive, electronic and thermal properties difficult, but computational GO evidence suggests facetted and off-centre cores play a major role for the bulk-immiscible Cu–Ag clusters, as well as for many other binary clusters [62, 209–211]. Beyond this size range, meaningful GO becomes too computationally expensive, even for the simplest of empirical potentials and the realm of dynamical simulations begins.

## 7.2 *Noble Metal Clusters*

A wide range of bimetallic noble metal clusters on the subnano- to nanoscale have been studied in the past 15 years, particularly with EA, PEW and BHMC GO. This work has resulted in the development of a broad picture of the likely geometric motifs and mixing patterns. In this section we consider some examples of the noble metal-containing clusters, which are important from their varied catalytic uses. Some authors have mixed together various noble metal bimetallics, treating several pairs with the same GO approach, in order to create a library of structures, and an overview of the segregation properties. One good example is the study of Rossi et al., in which size matched binary clusters of 34 and 38 atoms, including Ag–Pd, Pt–Pd and Ag–Au [212], and size mismatched particles, Cu–Ag, Au–Cu and Ag–Ni [81] were investigated within the EA approach with Gupta potentials. For the mismatched clusters, core-shell plh GM were found to be abundant, with magic numbers for Ag–Ni and Cu–Ag at 34 atoms (where seven interpenetrating icosahedra form a high symmetry geometry), and 38 atoms, for compositions involving a complete shell of silver. For Cu–Au, these plh motifs were less common, as gold supports the necessary strain in surface bonding less well than silver. For the size matched clusters, the preponderance of pIh structures was reduced still. This result highlights the importance of balancing the atomic radii, strain and bulk mixing preference in binary nanoalloys. It is noted, as is found to be common in binary clusters, that the predominant structural motifs favoured by a mixed cluster bear no resemblance to those of the pure, monometallic clusters of the same size, even resulting in an unexpected class of pIh structures, which were confirmed in the 30–40 atom size range with DFT calculations for Ag-Ni and Cu–Ag clusters.

Other authors have considered particular pairs of elements for their studies. We will divide these works into two sections: group 10 bimetallics, and mixed clusters with a group 10 component.

### 7.2.1 Group 10 Bimetallics

These metals represent the catalytically active set of elements Ni, Pd and Pt, which variously have electronic configurations ranging from $d^{10}s^0$ (closed shell), to $d^9s^1$ (open shell d and s orbitals) and $d^8s^2$ (open shell d), depending on cluster size and geometry. This relatively broad range of electronic structures lead to geometric effects in clusters and in the complexes adopted by group 10 metals, as the d orbitals are filled in a range of symmetries, including for example, the propensity towards square planar motifs [213]. Magnetism, commonly observed in 3d metals, and occasionally found in heavier metal and bimetallic clusters [214], additionally complicates the structure prediction of Ni-containing systems. However, the energetic and geometric arguments which go some way to predicting chemical ordering in coinage metals is seen to apply frequently to the group 10 metals.

EP GO for Pt–Pd clusters modelled with the Gupta potential has been shown to be sensitive to the choice of parameterisation. Massen et al. [215] used the BCGA for locate GM for Pt–Pd clusters up to 56 atoms, noting that with three parameterisations of the potential, three classes of GM were found. One, in which $Pt_{core}Pd_{shell}$ clusters predominated, one which favoured phase separation into Janus-like motifs, and one which gave cubic, mixed phases. The combination of surface energies and homometallic cohesive energies suggests a segregation of Pd to the shell of a core-shell particle, which was observed in the first parameterisation. It is of note that the authors, along with those of other studies, found a larger number of decahedral structures for Pt–Pd clusters than are found for either the pure clusters, or other late group bimetallic clusters in the <50 atom range [94]. This result was found to apply for the 98 atom cluster, for which the majority of compositions were found to be based on the Marks decahedron[216]. Interestingly, the excess energy, which is a measure of the stability of the GM of a mixed composition with respect to its monometallic counterparts, was observed to be optimised at compositions approaching 1:1 Pt:Pd ratios. It was further predicted that the Leary tetrahedron dominates the structural landscape in these highly stable compositions. With the FCEM/CBEV method, Rubinovitch and Polak investigated the preferred chemical ordering of the 923 atom Pt–Pd cuboctahedron [60], finding mixed surface faces, with dominant subsurface location of Pd. Palladium was observed to preferentially occupy vertex and edge sites, in agreement with the lower surface energy of Pd. Barcaro et al. utilised a similar approach, coupled to a shell model for homotop optimisation of magic number truncated octahedral clusters up to 201 atoms [217], for which they observed a complex pattern of ordering. Pd–Pd bonds were minimised by the creation of a patchy, multishell structure. In agreement with Rubinovitch, platinum was found to occupy surface (111) facets even at low mole fractions.

While the surface and cohesive energies suggest a clear segregation, the coordination-dependent binding energies vary in a complex manner for Pt–Pd, resulting in a difficult binary system to unambiguously describe. The relative size of (111) and (100) facets will play a role for large particles, but the result of the various GO investigations is that surface mixing in Pt–Pd clusters is favourable, with Pt-rich cores and Pd-rich subsurface occupation expected.

Pd–Ni clusters exhibit a significant size mismatch, of approximately 10 %, with nickel the smaller atom, while their surface energies are similar. The cohesive energy of Ni is higher than that of Pd. The prediction that palladium should segregate to the surface of a core-shell cluster has been tested by few GO investigations. Calculations on the electronic structure of such clusters have shown an enhancement of the magnetic moment of nickel, when surrounded by a core of palladium. Wang et al. employed a local spin density functional method to elucidate the electronic properties of $Ni_{core}Pd_{shell}$ clusters of up to 55 atoms [218], finding a ferromagnetic coupling between Pd and Ni, whose value depended on the size of the Pd shell. Guevara et al. later extended this analysis to closed (geometric) shell clusters of up to 561 atoms [219], observing that hybridisation between core Ni and surface Pd atoms enhances the magnetic moment of Ni, and that this effect is diminished at higher Ni mole fractions, due to the reduction in Ni–Pd bonds. These results suggest that the coupling of structure and chemical ordering to the electronic properties is very important in Pd–Ni clusters, and thus that DFT GO methods would be beneficial to capture the physics of the system.

Pt–Ni clusters have proven more enigmatic than other nickel-containing particles, as unlike the case of Pd–Ni, the factors usually used to predict the GM chemical ordering lead to erroneous results. While platinum has the greater cohesive energy and nickel, the lower surface energy, it has been observed experimentally and theoretically that core enrichment of nickel occurs. While not strictly a GO method, Wang et al. performed MC simulations for large fcc Pt–Ni clusters at $Pt_3Ni_1$ and $Pt_1Ni_1$ compositions in order to locate favourable structures and chemical ordering [220]. They observed that for clusters of below 2406 atoms, there exists a multishell arrangement of alternating Ni and Pt layers, with a surface of predominantly Pt. For larger sizes, a core shell structure is recovered, with a mixed Ni/Pt core and a Pt shell.

### 7.2.2 Noble Metal Mixed Clusters

Much work has been performed on mixed clusters containing noble elements, especially when alloyed with other late group metals. Earlier groups of the periodic table are less well represented, however, owing to a confluence of factors. Open shell electronic structure, along with the wide range of available oxidation states make the prediction of structure difficult, in tandem with the computational difficulty in attaining converged electronic states. The magnetic ordering of 3d metals add a further degree of computational freedom, demanding high level electronic structure methods of at least the quality of DFT. As a result, reliable GO has only been performed for ultrasmall cluster sizes, and the gap between theory and experiment remains large in most cases. We will mention a few key results from this sea of possible binary metal pairs.

Palladium–gold clusters have been investigated with GO methods, as their importance in catalysis necessitates the accurate determination of structure-activity relationships. Pittaway et al. [221] applied a two phase BCGA EP–GO + DFT reoptimisation procedure to the GO of $Pd_n$–$Au_n$ clusters up to 50 atoms, finding a sensitive dependence of parameterisation on the mixing propensity. Recently Bruma et al. [222] applied this procedure to find the GM over a range of compositions for the 98 atom Au–Pd alloy cluster, which is the magic size for the Leary tetrahedron. predicted to be the GM for silver with the Sutton-Chen potential, and with the LJ potential [223]. In this study, the Leary tetrahedron was observed to be suboptimal, when compared to fcc and Marks decahedral motifs, in agreement with experimental results. Recently the same authors applied a direct DFT BHMC GO to locate the GM of the $Au_{24}Pd_1$ cluster [224], finding an endohedrally doped cage structure. Charge donation from Pd to Au was proposed to activate the gold cage towards subsequent catalytic reactions. This screening procedure is common for complex systems, where the cost of a full GO at the DFT level is prohibitively expensive, but suffers from the drawback of incompleteness. Any structures which are unstable at the Gupta level of theory do not pass the screen for the DFT calculation, and are excluded, thus artificially removing geometries which may in fact become stable at the higher level. Any multiple phase GO process therefore includes the biases of both energy models. We draw particular attention to the Gupta + DFT method, because its use has been so ubiquitous in the GO of bimetallic clusters.

Pt/Au clusters have been studied due to their catalytic activities, with clusters of both single metal counterparts known for their CO oxidation capabilities [225–227] (as has additionally been shown for Pd clusters [228]). Ultrasmall platinum clusters have been applied to propane dehydrogenation recently [229], showing drastically higher turnover frequencies than larger clusters, while ultrasmall gold clusters have been the main focus of surface deposited cluster GO over the last decade [48, 230], along with experimental evidence of impressive activity towards alkyne hydration catalysis [231]. As depicted in Fig. 11, Pt–Au clusters up to 101 atoms display a wide array of stable structural motifs, which depend sensitively on dopant loading [232]. Alloying of gold particles has been investigated in the subnanometre range by many authors [45, 233–239]. In early instances of bimetallic DFT GO with BHMC, Gao et al. [45, 239] determined the GM structures of neutral M–$Au_n$ clusters, for the range of M = Pd, Rh, Ru, Mn, Re, W, V, Nb, Ta, Zr, Sc, Y, Mg, Ca, Sr, Na and K, with $n$ selected in each case to give a cluster which conformed to the eighteen electron rule. This, in practice, included $n$ from 8 to 17 gold atoms. The authors noted that all GM for $n > 8$ exhibited endohedrally doped gold cages, of varying symmetries, and that high CO oxidation activity is predicted [240]. Most of these cage structure were found to display large HOMO-LUMO gaps, implying particular chemical stability, as noted by Pyykkö and Runeberg for the thirteen-atom gold cluster [241].

Ferrando et al. [242] studied binary clusters of Ag–Pd, Ag–Co, Au–Cu, Au–Ni, and Au–Co for selected sizes up to 45 atoms with a two phase Gupta GO + DFT reoptimisation procedure, finding striking results for the cobalt-containing particles. The pIh structures which dominate the other late transition metal clusters are destabilised when cobalt is included, due to unfavourable magnetic interactions in the compressed core. This leads to a preference for mixed compositions, which is in opposition to the immiscibility known for bulk Ag–Co and Au–Co. Finite size effects are thus found to play a role in the structure and chemical ordering predictions of small cobalt-based clusters.

Negreiros et al. [52] applied the two-phase method to GO for Ag–Pd clusters of selected sizes between 38 and 100 atoms, in order to test the reparameterisation of a Gupta potential with DFT fitting. GO involved a BHMC structure search phase, followed by a BHMC homotop optimisation, with DFT reoptimisation of low lying motifs. Decahedra are found to be most stable for 100 atoms, with capped icosahedra at 60 atoms, and a combination of truncated octahedra and icosahedra at 38 atoms. The stability of pIh motifs was suppressed on refitting the potentials, which displays a divergence between standard Gupta potentials and DFT methods for this system. For larger clusters, Tang et al. [59] recently applied a Metropolis MC simulation to

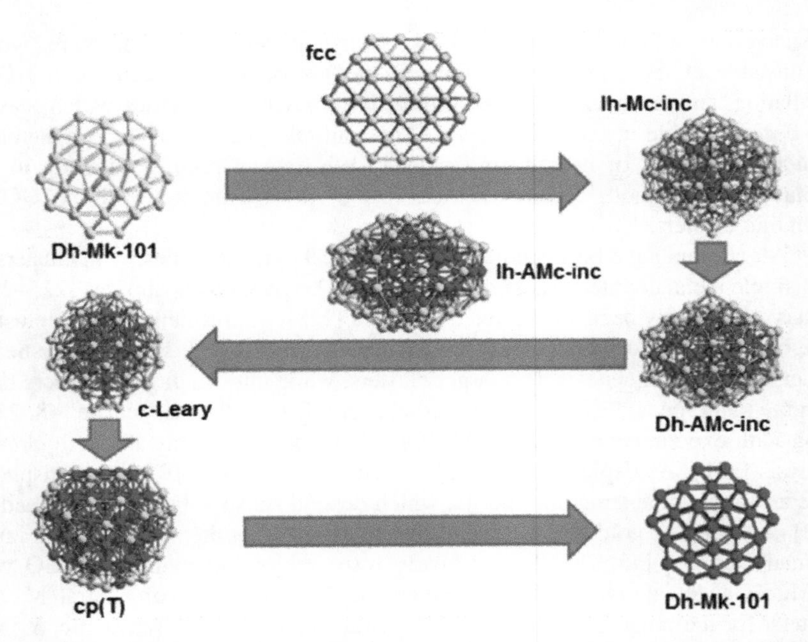


**Fig. 11** Transition of global stability of the Au $_{(101-n)}$–$Pt_n$ cluster. Structures are Marks decahedra (Dh–Mk), fcc, Mackay icosahedra (Ih–Mc), Anti-Mackay growth on a decahedral core (Dh–AMc), Anti-Mackay icosahedron (Ih–AMc), capped Leary tetrahedron (c-Leary) and close-packed tetrahedron (cp(T)). Reprinted from Dessens-Felix et al. [232], the Journal of Physical Chemistry C. **117** 2013, pp. 20967. Copyright 2013 American Chemical Society

Ag–Pd particles of selected sizes between 586 and 9201, finding a size and Ag mole fraction-dependent transition in GM chemical ordering, from $Pd_{core}(AgPd)_{shell}$ to $(AgPd)_{core}Ag_{shell}$. Complex surface (111) facetting is observed, in which Ag-rich surface occupation tends to isolate Pd atoms on the faces, due to a competition between exothermic hetero-bonding and preferential Ag surface location.

## 7.3 Main Group-Based Clusters

Structure prediction of main group clusters has garnered much interest since the advent of electronic structure calculation GO methods. Experimental deflection measurements of such clusters require accurate determination of electric and magnetic dipole moments. Indeed, matching the values for these properties to those of experiment has been used as a means to verify the structures found with GO [243, 244], in much the same way as the use of time-dependent DFT calculations to match optical spectra for coinage metal nanoalloys [202, 203].

Rohrmann, Schwerdtfeger and Schäfer employed a combined experimental/theoretical method to determine the cluster geometries of magnetic main group-doped clusters Mn–$Sn_n$ [245], in which Stern-Gerlach magnetic beam deflection measurements were matched with magnetic dipole moments calculated from GM structures, which in turn were generated with an EA DFT-GO search. Similar computational methods were also applied to resolve the structures in electrostatic deflection experiments on neutral $Pb_n$ clusters, for n = 7–18 [246], $Si_8$ and $Si_{11}$ clusters [244] and $Sn_m$–$Bi_n$ ($m$ = 5–13, $n$ = 1–2) [243]. In this case, electric dipole moments and polarisabilities were the structure-sensitive property used to compare directly with experiment.

Ultrasmall Ru–Sn clusters were recently investigated for the effect on structure and catalytic activity of doping and size, by Paz-Borbón et al. [115]. A BHMC DFT GO search for the GM of $Ru_n$, $Sn_n$ and $(Ru–Sn)_n$, with $n$ = 12, and $m$ = 14 was performed, followed by calculations of transition barrier heights for ethylene hydrogenation. The GM structures, along with their corresponding binding energies are given in Fig. 12. It was noted that doping tin into ruthenium clusters drastically altered their shape, spin preference and reactive properties—lowering the rate-determining barrier. Very recently Huang et al. [247] have discovered the ground state structures and chemical ordering of a class of three-shell Matryoshka clusters following the pattern $A_1B_{12}A_{20}$, where A = Pb or Sn, and B = Mn, Cd, Zn or Mg. They used an EA DFT GO approach, observing particularly stable icosahedral superatom structures, in agreement with the 108 electron rule found for other bimetallic multishell clusters.

Cu–Sn clusters have been globally optimised within a DFT/tabu search method by Fournier [248], displaying a preference for $Cu_{core}Sn_{shell}$, along with the geometric shell closing stability common for core-shell bimetallic clusters. The same

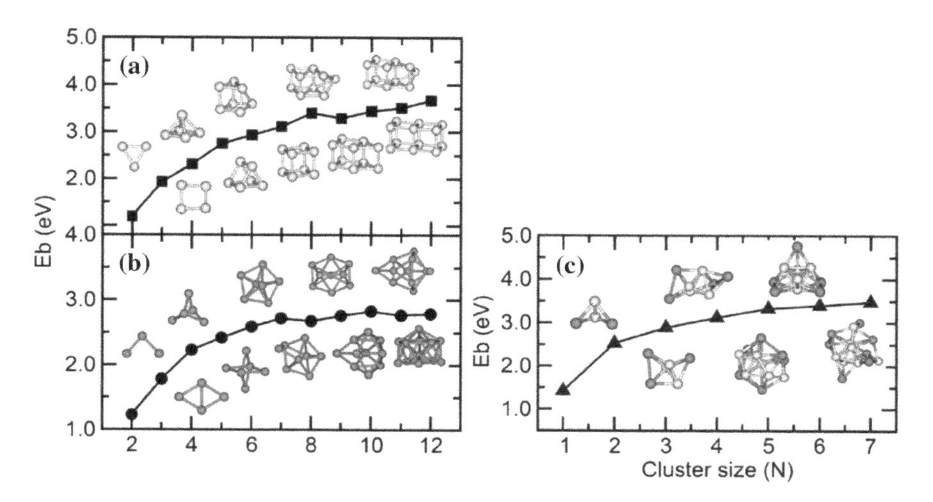


**Fig. 12** Binding energies of GM structures for Ru, Sn and Ru–Sn clusters with BHMC DFT GO. Reproduced from Ref [115] with permission of The Royal Society of Chemistry

author applied this tabu search to the location of Pb–Sn clusters of each composition and size between seven and twelve atoms [249]. The similarity in mono-elemental GM meant that each composition followed a simple substitution pattern, in which some segregation was observed. The authors stress that the application of spin-orbit (SO) coupling effects is important, and so a two phase approach with SO applied after structure prediction will not yield accurate results for lead-based clusters.

Larger binary and ternary alkali metal clusters have been treated by Aguado and Lopez [250, 251], with ternary $Li_{13}Na_{30}Cs_{12}$ particles showing an onion-layer core-shell chemical order, with a preference for pIh geometries. The larger, $Li_{13}Na_{30}Cs_{12}$ cluster was additionally considered, with an orbital-free DFT-MD approach, obtaining perfect core-shell segregation, with a complete Cs shell. It is observed that the large lattice mismatch between Li and Cs generates a richer structural landscape in the case of $Li_{55}Cs_{42}$, while the binary and ternary cluster display an enhanced thermal stability, induced by symmetry-breaking in the vibrational spectrum due to the varying atomic masses of elements in each shell. GO and thermal analysis of several magic number binary clusters, M–$Na_{54}$ (M = Li, Cs [252], K [253]), were considered with the same approach, with Cs occupying surface sites, and Li doping into the core. A two-phase Gupta + DFT method was used for the GO of 55 atom K–Na and Li–Na clusters [254]. In both systems, it was observed that for compositions approaching 1: 1, the icosahedral GM is replaced by pIh motifs with prolate structures, in agreement with the Jellium model prediction for simple metal clusters. This competition between geometric shell closing and Jellium distortions is thus found to be reordered by the presence of dopant atoms, and is composition-dependent.

It is noted that while main group clusters are found to exhibit a broader range of structural isomers, and as such present a more challenging task to unbiased GO approaches, several features are found to be in common with simpler bimetallic clusters. The geometric and electronic shell closing effects known originally for silicon, and used widely to explain high symmetry, high stability motifs in coinage and noble metal clusters, are found in main group systems, and thus still maintain some predictive power. The coupling of magnetic and non-magnetic elements together into binary clusters remains a complex issue, in which enhancement and depletion of magnetic moments may be observed, as found in Ni–Pt and Ni–Pd alloy systems. The common chemical ordering patterns used to describe late transition metals are still useful qualitative tools for these more diverse clusters. However, in main group-based systems care must be taken, as the omnidirectionality of bonding becomes a poorer approximation as systems become more covalent in type. Additionally, the need to satisfy coordination is a less straightforward measure of energetic stability. Despite these seemingly profound difficulties, the number and accuracy of GO investigations on main group-based binary systems highlights the power of the computational methods.

Many other binary pairs have been treated for varying compositions and sizes with GO methods, predominantly in a two phase EP/DFT manner. For a comprehensive description of experimental and computational GO results for bimetallic clusters, we recommend the review article of Ferrando et al. [1]. For the study of ultrasmall clusters, which is primarily the domain of DFT GO, we refer the reader to the recent review by Heiles and Johnston [255].

# 8 Beyond Isolated, Bare Clusters

All the methods thus far have been used to treat isolated clusters, and as such, model the behaviour of systems in vacuum, without the influence of substrates, complexing ligands, carrier gases or other clusters. These simplifications are sufficient for a first approximation to experimental behaviour, but clearly fall short of describing the nature of most situations in which clusters will be employed in practice. Historically, such features have been treated according to a two-phase procedure. The general issue arises, that structures which do not lie in the set of competitive minima in one scenario, will be erroneously excluded in the next. Nevertheless, many studies, indeed most to the present day, use such a multi-stage process. The extension to nanoalloys is the next step for these methods, and require the use of additional techniques to maximise efficiency as configuration space grows.

## *8.1 Surface-Supported Clusters*

The mode in which clusters are usually found in application and experiment, is supported on a surface. The choice of surface has a profound impact on the chemical properties of the cluster, which are due in turn, to their geometric and electronic structure. In the field of heterogeneous catalysis, it is well known that the effect of the support has a dramatic effect on cluster mobility, and thus its sintering resistance, as well as activity as a catalyst. [256, 257]. As a result, the prediction of cluster structure upon surfaces has become a major area of computational research, with unbiased GO a significant part. In Hartke's review of cluster GO with EAs, the state of the art as of 2004 was described [67]. He notes that "this, [surface GO] setup has been addressed in very few EA applications", and "EA applications in this area have only just begun." Since then, a vast amount of work has been performed, both on the development of efficient surface-cluster GO [77, 78, 258], and on the range of systems to which it has been applied, including binary nanoalloys clusters. We will focus primarily on the binary cluster work, though this subject deserves a review of its own.

The studies may be categorised into two main classes:

(1) Local reoptimisation of gas phase cluster structures, subsequently deposited on a rigid support
(2) Direct cluster GO on a support (rigid or flexible)

The first class makes up the vast majority of investigations, as it utilises existing results for gas phase cluster structures. This procedure suffers from the same drawbacks as any two phase GO approach, but they are particularly acute in this case, as a surface dramatically alters the range of potential structural isomers available to a cluster, whilst introducing an additional class of bonds unavailable to the isolated cluster. However, situations in which this procedure is valid exist, including the deposition of ultrasmall clusters, for whom the number of isomers is small enough that it is possible to avoid missing any, with a sufficiently broad library of putative GM as input [259–268]. Another example is the case of ultrasoft landing, in which it may be assumed that the surface should not interact strongly enough with the cluster to cause significant reordering of minima energies. This ultrasoft landing method has been used for noble metal clusters, such as small palladium clusters on alumina [269] and metal oxides, for example vanadia upon silica/Mo(112) [270].

The second class, modelling growth or annealing of the cluster upon the surface, in which a cluster is truly subjected to a GO process upon the surface has been exploited by several authors. Early attempts included a lattice model for surface sites, encoded in binary strings for LJ clusters [272], and a phenotype method for metal ad-atom clusters with an embedded atom potential [273]. Goniakowski et al. used PEW and BHMC to determine GM structures for Ag, Au, Pd and Pt clusters of selected sizes up to 500 atoms with a Gupta potential [274–276]. For all metals, it was observed that there is a size-dependent transition in global stability from cube-on-cube (100) structures to fcc (111) motifs, which comes at smaller sizes for those with larger lattic mismatch with the surface (Pt and Pd). Later, Ismail et al. [271] employed a two-phase GO for 30 and 40 atom Au–Pd clusters at selected compositions. The authors used a Gupta potential to locate several low-lying structures for the cluster upon a fixed MgO (100) slab, with a BHMC algorithm which employed SHAKE and exchange moves, followed by a DFT reoptimisation of promising candidate structures. It was observed that the clusters preferred to adopt the core-shell type motifs favoured by the isolated Au–Pd clusters of that size, but the presence of the support induced a build-up of Pd atoms at the interface, in order to maximise the stronger $Pd_{cluster}$–$O_{surf}$ bonds. This is shown for the 40 atom clusters in Fig. 13.

As in the case of ultrasmall isolated clusters, the GO of the very smallest metallic particles upon surfaces requires the accurate capture of electronic effects, such as charge transfer, or the “metal-on-top effect” [277–279]. The solution is once again, to perform DFT-GO at the surface. This expense, particularly when additionally considering the relaxation of the slab, limits investigations to the sub-nanoscale range. However, recent experimental work has shown, particularly for coinage and noble metal clusters, that very high catalytic activities and surprisingly sintering resistance may be observed for such minute clusters, even outperforming larger nanoparticles [229, 231, 280, 281]. There is, therefore, a convenient synergy between systems which must be treated with sophisticated electronic structure

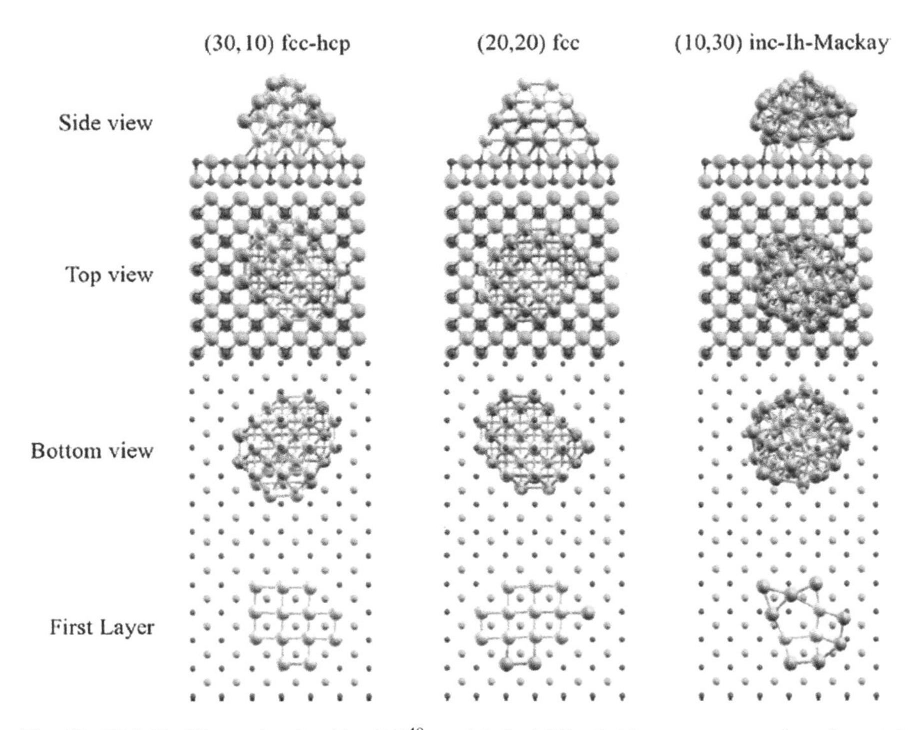


**Fig. 13** BHMC GO results for (Au–Pd)[40] on MgO (100). Cubic structures and preferential segregation of silver to maximise epitaxy are determined for the GM. Reprinted from Ismail et al. [271], in the Journal of Physical Chemistry C. 117 2013, pp. 293. Copyright 2013 American Chemical Society

methods, and their relevance to experiment. In the following section, we sample the recent work of several authors who have extended their surface GO methods to include direct DFT-GO.

### 8.1.1 Density Functional Surface GO

Fortunelli has implemented BHMC DFT GO for clusters upon surfaces [282–284], finding the GM structures of $Ag_nPd_1$ (n = 1–8) clusters bound over $F_s$-centre defects on MgO(100). It was noted that there is a transition from 2D to 3D motifs at $Ag_5Pd_1$, above which pentagonal bipryramidal structures are preferred, while the atom anchored to the defect site is always palladium. The closed-shell, complete pentagonal bipyramid was found to be a magic number cluster, with remarkable stability, due to the confluence of electronic and geometric stabilisation. The same authors remark in a related paper for $Ag_n$ clusters on $F_s$-MgO(100) [282] that "the DF-BH approach is practically unfeasible in the case of systems composed by more than few tens of atoms", which may still be considered largely the case.

Vilhelmsen and Hammer applied a direct EA DFT GO at the surface for sub-nanometre clusters of gold upon defect MgO [48], in which the upper surface layers were free to locally relax. They found a number of more stable $Au_8$ clusters than previously reported, implying the need for an unbiased GO process, even for the smallest of clusters. Extended to $Au_{24}$ upon $TiO_2$ [79], the authors note an unusual interfacial layer of oxygen, which was unlikely to be predicted from a biased search. More recently the same group has developed and benchmarked the EA to more efficiently perform direct DFT-GO [77], including a series of new move-classes, specific to the surface-bound regime. Mutations which were found to improve the convergence to the GM include a rotation of the cluster about the surface normal vector and a symmetrisation operator which reflects the image through a plane parallel to the surface normal. Several of the systems for which this algorithm has been applied are displayed in Fig. 14.

The BCGA has recently been extended in a similar manner, to include direct, surface-bound DFT–GO [258], and was first applied to the structures of bimetallic Pd–Ag and Pt–Pd tetramer clusters upon MgO(100). It was observed that the preference for particular binding sites for a CO molecule upon the clusters was governed by the localised nature of the charge, while the pre-organisation of the M–CO bonding orbital found for a single atom was lost in the mixed clusters. This result highlights both the important of accurate treatment of the electronic properties of clusters on supports, in addition to the optimisation of chemical ordering.

The accurate treatment of the surface, including GO analysis of its structure, rather than of the cluster, is not an area of focus for the current chapter. For work in this area, we refer the reader to articles by Sierka et al. [285] and Chuang et al. [286]. However, the hybrid ab initio genetic algorithm (HAGA) of Sierka is capable of isolated cluster GO, in addition to surface GO, in which the surface itself is the object of the GO process [78]. These two regimes overlap in the case where metal and metal oxide adsorbates are considered part of the surface to be optimised, and thus this method may be considered a DFT surface GO method in the same manner as those described above. HAGA has been used to locate the GM for low dimensional thin films of silica and alumina upon Mo(112) surfaces, directly at the DFT level, for example [78].

A view of the next stage in computational modelling of low-lying structures of supported clusters is provided by Negreiros et al. [287], who have introduced the so-called reactive global optimisation (RGO) scheme. RGO is in effect a DFT GO method capable of including kinetic information, in order to build up the picture of a set of possible reaction pathways. The method aims to locate low energy local minima of clusters upon surfaces according to a search which is limited to energy barriers defined for the reaction in question. By rejecting unprofitable reaction pathways, introducing adsorbates to the cluster, and taking timesteps in a sequential manner, the energy landscape of a whole reaction may be mapped out, according to limits of temperature, adsorbate gas pressure, time and activation energy which are user-defined. The strength of this procedure lies in its deviation from the strictly energetic view of GO. By adapting the structure prediction according to kinetic constraints, and including realistic operational conditions, the RGO improves upon

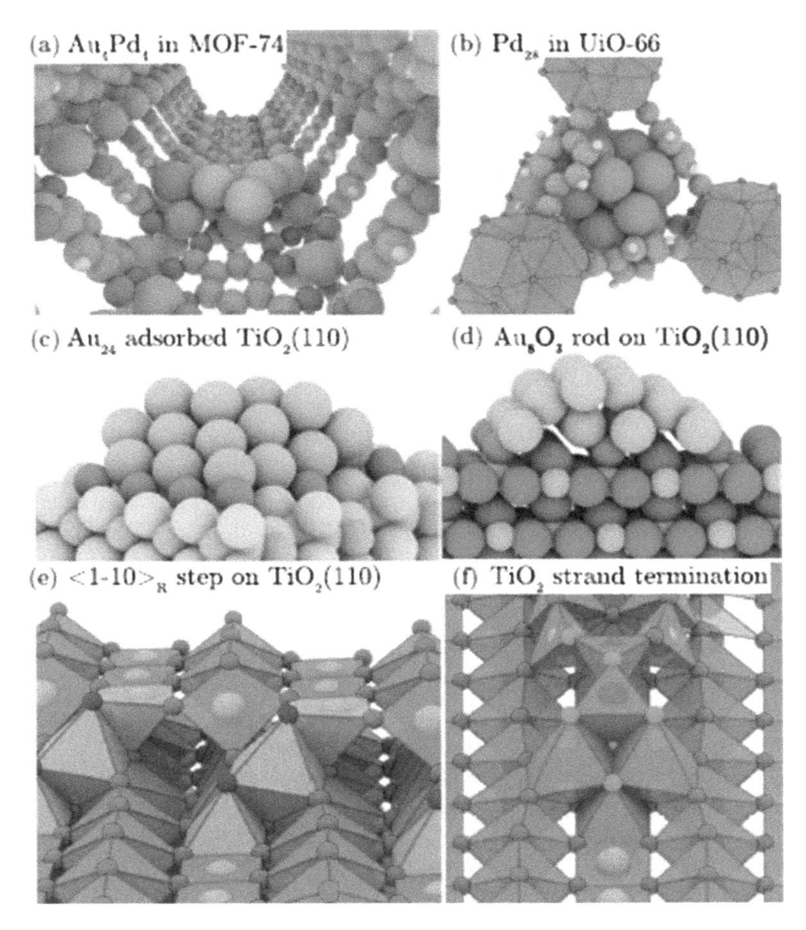


**Fig. 14** Example systems upon which the DFT-EA GO method of Vilhelmsen and Hammer has been applied recently. Reprinted with permission from Vilhelmsen et al. [77] in the Journal of Chemical Physics, 141, 044711. Copyright 2014, AIP Publishing LLC

static GO for any experimental scenario where the distribution of reaction products is not governed solely by the energies of the final structures. The method has so far been applied to propylene oxidation by mono and bimetallic Ag/Au trimer clusters on MgO(100) [280, 287], and CO oxidation over the same supported clusters [288].

## 8.2 *Passivated Clusters*

The study of ligand passivated clusters has been dominated by gold-thiol interactions, theoretically [289–293] and experimentally [294–297]. The question of cluster size, ligand chain length, and preferred binding mode, on the overall stability and structure of such systems has been debated since the 1990s. Early work involved the use of empirical potentials to model the effect of cluster environment on the orientation of the thiols, with Luedtke and Landman predicting "bundling" of the ligands along preferred directions [298], which was manifested in isolated clusters and superlattices, but not when bound upon a surface. Further, they found that magic sizes appear, which dominate the distribution of clusters in the 1–2 nm range, and consist of Marks decahedral motifs [289]. Wilson and Johnston applied the Murrell-Mottram potential to a series of high symmetry gold clusters [178], between 38 and 55 atoms, to which a monolayer of thiol groups are attached, in an implicit thiol model. A crossover in total stability between the icosahedron and fcc geometries is observed to take place.

Experimental studies of small gold-thiol clusters exhibit spectra which imply the existence of structurally ordered metal cores [295–297]. The ultimate determination of these structures for selected, notable stable sizes, became a focus of many studies, often performed with DFT calculations. $Au_{102}(RS)_{44}$ was determined [300], shortly followed theoretically by that of $Au_{25}(RS)_{18}$ [290], both of which found to obey the "divide and protect" concept of Häkkinen et al. [301] From the point of view of GO, Xiang et al. have developed a direct DFT EA method [299], which they applied to the $Ag_7^-$ cluster, ligated with $(SCH_3)_4$ or $(DMSA)_4$. The procedure involved performing mating and mutation steps on the metal cluster core, with only one ligand atom bound, followed by the re-introduction of the remaining ligand chain for local geometry optimisation. As a result, the usual phenotypic EA operators may be used, while the ligand is treated by the DFT calculations on the resulting structure. Steric hindrance between ligand chains is treated by inserting a step between the EA operators and the local optimisation, in which ligand chains are reoriented by an MC run, to minimise repulsion. The putative GM structure found with this method is shown in Fig. 15.

There is, however very little by way of computational GO for passivated clusters, despite the clear need to extend the study of such passivated clusters with an unbiased GO approach. The complexity of multiple types of bond, and the inclusion of several elements, drastically increases the computational cost, beyond the means of most current methods. To our knowledge there have been no formal GO studies of binary passivated metal clusters, although the machinery for such a study is in place, as evidenced by the study of Xiang et al. [299]. Theoretical work involving silver and copper dopants into $Au_{25}(RS)_{18}$ has recently been performed by Aikens et al. [302] and Guidez et al. [303], as well as joint DFT/experimental work with $Cu_sAu_{(25-x)}(RS)_{18}$ and $Ag_xAu_{(25-x)}(RS)_{18}$ by Negisihi et al. [304–306] Palladium doped gold thiolate clusters have been explored by Kacprzak et al. [307], Negishi et al. [308, 309] and Yixiang et al. [44], while a broader range of passivated clusters of gold, have also been investigated from both theoretical and experimental perspectives [310]. Where there are metallic clusters of unknown structure, the need for binary cluster GO follows closely behind.

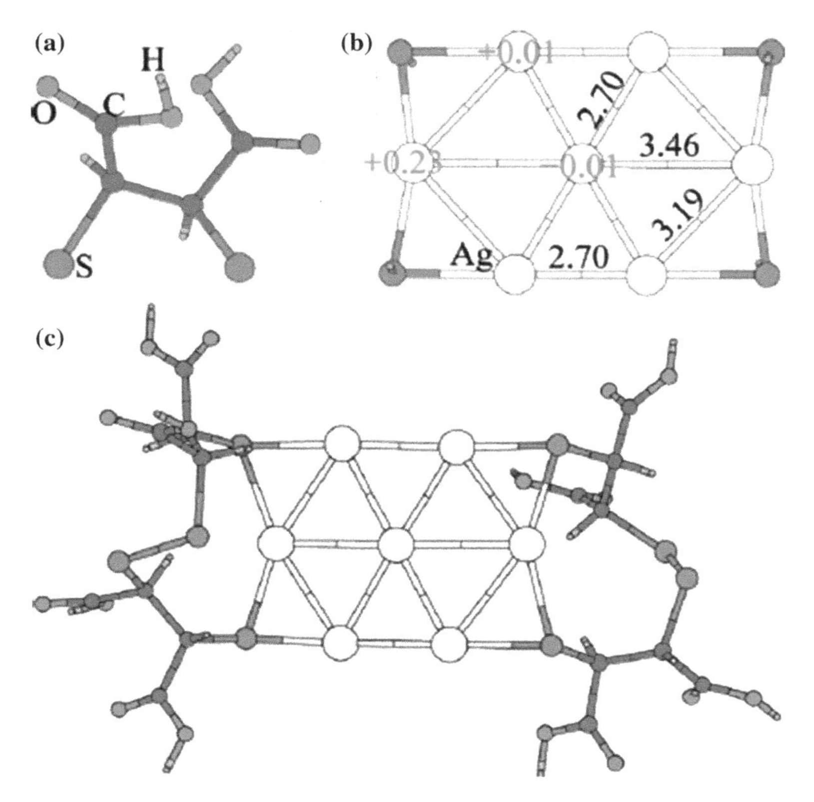


**Fig. 15** Ground state structure of $Ag_7(DMSA)^-$. Bond lengths and charges are overlaid. **a** is the DMSA ligand, **b** is the GM determined from the EA, **c** is resulting structure after reconstruction of the full, passivated system. Reprinted from Xiang et al. [299], in the Journal of the American Chemical Society. **132** 2010, pp. 7355. Copyright 2010 American Chemical Society

# 9 Higher Order Nanoalloys

From binary systems, the natural inclination is to extend to high order nanoalloys. There is good reason to be interested in such systems, particularly in the fields of mineralogy and materials science. An industrial justification comes from aerospace technology, where research into quarternary and quinary alloys is already extant, for the design of alloy materials, highly optimised for their thermal and structural properties.

Many of the improvements available for the GO of binary clusters apply to ternary and higher systems with little modification. Exchange moves, two-phase homotop searching and core-to-shell crossover steps in EAs are all generalisable. There is little published work on such clusters to date. One reason is simply that the combinatorial expansion of possible homotops proves fatal to the unambiguous prediction of chemical ordering. Another is the geometric part of the energy landscape becomes significantly more complex. The reduction in symmetry which is likely to result from including more metal elements means a more heterogeneous landscape with a larger number of minima, which are more difficult to claim to have been exhaustively searched. However, the more sophisticated recent chemical ordering optimization techniques, such as biminimisation [181] suggest orders of magnitude may be saved in computing cost with an intelligent choice of homotop moves, extending the complexity of systems available for investigation significantly. We therefore expect great steps to be made in this direction in coming years.

For three-component metal oxide clusters, Woodley applied an EA to $(MgAl_2O_4)_n$, $(Al_2SiO_5)_n$, $(Mg_2SiO_4)_n$, and $(MgSiO_3)_n$ clusters, for n = 1–6 [90], using a Born pairwise potential model. The performance of mutation moves, in particular ion exchange steps, is investigated, showing that exchange between anions and cations was more beneficial that merely exchanging cations. Additionally, the charge and size of the pair of cations is a useful descriptor of the benefit for ion exchange mutations, with similar cations exchanging more successfully. The effects of exchange move type, and their sensitivities to cluster size, and which other mutation steps to which they are coupled, says much about the fundamental principles underlying the search over a multimetallic oxide cluster energy landscape.

Turning towards multinary metal clusters, Cheng et al. [311] used a Gupta potential to study the trimetallic Au–Ag–Cu cluster with Metropolis MC simulations, for Ih magic number sizes, 147, 309, 561 and 923 atoms. The structural search was thus simplified, due to the structure-seeking nature of such clusters. The preference for chemical ordering, which was captured by the inclusion of an atom exchange moveclass, showed a clear preference for a silver shell, and a copper core, with gold atoms in intermediate positions with a broad radial distribution. For compositions in which the silver was insufficient to create a complete core, the gold intermixed, in line with expectations based on surface energy and atomic radius. As mentioned previously, ternary alkali metal nanoalloys have been considered by Aguado and López, for selected monovalent "simple" metal clusters, containing Li, Na and Cs sizes up to 55 atoms.

For test systems which employ variants of LJ potentials, Wu and colleagues [96] have applied a ternary TLJ model to the study of 38 and 55 atom Ar–Kr–Xe clusters, while Dieterich and Hartke have gone further, treating selected ternary and quinary clusters [99] for selected sizes up to 55 atoms, with LJ potentials fitted to high level ab-initio data, as shown in Fig. 16. They note that while binary mixtures of approximately 1:1 composition may retain similar structures to their homo-atomic counterparts, small amounts of ternary atom doping may be sufficient to significantly alter the geometric preference. For $Ar_{19}$ $Kr_{19}$, the familiar fcc truncated octahedron remains the GM, as for $Ar_{38}$ and $Kr_{38}$, whereas $Ar_{18}Kr_{19}Xe$ exhibits a dramatic destabilisation of the fcc structure with respect to an incomplete anti-Mackay icosahedral motif. These studies highlight the limitations of current methods, by considering simple empirical potentials and selected small cluster sizes, which are known to exhibit high symmetry structures. We know of no GO studies of clusters of metallic elements of higher than ternary order so far,

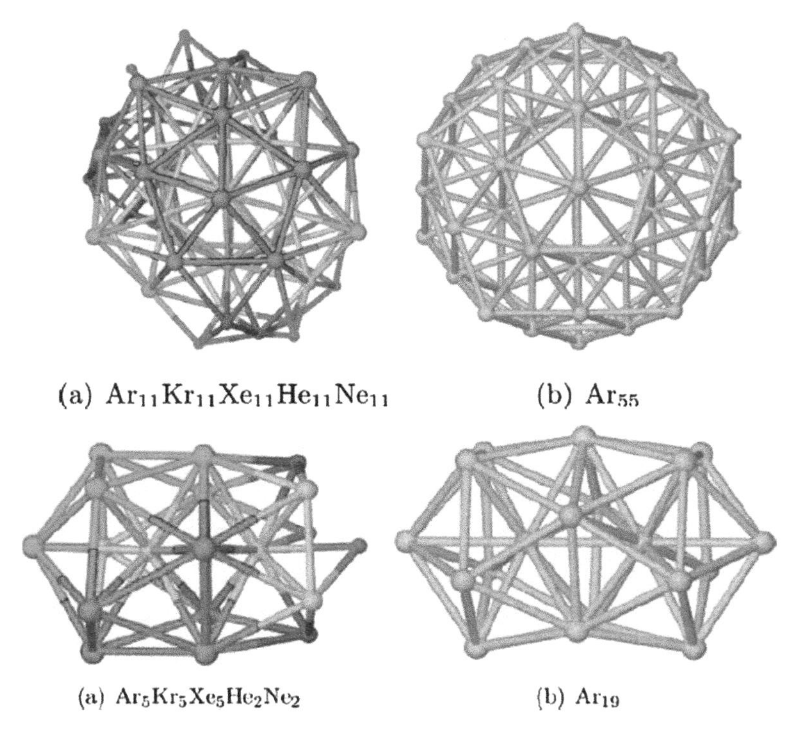


**Fig. 16** GM structures of quinary rare-gas clusters and their mono-elemental counterparts, calculated with the LJ potential. From Dieterich et al. [99], in the Journal of Computational Chemistry. Copyright 2011 by John Wiley & Sons, Inc. Reprinted by permission of John Wiley & Sons, Inc

though, as in the case of passivated clusters, the computational apparatus is now available.

## 10 Conclusions and Future Outlook

In this chapter we have outlined the various methods most commonly applied to the global optimisation of clusters, with a particular focus on those which have proven themselves capable of treating binary systems. The additional complexity of the binary cluster energy landscape, primarily the issue of permutational isomerisation has been tackled by many of these methods, either by the development and inclusion of appropriate moveclasses, or by the use of two-phase optimisation methods. We have covered several recent instances of these homotop optimisation strategies, ranging from simple exchange moves, to in-depth transition path sampling on the full 3*N*-dimensional energy landscape. The overall view of binary cluster global optimisation has, over the last 15–20 years grown in scope, as the sophistication of techniques and the available computing power have both grown, extending the range of sizes and elements which can be treated. With empirical potentials routinely aiming to locate global minima for binary clusters in the several hundred atom regime, the gap between theory and experiment is rapidly narrowing.

While ever larger clusters, with ever more demanding chemical ordering patterns are coming under the remit of global optimisation, in the past five to ten years there has also been a resurgence of interest in the very smallest of clusters, which has come about for several good reasons. The most important is the understanding of the unique properties exhibited by clusters of fewer than twenty atoms in experiments. Indeed, with recent results in the field of heterogeneous catalysis for gold, platinum, silver and more, it appears that the sub-nanometre cluster regime forms a class of its own, distinct from that of the nanoparticle. The accurate determination of shape and properties which comes from computational global optimisation of these tiny particles has never been so important, and the development of DFT-GO strategies has been vital in keeping up with the needs of experimentalists in this area. The field is ripe for re-evaluation of our understanding of the ultrasmall metallic clusters, with binary clusters an important part of this effort.

Looking forward, there are a number of areas in which the challenges of accurate global optimisation are still not satisfactorily met. We have highlighted some of those which we think are crucial for the next few years throughout this chapter. Accurately predicting the chemical ordering, structures and ligand-ligand effects of passivated metal clusters is an area which could become a major item in the theoretician's toolkit in coming years. The overlap between thermodynamics and kinetics in surface-bound cluster science is another area which is growing. Several authors have begun to take information gained from structure prediction further than simply the location of good structures. Simulation methods which utilise GO as part of a kinetic analysis of likely reaction pathways, and those which consider experimental information, such as the presence of impassable barriers, or entropic considerations are all available at present. We expect the adoption of these techniques, such that the theoretician can more closely connect to the specific experiment of interest will become a more mainstream concept in cluster science in coming years. This has already been performed for the smallest of binary metal clusters, and the availability of reliable surface-supported cluster optimisation algorithms has been a vital component of this advancement.

Broadening the horizons of bimetallic cluster GO will require the effective combination of many of the algorithms available today, as it has been known for many years that no one technique is able to outperform all others over a sufficiently broad range of systems. A significant forward step in the improvement of GO methods demands the dedicated combination, comparison and collaboration between the many varied, and subtle techniques currently in existence.

**Acknowledgments** R.L.J. acknowledges members of his research group and collaborators, past and present, for their contributions to his research in the area of nanoalloy GO and development of GO strategies. C.J.H. wishes to acknowledge the School of Chemistry at the University of

Birmingham, and the EPSRC for funding his PhD studies in the Johnston group. Both authors are grateful to Mark Oakley for supplying data on combined GO approaches, and acknowledge support from EU COST Action MP0903:NANOALLOY and EPSRC Critical Mass Grant (EP/J010804/1) "TOUCAN: TOwards an Understanding of CAtalysis on Nanoalloys".

## References

1. Ferrando R, Jellinek J, Johnston RL (2008) Chem Rev 108:845
2. Wales DJ (2003) Energy landscapes with applications to clusters, glasses and biomolecules. Cambridge University Press
3. Baletto F, Ferrando R (2005) Rev Mod Phys 77(1):371–423
4. Calvo F (In Press, 2015) Phys Chem Chem Phys
5. Leary RH (2000) J Glob Optim 18(4):367–383
6. Cheng L, Feng Y, Yang J, Yang J (2009) J Chem Phys 130:214112
7. Nocedal J, Wright S (2006) Numerical optimization, 2nd edn. Springer
8. Press WH, Teukolsky SA, Vetterling WT, Flattery BP (2007) Numerical recipes: the art of scientific computing, 3rd edn. Cambridge University Press
9. Liotard DA (1992) Int J Quant Chem 44:723–741
10. Schlegel HB (2011) Wiley Interdiscip Rev Comput Mol Sci 1:790–809
11. Tsai CJ, Jordan KD (1993) J Phys Chem 97:11227–11237
12. Stillinger FH (1999) Phys Rev B 59:48–51
13. Sutton AP, Chen J (1990) Philos Mag Lett 61:139
14. Doye JPK, Wales DJ. New J Phys 22:733–744, 1198
15. Cleri F, Rosato V (1993) Phys Rev B 48:22
16. Doye JPK (2006) Comp Mater Sci 35:227–231
17. Darby S, Mortimer-Jones TV, Johnston RL, Roberts C (2002) J Chem Phys 116:1536
18. Baletto F, Mottet C, Ferrando R (2003) Phys Rev Lett 90(13):135504
19. Rossi G, Rapallo A, Mottet C, Fortunelli A, Baletto F, Ferrando R. Phys Rev Lett
20. Doye JPK, Hendy SC (2003) Eur Phys J D 22:99–107
21. Doye JPK (2003) J Chem Phys 119:1136–1147
22. Morse PM (1929) Phys Rev 34:57
23. Doye JPK, Wales DJ, Berry RS (1996) J Chem Phys 103:4234–4249
24. Doye JPK, Wales DJ (1997) Z Phys D 40:194
25. Lennard-Jones JE (1924) Proc R Soc Lond A 106(738):463–477
26. Xiang Y, Jiang H, Cai W, Shao X (2004) J Phys Chem A 108:3586–3592
27. Deaven DM, Ho KM (1995) Phys Rev Lett 75:288–291
28. Wales DJW, Doye JPK (1997) J Phys Chem A 101:5111
29. Wolf MD, Landman U (1998) J Phys Chem A 102:6129–6137
30. Atanasov I, Ferrando R, Johnston RL (2014) J Phys Condens Mater 26:275–301
31. Barcaro G, Sementa L, Fortunelli A (2014) Phys Chem Chem Phys 16:24256–24265
32. Kresse G, Furthmüller J (1996) Comput Mater Sci 6:15–50
33. Giannozzi P, Baroni S, Bonini N, Calandra M, Car R, Cavazzoni C, Ceresoli D, Chiarotti GL, Concoccioni M, Dabo I, Dal Corso A, de Gironcoli S, Fabris S, Fratesi G, Gebauer R, Gerstmann U, Gougoussis C, Kokalj A, Lazzeri M, Martin-Samos L, Marzari N, Mauri F, Mazzarello R, Paolini S, Pasquarello A, Paulatto L, Sbraccia C, Scandolo S, Sclauzero G, Seitsonen AP, Smogunov A, Umari P, Wentcovitch RM (2009) J Phys Condens Matter 21:395502
34. Valiev M, Bylaska EJ, Govind N, Kowalski K, Straatsma TP, van Dam HJJ, Wang D, Nieplocha J, Aprá E, Windus TL, de Jong WA (2010) Comput Phys Commun 181:1477
35. Rogan J, Ramirez M, Varas A, Kiwi M (2013) Comp Theor Chem 1021:155–163
36. Heiles S, Johnston RL (2013) Int J Quant Chem 113:2091–2109

37. Bartlett RJ, Musial M (2007) Rev Mod Phys 79(1):291–352
38. Möller C, Plesset MS (1934) Phys Rev 46:618–622
39. Archibong EF, St-Amant A (1998) J Chem Phys 109(3):962–972
40. Pakiari AH, Jamshidi Z (2010) J Phys Chem A 114:9212–9221
41. Osorio E, Villalobos V, Santos JC, Donald KJ, Merino G, Tiznado W (2012) Chem Phys Lett 522:67–71
42. Tam NM, Tai TB, Ngan VT, Nguyen MT (2013) J Phys Chem A 117:6867–6882
43. David J, Guerra D, Hadad CZ, Restrepo A (2010) J Phys Chem A 114:10726–10731
44. Yixiang Q, Jia L, Shugang W (2010) Acta Chim Sinica 68(7):611–616
45. Gao Y, Bulusu S, Zeng XC (2005) J Am Chem Soc 127:15680–15681
46. Aprá E, Ferrando R, Fortunelli A (2006) Phys Rev B 73:205414
47. Pal R, Wang L-M, Huang W, Wang L-S, Zeng XC (2011) J Chem Phys 134:054306
48. Vilhelmsen L, Hammer B (2012) Phys Rev Lett 108:126101
49. Heiles S, Logsdail AJ, Schäfer R, Johnston RL (2012) Nanoscale 4(4):1109–1115
50. Barcaro G, Fortunelli A, Rossi G, Nita F, Ferrando R (2006) J Phys Chem B 110:23197–23203
51. Paz Borbón LO, Johnston RL, Barcaro G, Fortunelli A (2008) J Chem Phys 128:134–517
52. Negreiros F, Kuntová Z, Barcaro G, Rossi G, Ferrando R, Fortunelli A (2010) J Chem Phys 132:234703
53. Nuñez S, Johnston RL (2010) J Phys Chem C 114:13255
54. Toai TJ, Rossi G, Ferrando R (2007) Faraday Discuss 138:49–58
55. Raoult B, Farges J, Deferaudy MF, Torchet G (1989) Philos Mag 60(6):881–906
56. Cleveland CL, Landman U (1991) J Chem Phys 94(11):7376–7396
57. Krainyukova NV (2007) Eur Phys J D 43:45–48
58. Duan Z, Wang GJ (2011) Phys Condens Mater 23:475301
59. Tang J, Deng L, Deng H, Xiao S, Zhang X, Hu W (2014) J Phys Chem C 118:27850–27860
60. Rubinovitch L, Polak M (2009) Phys Rev B 80:045404
61. Tigger H, Rubinovitch L, Polak M (2012) J Phys Chem C 116:26000–26005
62. Laasonen K, Panizon E, Ferrando R (2013) J Phys Chem C 117:26405–26413
63. Shim J-H, Lee BJ, Cho YW (2002) Surf Sci 512:262–268
64. Qin Y, Fichthorn KA (2003) J Chem Phys 119:9745–9754
65. Cheng S, Grest GS (2013) J Chem Phys 138:064701
66. Leardi RJ (2001) Chemometrics 15:559–569
67. Hartke B (2004) In: Applications of evolutionary computation in chemistry. Springer
68. Hartke B (2011) Wiley Interdiscip Rev Comput Mol Sci 1(6):879–887
69. Johnston RL (2003) Dalton Trans 22:4193–4207
70. Sun J, Lai C-H., Wu X-J (2012) Particle swarm optimization: classical and quantum perspectives, 1st edn. Taylor and Francis
71. Yang XS (ed) (2014) Nature-inspired optimization algorithms, 1st edn. Elsevier
72. Bielli M, Caramia M, Carotenuto P (2002) Transp Res Part C 10:19–34
73. Hornby GS, Globus A, Linden DS, Lohn JD (2006) In: Proceedings of 2006 American institute of aeronautics and astronautics conference on space, pp 19–21
74. Woodley SM, Battle PD, Gale JD, Catlow CRA (2004) Phys Chem Chem Phys 6:1815–1822
75. Woodley SM (2004) Phys Chem Chem Phys 6:1823–1829
76. Hartke B (1993) J Phys Chem 97:9973–9976
77. Vilhelmsen L, Hammer B (2014) J Chem Phys 141:044701
78. Sierka M (2010) Prog Surf Sci 85:398–434
79. Vilhelmsen L, Hammer B (2013) J Chem Phys 139:204701
80. Shayeghi A, Götz D, Davis JBA, Schäfer R, Johnston RL (2015) Phys Chem Chem Phys 17:2014–2122
81. Rappallo A, Rossi G, Ferrando R, Fortunelli A, Curley BC, Lloyd LD, Tarbuck GM, Johnston RL (2005) J Chem Phys 122:194308
82. Deaven DM, Tit N, Morris JR, Ho KM (1996) Chem Phys Lett 256:195–200
83. Goedecker S (2004) J Chem Phys 120(21):9911–9917

84. Rossi G, Ferrando R (2006) Chem Phys Lett 423:17–22
85. Lee J, Lee I-H, Lee J (2003) Phys Rev Lett 91(8):080201
86. Stillinger FH, Weber TA (1988) J Stat Phys 52(5–6):1429–1445
87. Li Y, Scheraga HA (1987) Proc Natl Acad Sci 84:6611–6615
88. Lamarck J-BP (1778) Flore françoise, 1st edn. L'Imprimerie Royale
89. Darwin CR (1859) On the origin of species, 1st edn. John Murray
90. Woodley SM, Catlow CRA (2009) Comput Mater Sci 45:84–95
91. Pillardy J, Piela L (1995) J Phys Chem 99:11805–11812
92. Rossi G, Ferrando R (2009) J Phys Condens Mater 21:084–201
93. Lysgaard S, Landis DD, Bligaard T, Vegge T (2014) Top Catal 57:33–39
94. Ferrando R, Fortunelli A, Johnston RL (2007) Phys Chem Chem Phys 10:640–649
95. Marques JMC, Pereira FB (2010) Chem Phys Lett 485:211–216
96. Wu X, Huang C, Sun Y, Wu G (2013) Chem Phys 415:69–75
97. Wu X, Cai W, Shao X (2009) J Comput Chem 30:1992–2000
98. Wu X, Wu G, Chen Y, Qiao Y (2011) J Phys Chem A 115:13316–13323
99. Dieterich JM, Hartke B (2011) J Comput Chem 32:1377–1385
100. Kennedy J, Eberhardt R (1995) Proceedings of IEEE International Conference on Neural Networks, pp 1942–1948
101. Call ST, Zubarev DU, Boldyrev AI (2007) J Comput Chem 28:1177–1186
102. Zhou JC, Li WJ, Zhu JB (2008) Mater Sci Technol 24(7):870–874
103. Wang Y, Lv J, Zhu L, Ma Y (2010) Phys Rev B 82:094116
104. Wang Y, Miao M, Lv J, Zhu L, Lin K, Liu H, Ma Y (2012) J Chem Phys 137:224108
105. Lv J, Wang Y, Zhu L, Ma Y (2012) J Chem Phys 137:084104
106. Karaboga D, Basturk B (2007) J Glob Optim 39:459–471
107. von Rudorff GF, Wehmeyer C, Sebastiani D (2014) Comput Phys Commun 185:1639–1646
108. Wehmeyer C, von Rudorff GF, Wolf S, Kabbe G, Schärf D, Kühne TD, Sebastiani D (2012) J Chem Phys 137:194110
109. Dorigo M, Stützle T (2004) Ant colony optimization, 1st edn. Bradford Company
110. Oakley MT, Richardson EG, Carr H, Johnston RL (2013) EEE/ACM Trans Comput Biol Bioinf 10(6):1548–1552
111. Lourenco N, Pereira FB. 14th international conference on genetic and evolutionary computation conference (GECCO), pp 41–48
112. Tsutsui S. In: Ant colony optimization and swarm intelligence, 1st edn. Springer
113. Metropolis N, Rosenbluth AW, Rosenbluth MN, Teller AN, Teller E (1953) J Chem Phys 21:1087–1092
114. Doye JPK, Meyer L (2005) Phys Rev Lett 95:063401
115. Paz-Borbón LO, Hellman A, Thomas JM, Grönbeck H (2013) Phys Chem Chem Phys 15:9694–9700
116. Zhan L, Chen JZY, Liu W-K, Lai SK (2005) J Chem Phys 122:244707
117. Noya EG, Doye JPK, Wales DJ, Aguado A (2006) Eur Phys J D 43(1–3):47–60
118. Huang W, Lai X, Xu R (2011) Chem Phys Lett 507:199–202
119. Sicher M, Mohr S, Goedecker S (2011) J Chem Phys 134:044106
120. Kolossvary I, Bowers KJ (2010) Phys Rev E 82:056711
121. Bulusu S, Zeng XC (2006) J Chem Phys 125:154303
122. Bulusu S, Zeng XC (2006) Proc Natl Acad Sci 103(22):8326–8330
123. Tao Y, Ruchu X, Wenqi H (2011) J Chem Inf Model 51:572–577
124. Hartke B (1999) J Comput Chem 20(16):1752–1759
125. Takeuchi H (2006) J Chem Inf Model 46:2066–2070
126. Heidari I, De S, Ghazi SM, Goedecker S, Kanhere DG (2011) J Phys Chem A 115:12307–12314
127. Glover F (1989) Orsa J Comput 1:190–206
128. Kim HG, Choi SK, Lee HM (2008) J Chem Phys 128:144702
129. Neirotti JP, Calvo F, Freeman DL, Doll JD (2000) J Chem Phys 112(23):10340–10349

130. Miller MA, Bonhommeau DA, Heard CJ, Shin Y, Spezia R, Gaigeot M-P (2012) J Phys Condens Mater 24:284130
131. Calvo F, Yurtesver E (2004) Phys Rev B 70:045423
132. Jellinek J, Garzón ILZ (1991) Phys D Atom Mol Cl. 20(1–4):239–242
133. Garzón IL, Jellinek JZ (1993) Phys D Atom Mol Cl. 26(1–4):316–318
134. Garzón IL, Posada-Amarillas A (1996) Phys Rev B 54(16):11796–11802
135. Rodriguez-López JL, Montejano-Carrizales JM, José-Yacamán M (2003) Appl Surf Sci 219:56–63
136. Reyes-Nava JA, Rodriguez-López JL, Pal U (2009) Phys Rev B 80:161412
137. Tan TL, Wang L-L, Johnson DD, Bai K (2012) Nano Lett 12:4875–4880
138. Sibani P, Schön JC, Salamon P, Andersson JO (1993) Europhyhs Lett 22:479–485
139. Schön JC, Putz H, Jansen M (1996) J Phys Condens Mater 8:143
140. Neelamraju S, Schön JC, Doll K, Jansen M (2012) Phys Chem Chem Phys 14:1223
141. Doll K, Schön JC, Jansen M (2010) J Chem Phys 133:024107
142. Pacheco-Contreras R, Dessens-Félix M, Borbón-González DJ, Paz-Borbón LO, Johnston RL, Schön JC, Posada-Amarillas A (2012) J Phys Chem A 116:5235
143. Pacheco-Contreras R, Borbón-González DJ, Dessens-Félix M, Paz-Borbón LO, Johnston RL, Schön JC, Jansen M, Posada-Amarillas A (2013) RSC Adv 3:11571–11579
144. Heard CJ, Schön JC, Johnston RL (In press, 2015) Chem Phys Chem
145. Schön JC, Wevers MAC, Jansen M (2003) J Phys Condens Mater 15:5479
146. Wevers MAC. Schön JC, Jansen M (1999) J Phys Condens Mater 11:6487
147. Schön JC, Cancarevic ZP, Hannemann A, Jansen M (2008) J Chem Phys 128:194712
148. Zagorac D, Schön JC, Jansen M (2012) J Phys Chem C 116(31):16726
149. Calvo F, Fortunelli A, Negreiros F, Wales DJ (2013) J Chem Phys 139:111102
150. Miller MA, Doye JPK, Wales DJ (1999) Phys Rev E 60(4):3701–3718
151. Wales DJ (2002) Mol Phys 100:3285–3306
152. Carr J, Wales DJ (2005) J Chem Phys 123:234901
153. Chakrabarti D, Fejer S, Wales DJ (2009) Proc Natl Acad Sci 106:20164–20167
154. Oakley MT, Johnston RLJ (2013) Chem Theory Comput 9:650–657
155. Wales DJ. OPTIM: a program for optimizing geometries and calculating reaction pathways. www.wales.ch.cam.ac.uk/OPTIM
156. Trygubenko S, Wales DJ (2004) J Chem Phys 120(5):2082–2094
157. Sheppard D, Terrell R, Henkelman G (2008) J Chem Phys 128:134106
158. Shao XG, Cheng LJ, Cai WS (2004) J Comput Chem 25:1693–1698
159. Wu X, Cai WS, Shao XG (2009) Chem Phys 363(1–3):72–77
160. Lai XJ, Huang WQ, Xu RCJ (2011) Chem Phys A 115(20):5021–5026
161. Wu X, Cheng W (2014) J Chem Phys 141:124110
162. Shao X, Liu X, Cai WSJ (2005) Chem Theor Comput 1:762–768
163. Shao XG, Wu X, Cai WS (2010) J Phys Chem A 114(1):29–36
164. Lai X, Xu R, Huang W (2011) J Chem Phys 135:164109
165. Laio A, Parrinello M (2002) Proc Natl Acad Sci 99:12562–12566
166. Santarossa G, Vargas A, Ianuzzi M, Baiker A (2010) Phys Rev B 81:174205
167. Tribello GA, Ceriotti M, Parrinello M (2010) Proc Natl Acad Sci 107(41):17509–17514
168. Tribello GA, Cuny J, Eshet H, Parrinello M (2011) J Chem Phys 135:114109
169. Calvo F (2010) Phys Rev E 82:046703
170. Pavan L, Di Paola C, Baletto F (2013) Eur Phys J D 67:24
171. Zhai Y, Laio A, Tosatti E, Gong X-G (2011) J Am Chem Soc 133:2535–2540
172. Wang H-Q, Kuang X-Y, Li H-F (2012) Phys Chem Chem Phys 10:5156–5165
173. Calvo F, Cottancin E, Broyer M (2008) Phys Rev B 77:121406
174. Owicki JC, Scheraga HA (1977) Chem Phys Lett 47(3):600–602
175. Faller R, de Pablo JJ (2003) J Chem Phys 119(8):4405–4408
176. Shao GF, Wang TN, Liu TD, Chen JR, Zheng JW, Wen YH (2015) Comput Phys Commun 186:11–18
177. Fortunelli A, Velasco AM (1999) J Mol Struct 487(3):251–266

178. Wilson NT, Johnston RL (2002) J Mater Chem 12:2913–2922
179. Paz-Borbón LO, Johnston RL, Barcaro G, Fortunelli A (2009) Eur Phys J D 52(1–3):131–134
180. Schebarchov D, Wales DJ (2013) J Chem Phys 139:221101
181. Schebarchov D, Wales DJ (2014) Phys Rev Lett 113:156102
182. Cheng D, Huang S, Wang W (2006) Eur Phys J D 39:41–48
183. Cheng D, Huang S, Wang W (2006) Phys Rev B 74:064117
184. Cheng D, Wang W, Huang S (2006) J Phys Chem B 110:16193–16196
185. Davis JBA, Johnston RL, Rubinovitch L, Polak M (2014) J Chem Phys 101(22):224307
186. Kozlov SM, Kovács G, Ferrando R, Neyman KM. Unpublished Work, 2015
187. Liu TD, Fan TE, Shao GF, Zheng JW, Wen Y-H (2014) Phys Lett A 378:2965–2972
188. Oakley MT. Personal communication
189. Hsu PJ, Lai SK (2006) J Chem Phys 124:044711
190. Schwerdtfeger P (2002) Heteroat Chem 13(6):578–584
191. Pyykkö P (2012) Ann Rev Phys Chem 63:45–64
192. Häkkinen H, Moseler M, Landman U (2002) Phys Rev Lett 89:033401
193. Kittel C (2005) Introduction to solid state physics, 8th edn. Wiley
194. Slater JC (1964) J Chem Phys 41(10):3191–3205
195. Tyson WR, Miller WA (1977) Surf Sci 62:267–276
196. deBoer FR, Boom R, Mattens WCM, Niessen AR (1988) Cohesion in metals, 1st edn. North Holland
197. Bonaçiç-Koutecký V, Burda J, Mitriç R, Ge M, Zampella G, Fantucci P (2002) J Chem Phys 117:3120
198. Weis P, Welz O, Vollmer E, Kappes MM (2004) J Chem Phys 120:677
199. Zhao JF, Zeng Z (2006) J Chem Phys 125:014303
200. Hong L, Wang H, Cheng J, Huang X, Sai L, Zhao J (2012) Comput. Theor Chem 993:36–44
201. Rohrdanz MA, Martins KM, Herbert JM (2009) J Chem Phys 130:054112
202. Shayeghi A, Johnston RL, Schäfer R (2013) Phys Chem Chem Phys 15(45):19715–19723
203. Shayeghi A, Heard CJ, Johnston RL, Schäfer R (2014) J Chem Phys 140:054312
204. Heard CJ, Johnston RL (2013) Eur Phys J D 65:1–6
205. Wang LM, Pal R, Huang W, Zeng XC, Wang L-S (2010) J Chem Phys 132:114306
206. Lang SM, Claes P, Cuong NT, Nguyen MT, Lievens P, Janssens E (2011) J Chem Phys 135:224305
207. Molayem M, Grigoryan VG, Springborg M (2011) J Phys Chem C 115:22148–22162
208. Rapallo A, Rossi G, Ferrando R, Curley BC, Lloyd LD, Tarbuck GM, Johnston RL (2005) J Chem Phys 122:194308
209. Langlois C, Li ZL, Yuan J, Alloyeau D, Nelayah J, Bochiccio D, Ferrando R, Ricolleau C (2012) Nanoscale 4:3381–3388
210. Panizon E, Bochicchio D, Rossi G, Ferrando R (2014) Chem Mater 26:3354–3356
211. Ferrando RJ (2015) Phys Condens Mater 27:013003
212. Rossi G, Ferrando R, Rappallo A, Fortunelli A, Curley BC, Lloyd LD, Johnston RL (2005) J Chem Phys 122:194309
213. Greenwood NN, Earnshaw A (1997) Chemistry of the elements, 2nd edn. Butterworth-Heinemann
214. Albert K, Neyman KM, Pacchioni G, Rösch N (1996) Inorg Chem 35:7370–7376
215. Massen C, Mortimer-Jones TV, Johnston RLJ (2002) Chem Soc Dalton Trans 23:4375–4388
216. Paz-Borbón LO, Mortimer-Jones TV, Johnston RL, Posada-Amarillas A, Barcaro G, Fortunelli A (2007) Phys Chem Chem Phys 9:5202–5208
217. Barcaro G, Fortunelli A, Polak M, Rubinovitch L (2011) Nano Lett 11:1766–1769
218. Wang Q, Sun Q, Yu JZ, Hashi Y, Kawazoe Y (2000) Phys Lett A 267:394–402
219. Guevara J, Llois AM, Aguilera-Granja F, Montejano-Carrizales JM (2004) Phys B 354:300–302
220. Wang G, Van Hove MA, Ross PN, Baskes MI (2005) J Chem Phys 122:024706

221. Pittaway F, Paz-Borbón LO, Johnston RL, Arslan H, Ferrando R, Mottet C, Barcaro G, Fortunelli A (2009) J Phys Chem C 113:9141–9152
222. Bruma A, Ismail R, Paz-Borbón LO, Arslan H, Barcaro G, Fortunelli A, Li ZY, Johnston RL (2013) Nanoscale 5:646–652
223. Leary RH, Doye JPK (1999) Phys Rev E 60(6):6320–6322
224. Bruma A, Negreiros FR, Xie S, Tsukuda T, Johnston RL, Fortunelli A, Li ZY (2013) Nanoscale 5:9620–9625
225. Haruta M, Yamada N, Kobayashi T, Iijima S (1989) J Catal 115(2):301–309
226. Falsig H, Hvolbök B, Kristensen IS, Jiang T, Bligaard T, Christensen CL, Nørskov JK (2008) Angew Chem Int Ed 47:4835–4839
227. Allian AD, Takanabe K, Fujdala KL, Hao X, Truex TJ, Cai J, Buda C, Neurock M, Iglesia E (2008) J Am Chem Soc 133:4498–4517
228. Kane MD, Roberts FS, Anderson SL (2014) Int J Mass Spectrom 370:1–15
229. Vajda S, Pellin MJ, Greeley JP, Marshall CL, Curtiss LA, Ballentine GA, Elam JW, Catillon-Mucherie S, Redfern PC, Mehmood F, Zapol P (2009) Nat Mat 8:213–216
230. Häkkinen H, Abbet S, Sanchez A, Heiz U, Landman U (2003) Angew Chem Int Ed 42 (11):1297–1300
231. Oliver-Meseguer J, Cbrero-Antonio JR, Dominguez I, Leyba-Perez A, Corma A (2012) Science 338:1452–1455
232. Dessens-Félix M, Pacheco-Contreras R, Barcaro G, Sementa L, Fortunelli A, Posada-Amarillas A (2013) J Phys Chem C 117:20967–20974
233. Hossain D, Pittman CU Jr, Gwaltney SR (2014) J Inorg Organomet Polym 24:241–249
234. Li HF, Wang HQ (2014) Phys Chem Chem Phys 16:244–254
235. Die D, Kuang X-Y, Zhu B, Guo J-J (2011) Phys B 406:3160–3165
236. Die D, Zeng BX, Wang H, Du Q (2013) Comput. Theor Chem 1025:67–73
237. Zhang F, Fa W (2012) Phys Lett A 376:1612–1616
238. Pal R, Wang L-M, Huang W, Wang L-S, Zeng XC (2009) J Am Chem Soc 131:3396–3404
239. Gao Y, Bulusu S, Zeng XC (2006) Chem Phys Chem 7:2275–2278
240. Gao Y, Shao N, Bulusu S, Zeng XC (2008) J Phys Chem C 112:8234–8238
241. Pyykkö P, Runeberg N (2002) Anger Chem Int Ed 41(12):2174–2176
242. Ferrando R, Fortunelli A, Rossi G (2005) Phys Rev B 72:085449
243. Heiles S, Johnston RL, Schäfer R (2012) J Phys Chem A 116:7756–7764
244. Götz DA, Heiles S, Johnston RL, Schäfer R (2012) J Chem Phys 136:186101
245. Rohrmann U, Schwerdtfeger P, Schäfer R (2014) Phys Chem Chem Phys 16:23952–23966
246. Götz DA, Shayeghi A, Johnston RL, Schwerdtfeger P, Schäfer R (2014) J Chem Phys 140:164313
247. Huang X, Zhao J, Su Y, Chen Z, King RB (2014) Sci Rep 4:6915
248. Fournier R (2010) Can J Chem 88(11):1071–1078
249. Orel S, Fournier R (2013) J Chem Phys 138(6):064306
250. Aguado A, López JMJ (2004) Chem Theor Comput 1(2):299–306
251. Aguado A, López JM (2005) Phys Rev B 72:205420
252. Aguado A, López JM (2004) J Phys Chem B 108:11722–11731
253. Aguado A, López JM (2006) Comput Mater Sci 35:174–178
254. Aguado A, López JM (2010) J Chem Phys 133:094302
255. Heiles S, Johnston RL (2013) Int J Quant Chem 113(18):2091
256. Ferguson GA, Mehmood F, Rankin RB, Greeley JP, Vajda S, Curtiss LA (2012) Top Catal 55(5–6):353–365
257. Henry CR (1998) Surf Sci Rep 31(7–8):235–325
258. Heard CJ, Heiles S, Vajda S, Johnston RL (2014) Nanoscale 6:11777–11788
259. Yudanov IV, Vent S, Neyman K, Pacchioni G, Rösch N (1997) Chem Phys Lett 275(3–4):245–252
260. Yudanov I, Pacchioni G, Neyman K, Rösch N. J Phys Chem B
261. Neyman KM, Vent S, Rösch N, Pacchioni G (1999) Top Catal 9:153–161
262. Matveev AV, Neyman KM, Yudanov IB, Rösch N

263. Matveev AV, Neyman K, Yudanov IV, Rösch N (1999) Surf Sci 426:123–139
264. Ferrari AM, Xiao C, Neyman KM, Pacchioni G, Rösch N (1998) Phys Chem Chem Phys 1:4655–4661
265. Nasluzov VA, Rivanenkov VV, Shor AM, Neyman K, Rösch N (2003) Chem Phys Lett 374:487–495
266. Yoon B, Häkkinen H, Landman U, Wörz AS, Antonietti J-M, Abbet S, Judai K, Heiz U (2005) Science 307:403–407
267. Inntam C, Moskaleva LV, Yudanov IV, Neyman KM, Rösch N (2006) Chem Phys Lett 417:515–520
268. Neyman KM, Inntam C, Moskaleva LV, Rösch N (2007) Chem Eur J 13:277–286
269. Heard CJ, Vajda S, Johnston RL (2014) J Phys Chem C 118:3581–3589
270. Todorova TK, Döbler J, Sierka M, Sauer J (2009) J Phys Chem C 113:8336–8342
271. Ismail R, Ferrando R, Johnston RL (2012) J Phys Chem C 117:293–301
272. Miyazaki K, Inoue T (2002) Surf Sci 501:93–101
273. Zhuang J, Kojima T, Zhang W, Liu L, Zhao L, Li Y (2002) Phys Rev B 65:045411
274. Ferrando R, Rossi G, Levi AC, Kuntová Z, Nita F, Jelea A, Mottet C, Barcaro G, Fortunelli A, Goniakowski J (2009) J Chem Phys 130:174703
275. Goniakowski J, Jelea A, Mottet C, Barcaro G, Fortunelli A, Kuntová Z, Nita F, Levi AC, Rossi G, Ferrando R (2009) J Chem Phys 130:174703
276. Kozlov S, Aleksandrov HA, Goniakowski J, Neyman KM (2013) J Chem Phys 139:084701
277. Barcaro G, Fortunelli A (2005) J Chem Theory Comput 1:972–985
278. Barcaro G, Fortunelli A, Rossi G, Nita F, Ferrando R (2007) Phys Rev Lett 98:156101
279. Atanasov I, Barcaro G, Negreiros FR, Fortunelli A, Johnston RL (2013) J Chem Phys 138:224703
280. Lei Y, Mehmood F, Lee S, Greeley B, Lee S, Seifert S, Winans RE, Elam JW, Meyer RJ, Redfern PC, Teschner D, Shlögl R, Pellin MJ, Curtiss LA, Vajda S (2010) Science (9):224–228
281. Kwon G, Ferguson GA, Heard CJ, Tyo EC, Yin C, DeBartolo J, Sönke S, Winans RRE, Kropf AJ, Greely JP, Johnston RL, Curtiss LA, Pellin MJ, Vajda S (2013) ACS Nano 7:5808–5817
282. Barcaro G, Fortunelli A (2007) Faraday Discuss 138:37–47
283. Barcaro G, Fortunelli A (2007) New J Phys 9(22):1–17
284. Barcaro G, Fortunelli A (2007) J Phys Chem C 111:11384–11389
285. Sierka M, Todorova TK, Sauer J, Kaya S, Stacchiola D, Wiessenrieder J, Shaikhutdinov S, Freund H-J (2007) J Chem Phys 126:234710
286. Chuang FC, Ciobanu CV, Shenoy VB, Wang CZ, Ho KM (2004) Surf Sci Lett 573:375–381
287. Negreiros FR, Sementa L, Barcaro G, Vajda S, Aprá E, Fortunelli A (2012) ACD Catal 2 (9):1860–1864
288. Negreiros FR, Sementa L, Barcaro G, Vajda S, Aprá E, Fortunelli A (2012) ACS Catal 2:1860–1864
289. Cleveland CL, Landman U, Schaaff TG, Shafigullin MN, Stephens PW, Whetten RL (1997) Phys Rev Lett 79(10):1873–1876
290. Akola J, Walter M, Whetten RL, Häkinnen H, Grönbeck H (2008) J Am Chem Soc 130:3756–3757
291. López-Acevedo O, Kacprzak KA, Akola J, Häkkinen H (2010) Nat. Chem 2:329–334
292. Goh J-Q, Malola S, Häkkinen H, Akola J (2013) J Phys Chem C 117:22079–22086
293. Weissker H-C, Escobar HB, Thanthirige VD, Kwak K, Lee D, Ramakrishna G, Whetten RL, López-Lozano X (2014) Nat. Comm. 5:3785
294. Quinn BM, Liljeroth P, Ruiz V, Laaksonen T, Kontturi K (2003) J Am Chem Soc 125:6644–6645
295. Negishi Y, Chaki NK, Shichibu Y, Whetten RL, Tsukuda T (2007) J Am Chem Soc 129:11322–11323
296. Price RC, Whetten RL (2005) J Am Chem Soc 127:13750–13751

297. Tracy JB, Crowe MC, Parker JF, Hampe O, Fields-Zinna CA, Dass A, Murray RW (2007) J Am Chem Soc 129:16209–16215
298. Luedtke WD, Landman U (1996) J Phys Chem 100(32):13323–13329
299. Xiang H, Wei S-H, Gong X (2010) J Am Chem Soc 132:7355–7360
300. Jadzinsky PD, Calero G, Ackerson CJ, Bushnell DA, Kornberg RD (2007) Science 318:430–433
301. Häkkinen H, Walter M, Grönbeck H (2006) J Phys Chem B 110:8827–9931
302. Kumara C, Aikens CM, Dass A (2014) Chem Phys Lett 5:461–466
303. Guidez EB, Mäkinen V, Häkkinen H, Aikens CM (2012) J Phys Chem C 116:20617–20624
304. Negishi Y, Munakata K, Ohgake W, Nobusada K (2012) J Phys Chem Lett 3:2209–2214
305. Kurashige W, Munakata K, Nobusada K, Negishi Y (2013) Chem Commun 49:5447–5449
306. Yamazoe S, Kurashige W, Nobusada K, Negishi Y, Tsukuda T (2014) J Phys Chem C 118 (43):25284–25290
307. Kacprzak KA, Lehtovaara L, Akola J, López-Acevedo O, Häkkinen H (2009) Phys Chem Chem Phys 11:7123–7129
308. Negishi Y, Kurashige W, Kobayashi Y, Yamazoe S, Kojima N, Seto M, Tsukuda T (2013) J Phys Chem Lett 4:3579–3583
309. Niihori Y, Kurashige W, Matsuzaki M, Negishi Y (2013) Nanoscale 5:508–512
310. Akola J, Kacprzak KA, López-Avecedo O, Walter M, Grönbeck G, Häkkinen H (2010) J Phys Chem C 114:15986–15944
311. Cheng D, Liu X, Cao D, Wang W, Huang S (2007) Nanotech 18:475702

# Structural Identification of Doped Silicon Clusters

**Yejun Li, André Fielicke, Peter Lievens and Ewald Janssens**

**Abstract** In this chapter we review recent research on the structural identification of isolated doped silicon clusters by combining state-of-the-art experiments and computational modelling using the density functional theory formalism. The experimental techniques include chemical probe mass spectrometric methods, infrared action spectroscopy, photoelectron spectroscopy, and x-ray absorption spectroscopy. Coinage metal elements, transition metals with an incomplete *d* sub-shell, lanthanides, and non-metallic main-group elements are considered as dopant atoms. The growth mechanisms of the doped silicon clusters are described with particular emphasis on the formation of endohedral cages. Specific species that may be considered as building blocks in future nano-structured materials and devices are highlighted, thereby exploiting their unique structural, electronic, or magnetic properties.

**Keywords** Silicon clusters · Growth mechanism · Dopants in atomic clusters · Molecular beams · Mass spectrometry · Infrared multiple photon dissociation spectroscopy · Photoelectron-spectroscopy · X-ray absorption spectroscopy

Y. Li · P. Lievens · E. Janssens (✉)
Laboratory of Solid State Physics and Magnetism, KU Leuven,
Celestijnenlaan 200d, Leuven 3001, Belgium
e-mail: ewald.janssens@kuleuven.be

Y. Li
e-mail: liyejun-mse@163.com

A. Fielicke
Institut für Optik und Atomare Physik, Technische Universität Berlin,
Hardenbergstrasse 36, 10623 Berlin, Germany



M.T. Nguyen and B. Kiran (eds.), *Clusters*, Challenges
and Advances in Computational Chemistry and Physics 23,
DOI 10.1007/978-3-319-48918-6_2

# 1 Introduction

The experimental study of small silicon clusters started about 30 years ago with the advent of cluster sources [1–3] and was arguably often motivated by the downsizing of silicon based devices. The properties of silicon clusters are, as for many other atomic clusters, known to vary strongly with size and can be very different from bulk. Notable examples are their size dependent structural motives that are dissimilar from pieces of bulk silicon, such as prolate shaped cationic silicon clusters composed of 20–30 atoms and spherical shapes for larger clusters [4]. Also the photofragmentation behavior of silicon clusters is peculiar with relatively small clusters losing relatively large $Si_6$ and $Si_{10}$ fragments, as opposed to the more common atom-by-atom evaporation for other types of clusters such as metals [1, 5]. Unlike the isolobal carbon that forms $sp^2$ hybridized fullerenes and carbon nanotubes, silicon favors $sp^3$ hybridization [6], which leads to rather asymmetric and reactive structures for pure silicon clusters.

It was shown in the last decades that doping can be used to modify and optimize the structural, electronic, chemical, and magnetic properties of silicon clusters [7, 8]. In particular, doping with transition metals can induce the formation of stable and unreactive cages of high symmetry, which may have appealing properties [9–13]. Following the first experimental realization of transition metal *(TM)* doped silicon clusters and the mass spectrometric discovery of the particular stability of $Si_{15}TM^+$ and $Si_{16}TM^+$ ($TM$ = Cr, Mo, W) clusters by Beck [14], a large number of theoretical and experimental studies have been devoted to the structures and properties of endohedrally doped silicon clusters. Famous examples include the predictions of icosahedral $Si_{20}Zr$ [9], hexagonal prism structures for $Si_{12}W$ [10] and $Si_{12}Cr$ [15], fullerene-like $Si_{16}Zr$ [11], Frank-Kasper polyhedral $Si_{16}Ti$ [11], and cubic $Si_8Be$ [16]. These results opened prospects for the production in large quantities of size selected silicon clusters with tailored properties, what is of utmost relevance for the use of silicon clusters as building blocks in future nanostructured devices.

The interest in doped silicon clusters has been so high that almost every element of the periodic table has been considered for doping. Nonetheless, identification of the geometric and electronic structures of doped, and even of bare, silicon clusters still is a great challenge for spectroscopic and theoretical studies alike. For example, it took many years before a consensus was reached on the geometry of $Si_6$ [17–19] and the lowest energy structure of $Si_8^+$ was only recently identified [20].

The reason why a dopant can stabilize a high-symmetry silicon cage is still subject of debate. Arguments either refer to bonding between the valence orbitals of the dopant with the silicon cage [9, 21] or to closed electronic shell structures [22] and compact geometries. Electron counting rules were also applied to explain the magic numbers of silicon clusters doped with transition metals. For example, $Si_{12}Cr$ and $Si_{16}Ti$ were claimed to be more stable than their neighboring sizes because of a 18 and 20 electron shell closure, respectively [13, 15]. On the other hand the enhanced stability of $Si_{16}Ti$ also could be explained by bonding arguments with a combination of 3*d* electrons of the Ti with electrons of the Si cages forming a bonding orbital at the HOMO level of the cluster [23].

In spite of intensive efforts, the influence of different dopants on the structures of the host silicon clusters, in particular on the growth mechanisms that govern the evolution of their physical and chemical properties, is still poorly understood. As a matter of fact, both experiment and theory have been facing difficulties and challenges to investigate the structures of clusters. Computationally, many energetically low-lying isomers may exist for a given cluster size, with small energy differences that are within the accuracy of contemporary methods, in particular for density functional theory (DFT). Hence, it is clear that theory alone does not suffice to accurately predict the most stable geometry of a cluster. Concurrently, experimental studies of clusters in gas phase are far from trivial and there is no single experimental method that straightforwardly probes the geometry and electronic structure of clusters. Chemical probe techniques based on mass spectrometry have been developed and various molecules ($H_2O$, $O_2$) or atoms (H, Ar) have been used to probe the transition from exohedral, where the dopant is exposed on the surface of the silicon cluster, to endohedrally doped silicon clusters, where the dopant resides inside a silicon cage [10, 24–26]. These techniques, however, do not provide detailed structural information of the clusters. In the last years, infrared (IR) spectroscopy has proven to be an invaluable technique to determine the structure of clusters in the gas phase, since it is very sensitive to the cluster's internal structure with molecular vibrations reflecting the arrangements of the atoms [27]. The growth patterns of several neutral and cationic transition metal, coinage metal, and non-metal main-group doped silicon clusters have been identified by IR action spectroscopy in combination with DFT calculations [28–33]. Photoelectron spectroscopy (PES) has commonly been used to study the electronic structure of anionic doped silicon clusters, yielding information about electron detachment energies, and in combination with DFT was also used for structural identification [24]. Extensive research efforts by PES were undertaken on transition metal and lanthanide metal atom doped silicon clusters [13, 34–40]. PES probes the electronic signature of valence electrons, which is done in the visible or ultraviolet spectral range and can be performed by standard laboratory lasers. To access the deeper valence bands or core levels, photon energies in the extreme ultraviolet to soft x-ray range are needed, requiring synchrotron radiation facilities or free-electron lasers as light sources. X-ray absorption (XAS) and x-ray magnetic circular dichroism (XMCD) spectroscopy, concentrating on the $2p \rightarrow 3d$ excitation energies of 3*d* transition metal dopants, were used to study the electronic and magnetic properties of doped cationic silicon clusters [26]. XMCD for instance revealed a coordination-driven magnetic-to-nonmagnetic transition in manganese doped $Si_nMn^+$ clusters [26].

The current chapter focuses on the evolution of geometric and electronic structures of doped silicon clusters as obtained by state-of-the-art experiments in combination with DFT calculations. In view of the extensive amount of literature, we mainly concentrate on recent studies dealing with coinage metal, transition metals with open atomic *d* sub-shells, lanthanides, and non-metallic main-group elements as dopants. It should be mentioned that there also are plenty of studies on silicon clusters doped with alkali atoms and other main group elements [16, 41–48], which are out of the scope of the present review.

The experimental techniques that have extensively been applied for the structural study of doped silicon clusters in the gas phase are briefly introduced in Sect. 2 with comments on the strengths and limitations of each technique. In Sect. 3, the effects of the different dopants on the growth pattern are discussed on the basis of recent experimental and theoretical results. Also the electronic and magnetic properties of the selected clusters are commented on. Conclusions and an outlook for future work are presented in the final Sect. 4.

# 2 Experimental Techniques to Study Doped Si Clusters in the Gas Phase

## *2.1 Mass Spectrometric Techniques*

The first reported experimental investigation of doped silicon clusters is a mass spectrometric study of transition metal doped silicon clusters by Beck [14]. After, strikingly high abundances were observed for $Si_{15}TM^+$ and $Si_{16}TM^+$ with $TM$ = Cr, Mo, and W, while most other sizes had low abundances. These mass spectra thus revealed the particular stability of certain doped silicon clusters. Beck suggested that the transition metal atom acts as a seed and silicon atoms form a shell around the metal dopant. The work of Hiura et al. constitutes another landmark in the investigation of doped silicon clusters [10]. $Si_nTMH_x{}^+$ species were produced in an ion trap via the reaction of 5*d* transition metal ions with silane $SiH_4$ gas [10]. For long ion trapping times, surprisingly the end products are dehydrogenated doped silicon clusters with a maximal size *n* that depends on the transition metal ion. More specifically, for $TM^+$ with $TM$ = Hf, Ta, W, Re, and Ir, the end products were dehydrogenated $Si_nTM^+$ clusters with $n$ = 14, 13, 12, 11, and 9, respectively. It was predicted that those clusters have endohedral structures with the metal ion stabilizing a Si polyhedral cage, indeed confirmed by ab initio calculations for $Si_{12}W$ [10].

Later, mass spectrometric investigations of Cr, Mn, Cu, and Zn doped silicon clusters by Neukermans et al. demonstrated the enhanced stability of $Si_{15,16}Cr^+$, $Si_{15,16}Mn^+$ and $Si_{10}Cu^+$ [49] relative to neighboring sizes. Ohara et al. studied anionic silicon clusters doped with single Ti, Hf, Mo, and W atoms [35]. Also in that work, predominant formation of doped clusters containing 15 and 16 silicon atoms was found independent of the dopant atom, pointing to a geometric stabilization. To better understand the role of the number of valence electrons for the stability of the clusters, a systematical investigation of cationic, neutral, and anionic $Si_nTM^{+/0/-}$ ($TM$ = Sc, Y, Lu, Ti, Zr, Hf, V, Nb, and Ta) was conducted by Nakajima and coworkers [13, 34, 50]. Particular enhanced intensities were detected for the isoelectronic $Si_{16}Sc^-$, $Si_{16}Ti^0$, and $Si_{16}V^+$ species [13]. It was conjectured that these clusters have closed electronic shell structures with 20 itinerant electrons: four from the dopant 3*d* orbitals and one electron from each $3p_z$ orbital of the $sp^2$-hybridized silicon atoms that are arranged in a caged structure.

While intensities in mass spectrometric studies are useful to obtain information about relative stabilities of certain cluster sizes, the combination of chemical probes with mass spectrometry can provide additional information about the position of the dopant atom. Nakajima and coworkers demonstrated that an exohedral transition metal dopant atom is a reactive site for water [24, 34, 35, 50]. Large doped Si clusters, that presumably adopt a caged structure, show a low reactivity upon exposure to $H_2O$, while exohedrally doped clusters readily interact with water. Their studies of transition metal doped $Si_nTM$ ($TM$ = Sc, Ti, V, Y, Zr, Nb, Mo, Hf, Ta, and W) and lanthanide doped $Si_nLn$ ($Ln$ = Tb, Ho, and Lu) clusters provided element and charge state dependent threshold sizes for the reaction with water vapor. Similarly, Lau and coworkers observed a size dependent depletion of singly doped $Si_nMn^+$ ($n$ = 7–14) in the presence of oxygen, which indicated the exohedral to endohedral transition at $n$ = 10 [26].

Janssens et al. proved that not only the reactivity with molecules such as $H_2O$, but also the physisorption of argon atoms, can be used as a structural probe for endohedral doping [25]. At 80 K, argon does not attach to elemental silicon clusters but only to surface-located transition-metal atoms. Critical sizes for Ar attachment were found to depend on the size of the transition metal dopant atoms for $Si_nTM^+$ with $TM$ = Ti, V, Cr, Mn, Co, and Cu [25, 51]. As illustrated in Fig. 1, no or negligible Ar physisorption is observed for sizes larger than $Si_{12}Ti^+$, $Si_{11}V^+$, $Si_{10}Cr^+$, $Si_{10}Mn^+$, $Si_7Co^+$, and $Si_{11}Cu^+$. Agreement with theoretical predictions confirms that these critical sizes for Ar adsorption are indeed related to the formation of endohedral clusters [25]. Furthermore, also doubly doped $Si_nTM_2^+$ clusters show a critical size for Ar adsorption, which might indicate that larger endohedral caged molecules are formed, eventually to be considered as seeds for the growth of metal-doped silicon nanorods [52, 53].

It is obvious that these mass spectrometric techniques alone cannot provide detailed structural information. For example, Ar tagging experiments can hint towards cage formation of $Si_nV^+$ from $n$ = 11 onwards but cannot differentiate between trigonal and hexagonal prism based structures [28]. More detailed structural and electronic information of doped silicon clusters could be obtained by infrared spectroscopy, photoelectron spectroscopy, or x-ray spectroscopy, as is described in the subsequent sections.

## 2.2 *Infrared Spectroscopy*

IR spectroscopy is a powerful technique to determine the structure of clusters. The vibrational modes of a cluster, with energies corresponding to far-IR photons, are very sensitive to forces between the atoms and to the structural arrangement of

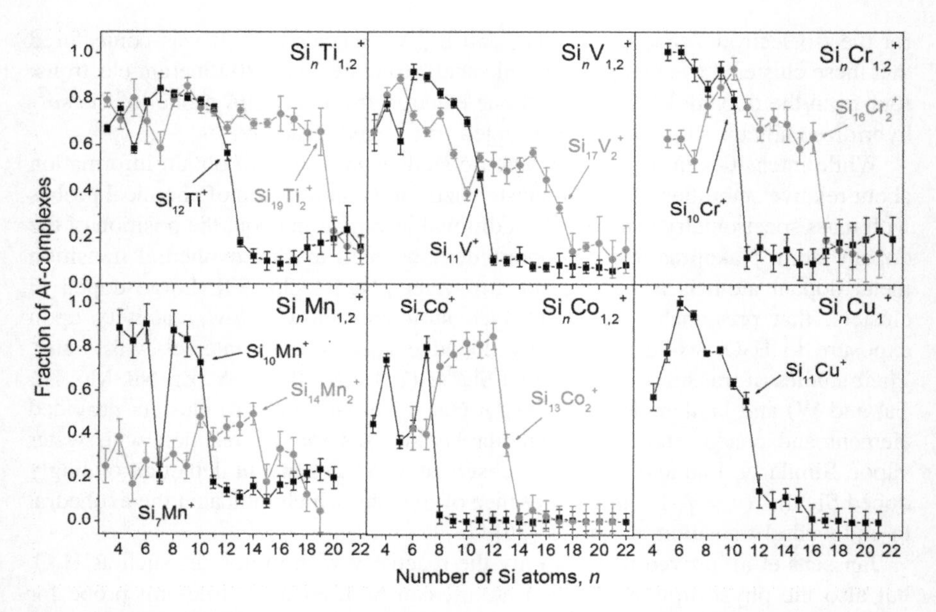


**Fig. 1** Fraction of argon complexes formed for $Si_nTM_{1,2}^+$ ($TM$ = Ti, V, Cr, Mn, Co, Cu) as function of the cluster size. A dopant dependent critical size, beyond which the argon-complex formation is unlikely, is found for both the singly and doubly *TM* doped species. The critical size likely corresponds to the onset of cage formation. The absence of Ar attachment on $Si_7Mn^+$ is not caused by encapsulation of the Mn dopant atom but to a very weak interaction between Mn and Ar in $Si_7Mn^+{\cdot}Ar$, which is related to a particular electronic shielding effect of the dopant *s*-electrons at this size as discussed in detail in [51]

atoms in a cluster. Conventional IR spectroscopy measures absorption spectra via the wavelength dependence of the transmittance. For sensitivity reasons it is required that a sufficient amount of material is sampled, which makes the technique inappropriate for low-density media such as clusters in the gas phase. In addition, convential IR spectroscopy is not size selective and the spectra of different species in the molecular beam cannot be disentangled, unless prior size selection is made (remark that a cluster source produces a distribution of sizes and compositions). Both mass selectivity and high sensitivity can be obtained by combining IR excitation with mass spectrometry. The reaction of a cluster in reponse to photon absorption is then detected mass spectrometrically in so-called action spectroscopy. Such action can be either a change of mass, cluster dissociation, or a change in its charge state, cluster ionization. The last years basically four experimental techniques for IR action spectroscopy on clusters in the gas phase have been developed (see Fig. 2): IR resonance enhanced multiple photon ionization (IR-REMPI), IR multiple photon dissociation (IR-MPD), infrared-ultraviolet two-color ionization (IR-UV2CI) , and IR multiple photon dissociation spectroscopy of cluster–rare gas complexes [54]. These techniques have in common that a vibrational spectrum is obtained by recording a mass-selective ion yield as a function of the IR radiation

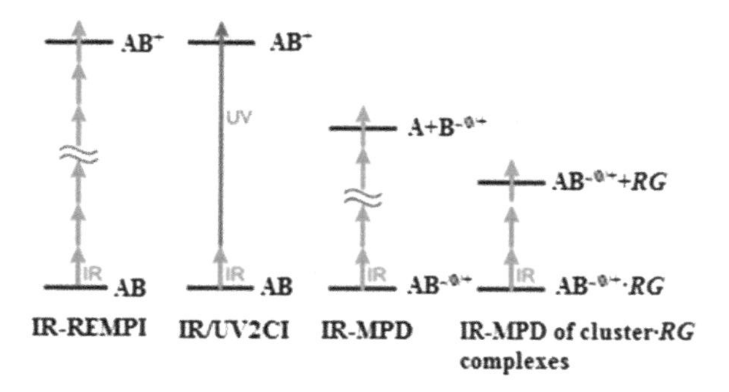


**Fig. 2** Schematic presentation of four experimental approaches for IR action spectroscopy of atomic clusters, AB, in the gas phase: IR resonance enhanced multiple photon ionization (IR-REMPI), IR/UV two-color ionization (IR-UV2CI), IR multiple photon dissociation (IR-MPD), and IR multiple photon dissociation of cluster–rare gas complexes. The first two techniques are used for neutral clusters. The last two can be applied both on charged and neutral clusters. In the case of neutral clusters postionization is needed. (Adapted from Asmis et al. (2012) Phys Chem Phys Chem 14:9270)

wavelength and that they require an intense and tunable IR laser source. Of these four techniques IR-UV2CI and IR-MPD of cluster—rare gas complexes have been used to characterize neutral and cationic doped silicon clusters. Those experiments on doped silicon clusters were all done at the Free Electron Laser for Infrared eXperiments (FELIX), Nieuwegein, the Netherlands. Recently this facility was moved to the Radboud University Nijmegen [55].

Silicon clusters are strongly bound with typical bond energies of 2–4 eV [56], while the vibrational transitions of these clusters are typically in the 50–500 $cm^{-1}$ range, corresponding to photon energies of 6–60 meV. Many IR photons are required to dissociate pure and doped silicon clusters, which means the action spectroscopy will not be a coherent multiphoton process. Intramolecular vibrational redistribution of energy is taking place in between successive photon absorption event. Also, given the low absorption probability, even with the available high intensity free electron laser sources, dissociation of silicon clusters is challenging. The so-called messenger atom technique, where weakly bound cluster—ligand complexes are probed instead of pure clusters, can overcome the need of many photons [57, 58]. Such a ligand is typically an inert gas atom, which can be attached to the silicon clusters by adding a small fraction of the messenger gas to the carrier gas in the cluster source and cooling the source to cryogenic temperatures. The loosely bound ligands are supposed to have a negligible influence on the geometry of the clusters. This assumption has to be confirmed ex post and does not always hold [59]. Following absorption of a few IR photons the cluster—rare gas atom bond will break. The light absorption can then be monitored mass spectrometrically by depletion of the complex or alternatively by the increase of the bare cluster abundance.

Structural identification of the clusters can subsequently be obtained by comparison of the experimental IR spectra with computed spectra for different structural isomers. DFT calculations have been the most used computational tool for doped silicon clusters. An example is shown in Fig. 3 for $Si_nAu^+$ clusters, where the comparison of the experimental and the calculated spectra of the obtained lowest energy isomers for $n = 4, 6, 8, 9, 11$ is depicted [31]. The experimental action spectra were obtained by IR-MPD of $Si_nAu^+{\cdot}Ar$ or Xe complexes, while DFT calculations were employed using the BP86 functional in combination with the SVP basis set for the Si atoms and the SDD pseudopotential for Au. The rare gas ligand was not included in the computations. The observation that the experimental spectra are well reproduced by the calculations, even for the small features, thus confirms the minor influence of the messenger atom on the cluster's vibrational spectrum in this case. The structures are found to be planar for $Si_4Au^+$ and three-dimensional for larger sizes. $Si_6Au^+$ is a distorted octahedron, while $Si_8Au^+$ and $Si_9Au^+$ have edge-capped pentagonal bipyramidal structures [31]. However, it should be noted that the structural assignment is not always straightforward. On one hand, the anharmonicity of the potential energy and finite temperature effects are normally not taken into account in the calculations, which will result in band shifts and can have a significant effect on the intensities of bands. On the other hand, DFT calculations have difficulties to properly describe the cluster—rare gas bond. Furthermore, recent kinetic Monte Carlo simulations also demonstrated that the multiple photon character of the action spectroscopy influences the exact frequency, intensity, and width of the adsorption modes [60]. Therefore, a perfect match between theory and experiment cannot be expected. Definite assignments are even more challenging for larger-sized clusters, mainly due to the emergence of many low energy isomers, of which several may have harmonic IR spectra in reasonable correspondence with the experiment. This is illustrated in Fig. 3 for $Si_{11}Au^+$, which has two isomers, with a relative energy of less than 0.2 eV with respect to the obtained ground state. All three could exist in the molecular beam as the main features in their IR spectra show a reasonable agreement with the experimental action spectrum.

The messenger atom technique is not restricted to ionic species only [29]. If the clusters are postionized by a single photon with an energy above the cluster's ionization potential, IR spectroscopy of neutral cluster—rare gas complexes can be done in an analogous way as for ionic species. This was illustrated by Claes et al. for small neutral $Si_nV$ and $Si_nMn$ ($n = 6$–$9$) clusters tagged with Xe [29].

The inherent disadvantage of the messenger atom techniques is the possible perturbation of the cluster's structure by the adsorbed ligand, even if weakly bound rare gas atoms are used. Also in specific cases, especially for neutral clusters, it is quite difficult to create cluster—rare gas complexes, even at low temperatures. An alternative method for neutral clusters, without the need for formation of messenger atom complexes, is the IR-UV2CI technique [17, 61]. This generally applicable technique combines IR excitation with near threshold photoionization. It relies on the absorption of a single (or a few) IR photons prior to absorption of a UV photon, which has a photon energy just below the ionization energy, to lift the total internal

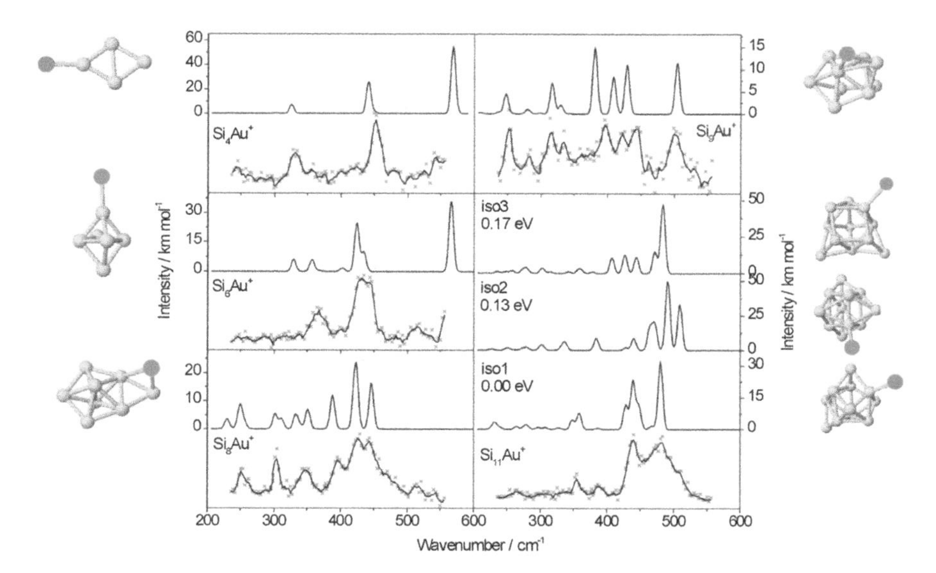


**Fig. 3** Experimental IR-MPD spectra (*lower traces*) of $Si_nAu^+{\cdot}Ar$ ($n$ = 4, 6, 9) and $Si_nAu^+{\cdot}Xe$ ($n$ = 8, 11), the calculated harmonic single photon adsorption spectra (*full lines*, traces above the experimental data), and the geometric structures (*left* and *right*) of the obtained lowest-energy isomers. The *crosses* are the original data points, while the *full lines* correspond to three-point running averages

energy of the species above the ionization threshold. The direct photoionization generally prevails over the slower statistical fragmentation process. The formed ions can be sensitively detected by means of mass spectrometry. Scanning the energy of the IR photons changes the ionization efficiency and the recorded ion intensity reflects the IR absorption spectrum of the corresponding neutral clusters. This is illustrated by the mass spectrum in Fig. 4 for $Si_nCo$ clusters. Species with a vertical ionization energy close to (or slightly above) the photon energy of the ionization laser (such as $Si_nCo$ with $n$ = 10–12) have a low intensity in the mass spectra (Fig. 4a), but show enhanced abundance if they are excited by resonant absorption of one or more IR photons prior to interaction with UV radiation (Fig. 4b). By comparison of the IR-UV2CI spectra of $Si_nCo$ ($n$ = 10–12) with simulated IR spectra for different structural isomers (not shown), endohedral geometries were assigned to those clusters [61].

## 2.3 *Photoelectron Spectroscopy*

PES is a commonly used technique to study the electronic structures and chemical bonding of gas-phase atomic clusters. The basic physical idea behind PES is simple: a cluster is excited by absorption of a single photon with a photon energy

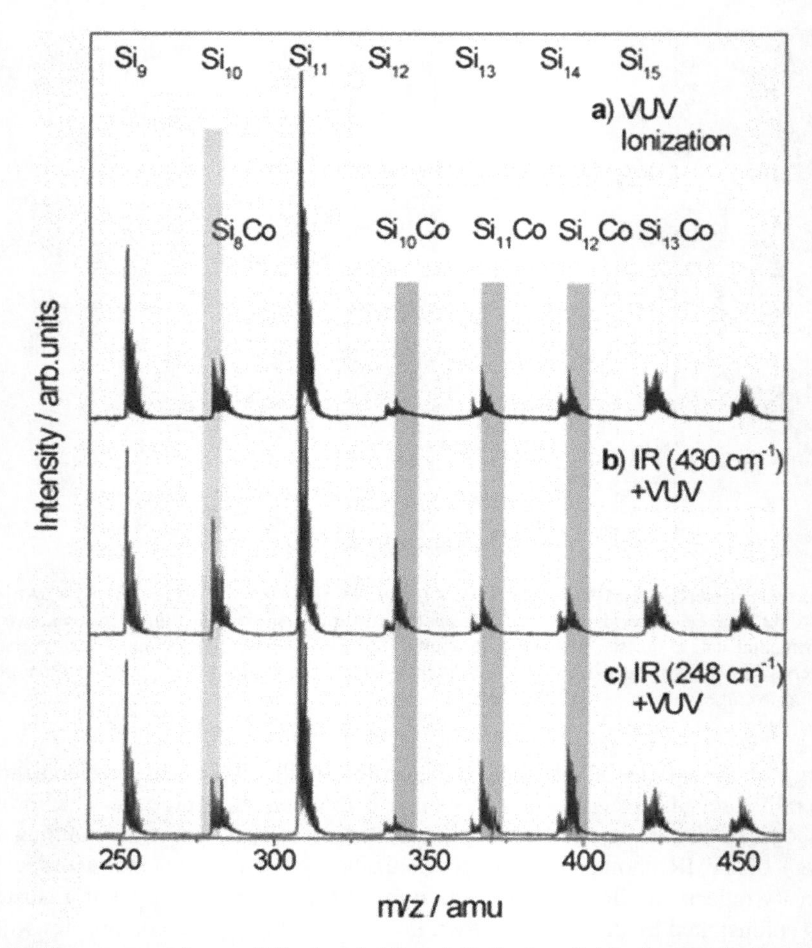


**Fig. 4** Mass spectra of neutral $Si_n$ ($n$ = 9–16) and $Si_nCo$ ($n$ = 7–14) clusters obtained under different ionization conditions. **a** Mass spectrum obtained by ionization with 7.87 eV UV photons solely. **b** Prior irradiation of the neutral cluster distribution with intense IR light at 430 $cm^{-1}$ significantly increases the signal of $Si_{10}$ and $Si_{10}Co$, while the rest of the mass spectrum is nearly unchanged. **c** Irradiation with IR light of 248 $cm^{-1}$ followed by UV ionization enhances the intensities of $Si_{11}Co$ and $Si_{12}Co$. (Reproduced from Li Y et al. (2014) J Phys Chem A 37:8198)

exceeding the binding energy of the least bound electron, which results in electron detachment. The remaining part of the incoming photon energy is shared by the ionized cluster and the photoelectron. Since the mass of the cluster is much larger than that of the electron, most excess energy is converted into kinetic energy of the emitted electron. Hence, measuring the intensity of photoelectrons as a function of their kinetic energy provides a spectrum of the occupied electronic energy levels of the cluster.

PES is usually performed on anionic clusters, since the charge is required for size selection and anions usually have low valence electron binding energies that are accessible by commercial lasers, such as the 2nd, 3rd or 4th harmonics of a Nd: YAG laser. An anion photoelectron spectrum provides the adiabatic detachment energy (ADE) and the vertical detachment energy (VDE), as well as the electronic excitation energies, which constitute an electronic fingerprint of the cluster. PES on anionic clusters also yields information about the corresponding neutral species. Since the detachment process is fast in comparison to structural arrangement of the nuclei upon removal of the electron, anion photoelectron spectroscopy probes the electronic structure of a neutral cluster in the ground state geometry of its anion. The resulting neutral cluster can hereby either be in the vibronic ground state or in a vibronically excited state. For example the adiabatic electron affinity ($EA_a$) of $Si_nEu$ ($n = 3$–17) clusters was inferred from their photoelectron spectra [37]. For small sizes $n = 3$–11 the $EA_a$ is relatively small and ranges from 1.45–2.2 eV. However, it shows an abrupt increase ($EA_a > 2.5$ eV) from $n = 12$ onwards, which has been related to geometric rearrangements from exohedral to endohedral structures [37].

In addition to the valuable knowledge about the electronic states, PES can be used for structural assignment if changes in the geometry lead to changes in the electronic levels. The photoelectron spectrum of various isomers can be simulated by means of quantum chemical calculations if the generalized Koopman's theorem is applied to predict vertical detachment energies of different occupied orbitals. The cluster structure can then be assigned through comparison of the PES spectrum with simulated density of states (DOS) of candidate isomeric structures.

Figure 5 depicts the photoelectron spectra of $Si_{12}V_n^-$ ($n = 1, 2, 3$) clusters as recorded by Zheng and coworkers using the 4th harmonic (4.66 eV) of a Nd:YAG laser [38]. The photoelectrons are energy-analyzed by a magnetic bottle photoelectron spectrometer. The experimental photoelectron spectra of $Si_{12}V_n^-$ are well reproduced by DFT computations applying the generalized Koopman's theorem. With the structures identified, the computations of the electronic structures were used to infer other physical properties. It was found that $Si_{12}V_3^-$ shows a total spin of 4 $\mu_B$ with a ferromagnetic coupling between the V dopants, while $Si_{12}V^-$ and $Si_{12}V_2^-$ favor low spin states.

## 2.4 X-Ray Absorption and Magnetic Circular Dichroism Spectroscopy

While the valence electron spectra of size-selected anionic clusters can be obtained by PES using commercial laser systems in the visible or UV spectral range, access to deeper core electron levels requires photon energies in the extreme ultraviolet to soft X-ray range. X-ray absorption experiments can be performed at synchrotrons and provide element specific electronic and magnetic information, which is particularly valuable for the study of doped clusters. They are, however, challenging

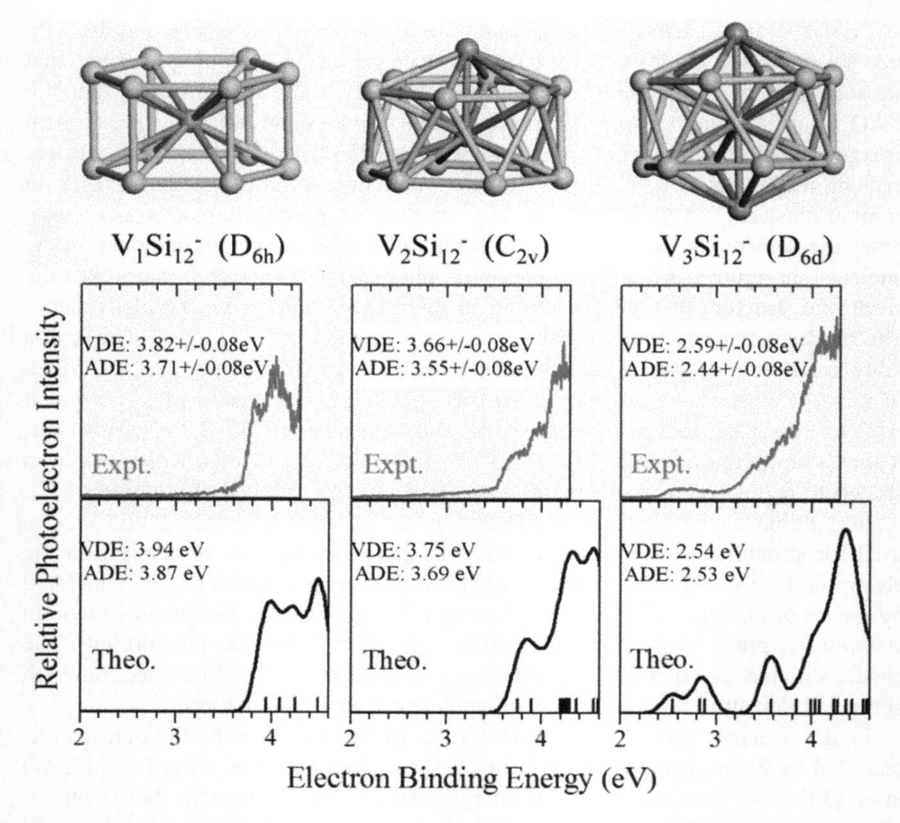


**Fig. 5** Photoelectron spectra of $Si_{12}V_m^-$ ($m$ = 1, 2, 3) clusters. *Upper* geometrical structures; *middle* experiment; *lower* theoretical simulation using DFT and a uniform Gaussian broadening of 0.1 eV for each vertical detachment process. The individual calculated energy levels are labelled by *short vertical lines*. Both the measured VDE and ADE are well reproduced by the simulated values for the obtained lowest energy isomers. (Adapted from Huang X et al. (2014) Nanoscale 6:14617)

for clusters in the gas phase given the low target density and the required high photon fluence [62].

Opposed to PES and IRMPD experiments, which are often done in molecular beams, X-ray experiments on clusters in the gas phase are usually performed in an ion trap to ensure a high density of size selected clusters. Size selected trapped clusters are electronically excited by a collinear beam of tunable soft X-ray radiation and will undergo fragmentation and multiple ionization steps. Ions are subsequently extracted from the ion trap and analyzed in a time-of-flight mass spectrometer. X-ray absorption spectra are obtained by monitoring the yield of cluster fragment ions as a function of incident photon energy. For XMCD spectroscopy low temperatures and high magnetic fields are required to align the magnetic moment of the clusters. This can be done by a superconducting solenoid in combination with cryogenic cooling of the ion trap and a buffer gas for thermal equilibration of the clusters [26]. Ion yield spectra are then recorded for parallel and antiparallel alignment of the photon helicity and magnetic field.

The basic principle of XAS is the absorption of an X-ray photon by a single atom, exciting a core electron. The excited core electron either goes into an empty valence state (resonant photoabsorption) or, if the photon energy is sufficient, into a continuum state above the ionization energy (non-resonant photoionization). A core hole is created in both cases, making XAS a local and element specific method due to the strong localization of the created core hole and the characteristic binding energy of the core electrons. Moreover, for resonant photoabsorption experiments also valence electron information is obtained. Indeed, because of the dipole selection rule $\Delta l = \pm 1$ one can selectively probe specific orbital angular momentum states of the empty valence electron levels.

Quantitative magnetic information of the clusters can be obtained by XMCD spectroscopy. Hereto the ion trap is placed inside a high magnetic field (for instance provided by a superconducting solenoid), so that their magnetic moments are (partly) aligned. Circularly polarized x-ray radiation is then used to measure the differential X-ray absorption spectra for left- and right- handed circularly polarized X-ray beams. Using magneto-optical sum rules both element specific orbital and spin magnetic moment information can be extracted from the dichroic signal [63]. In this case, the dipole selection rule is expanded with $\Delta m_l = +1$ and $\Delta m_l = -1$ for right and left circularly polarized radiation, respectively. For an unequal occupation of the $m_l$ states, the absorption of right and left circularly polarized light will be different. Note that also magnetic deflection experiments can be used to investigate cluster magnetism, in particular total magnetic moments of neutral clusters. We are, however, not aware of any magnetic deflection experiment on doped silicon clusters.

XAS and XMCD have been used to study transition metal doped silicon clusters by Lau and coworkers [26, 64, 65]. Using XAS the electronic structure of the *TM* dopant in $Si_nV^+$ ($n$ = 14–18), $Si_{16}Ti^+$, and $Si_{16}Cr^+$ was probed via $2p \rightarrow 3d$ transitions, i.e., at the $L_{2,3}$ edges [64]. Figure 6 shows that $L_{2,3}$ X-ray absorption spectra of $Si_{16}Ti^+$, $Si_{16}V^+$, and $Si_{16}Cr^+$ clusters and those of isolated $Ti^+$, $V^+$, and $Cr^+$ ions. The XAS spectra of the isolated ions (upper panel of Fig. 6) have multiplet electronic structures characteristic for each element, and reflect the different $3d$ orbital occupancies of $Ti^+$, $V^+$, and $Cr^+$. However, these differences disappear in case the corresponding atoms are encapsulated in $Si_{16}$ cages. $Si_{16}Ti^+$, $Si_{16}V^+$, and $Si_{16}Cr^+$ endohedral clusters exhibit a very similar fine structure in their X-ray absorption spectra (lower panel of Fig. 6). The relative excitation energies, both the intense features at 0 and 2.1 eV and the less intense ones at 1.1 and 5.0 eV, are practically identical for the three transition metal doped silicon clusters. These similarities imply nearly identical local electronic states of the transition metal dopant atoms, which further indicates a very similar structural environment, as XAS is very sensitive to both [64].

XMCD was used to study $Si_nMn^+$ ($n$ = 7–14) via the $L_{2,3}$ local $2p \rightarrow 3d$ excitations of the Mn dopant [26]. A clear correlation of the magnetic moment with the

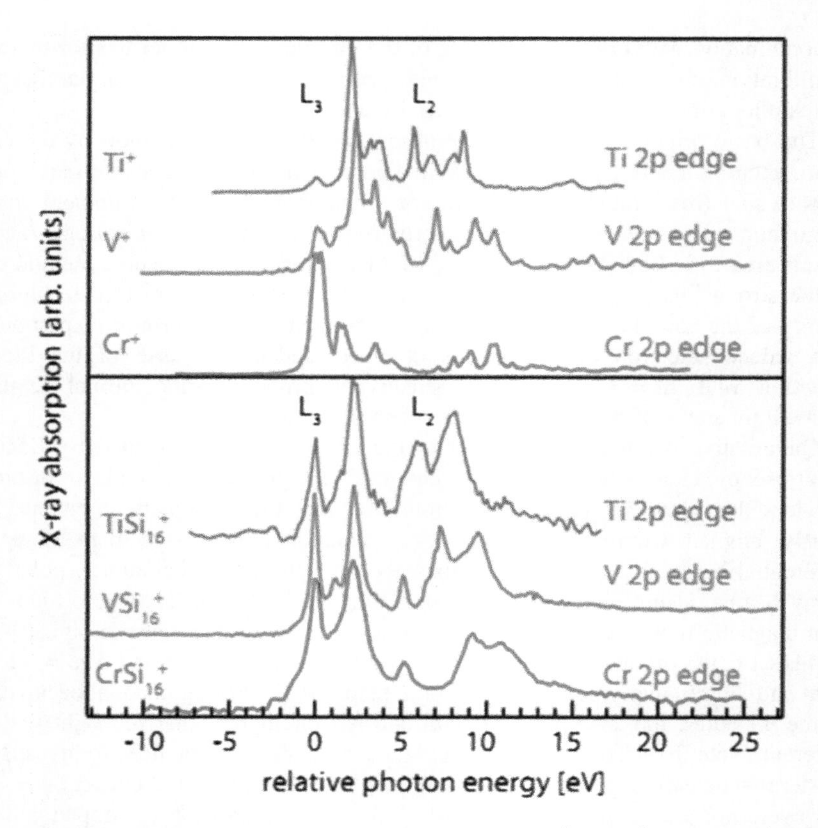


**Fig. 6** Transition metal $L_{2,3}$ X-ray absorption spectra of $Si_{16}Ti^+$, $Si_{16}V^+$, and $Si_{16}Cr^+$ clusters (*lower panel*) and of single $Ti^+$, $V^+$, and $Cr^+$ ions (*upper panel*). The spectra are aligned at the first peak position. While the bare ions have different spectra, the relative excitation energies of transition metal dopants in silicon clusters are nearly identical. This surprising fact indicates almost identical local electronic structure at the dopant atom site. Relative positions of the $L_2$ edge shift because of decreasing $2p$ spin-orbit splitting from chromium to titanium. (Reproduced from Lau JT et al. (2009) Phys Rev A 79:053201)

manganese coordination number and nearest-neighbor distance was found. This work indicated quenching of the Mn magnetic moment at coordination numbers corresponding to those in bulk silicon as a consequence of 3*d* electron delocalization because of the strong interaction between Mn and the silicon cage. The correlation of the magnetic moment and the weighted coordination number provides guidelines to the stabilization of high-spin states in dilute manganese-doped silicon [26].

It was also shown that direct and resonant core-level photoionization spectroscopy in combination with valence band photoionization curves can be used to accurately quantify HOMO–LUMO gaps of size-selected clusters via a Born-Haber cycle [65]. In particular the size-evolution of the HOMO–LUMO gaps of $Si_nV^+$

($n$ = 15–17) was obtained by this novel method, confirming the special electronic stability of $Si_{16}V^+$. The method was claimed to be widely applicable and expands the range of current techniques for the determination of band gaps to ultra-dilute samples and electron binding energies in the vacuum UV spectral range.

# 3 Dopant Dependent Growth Mechanisms and Properties

## 3.1 Coinage Metal Dopants ($kd^{10}(k + 1)s^1$)

Coinage metal (Cu, Ag, and Au) doped silicon clusters have been studied extensively by DFT calculations. For instance, Xiao et al. showed that the most stable isomers of $Si_nCu$ ($n$ = 4, 6, 8, and 10) have either Si frameworks similar to the structures of the ground state or low-lying isomers of bare $Si_n$ or that Cu is making substitutions in $Si_{n+1}$ [66]. It was also demonstrated that the structure of the Si framework in $Si_nCu$ is largely determined by the Si–Si interactions, because they are stronger than the Si–Cu interactions. However, several computational predictions of the structures of coinage metal doped silicon clusters are in disagreement with one another, especially concerning the dopant induced cage formation. For example, Chuang et al. predicted by first-principles calculations at the PBE/PAW level that $Si_nAg$ clusters ($n$ = 1–13) have exohedral structures with the Ag dopant atom taking a capping position [67], while another computational study (using B3LYP/3-21G*) of $Si_nAg$ ($n$ = 1–15) by Ziella et al. indicated endohedral geometries for $n > 10$ [68]. Conflicting computational predictions were also made for neutral $Si_nAu$ clusters. Wang et al. studied (at B3PW91/LanL2DZ and PW91/DNP levels) the configurations, stability, and electronic structure of neutral $Si_nAu$ ($n$ = 1–16) clusters and found that the Au dopant moves to an interior site in $Si_{11}Au$ and that $Si_{12}Au$ adopts an endohedral structure [69], while Chuang et al. (PBE/PAW) obtained exohedral geometries for neutral $Si_nAu$ ($n$ = 1–16) [70].

The geometry of the coinage metal doped silicon clusters has extensively been investigated by IR-MPD spectroscopy of cluster–rare gas complexes in combination with DFT calculations [30, 31, 33, 71]. The growth mechanisms, assigned on the basis of this combined experimental and computational work, of $Si_nCu^+$ ($n$ = 2–11), $Si_nAg^+$ ($n$ = 6–15), and $Si_nAu^+$ ($n$ = 2–15) clusters are summarized in Fig. 7. In view of the similar electronic structure ($kd^{10}(k + 1)s^1$) of the coinage metal atoms, one may expect that they have a similar influence on the geometry of the silicon clusters. There are indeed similarities in the growth mechanisms of $Si_nCu^+$, $Si_nAg^+$, and $Si_nAu^+$ clusters. It is found that all studied sizes have exohedral structures, in which the dopant atoms favor adsorption on bare $Si_n^+$ clusters rather than taking substitutional positions as is the case in silicon clusters doped by transition metal elements like V or Mn [32, 33]. This difference must be related to the occupancy of the valence $d$ orbitals. Also the silicon building blocks in $Si_nCu^+$,

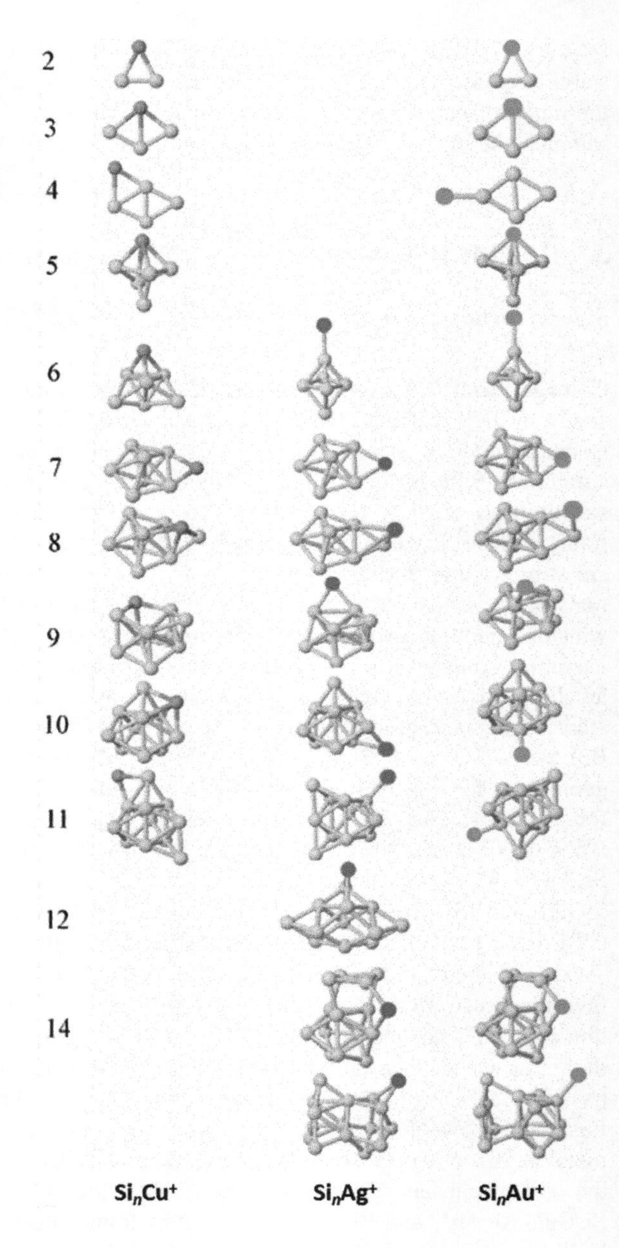


**Fig. 7** Structural evolution for $Si_nCu^+$ (*left*), $Si_nAg^+$ (*middle*), and $Si_nAu^+$ clusters (*right*). The structures are assigned based on comparison of IR-MPD spectra, measured on the corresponding cluster—rare gas complexes, with harmonic vibrational spectra calculated for different structural isomers by DFT. (Reproduced from Li Y et al. (2015) J Phys Chem C 119: 10896-10903)

$Si_nAg^+$, and $Si_nAu^+$ are similar: the clusters are based on a pentagonal bipyramid for $n$ = 7–9, while a transition to a trigonal prism motif appears from $n$ = 10.

There are, however, also differences between the structures of Cu, Ag, and Au doped silicon clusters. $Si_nAg^+$ and $Si_nAu^+$ have similar structures for the smallest sizes ($n$ = 6–8) and for $n$ = 14, while the dopants cap a different position for $n$ = 9 and 10, or the doped clusters adopt different Si frameworks ($n$ = 11 and 15). In general, there seems to be a trend that Cu likes facial, Ag edge, and Au corner positions (see Fig. 7) [31]. It was shown that the *d* orbital occupancy of the coinage metal dopant plays an important role in the binding site. Natural electron configuration analysis gives that the number of electrons in 4*d* orbitals of Ag in $Si_nAg^+$ is slightly higher (9.9) than that in Cu 3*d* orbitals of $Si_nCu^+$ (9.8) [30]. Because of this difference in occupancy, Ag prefers to add to the bare Si clusters in a lower coordinated position than Cu. The same may hold for Au in $Si_nAu^+$ clusters. The coinage metal dopants thus add to the bare $Si_n$ clusters with coordination numbers decreasing from Cu over Ag to Au.

Formation of endohedrally doped cages is experimentally not observed for coinage metal doped silicon clusters, until the largest sizes studied, which is $n$ = 15 for $Si_nAg^+$ and $Si_nAu^+$ (see Fig. 7). For $Si_nCu^+$ the situation is less clear. The IR spectra of cluster–rare gas complexes demonstrated that for $n \leq 11$ no caged structures are formed, but larger sizes were not studied [33]. Larger sizes might be endohedral as predicted computationally [66, 72] and as implied from mass spectrometric experiments, which showed that $Si_{11}Cu^+$ is the critical size for Ar attachment hinting that the Cu dopant atom may take an interior position in larger $Si_nCu^+$ clusters [25].

The atomic radius of the dopant atom has been shown to play an important role in determining the critical size for cage formation of the transition metal doped silicon clusters [25, 73]. It may also have a significant influence on the binding site of the coinage metal dopant. Au and Ag have similar atomic radii (0.174 and 0.165 nm), which are larger than that of Cu (0.145 nm), therefore, they prefer a more facial (low coordinated) position on the Si clusters and more Si atoms are needed to encapsulate them. The similar growth patterns of $Si_nAg^+$ and $Si_nAu^+$, without cage formation up to $n$ = 15, indicate that not only the atomic radius of the dopant atom but also the occupancy of the valence *d* orbitals and the orbital hybridization between the dopant atom and Si atoms are important in the formation of endohedral structures [74].

Aside from the IR spectroscopy work on cationic coinage metal doped Si clusters, PES investigations were performed on anionic coinage metal doped Si clusters. Zheng and coworkers conducted a combined PES and DFT study on the electronic and geometrical structures of $Si_nCu^-$ ($n$ = 4–18) [75] and $Si_nAg^-$ ($n$ = 6–12) [40]. They tentatively assigned the structures of these clusters based on the comparison between the experimental and calculated photoelectron spectra, vertical detachment energies, and adiabatic detachment energies. It was found that $Si_nCu^-$ ($n > 12$) clusters have endohedral structures. This critical size is consistent with the above mentioned Ar physisorption experiment on cationic $Si_nCu^+$ clusters [25]. The $Si_nAg^-$ ($n$ = 6–12) clusters are predicted to all have exohedral structures. Most structurally assigned anionic clusters are different from the corresponding cationic $Si_nAg^+$ and $Si_nCu^+$ clusters, although they have in common that the Ag dopant seems to take a lower coordinated position than the Cu dopant.

## 3.2 Transition Metal Dopants ($kd^x$ 0 < x< 10)

Among all elements in the period table, transition metal atoms and in particular 3*d* transition metal atoms have been used most frequently as dopants in silicon clusters. We therefore do not intend to give an exhaustive overview and concentrate on some recent combined experimental and computational results.

Mass spectrometric Ar tagging experiments predicted that $Si_nV^+$ adopts endohedral cage structures from $n = 12$ onwards [25]. The geometry of cationic vanadium doped silicon clusters, $Si_nV^+$ ($n = 4$–16), was investigated in more detail by IR-MPD in combination with DFT calculations [33, 71]. The IR spectra of $Si_nV^+$ ($n = 4$–11) were measured on the corresponding $Si_nV^+$•Ar rare gas complexes, which could be formed for the exohedrally doped silicon clusters [33]. Different from coinage metal dopants, the V substitutes a highly coordinated silicon atom of the pure cationic bare silicon clusters. The different growth pattern of $Si_nV^+$ and $Si_nCu^+$ clusters reflects the role of the dopant's 3*d* orbital occupancy on the binding. The natural electronic configurations obtained by the natural bond orbital (NBO) method indicated that the number of electrons in the V 3*d* orbitals varied from 3.3 ($n = 6$) to 5.6 ($n = 11$) in $Si_nV^+$. To probe endohedral $Si_nV^+$ ($n = 12$–16) clusters by IR-MPD, Xe was used as messenger atom, which does bind to Si due to its higher polarizability than Ar [28]. The assigned $Si_nV^+$ ($n = 12$–16) clusters have Si frameworks that are significantly different from bare Si clusters. $Si_nV^+$ ($n = 12$–14) have hexagonal prism based structures, $Si_{15}V^+$ is found to be built up from pentagons and rhombuses, and $Si_{16}V^+$ has a slightly distorted Frank-Kasper polyhedral geometry.

$Si_{16}V^+$ is one of the best studied doped silicon clusters in the literature and was computationally predicted to have a symmetric Frank-Kasper polyhedral structure [12, 23, 76]. Its stability was rationalized by means of an electron counting rule. Each Si atom binds to the V dopant and contributes one delocalized electron to the electron count. With the five valence electrons of V and the cationic state of the cluster, this results in a 20 electron closed shell system [77]. Its broad spectral lines in the IR-MPD experiments were initially suggested to be related to the fluxional behavior of $Si_{16}V^+$ resulting from rapid interchange of the atomic positions between a slightly distorted and a perfect Frank-Kasper polyhedral geometry corresponding to the ground state (**iso1**-$T$) and a transition state (**TS**-$T_d$), respectively (see Fig. 8). Molecular dynamics simulations revealed that the Si cage is highly dynamic showing movement and interchange within a quasi-liquid Si shell [28]. The high symmetry of cold $Si_{16}V^+$ was confirmed by Lau et al. through X-ray absorption spectroscopy at the transition metal $L_{2,3}$ edges, which revealed a nearly identical local electronic structure of the dopant atoms in $Si_{16}Ti^+$, $Si_{16}V^+$, and $Si_{16}Cr^+$ in spite of a different number of valence electrons [64]. The HOMO–LUMO gaps of $Si_nV^+$ ($n = 15$–17), derived by a combination of direct and resonant core-level photoionization spectroscopy with valence band photoionization curves, showed a striking size-dependence and gave further evidence of the special electronic structure and stability of $Si_{16}V^+$ [65].

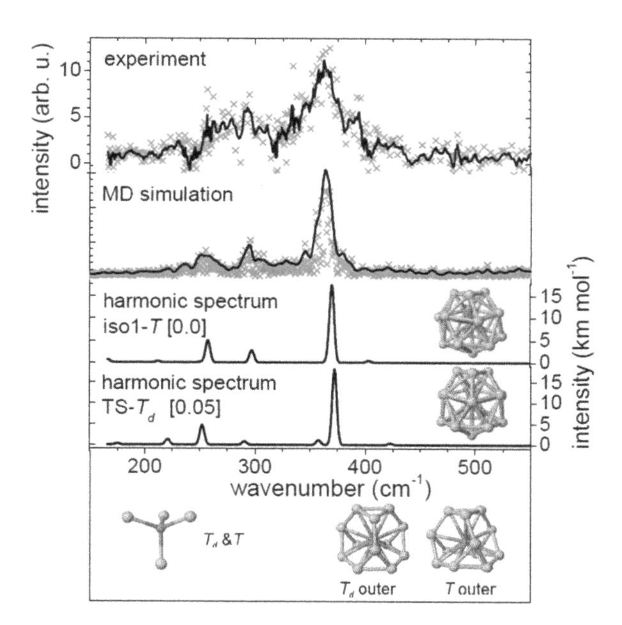


**Fig. 8** IR-MPD spectrum of $Si_{16}V^{+}$•Xe (*upper trace*). The 2nd trace shows the IR spectrum obtained from the molecular dynamics (MD) simulation starting from **iso1**-$T$ (*crosses*) and a spectrum that is obtained by overlaying the discrete points with a 5 $cm^{-1}$ FWHM Gaussian profile (*line*). Computed harmonic vibrational spectra of **iso1**-$T$ and **TS**-$T_d$ are given in the 3rd and 4th traces. Relative energies in eV at the BP86/6-311 + G(d) level are given in brackets. The structures shown in the *lower trace* represent the *inner* tetrahedral and *outer shells* of **TS**-$T_d$ and **iso1**-$T$. Relaxing from $T_d$ to $T$ symmetry, the *inner shell* remains essentially unchanged while the *outer shell* is distorted at the symmetrical planes. The small barrier over **TS**-$T_d$ between degenerate **iso1**-$T$ states suggests fluxional rearrangement. The MD simulated trajectory supports the interpretation of $Si_{16}V^{+}$ undergoing rapid transitions between nearly degenerate local minima. This fluxional behavior leads to a broadening of the IR absorption bands. (Reproduced from Claes P et al. (2011) Phys Rev Lett 107:173401)

Bowen and coworkers measured photoelectron spectra of $Si_nCr^{-}$ ($n$ = 8–12) following excitation by 355 nm photons [78]. They speculated that the onset of cage formation may be at $n$ = 8. The high VDE found for $Si_{12}Cr^{-}$ is consistent with the theoretical prediction of an enhanced stability for this cluster [15]. A combined PES and DFT study of anionic Cr doped silicon $Si_nCr^{-}$ ($n$ = 3–12) refined the earlier prediction and the structural transition from exohedral to endohedral was shown to occur at $n$ = 10 [39]. This critical size is close to the prediction for cationic $Si_nCr^{+}$ based on mass spectrometric Ar physisorption experiments, where $Si_{11}Cr^{+}$ is found as the smallest endohedral $Si_nCr^{+}$ cluster (see Fig. 1). Similar to Mn doped silicon clusters, the magnetic properties of $Si_nCr^{-}$ correlate with their geometric structures. Computations found high magnetic moments (3–5 $\mu_B$) for small ($n$ = 3–9) exohedral structures, while larger ($n$ = 10–12) endohedral sizes have low local magnetic moments (1 $\mu_B$). Also the local charges on the Cr atom abruptly increase from $n = 10$ onwards. $Si_{12}Cr^-$ was predicted to have a $D_{3d}$ structure with the Cr atom encapsulated in a $Si_{12}$ hexagonal prism cage and enhanced stability [39], in agreement with earlier work [15].

Similar approaches were used to study $Si_nMn^+$ ($n$ = 6–10, 12–14, and 16) clusters [32]. The Mn dopant favors substitution of a Si atom in $Si_{n+1}{}^+$ cations for small, exohedral $Si_nMn^+$ ($n$ = 6–10), while the endohedral $Si_nMn^+$ ($n$ = 12–14, and 16) clusters have fullerene-like structures. Opposed to several other 3*d* transition-metal dopants, the Mn dopants in small $Si_nMn^+$ ($n$ = 6–10) clusters have, according to natural population analysis on the experimentally assigned isomers, high local magnetic moments. The atomic charges on the Mn dopant are around +1 e for all the considered clusters with electron populations of around 0.3 e in the Mn 4*s* and 5.6 e in the Mn 3*d* atomic orbitals. $Si_7Mn^+$ is a remarkable exception in which both the 4*s* and the 3*d* Mn atomic orbitals are half filled [51]. The electronic and magnetic properties of the $Si_nMn^+$ ($n$ = 7–14) clusters were analyzed in detail by a combined experimental (XAS and XMCD spectroscopy) and theoretical study [26]. In line with the predictions above, magnetic moment quenching or better a transition from high spin to low spin state was found to coincide with the exohedral to endohedral structural transition around $Si_{11}Mn^+$. The quenching of the magnetic moment is due to the Mn 3*d* electron delocalization because of the strong interaction of the Mn dopant with the silicon cage in case the dopant is highly coordinated.

To investigate the effect of charge on the geometrical and electronic properties of Mn and V doped silicon clusters, a combined experimental and theoretical study was carried out on $Si_nTM^{0/+}$ ($TM$ = V, Mn; $n$ = 6–9) [29]. In general, the effect of the charge state was found to be small. Neutral and cationic $Si_nV^{0/+}$ and $Si_nMn^{0/+}$ clusters have similar geometrical structures, although the positions of the capping atoms are different for some sizes. This similarity is also reflected in the electronic structure with the additional electron in the neutral clusters occupying the majority spin LUMO of the corresponding cation. The analysis of the NBO charge on the transition metal atom showed that most of the additional electron charge in the neutral clusters is distributed over the silicon framework. In contrast, a PES study by Zheng and coworkers [79], in which the most likely structures of these clusters were identified by comparing calculated and experimental VDEs, showed that the structures of anionic $Si_nV^-$ ($n$ = 3–6) clusters are different from their cationic and neutral counterparts.

The structures of cobalt doped neutral silicon clusters, $Si_nCo$ ($n$ = 10–12), were recently investigated by IR-UV2CI experiments [61]. These clusters prefer caged structures in which the silicon frameworks are double-layered. Electronic structure analysis indicated that the clusters are stabilized by an ionic interaction between the Co dopant atom and the silicon cage due to the charge transfer from the silicon valence *sp* orbitals to the cobalt 3*d* orbitals. The DOS revealed a pronounced electronic shell structure. For instance, $Si_{10}Co$ has 49 valence electrons (4 for each Si atom and 9 for Co). By comparison of the molecular orbitals with wave functions of a single free electron in a square well potential with a spherical shape, the level sequence of the occupied electronic states in $Si_{10}Co$ can be described as

$(1S)^2(1P)^6(1\ D_a)^6(1D_b)^4(2S)^2(2\ D_a)^2(1F_a)^2(2D_b)^8\ (1F_b)^6(1F_c)^6(2P)^5$. The subscripts a, b, and c indicate that the degeneracy of the high angular momentum states (1D, 2D, and 1F) is lifted because of crystal field splitting related to the distortion from spherical symmetry of the cluster. The most important difference with the energy level sequence of free electrons in a square well potential is the lowering of the 2D level. Examination of the 2D molecular orbitals shows that they are mainly composed of the Co 3*d* atomic orbitals, representing the strong hybridization between the central Co atom and the surrounding Si cage. In the same study it was also shown that the strong hybridization between the Co dopant atom and the silicon host quenches the local magnetic moment on the encapsulated Co atom [61].

The particular stability of the $Si_{16}TM^+$ clusters was systematically investigated by Nakajima and coworkers using mass spectrometry, anion photoelectron spectroscopy, adsorption reactivity, and theoretical calculations [13, 50]. Enhanced abundances were found for $Si_{16}Sc^-$, $Si_{16}Ti$, $Si_{16}V^+$, $Si_{16}Nb^+$, and $Si_{16}Ta^+$ clusters in addition to high threshold energies for electron detachment, large HOMO–LUMO gaps, and abrupt drops in reactivity towards $H_2O$ adsorption from $n = 16$ onwards. X-ray absorption spectroscopy experiments by Lau et al. predicted highly symmetric silicon cage structures for $Si_{16}Ti^+$, $Si_{16}V^+$, and $Si_{16}Cr^+$ [64], indicating that the interaction with the silicon cage determines the local electronic structure of the dopants, while the dopants induce a geometric rearrangement of the silicon clusters.

Theoretical investigations have been carried out for some other transition metal doped silicon clusters. A review of all theoretical work on transition metal doped silicon clusters is beyond the scope of this chapter, but we will briefly discuss a selection of recent results. A DFT study by Ma et al. has shown that the equilibrium site of the Fe dopants in $Si_nFe$ ($n = 2$–$4$) gradually moves from convex, over surface, to concave sites as the number of Si atoms increases [74]. From $n = 10$ onwards the Fe atom is at the center of a Si outer frame, forming metal-encapsulated Si cages [74]. For neutral $Si_nFe$ ($n = 1$–$8$) clusters, Liu et al. predicted trigonal, tetragonal, capped tetragonal, capped pentagonal, and combined tetragonal bipyramids as ground state structures for $n = 4$–$8$ [80]. Although the located ground states are essentially similar for both neutral and cationic $Si_nFe$ ($n = 1$–$8$) clusters, the structures of their anionic counterparts are significantly deformed. Endohedral larger neutral $Si_nFe$ ($n \geq 9$) clusters were recently studied by Chauhan et al. [21]. They concluded that the stability of those clusters was determined in the first place by structural arguments such as compact geometries and high symmetric cages, allowing a better mixing of Fe 3d and Si 3p states, rather than by electron counting rules. According to Deng et al., small $Si_nTi$ ($n = 1$–$8$) clusters have pyramidal structures: trigonal pyramid ($n = 4$), trigonal bipyramid ($n = 5$), square bipyramid ($n = 6$), pentagonal bipyramid ($n = 7, 8$), and they follow a growth pattern where the Si atom prefers to cap $Si_{n-1}Ti$ to form $Si_nTi$ [81]. For $Si_nTM$ ($n = 8$–$16$) with $TM$ = Ti, Zr, and Hf, Kawamura et al. predicted basketlike structures for $n = 8$–$12$ and endohedral structures for $n = 13$–$16$ [82]. The critical size for cage formation and the magic cluster $n = 16$ are also consistent with the experiments [35]. For the 4*d* transition metal yttrium, Jaiswal et al. have shown a peculiar size dependence of the structures of anionic $Si_nY^-$ ($n = 4$–20) clusters [83]. For sizes $n = 8$, 10–15, 18, 19, the yttrium acts as a linker between two silicon sub-clusters, while for some other sizes ($n = 16$, 17, 20) the Y dopant is endohedrally encapsulated.

Besides the work on the singly doped silicon clusters, multiply doped silicon clusters recently also attracted attention. Questions of particular interest are how the additional *TM* dopants can alter the structure of the Si framework and how the local magnetic moments on the dopant atoms are coupled to one another. A PES study by Zheng and coworkers, as mentioned in Sect. 2.3, showed that $Si_{12}V^-$ has a V-centered hexagonal prism structure [38], which is slightly different from the distorted hexagonal prism of $Si_{12}V^+$ [28]. Addition of a second V dopant leads to a V capped hexagonal antiprism for $Si_{12}V_2^-$. A third V atom caps $Si_{12}V_2^-$ to form a bicapped hexagonal antiprism wheel-like structure for $Si_{12}V_3^-$. For $Si_{20}V_2^-$ those researchers predicted an elongated dodecahedron cage structure with a $V_2$ unit encapsulated inside the cage [84]. The strong V–V interaction makes $V_2$ small enough to fit into a dodecahedral $Si_{20}$ cage. In a recent study on $Si_nV_3^-$ ($n = 3$–14) [85], a good agreement between experimental and calculated vertical and adiabatic detachment energies was found, which serves as support for the correctness of the calculated cluster geometries. They also predicted that single V doped anionic silicon clusters are encapsulated from $n = 11$ onwards.

Computationally, Khanna and coworkers studied $Si_nTM_2$ (*TM* = Fe, Co, Ni, Cr, and Mn; $n = 1$–8) clusters by DFT calculations [86, 87]. It was shown that these clusters display a variety of magnetic configurations with varying magnetic moment and different magnetic coupling between the two transition metal atoms depending on the cluster size and charge state. The coupling between the dopants in $Si_nFe_2$ and $Si_nMn_2$ clusters are mostly found to be ferromagnetic with large moments on the dopants, while $Si_nNi_2$ clusters are predicted to be nonmagnetic for most sizes. Most of $Si_nCo_2$ clusters have ferromagnetic ground states, but in general the magnetic moments on Co are smaller than those on the Fe and Mn dopants in silicon clusters. In cationic $Si_nCo_2^+$ ($n = 8$–12) the magnetic coupling between the Co atoms was shown to the depend on the Co–Co distance [88]. Local magnetic moments on Cr in small $Si_nCr_2$ ($n = 1$–5) clusters are found to be coupled ferromagnetic or anti-ferromagnetic, while larger sizes ($n = 6$–8) are nonmagnetic. Recently a non-magnetic, triple ring tubular structure with a $Mn_2$ dimer inside the symmetric Si skeleton was reported for $Si_{15}Mn_2$ [89]. An interesting growth pattern is found for $Si_nPd_2$ ($n = 10$–20) by Zhao et al. Most clusters are based on the pentagonal prism structure of $Si_{10}Pd$ [90]. In $Si_nPd_2$ ($n = 10$–15) one Pd dopant is located at a facial position and the other Pd dopant is inside the silicon framework, while for larger clusters ($n = 16$–20) both Pd atoms are encapsulated.

### 3.3 *Lanthanide Dopants (4f$^x$ 0 < x < 14)*

Lanthanide (*Ln*) doped silicon clusters are of particular interest for applications given the limited binding interaction of localized 4*f* electrons with the environment, which may imply that the lanthanide dopants maintain large magnetic moments in silicon clusters. Experimentally, lanthanide doped silicon clusters mainly have been investigated by PES. Nakajima and coworkers conducted the first study on terbium doped anionic silicon clusters, $Si_nTb^-$ ($n$ = 6–16) [24]. The clusters were excited using the 266 nm (4.66 eV) 4th harmonic output of a Nd:YAG laser and remarkably high electron affinities were observed for $Si_{10}Tb^-$ and $Si_{11}Tb^-$, which was interpreted as support for the onset of cage formation. The reduced reactivity of $Si_nTb^-$ at $n$ = 10–16 towards $H_2O$ is further evidence that the Tb atom is located inside $Si_n$ cages.

Bowen and coworkers investigated $Si_nEu^-$ ($n$ = 3–17) and revealed pronounced electronic rearrangement in the $n$ = 10–12 size range with a remarkable increase of the electron affinities from $Si_{11}Eu^-$ (1.9 eV) to $Si_{12}Eu^-$ (2.8 eV) [37]. Such enhancement of the electron affinity is very similar to what was seen for Tb doped Si clusters before and, again, presumably reflects a geometric rearrangement due to the encapsulation of the Eu dopant atom. That one atom more is needed for cage formation in $Si_nEu^-$ can be explained by the larger atomic radius of Eu (0.185 nm) than that of Tb (0.175 nm) and thus more silicon atoms are required to encapsulate Eu [91]. Differences between the photoelectron spectra of $Si_nTb^-$ and $Si_nEu^-$ were attributed to different oxidation states, Tb(III) and Eu(II), that the dopant atoms adopt in these clusters [37]. DFT calculations by Zhao et al. also predicted that, starting from $n$ = 12, the Eu atom is located in a Si cage [92]. Interestingly, it was also found that the magnetic moments on Eu do not quench in those cages because the 4*f* electrons of the Eu atom do not interact strongly with the silicon cage. An extensive computational study of $Si_nEu$ ($n$ = 3–11) was recently reported by Yang et al. [93].

Bowen and coworkers later carried out a more systematic PES study to understand the electronic structures of different lanthanide dopants in silicon clusters: $Si_nLn^-$ ($n$ = 3–13; *Ln* = Ho, Gd, Pr, Sm, Eu, and Yb) [36]. These spectra are presented in Fig. 9. On the basis of their appearance, the *Ln* doped silicon systems can be categorized either into one of the two distinct categories A and B or in the intermediate category AB. The photoelectron spectra of $Si_nEu^-$ and $Si_nYb^-$ ($n$ 12) have a common feature with a low electron binding energy and form category A. The photoelectron spectra of Ho, Gd, and Pr doped anionic Si clusters lack low energy electron binding peaks and form category B. Most interestingly, $Si_nSm^-$ is an exception (intermediate category AB), because the small sizes $n \leq 6$ and $n = 9$ have the same low energy feature as the clusters in category A, while larger sizes $n \geq 11$ have spectra similar to the clusters in category B. Intermediate sizes of $Si_nSm^-$ ($n$ = 7, 8, and 10) even show features of both A and B. The different electronic features of the different lanthanide doped silicon clusters were explained by different oxidation states adopted by the lanthanide atoms. It was also claimed

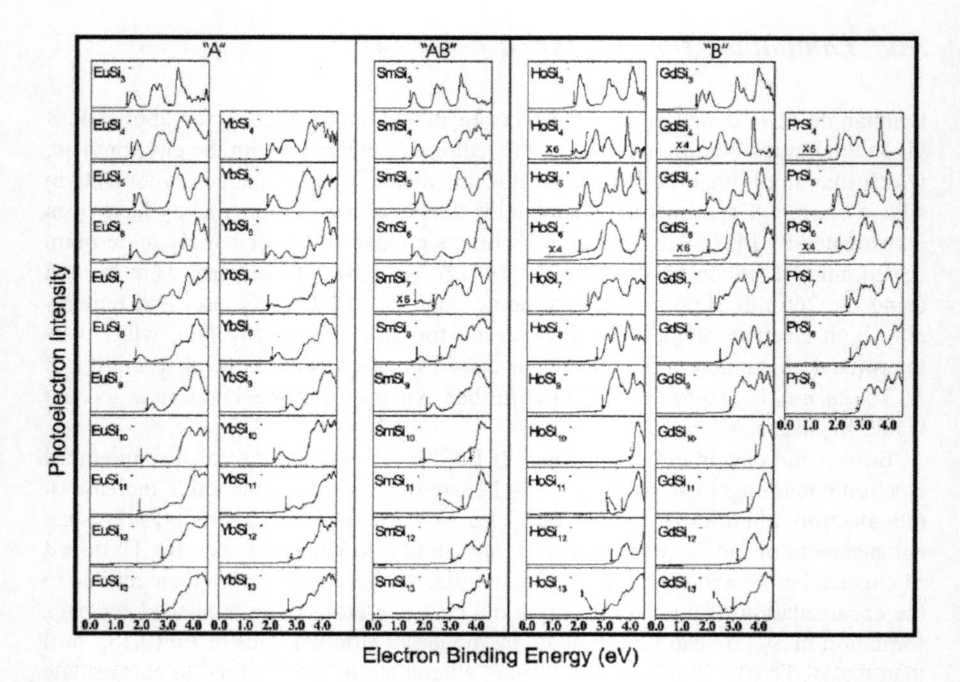


**Fig. 9** Photoelectron spectra of $Si_nLn^-$ ($3 \leq n \leq 13$) anionic clusters recorded following excitation with 266 nm photons. The *vertical arrows* indicate the threshold energies from which the adiabatic electron affinities ($EA_a$) of the clusters are inferred. Due to the presence of two isomers for $Si_nSm^-$ ($n$ = 7, 9, 10) two *vertical arrows* are drawn in their photoelectron spectra approximately pointing at the threshold energies of each isomer. Peaks corresponding to an impurity are marked with "*I*". The categories *A*, *B*, and *AB* are explained in the text. (Reproduced from Grubisic A et al. (2009) J Am Chem Soc 131:10783)

that based on the adopted oxidation state of the lanthanide atoms it should be possible to predict the likely electronic structures of other $Si_nLn^-$ clusters.

Plenty of computational work has been done on lanthanide doped silicon clusters. Neutral and anionic lanthanide doped silicon clusters $Si_6Ln^{0/-}$ (*Ln* = La, Ce, Pr, Eu, Gd, Ho, Tb, Yb, and Lu) have been investigated by DFT [94–96]. It was shown that all anionic $Si_6Ln^-$ clusters, except $Si_6Yb^-$, prefer structures with the *Ln* dopant atom located on top of a pentagonal bipyramid. On the other hand, in all neutral $Si_6Ln$ clusters and in the anionic $Si_6Yb^-$, the dopant prefers an equatorial position in the pentagonal bipyramidal structure. In general, the magnetic moments of the dopant atoms in these clusters remain largely localized and the atomic-like magnetism is maintained in the doped clusters. Guo et al. have shown that caged lanthanide doped silicon clusters, $Si_{16}Ln$ (*Ln* = La, Ce, Pr, Nd, Sm, Eu, Gd, Tm, Yb, and Lu), favor fullerene-like over Frank-Kasper structures, which is the most stable structure for many transition metal doped clusters [97]. They found large spin magnetic moments on late lanthanide dopants (Eu: 5.85 $\mu_B$; Gd: 6.81 $\mu_B$), while some others (Pr, Nd, Sm, and Tm) have large orbibal moments. In addition, for dopants in which the lanthanide 4*f* atomic orbitals are less than half-filled (Pr, Nd, and Sm) the orbital and spin moments are antiparallel to one another, but when the 4*f* atomic orbitals are more than half filled (Tm) the orbital and spin moments are parallel. Also $Si_nHo$ ($n$ = 1–20) has been studied computationally. Liu et al. predicted that the Ho dopant replaces a Si atom in the pure $Si_{n+1}$ clusters for most small $Si_nHo$ ($n$ = 1–12) clusters with large local magnetic moments on Ho [98], while Zhao et al. suggested that the Ho dopant atom is encapsulated by a Si cage from $n$ = 15 onwards [99]. Mulliken population analysis showed charge transfer from the Ho dopant to the Si framework for small sizes ($n$ = 1–12), however, the calculated atomic polar tensor-based charges for larger $Si_nHo$ ($n$ = 12–20) indicate an inverted transfer [99]. Cao et al. found growth patterns for $Si_nLu$ ($n$ = 1–12) very similar to $Si_nHo$ clusters and the calculated ground state structures of $Si_nLu$ ($n$ = 1–12) are formed by substituting a Si atom of the pure $Si_{n+1}$ clusters with a Lu atom [100].

## 3.4 Non-metallic Main Group Dopants

Besides transition metal and lanthanide doped silicon clusters, also silicon clusters doped with main group elements have intensively been studied the last years. In this review we restrict to main group dopant elements that are not metallic in the bulk phase. In particular, $Si_nX$ with $X$ = B, C, O, F, N, and As will be discussed and the focus is on experimental studies.

Fielicke and coworkers obtained infrared spectra of neutral silicon carbide clusters, $Si_mC_n$ ($m + n = 6$, $n \leq 2$), using the IR-UV2CI technique [101]. Structural assignment of $Si_6$, $Si_5C$, and $Si_4C_2$ was accomplished by comparison between experimental IR spectra and calculated linear absorption spectra. Computations for the entire $Si_mC_n$ ($m + n = 6$, $n$ = 0–6) series predict a structural transition from chain-like $C_6$ to three-dimensional bipyramidal structures for $Si_6$ (see Fig. 10). This example demonstrates that the structural differences between

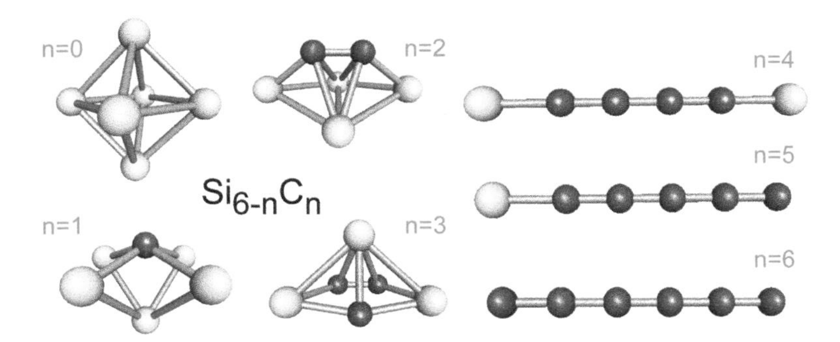


**Fig. 10** Evolution of the geometrical structures of neutral $Si_{6-n}C_n$ clusters ($n$ = 0–6). (Reproduced from Savoca M et al. (2013) J Phys Chem A 117:1158)

carbon, being flexible and exhibiting different types of $sp^n$ ($n$ = 1–3) hybridization, and silicon, preferring single bonds with $sp^3$ hybridization, are retained at the subnanometer scale. Furthermore, even for the smallest possible number of carbon atoms ($n$ = 2), carbon segregation is observed, which can be rationalized by chemical bonding arguments. The bond strength decreases along the C–C, C–Si, Si–Si series and Si has a high ability for multicenter bonding.

Nakajima et al. investigated $Si_mC_n^-$ ($1 \leq m \leq 7$ and $1 \leq n \leq 5$) cluster anions by PES. Similar spectra were found for $Si_mC^-$ ($3 \leq m \leq 7$) and pure $Si_{m+1}^-$, which was attributed to the isovalent electronic structure of Si and C atoms and suggested similar geometric structures. However, significant differences were found for $SiC_n^-$ and $C_{n+1}^-$ ($2 \leq n \leq 5$), which was explained by the change of the geometries [102].

Recent complementary experimental and theoretical studies dealt with the vibrational spectra and structures of neutral silicon clusters doped with first row elements $Si_6X$ with $X$ = Be, B, C, N, O [103, 104]. For $Si_6X$ ($X$ = B, C, O) calculated IR spectra could be compared to experimental spectra obtained by IR-UV2CI, while for the other clusters no experimental spectra were measured. Computations showed that the dopant atoms in $Si_6X$ have a negative net charge and the Si atoms act as electron donors. Their structures strongly depend on the nature of the dopant atom. In particular, Be, B, and C favor structures based on the $Si_7$ pentagonal bipyramid with substitution at an apex, while N and O doping cause complete structural rearrangement of the silicon cluster. Also vibrational spectra of Xe-tagged cationic silicon oxide clusters $Si_nO_m^+$ with $n$ = 3–5 and $m = n, n \pm 1$ have been obtained by IR-MPD [105]. For most sizes, the experimental IR spectra are consistent with the harmonic vibrational spectra of the calculated lowest energy isomer of the corresponding clusters (without considering the Xe messenger atom). In some cases ($Si_4O_3^+$, $Si_4O_5^+$, and $Si_5O_4^+$) it was found that Xe tagging changed the energetic order of the structural isomers, although the overall influence of the Xe adsorption on the cluster geometries and on the IR spectra is small. For the $Si_nO_m^+$ clusters, no simple sequential growth mechanism could be identified, but interesting structural motifs were found such as the $Si_2O_2$ rhombus, the $Si_3O_2$ pentagon, and the $Si_3O_3$ hexagon, which may be building blocks of larger $Si_nO_m^+$ clusters. For silicon monoxide $(SiO)_n^+$ ($n$ = 3–5), Garand et al. used loss of a SiO unit following infrared irradiation to record action spectra [106]. Although $(SiO)_4^+$ shows a structure different from its neutral counterpart and the presence of two isomers was suggested, neutral and cationic $(SiO)_3^{+/0}$ and $(SiO)_5^{+/0}$ have similar structures.

Besides these combined experimental and computational studies, there also exist a huge number of computational studies. For instance, Avaltroni et al. studied the stability of small endohedrally B and C doped silicon clusters [107]. The atomic radius, the electronegativity and the bonding pattern of the dopant atoms were found to be the main factors that determine the relative stability of the cluster. For $Si_8$ it was predicted that a B dopant could stabilize a small cubic caged structure. Work by Ngan and Nguyen showed that a closed electronic shell structure, with 34 delocalized valence electrons, is able to stabilize cubic caged $Si_8B^+$, $Si_8Be$, and

$Si_8C^{2+}$ isomers [16]. They found that the Si atoms in the cluster have $sp^3$ hydridization and that dangling bonds are saturated by the main group dopant atoms. Later studies of the same group identified the growth mechanism of boron doped silicon clusters $Si_nB$ ($n$ = 1–10) [108]. They found that $Si_nB$ can be formed by adding a Si atom to $Si_{n-1}B$ and the B dopant is encapsulated by a $Si_n$ cage from $n$ = 8 onwards, although the predicted lowest energy isomer of $Si_8B$ is based on a pentagonal bipyramidal motif and is not a cube. The high stability of $Si_9B^-$ and $Si_{10}B^+$ is claimed to originate from closed electronic shells and from spherical aromaticity. For C doped silicon clusters, a planar tetracoordinated carbon $Si_4C^{2+}$ structure with enhanced stability was predicted, which is a building block for $Si_9C$ [109]. Combined effects of the electron delocalization and geometrical constraint of the $Si_9$ and an additional electrostatic interaction also contribute to the cluster stabilization.

Also fluorine has been considered as dopant in silicon clusters. Zhang et al. conducted a systematical study of the geometries, stabilities, and electronic properties of $Si_nF$ ($n$ = 1–12) by first-principles calculations [110]. The located ground state structures of $Si_nF$ ($n$ = 1–12) are all exohedral and the F dopant was found to enhance the chemical reactivity and to reduce the stability of the pure Si clusters with smaller HOMO–LUMO gaps in the F doped clusters compared to the corresponding pure silicon clusters. They also found negative charges on the F dopant for the studied sizes. Arsenic doped silicon clusters, $Si_nAs$ ($n$ = 1–14), have been studied by Kodlaa et al. [111]. The As dopant seems to have a limited effect on the structures of the pure Si clusters and favors to adsorb to the pure $Si_n$ clusters at exohedral positions. Interestingly, enhanced metallic characteristics were found for the larger clusters, i.e., from $n$ = 4 onwards.

## 4 Summary and Outlook

In this chapter, we presented a review of recent progress on the structural identification of doped silicon clusters. Besides geometric structures, also electronic and magnetic properties of the clusters are briefly commented on. Growth patterns of the doped silicon clusters are discussed and clusters with appealing structural, electronic, or magnetic properties are identified. As dopant elements coinage metals, transition metals, lanthanides, and non-metallic main-group atoms were considered. Experimentally, the main techniques that have been applied in recent years for studying doped silicon clusters in the gas phase are chemical probe mass spectrometry methods, IR action spectroscopy, photoelectron spectroscopy, and x-ray absorption spectroscopy. Computationally most work makes use of the density functional theory formalism.

Specific doped silicon clusters with appealing properties may be used in future applications. However, it is important to realize that the current chapter dealt with isolated clusters in the gas phase. In devices, the interactions with the environment will affect the cluster's properties. Further extension of the work should deal with deposited size selected clusters, which have potential as building blocks of nano-assemblies in microelectronic devices. The choice of the substrate will be crucial, as the geometric and electronic properties of these clusters will be altered by interaction with the support. Recently, using scanning tunneling microscopy (STM), Nakajima and coworkers investigated the initial products created by the deposition of gas phase synthesized $Si_{16}Ta^+$, which is predicted to be very stable, on monolayer films of $C_{60}$ molecules. $Si_{16}Ta$–$C_{60}$ heterodimers are formed preferentially [112]. They also demonstrated that densely packed $Si_{16}Ta^+$ clusters can be immobilized onto $C_{60}$ terminated surfaces, while retaining their cage shape and positive charge [113]. Another example concerns materials based on $Si_{12}V^+$. Its symmetric hexagonal prism structure in the gas phase [28] inspired researchers to build networks with $Si_{12}V^+$ as building block. A first-principle calculation demonstrated that hexagonal porous and honeycomb-like frameworks build up from $Si_{12}V$ units with regularly and separately distributed V atoms are stable at room temperature [114]. The preferred magnetic coupling in both the hexagonal porous and honeycomb-like sheets is found to be ferromagnetic due to a free-electron-mediated mechanism. By using external strain, it was shown that the magnetic moments and the strength of the magnetic coupling with the sheets can be deliberately tuned [115], which is propitious for advanced applications, in particular for 2D silicon-based spintronic nanomaterials.

Another subject that needs further exploration is multiply doped silicon clusters, especially using magnetic dopants. In these systems, the magnetic coupling between the dopant atoms can be tuned by the size of the silicon host cage and by the interaction with the silicon which can be strongly size-dependent [38, 86, 87]. This approach could allow to control the magnetic properties in silicon clusters.

**Acknowledgments** We gratefully acknowledge all who contributed to our original papers on the structural identification of doped silicon clusters: Pieterjan Claes, Philipp Gruene, Marko Haertelt, Dan J. Harding, Jonathan T. Lyon, Gerard Meijer, Vu Thi Ngan, Minh Tho Nguyen, Nguyen Minh Tam, Alex Woodham, and Ludger Wöste. The authors are thankful to the Stichting voor Fundamenteel Onderzoek der Materie (FOM) in providing beam time on FELIX and highly appreciate the skillful assistance of the FELIX staff. This work is supported by the Research Foundation-Flanders (FWO), the KU Leuven Research Council (GOA 14/007), and the Deutsche Forschungsgemeinschaft within FOR 1282 (FI 893/4).

## References

1. Bloomfield LA, Freeman RR, Brown WL (1985) Photofragmentation of mass-resolved $Si_{2\text{-}12}^+$ clusters. Phys Rev Lett 54:2246–2249
2. Heath JR, Liu Y, O'Brien SC, Zhang QL, Curl RF, Tittel FK, Smalley RE (1985) Semiconductor cluster beams—one and two color ionization studies of $Si_x$ and $Ge_x$. J Chem Phys 83:5520–5526
3. Martin TP, Schaber H (1985) Mass spectra of Si, Ge, and Sn clusters. J Chem Phys 83:855–858

4. Shvartsburg AA, Hudgins RR, Dugourd P, Jarrold MF (2001) Structural information from ion mobility measurements: applications to semiconductor clusters. Chem Soc Rev 30:26–35
5. Zhang QL, Liu Y, Curl RF, Tittel FK, Smalley RE (1987) Photodissociation of semiconductor positive cluster ions. J Chem Phys 88:1670–1677
6. Rothlisberger U, Andreoni W, Parrinello M (1994) Structure of nanoscale silicon clusters. Phys Rev Lett 72:665–668
7. Veldeman N, Gruene P, Fielicke A, Ngan VT, Nguyen MT, Lievens P (2010) Endohedrally doped silicon clusters. In: Sattler KD (ed) Handbook of nanophysics: clusters and fullerenes. CRC Press Boca Raton
8. Kumar V (2007) Metal encapsulated clusters of silicon: silicon fullerenes and other polyhedral forms. In: Kumar V (ed) Nanosilicon. Elsevier London, pp 114–148
9. Jackson K, Nellermoe B (1996) $Zr@Si_{20}$: a strongly bound Si endohedral system. Chem Phys Lett 254:249
10. Hiura H, Miyazaki T, Kanayama T (2001) Formation of metal-encapsulating Si cage clusters. Phys Rev Lett 86:1733–1736
11. Kumar V, Kawazoe Y (2001) Metal-encapsulated fullerenelike and cubic caged clusters of silicon. Phys Rev Lett 87:045503
12. Reveles JU, Khanna SN (2006) Electronic counting rules for the stability of metal-silicon clusters. Phys Rev B 74:035435
13. Koyasu K, Akutsu M, Mitsui M, Nakajima A (2005) Selective formation of $MSi_{16}$ ($M$ = Sc, Ti, and V). J Am Chem Soc 127:4998–4999
14. Beck SM (1987) Studies of silicon cluster–metal atom compound formation in a supersonic molecular beam. J Chem Phys 87:4233–4234
15. Khanna SN, Rao BK, Jena P (2002) Magic numbers in metallo-inorganic clusters: chromium encapsulated in silicon cages. Phys Rev Lett 89:016803
16. Ngan VT, Nguyen MT (2010) The aromatic 8-electron cubic silicon clusters $Be@Si_8$, $B@Si_8^+$ and $C@Si_8^{2+}$. J Phys Chem A 114:7609–7615
17. Fielicke A, Lyon JT, Haertelt M, Meijer G, Claes P, de Haeck J, Lievens P (2009) Vibrational spectroscopy of neutral silicon clusters via far-IR-VUV two color ionization. J Chem Phys 131:171105
18. Raghavachari K (1986) Theoretical study of small silicon clusters: Equilibrium geometries and electronic structures of $Si_{2\text{-}7}$, $Si_{10}$. J Chem Phys 84:5672–5686
19. Zdetsis AD (2007) Fluxional and aromatic behavior in small magic silicon clusters: a full ab initio study of $Si_n$, $Si_n^{1-}$, $Si_n^{2-}$, and $Si_n^{1+}$, $n$ = 6, 10 clusters. J Chem Phys 127:014314
20. Lyon JT, Gruene P, Fielicke A, Meijer G, Janssens E, Claes P, Lievens P (2009) Structures of silicon cluster cations in the gas phase. J Am Chem Soc 131:1115–1121
21. Chauhan V, Abreu MB, Reber AC, Khanna SN (2015) Geometry controls the stability of $FeSi_{14}$. Phys Chem Chem Phys 17:15718–15724
22. Phi DN, Trung NT, Janssens E, Ngan VT (2016) Electron counting rules for transition metal-doped $Si_{12}$ clusters. Chem Phys Lett 643:103–108
23. Torres MB, Fernández EM, Balbás LC (2007) Theoretical study of isoelectronic $Si_nM$ clusters ($M = Sc^-$, Ti, $V^+$; $n$ = 14–18). Phys Rev B 75:205425
24. Ohara M, Miyajima K, Pramann A, Nakajima A, Kaya K (2002) Geometric and electronic structures of terbium-silicon mixed clusters ($TbSi_n$; $6 \leq n \leq 16$). J Phys Chem A 106:3702–3705
25. Janssens E, Gruene P, Meijer G, Wöste L, Lievens P, Fielicke A (2007) Argon physisorption as a structural probe for endohedral doped silicon clusters. Phys Rev Lett 99:063401
26. Zamudio-Bayer V, Leppert L, Hirsch K, Langenberg A, Rittmann J, Kossick M, Vogel M, Richter R, Terasaki A, Möller T, von Issendorff B, Kummel S, Lau JT (2013) Coordination-driven magnetic-to-nonmagnetic transition in manganese-doped silicon clusters. Phys Rev B 88:115425
27. Asmis KR, Fielicke A, von Helden G, Meijer G. Vibrational spectroscopy of gas-phase clusters and complexes. In: Woodruff DP (ed) Atomic cluster: from gas phase to deposited. Elsevier, Amsterdam

28. Claes P, Janssens E, Ngan VT, Gruene P, Lyon JT, Harding DJ, Fielicke A, Nguyen MT, Lievens P (2011) Structural identification of caged vanadium doped silicon clusters. Phys Rev Lett 107:173401
29. Claes P, Ngan VT, Haertelt M, Lyon JT, Fielicke A, Nguyen MT, Lievens P, Janssens E (2013) The structures of neutral transition metal doped silicon clusters, $Si_nX$ ($n$ = 6–9; X = V, Mn). J Chem Phys 138:194301
30. Li Y, Lyon JT, Woodham AP, Fielicke A, Janssens E (2014) The geometric structure of silver-doped silicon clusters. ChemPhysChem 15:328–336
31. Li Y, Lyon JT, Woodham AP, Lievens P, Fielicke A, Janssens E (2015) Structural identification of gold doped silicon clusters via far-infrared spectroscopy. J Phys Chem C 119:10896–10903
32. Ngan VT, Janssens E, Claes P, Lyon JT, Fielicke A, Nguyen MT, Lievens P (2012) High magnetic moments in manganese-doped silicon clusters. Chem Eur J 18:15788–15793
33. Ngan VT, Gruene P, Claes P, Janssens E, Fielicke A, Nguyen MT, Lievens P (2010) Disparate effects of Cu and V on structures of exohedral transition metal-doped silicon clusters: A combined far-infrared spectroscopic and computational study. J Am Chem Soc 132:15589–15602
34. Koyasu K, Atobe J, Akutsu M, Mitsui M, Nakajima A (2007) Electronic and geometric stabilities of clusters with transition metal encapsulated by silicon. J Phys Chem A 111:42–49
35. Ohara M, Koyasu K, Nakajima A, Kaya K (2003) Geometric and electronic structures of metal ($M$)-doped silicon clusters ($M$ = Ti, Hf, Mo and W). Chem Phys Lett 371:490–497
36. Grubisic A, Ko YJ, Wang H, Bowen KH (2009) Photoelectron spectroscopy of lanthanide-silicon cluster anions $LnSi_n^-$ ($3 \leq n \leq 13$; $Ln$ = Ho, Gd, Pr, Sm, Eu, Yb): Prospect for magnetic silicon-based clusters. J Am Chem Soc 131:10783–10790
37. Grubisic A, Wang H, Ko YJ, Bowen KH (2008) Photoelectron spectroscopy of europium-silicon cluster anions, $EuSi_n^-$ ($3 \leq n \leq 17$). J Chem Phys 129:054302
38. Huang X, Xu HG, Lu S, Su Y, King RB, Zhao J, Zheng W (2014) Discovery of a silicon-based ferromagnetic wheel structure in $V_xSi_{12}^-$ ($x$ = 1–3) clusters: Photoelectron spectroscopy and density functional theory investigation. Nanoscale 6:14617
39. Kong X, Xu HG, Zheng W (2012) Structures and magnetic properties of $CrSi_n^-$ ($n$ = 3–12) clusters: photoelectron spectroscopy and density functional calculations. J Chem Phys 137:064307
40. Kong XY, Deng XJ, Xu HG, Yang Z, Xu XL, Zheng W (2013) Photoelectron spectroscopy and density functional calculations of $AgSi_n^-$ ($n$ = 3–12) clusters. J Chem Phys 138:244312
41. Tai TB, Nguyen MT (2012) Electronic structure and thermochemical properties of silicon-doped lithium clusters $Li_nSi^{0/+}$, $n$ = 1–8: new insights on their stability. J Comp Chem 33:800–809
42. Tam NM, Tai TB, Ngan VT, Nguyen MT (2013) Structure, thermochemical properties, and growth sequence of aluminum-doped silicon clusters $Si_nAl_m$ ($n$ = 1–11, $m$ = 1–2) and their anions. J Phys Chem A 117:6867–6882
43. Majumder C, Kulshreshtha SK (2004) Influence of Al substitution on the atomic and electronic structure of Si clusters by density functional theory and molecular dynamics simulations. Phys Rev B 69:115432
44. Nigam S, Majumder C, Kulshreshtha SK (2006) Structural and electronic properties of $Si_n$, $Si_n^-$, and $PSi_{n-1}$ clusters ($2 \leq n \leq 13$): theoretical investigation based on ab initio molecular orbital theory. J Chem Phys 125:074303
45. Rabilloud F, Sporea C (2007) Ab initio investigation of structures and properties of mixed silicon-potassium $Si_nK_p$ and $Si_nK_p^+$ ($n \leq 6$, $p \leq 2$) clusters. J Comput Methods Sci Eng 7:273–286
46. Sporea C, Rabilloud F, Aubert-Frécon M (2007) Charge transfers in mixed silicon–alkali clusters and dipole moments. J Mol Struc-Theochem 802:85–90

47. Sporea C, Rabilloud F, Cosson X, Allouche AR, Aubert-Frécon M (2006) Theoretical study of mixed silicon—lithium clusters $Si_nLi_p^{(+)}$ ($n$ = 1–6, $p$ = 1–2). J Phys Chem A 110:6032–6038
48. De Haeck J, Bhattacharyya S, Le HT, Debruyne D, Tam NM, Ngan VT, Janssens E, Nguyen MT, Lievens P (2012) Ionization energies and structures of lithium doped silicon clusters. Phys Chem Chem Phys 14:8542–8550
49. Neukermans S, Wang X, Veldeman N, Janssens E, Silverans RE, Lievens P (2006) Mass spectrometric stability study of binary $MS_n$ clusters (S = Si, Ge, Sn, Pb, and $M$ = Cr, Mn, Cu, Zn). Int J Mass Spectrom 252:145–150
50. Koyasu K, Atobe J, Furuse S, Nakajima A (2008) Anion photoelectron spectroscopy of transition metal- and lanthanide metal-silicon clusters: $MSi_n^-$ ($n$ = 6–20). J Chem Phys 129:214301
51. Ngan VT, Janssens E, Claes P, Fielicke A, Nguyen MT, Lievens P (2015) Nature of the interaction between rare gas atoms and transition metal doped silicon clusters: the role of shielding effects. Phys Chem Chem Phys 17:17584–17591
52. Andriotis AN, Mpourmpakis G, Froudakis GE, Menon M (2002) Stabilization of Si-based cage clusters and nanotubes by encapsulation of transition metal atoms. New J Phys 4:78
53. Menon M, Andriotis AN, Froudakis GE (2002) Structure and stability of Ni-encapsulated Si nanotube. Nano Lett 2:301–304
54. Asmis KR (2012) Structure characterization of metal oxide clusters by vibrational spectroscopy: possibilities and prospects. Phys Chem Chem Phys 14:9270–9281
55. http://www.ru.nl/felix/
56. Raghavachari K, Logovinsky V (1985) Structure and bonding in small silicon clusters. Phys Rev Lett 55:2853
57. Okumura M, Yeh LI, Lee YT (1985) The vibrational predissociation spectroscopy of hydrogen clusters ions. J Chem Phys 83:3705
58. Okumura M, Yeh LI, Myers JK, Lee YT (1986) Infrared spectra of the cluster ions $H_7O_3^+{\cdot}H_2$ and $H_9O_4^+{\cdot}H_2$. J Chem Phys 85:2328
59. Savoca M, Langer J, Harding DJ, Dopfer O, Fielicke A (2013) Incipient chemical bond formation of Xe to a cationic silicon cluster: vibrational spectroscopy and structure of the $Si_4Xe^+$ complex. Chem Phys Lett 557:49–52
60. Calvo F, Li Y, Kiawi DM, Bakker JM, Parneix P, Janssens E (2015) Nonlinear effects in infrared action spectroscopy of silicon and vanadium oxide clusters: experiment and kinetic modeling. Phys Chem Chem Phys 17:25956–25967
61. Li Y, Tam NM, Claes P, Woodham AP, Lyon JT, Ngan VT, Nguyen MT, Lievens P, Fielicke A, Janssens E (2014) Structure assignment, electronic properties, and magnetism quenching of endohedrally doped neutral silicon clusters, $Si_nCo$ ($n$ = 10–12). J Phys Chem A 118:8198–8203
62. Hirsch K, Lau JT, Klar Ph, Langenberg A, Probst J, Rittmann J, Vogel M, Zamudio-Bayer V, Möller T, von Issendorff B (2009) X-ray spectroscopy on size-selected clusters in an ion trap: from the molecular limit to bulk properties. J Phys B: At Mol Opt Phys 42:154029
63. Peredkov S, Neeb M, Eberhardt W, Meyer J, Tombers M, Kampschulte H, Niedner-Schatteburg G (2011) Spin and orbital magnetic moments of free nanoparticles. Phys Rev Lett 107:233401
64. Lau JT, Hirsch K, Klar Ph, Langenberg A, Lofink F, Richter R, Rittmann J, Vogel M, Zamudio-Bayer V, Möller T, von Issendorff B (2009) X-ray spectroscopy reveals high symmetry and electronic shell structure of transition-metal-doped silicon clusters. Phys Rev A 79:053201
65. Lau JT, Vogel M, Langenberg A, Hirsch K, Rittmann J, Zamudio-Bayer V, Möller T, von Issendorff B (2011) Highest occupied molecular orbital–lowest unoccupied molecular orbital gaps of doped silicon clusters from core level spectroscopy. J Chem Phys 134:041102
66. Xiao C, Hagelberg F (2002) Geometric, energetic, and bonding properties of neutral and charged copper-doped silicon clusters. Phys Rev B 66:075425

67. Chuang FC, Hsieh YY, Hsu CC, Albao MA (2007) Geometries and stabilities of Ag-doped $Si_n$ ($n$ = 1-13) clusters: a first-principles study. J Chem Phys 127:144313
68. Ziella DH, Caputo MC, Provasi P (2010) Study of geometries and electronic properties of $AgSi_n$ cluster using DFT/TB. Int Quantum Chem 111:1680–1693
69. Wang J, Liu Y, Li YC (2010) Au@Sin: growth behavior, stability and electronic structure. Phys Lett A 374:2736–2742
70. Chuang FC, Hsu CC, Hsieh YY, Albao MA (2010) Atomic and electronic structures of $AuSi_n$ ($n$ = 1–16) clusters: a first-principles study. Chin J Phys 48:82–102
71. Gruene P, Fielicke A, Meijer M, Janssens E, Ngan VT, Nguyen MT, Lievens P (2008) Tuning the geometric structure by doping silicon clusters. ChemPhysChem 9:703–706
72. Hagelberg F, Xiao C, Lester WA (2003) Cagelike $Si_{12}$ clusters with endohedral Cu, Mo, and W metal atom impurities. Phys Rev B 67:035426
73. Guo LJ, Zhao GF, Gu YZ, Liu X, Zeng Z (2008) Density-functional investigation of metal-silicon cage clusters MSin (M = Sc, Ti, V, Cr, Mn, Fe Co, Ni, Cu, Zn; n = 8–16). Phys Rev B 77:195417
74. Ma L, Zhao JJ, Wang JG, Wang BL, Lu QL, Wang GH (2006) Growth behavior and magnetic properties of $Si_nFe$ ($n$ = 2–17) cluster. Phys Rev B 73:125439
75. Xu HG, Wu MM, Zhang ZG, Yuan J, Sun Q, Zheng W (2012) Photoelectron spectroscopy and density functional calculations of $CuSi_n^-$ ($n$ = 4–18) Clusters. J Chem Phys 136:104308
76. Palagin D, Gramzow M, Reuter K (2011) On the stability of "non-magic" endohedrally doped Si clusters: a first-principles sampling study of $MSi_{16}^+$ ($M$ = Ti, V, Cr). J Chem Phys 134:244705
77. Reveles JU, Khanna SN (2005) Nearly-free-electron gas in a silicon cage. Phys Rev B 72:165413
78. Zheng W, Nilles JM, Radisic D, Bowen KH (2005) Photoelectron spectroscopy of chromium-doped silicon cluster anions. J Chem Phys 122:071101
79. Xu HG, Zhang ZG, Feng Y, Yuan J, Zhao Y, Zheng W (2010) Vanadium-doped small silicon clusters: photoelectron spectroscopy and density-functional calculations. Chem Phys Lett 487:204–208
80. Liu Y, Li GL, Gao AM, Chen HY, Finlow D, Li QS (2011) The structures and properties of $FeSi_n/FeSi_n^+/FeSi_n^-$ ($n$ = 1–8) clusters. Eur Phys J D 64:27–35
81. Deng C, Zhou L, Li G, Chen H, Li Q (2012) Theoretical studies on the structures and stabilities of charged, titanium-doped, small silicon clusters, $TiSi_n^-/TiSi_n^+$ ($n$ = 1–8). J Clust Sci 23:975–993
82. Kawamura H, Kumar V, Kawazoe Y (2005) Growth behavior of metal-doped silicon clusters $Si_nM$ ($M$ = Ti, Zr, Hf; $n$ = 8–16). Phys Rev B 71:075423
83. Jaiswal S, Babar VP, Kumar V (2013) Growth behavior, electronic structure, and vibrational properties of $Si_nY$ anion clusters ($n$ = 4–20): Metal atom as linker and endohedral dopant. Phys Rev B 88:085412
84. Xu HG, Kong XY, Deng XJ, Zhang ZG, Zheng W (2014) Smallest fullerene-like silicon cage stabilized by a $V_2$ unit. J Chem Phys 140:024308
85. Huang X, Xu HG, Lu S, Su Y, King RB, Zhao J, Zheng W (2014) Discovery of a silicon-based ferromagnetic wheel structure in $V_xSi_{12}^-$ ($x$ = 1–3) clusters: photoelectron spectroscopy and density functional theory investigation. Nanoscale 6:14617
86. Robles R, Khanna SN, Castleman AW (2008) Stability and magnetic properties of $T_2Si_n$ ($T$ = Cr, Mn, $1 \leq n \leq 8$) clusters. Phys Rev B 77:235441
87. Robles R, Khanna SN (2009) Stable $T_2Si_n$ ($T$ = Fe Co, Ni, $1 \leq n \leq 8$) cluster motifs. J Chem Phys 130:164313
88. Li Y, Tam NM, Woodham AP, Lyon JT, Li Z, Lievens P, Fielicke A, Nguyen MT, Janssens E (2016) Structure dependent magnetic coupling in cobalt-doped silicon clusters. J Phys Chem C 120:19454–19460
89. Pham HT, Phan TT, Tam NM, Van Duong L, Pham-Ho MP, Nguyen MT (2015) $Mn_2@Si_{15}$: the smallest triple ring tubular silicon cluster. Phys Chem Chem Phys 17:17566–17570

90. Zhao RN, Han JG, Duan Y (2014) Density Functional theory investigations on the geometrical and electronic properties and growth patterns of $Si_n$ ($n$ = 10–20) clusters with bimetal $Pd_2$ impurities. Thin Solid Films 556:571–579
91. WebElements periodic table. http://www.webelements.com/
92. Zhao G, Sun J, Gu Y, Wang Y (2009) Density-functional study of structural, electronic, and magnetic properties of the $Si_nEu$ ($n$ = 1–13) clusters. J Chem Phys 131:114312
93. Yang J, Wang J, Hao Y (2015) Europium-doped silicon clusters $EuSi_n$ ($n$ = 3–11) and their anions: structures, thermochemistry, electron affinities, and magnetic moments. Theo Chem Acc 134:81
94. Li HF, Kuang XY, Wang HQ (2011) Probing the structural and electronic properties of lanthanide-metal-doped silicon clusters: $M@Si_6$ ($M$ = Pr, Gd, Ho). Phys Lett A 375:2836–2844
95. Wang HQ, Li HF (2014) A combined stochastic search and density functional theory study on the neutral and charged silicon-based clusters $MSi_6$ ($M$ = La, Ce, Yb and Lu). RSC Adv 4:29782–29793
96. Xu W, J WX, Xiao Y, Wang SG (2015) Stable structures of $LnSi_6$—and $LnSi_6$ clusters ($Ln$ = Pr, Eu, Gd, Tb, Yb), $C_{2v}$ or $C_{5v}$? Explanation of photoelectron spectra. Comp Theo Chem 1070: 1–8
97. Guo L, Zheng X, Zeng Z, Zhang C (2012) Spin orbital effect in lanthanides doped silicon cage clusters. Chem Phys Lett 550:134–137
98. Liu TG, Zhang WQ, Li YL (2013) First-principles study on the structure, electronic and magnetic properties of $HoSi_n$ ($n$ = 1–12, 20) clusters. Front Phys 9:210–218
99. Zhao RN, Han JG (2014) Geometrical stabilities and electronic properties of $Si_n$ ($n$ = 12–20) clusters with rare earth holmium impurity: a density functional investigation. RSC Adv 4:64410–64418
100. Cao TT, Zhao LX, Feng XJ, Lei YM, Luo YH (2008) Structural and electronic properties of $LuSi_n$ ($n$ = 1–12) clusters: A density functional theory investigation. J Mol Struc-Theochem 895:148–155
101. Savoca M, Lagutschenkov A, Langer J, Harding DJ, Fielicke A (2013) Vibrational spectra and structures of neutral $Si_mC_n$ clusters ($m + n = 6$): sequential doping of silicon clusters with carbon atoms. J Phys Chem A 117:1158–1163
102. Nakajima A, Taguwa T, Nakao K, Gomei M, Kishi R, Iwata S, Kaya K (1995) Photoelectron spectroscopy of silicon–carbon cluster anions ($Si_nC_m^-$). J Chem Phys 103:2050–2057
103. Truong NX, Savoca M, Harding DJ, Fielicke A, Dopfer O (2014) Vibrational spectra and structures of neutral $Si_6X$ clusters ($X$ = Be, B, C, N, O). Phys Chem Chem Phys 16:22364–22372
104. Truong NX, Haertelt M, Jaeger BKA, Gewinner S, Schöllkopf W, Fielicke A, Dopfer O (2016) Characterization of neutral boron-silicon clusters using infrared spectroscopy: the case of $Si_6B$. Int J Mass Spectrom 395:1–6
105. Savoca M, Langer J, Harding DJ, Palagin D, Reuter K, Dopfer O, Fielicke A (2014) Vibrational spectra and structures of bare and Xe-tagged cationic $Si_nO_m^+$ clusters. J Chem Phys 141:104313
106. Garand E, Goebbert D, Santambrogio G, Janssens E, Lievens P, Meijer G, Neumark DM, Asmis KR (2008) Vibrational spectra of small silicon monoxide cluster cations measured by infrared multiple photon dissociation spectroscopy. Phys Chem Chem Phys 10:1502–1506
107. Avaltroni F, Steinmann SN, Corminboeuf C (2012) How are small endohedral silicon clusters stabilized? Phys Chem Chem Phys 14:14842–14849
108. Tam NM, Tai TB, Nguyen MT (2012) Thermochemical parameters and growth mechanism of the boron-doped silicon clusters, $Si_nB^q$ with $n$ = 1–10 and $q$ = −1, 0, +1. J Phys Chem C 116:20086–20098
109. Tam NM, Ngan VT, Nguyen MT (2013) Planar tetracoordinate carbon stabilized by heavier congener cages: the $Si_9C$ and $Ge_9C$ clusters. Chem Phys Lett 595–596:272–276

110. Zhang S, Jiang HL, Wang P, Lu C, Li GQ, Zhang P (2013) Structures, stabilities, and electronic properties of F-doped $Si_n$ ($n$ = 1–12) clusters: Density functional theory investigation. Chin Phys B 22:123601
111. Kodlaa A, El-Taher S (2012) A DFT study on the structures and stabilities of As-doped $Si_{n-1}$ ($n$ = 2–15) clusters. Comp Theo Chem 992:134–141
112. Nakaya M, Nakaya M, Iwasa T, Tsunoyama H, Eguchi T, Nakajima A (2015) Heterodimerization via the covalent bonding of $Ta@Si_{16}$ nanoclusters and $C_{60}$ molecules. J Phys Chem C 119:10962–10968
113. Nakaya M, Iwasa T, Tsunoyama H, Eguchi T, Nakajima A (2014) Formation of a superatom monolayer using gas-phase-synthesized $Ta@Si_{16}$ nanocluster ions. Nanoscale 6:14702–14707
114. Liu Z, Wang X, Cai J, Zhu H (2015) Room-temperature ordered spin structures in cluster-assembled single $V@Si_{12}$ sheets. J Phys Chem C 119:1517–1523

# Structural Evolution, Vibrational Signatures and Energetics of Niobium Clusters from $Nb_2$ to $Nb_{20}$

**Pham Vu Nhat, Devashis Majumdar, Jerzy Leszczynski and Minh Tho Nguyen**

**Abstract** A comprehensive review on geometric and electronic structures, spectroscopic and energetic properties of small niobium clusters in the range from two to twenty atoms, $Nb_n$, n = 2–20, in three different charged states is presented including a systematic comparison of quantum chemical results with available experimental data to assign the lowest-lying structures of $Nb_n$ clusters and their IR spectra and some basic thermochemical parameters including total atomization (TAE) and dissociation ($D_e$) energies based on DFT and CCSD(T) results. Basic energetic properties including electron affinities, ionization energies, binding energies per atom, and stepwise dissociation energies are further discussed. Energetic parameters of small sizes often exhibit odd–even oscillations. Of the clusters considered, $Nb_2$, $Nb_4$, $Nb_8$ and $Nb_{10}$ were found to be magic as they hold the numbers of valence electrons corresponding to the closed-shell in the electron shells [1S/1P/2S/1D/1F.....]. $Nb_{10}$ has a spherically aromatic character, high chemical hrT high chemical hardness and large HOMO–LUMO gap. The open-shell $Nb_{15}$ system is also particularly stable and can form a highly symmetric structure in all charged states. For species with an encapsulated Nb atom, an electron density flow is present from the cage skeleton to the central atom, and the greater the charge involved the more stabilized the cluster is.

**Keywords** Niobium clusters · Structural evolution · Vibrational signatures · Infrared spectra · Electronic structure · Electron shells · Spherical aromaticity

P.V. Nhat
Department of Chemistry, Can Tho University, Can Tho, Vietnam
e-mail: nhat@ctu.edu.vn

D. Majumdar · J. Leszczynski
Interdisciplinary Center for Nanotoxicity, Department of Chemistry and Biochemistry, Jackson State University, Jackson, MS 39217, USA
e-mail: devashis@icnanotox.org

J. Leszczynski
e-mail: jerzy@icnanotox.org

M.T. Nguyen (✉)
Department of Chemistry, KU Leuven, B-3001 Leuven, Belgium
e-mail: minh.nguyen@kuleuven.be



M.T. Nguyen and B. Kiran (eds.), *Clusters*, Challenges and Advances in Computational Chemistry and Physics 23,
DOI 10.1007/978-3-319-48918-6_3

# 1 Introduction

The name *Niobium* came from Greek mythology: *Niobe*, the daughter of *Tantalus*, as niobium and tantalum are so closely related, always found together and very difficult to isolate or separate them from each other. Originally, the element niobium was named 'columbium' by Charles Hatchett, an English chemist, after he discovered in 1801 a new element from the mineral columbite [1]. In 1809, another English scientist, William Hyde Wollaston, analyzed both columbite and tantalite mineral specimens and claimed that columbium was actually the element tantalum [2]. This controversial issue was partly elucidated in 1844 by the German chemist Heinrich Rose when he argued that tantalum ores contain a second element and named it niobium [3]. The confusion was unequivocally resolved in 1864 when the Swedish chemist Christian Wilhelm Blomstrand was able to isolate the metallic niobium [4]. Both names columbium and niobium were used to identify this element until the International Union of Pure and Applied Chemistry (IUPAC) officially adopted in 1950 the name of niobium [5].

In its pure form, niobium is a lustrous, grey, ductile, paramagnetic metal. It becomes a superconductor when cooled below 9.25 K [6]. Although niobium is mostly used in production of high grade structural steel and superalloys, its most interesting applications appear to be in the field of superconductivity. The superconductive cavities of the electron accelerator machine built at the Thomas Jefferson National Accelerator Facility were made from pure niobium. Due to its high reactivity with nonmetal elements, addition of niobium to the steel induces formation of niobium carbide and niobium nitride and leads to improvement of grain refining, retardation of recrystallization, and precipitation hardening of the steel. Such doping increases the toughness and strength of the microalloyed steel [7]. Currently, metallic niobium is extracted from columbite and pyrochlore minerals, and mainly used as an alloying agent.

Studies in the clusters of the elements constitute an initial step in the understanding and interpreting experimental observations on the thermal, optical, magnetic and catalytic properties of nano-particles [8, 9]. Compared to other transition metals, the coinage metals in particular, relatively fewer studies have been devoted to either pure or doped niobium clusters. Reported experiments on $Nb_n$ mainly focused on their mass, optical, photoelectron and infrared spectroscopic properties, and thermochemical parameters including ionization energies (IE), electron affinities (EA) and total atomization energies (TAE), as well as their gas phase chemical reactivities. Knickelbein and Yang [10] measured the IEs for a large number of $Nb_n$ ($n$ = 2–76) using the photo-ionization efficiency (PIE) technique. Kietzmann et al. [11, 12] recorded the photo-electron spectra (PES) of small niobium cluster anions (from 3 to 25 atoms) using a laser vaporization technique. The ultraweak photo emission (UPE) spectra of the larger anions $Nb_n^-$ ($n$ = 4–200) were subsequently recorded by Wrigge et al. [13]. The plot of experimental IE values as a function of the size shows local maxima at $n$ = 8, 10 and 16. Photoemission studies also confirmed that these clusters have larger frontier orbital energy gaps than the other members. Such results are in line with the low chemical reactivities of $Nb_8$, $Nb_{10}$ and $Nb_{16}$ with small molecules, such as $H_2$, $N_2$ and CO [14, 15], and hence suggest that these systems have closed electronic shell structure in their ground state. Recently, Lapoutre et al. reported the IR spectra of neutral $Nb_n$ clusters ranging from $n = 5$–20 using thermionic emission as a probe [16].

Reactions of cationic $Nb_n^+$ ($n$ up to 28) with $H_2$, $H_2O$ and a variety of simple organic compounds were investigated [17, 18]. Ferro-electricity in neutral clusters $Nb_n$ with $n$ ranging from 2 to 150 was also recorded [19]. Optical absorption spectra of isolated Nb clusters from 7 to 20 atoms were measured from 334 to 614 nm via photo-depletion of $Nb_n$. Ar van der Waals complexes [20, 21].

Most of theoretical investigations to date focused on clusters smaller than $Nb_{10}$ and employed density functional theory (DFT) methods, and only a few used MO methods [22–24]. Earlier DFT calculations [25, 26] on neutral $Nb_n$ with $n$ up to 10 suggested that these clusters prefer the lowest possible spin state, except for $Nb_2$, and a close-packed growth behavior. Subsequent DFT study on neutral $Nb_n$ with $n = 2$–23 by Kumar and Kawazoe [27] also supported such observations. Fowler et al. [28] considered many low-lying states of the cationic and neutral clusters containing three and four Nb atoms using the hybrid functional B3LYP in conjunction with a relativistic compact effective core potential basis set (SBKJ). These authors employed additionally other MO methods, such as the $PMP_m$, MP4SDQ, CCSD and CCSD(T), and concluded that the B3LYP functional gives, in general, the most reliable predictions for both IEs and binding energies (BE). Structures of anionic clusters up to $n = 8$ were investigated by comparing experimental energetic results with the local spin density functional (LSD) calculations [11]. Recently, structures and related properties of small neutral, cationic, and anionic niobium clusters ranging from two to six atoms were revisited by Calaminici and Mejia-Olvera using the linear combination of Gaussian-type orbitals density functional theory (LCGTO-DFT) approach [29].

As with other transition metal containing compounds, niobium clusters are a rather difficult target for theoretical investigations, in part due to the presence of many unpaired $d$ electrons and their intrinsic problems. A crucial problem is that truncated wave functions for this type of systems often lead to an inaccurate energy ordering of the isomers and their electronic states. Furthermore, the lack of experimental information constitutes another source of uncertainty for spectroscopic parameters. When available, the assignment of experimental data is not straightforward. In principle, the most favored structure of a specific cluster could be identified by comparing the observed properties to the computed counterparts. A number of recent studies [30–32] thus combined experiment and theory to acquire complementary information, and thereby led to more reliable assignments. However, it sometime appears that the thermodynamically most stable form of a cluster may not be present under the experimental conditions used, due to some subtle kinetics. As a result, theoretical results for the lowest-lying structure could not match those obtained from measurements. For some small $Nb_n$ clusters, Fielicke et al. [31] assigned the structure for each cluster by a comparison between the experimental far IR spectra and the calculated ones, based on harmonic vibrational frequencies of different structural isomers.

In this context, we now review the available results on geometric structures and related properties of the simplest series of niobium clusters containing from two to twenty atoms, in both neutral and singly charged states (anions and cations). For this purpose, we used quantum chemical results as the main set of data, and then discuss the electronic structure and growth behavior by comparing the calculated IR spectra, electron affinities (EA), ionization energies (IE), and dissociation energies ($D_e$) to available experimental information. Since the identity of the ground state of several Nb oligomers is still not clear, attempts are made to determine the relevant structures and assign the corresponding spectral properties.

## 2 Equilibrium Structures and Vibrational Spectra of the Clusters

In the following sections, the equilibrium structures of $Nb_n$ clusters, along with their vibrational signatures, will be discussed in some details. Conventionally, the structures considered are designated by a number ***n*** going from 2 to 20, accompanied by a letter **A, B, C**.... The superscripts $^+$ and $^-$ obviously refer to the cation and anion, respectively.

The present review mostly discusses the computational results that were obtained in the single-reference framework of computations [33–35] which included density functional theory (DFT) and coupled cluster calculations (CCSD(T)). The results obtained through various multi-reference and other DFT methods would be mentioned (wherever available), but any critical comment on the result comparisons would be avoided, as the present article reviews only the specific results [34–37]. In DFT calculations the GGA functional BPW91 [36, 37] and effective core potential LanL2DZ basis set [38] and the correlation-consistent cc-pV*a*Z-PP ($a$ = D, T, Q) basis sets [39–41] were employed. Harmonic vibrational frequencies were also computed for each computed spin states to simulate their vibrational spectra. The initial calibrations were carried out using the coupled-cluster CCSD(T) method in conjunction with the cc-pVTZ-PP basis set for the equilibrium distance and energies of the dimer and its singly charged clusters. For larger systems up to $Nb_{12}$, improved relative energies were performed using single-point electronic energy calculations for quasi-degenerate states competing for the ground state using the coupled-cluster theory CCSD(T) and the cc-pV*a*Z-PP basis set on the basis of the BPW91/cc-pV*a*Z-PP ($a$ = D, T) optimized geometries.

The search for cluster structures was conducted using two different routes. In the first one, the initial geometries of a certain size $Nb_n$ were generated from the lowest-lying isomers of the smaller size $Nb_{n-1}$ by systematically adding an extra Nb atom at all possible positions. This procedure can also be called a *successive growth algorithm* [42]. In the second route, the reported results of other transition metal

**Table 1** Predictions of bond length ($R_e$, Å), dissociation energy ($D_e$, eV), ionization energy (IE, eV), electron affinity (EA, eV) and vibrational frequency ($\omega_e$, $cm^{-1}$) for gas phase $Nb_2$ and $Nb_2^+$, along with available experimental data

| Method | $Nb_2$ | | | | $Nb_2^+$ | | |
|---|---|---|---|---|---|---|---|
| | $R_e$ | $D_e$ | IE | $\omega_e$ | $R_e$ | $D_e$ | $\omega_e$ |
| BP86 | 2.083 | 4.91 | 6.47 | 452.3 | 2.039 | 5.72 | 469.6 |
| PBE | 2.079 | 4.77 | 6.31 | 453.6 | 2.036 | 5.62 | 470.1 |
| BPW91 | 2.081 | 4.34 | 6.25 | 453.1 | 2.040 | 5.26 | 456.4 |
| B3P86 | 2.049 | 3.45 | 6.57 | 485.0 | 2.002 | 4.46 | 505.9 |
| B3LYP | 2.063 | 3.64 | 6.13 | 471.7 | 2.017 | 4.46 | 519.9 |
| TPSS | 2.084 | 4.37 | 6.14 | 453.7 | 2.040 | 5.20 | 467.0 |
| TPSSh | 2.069 | 3.51 | 5.93 | 468.4 | 2.020 | 4.47 | 503.0 |
| M06 | 2.038 | 5.86 | 6.39 | 482.6 | 2.007 | 5.99 | 506.1 |
| CCSD (T) | 2.084 | 4.63 | 6.32 | 449.0 | 2.031 | 5.13 | 519.6 |
| Exptl. | 2.078[a] | 5.24 ± 0.26[b]<br>5.43 ± 0.01[c] | 6.20 ± 0.05[d]<br>6.37 ± 0.001[e] | 424.9[a] | | 5.87 ± 0.12[b]<br>5.94 ± 0.01[c] | 420 ± 3[f] |

Calculations were performed using the cc-pVQZ-PP basis set
[a]Taken from Ref. [44]
[b]Ref. [45]
[c]Ref. [46]
[d]Ref. [10]
[e]Ref. [47]
[f]Ref. [48]

clusters at similar sizes were used as a guide. Some thermochemical properties, including ionization energy (IE), electron affinity (EA) and atomization energy (TAE) were computed by employing the M06 functional [43]. Details of such computational approaches are available in Refs. [33–35].

One of the main points in this context of structural analysis is the calibration of the methods to generate reasonable results. The various methods used in the calibration of various parameters are shown in Table 1.

As it could be seen, the bond length deviations from experiment vary from 0.01 to 0.05 using GGA functionals, while hybrid functionals significantly underestimate the bond distance of $Nb_2$ as the B3LYP, B3P86 and M06 values are 0.015, 0.029 and 0.040 Å, respectively, shorter than the measured values.

Concerning vibrational frequencies, all methods considered tend to overestimate the experimental data (Table 1). The BPW91 functional overall has the most accurate performance for such parameter, having the absolute errors of 29 $cm^{-1}$ for $Nb_2$ and 36 $cm^{-1}$ for $Nb_2^+$. This functional is also reliable in computing IEs. The BPW91 value of 6.25 eV for IE($Nb_2$) agrees well with the experimental data of 6.20 ± 0.05 [10] and 6.37 ± 0.001 eV [47]. Although the geometric and spectroscopic information can sufficiently be established by DFT computations, most functionals yield inconsistent results for the dissociation energy ($D_e$). The B3LYP, B3P86 and TPSSh values for $D_e(Nb_2^+)$ are around 4.46–4.47 eV being much smaller than experimental values of 5.87 ± 0.12 [45] and 5.94 ± 0.01 eV [46]. GGA and meta-GGA methods, such as BP86, BPW91, PBE and TPSS, produce better results for such quantity but they still tend to underestimate the bond strength. Of the functionals tested, only the M06, in which corrections for medium- and long-range interactions are included, is found to describe well for the bond energy. The $D_e(Nb_2)$ = 5.86 and $D_e(Nb_2^+)$ = 5.99 eV computed using M06 functional are in better agreement with experiment (see Table 1).

The above discussion clearly shows the difficulty to judge on the absolute accuracy of a specific functional, as each naturally has its own advantages and drawbacks for different properties. Even though the comparison is only made for a simple species, the BPWP1 appears in general more reliable than other functionals in predicting molecular geometries, vibrational spectra and IEs, but it is likely to be deficient in computing the bond energy for which the M06 functional performs better. Therefore, we selected both BPW91 and M06 functionals for most results reviewed here. While the former is employed for predicting the lowest-energy structures and generating IR spectra, the latter is used for energetic quantities.

## 2.1 The Dimers

$Nb_2$ and its ions have extensively been investigated by both experimental and theoretical methods [44, 45, 49–52]. A unanimous conclusion which was long reached, is that $Nb_2$ is characterized by a triplet ground state with the orbital configuration $^3\Sigma_g^-: \ldots\pi_g^4 1\sigma_g^2 2\sigma_g^2 \delta_g^2$ where the two unpaired electrons occupy the doubly degenerate $\delta_g$ orbital. Such an occupancy gives rise to three possible states $^1\Sigma_g^+$, $^3\Sigma_g^-$, and $^1\Gamma_g$ in which the triplet $^3\Sigma_g^-$ is the lowest-energy state as expected by the Hund's rule.

The pure functionals BP86, BPW91, PBE and TPSS along with the hybrid meta-GGA M06 correctly reproduce the triplet $^3\Sigma_g^-$ ground state for $Nb_2$. The hybrid functionals B3LYP, B3P86 and TPSSh yield a singlet ground state, which contradicts to both experimental observations and high-level wavefunction theory calculations. This is in line with the prior finding that GGA and meta-GGA functionals are in general more consistent than hybrid approaches in treatments of compounds containing only transition elements [53].

The bond length of the neutral dimer was further computed using different functionals and CCSD(T) method in conjunction with the cc-pV*a*Z-PP, $a$ = D, T and Q, basis sets. The results are slightly changed with respect to the basis sets. Using the cc-pVQZ-PP basis set, the bond distance of the $^3\Sigma_g^-$ ground state of $Nb_2$ amounts from 2.04 (M06) to 2.08 Å (CCSD(T) and TPSS), as compared to the experimental value of 2.078 Å experimentally determined using rotationally resolved electronic spectroscopy [44]. The closed-shell singlet $^1\Sigma_g^+$ state is

~0.30 eV above the ground state by BPW91, which is larger than the values of 0.1–0.2 eV previously obtained from FOCI–FO(MR)CI techniques [52].

Concerning the cation, in agreement with earlier studies [52, 54], the high spin state associated with an orbital configuration of $^4\Sigma_g^-$ : $\ldots\pi_g^4 1\sigma_g^2\delta_g^2 2\sigma_g^1$ is confirmed to be the ground state with the doublet state ($^2\Gamma_g$) being 0.29 (BPW91) or 0.32 eV (CCSD(T) above it. Removal of an electron from $Nb_2$ ($^3\Sigma_g^-$) to form the $^4\Sigma_g^-$ cation shortens the equilibrium distance from 2.08 to 2.04 Å (BPW91). Hence, electron detachment results in a charge polarization between both nuclei, which induces an attraction.

The anion appears to be more sensitive with the methods employed. Most functionals tested predict that $Nb_2^-$ prefers a low spin ground state and possesses a shorter equilibrium distance as compared to the neutral, since one electron is added to the bonding $\delta_g$ MO. The ground state configuration of $Nb_2^-$ thus contains three electrons on the doubly degenerate orbital $\delta_g$ which gives rise to a single term $^2\Delta_g$. However, the BPW91 functional predicts the $^2\Delta_g$ state to be ~0.02 eV above the higher spin $^4\Sigma_u^-$ state ($1\sigma_g^2\pi_g^4 2\sigma_g^2\delta_g^2\sigma_u^1$). CCSD(T)/cc-pVQZ-PP calculations indicate that the quartet state is ~0.03 eV higher than the doublet state. Within the expected accuracy of these methods, both anionic states can thus be regarded as nearly degenerate. The lack of diffuse functions could induce an underestimation of the anion energy.

The ground state vibrational frequencies of $Nb_2$ and $Nb_2^+$ at different levels of theory are summarized in Table 1, along with experimental values. BPW91/cc–pVQZ–PP computations predict the harmonic frequencies of 453 and 456 $cm^{-1}$ for $Nb_2$ and $Nb_2^+$, respectively, which are compatible with the experimental values of 425 and 420 ± 3 $cm^{-1}$ [44, 48]. As far as we are aware, the experimental vibrational frequency for the anion $Nb_2^-$ is not available yet. Experimental bond lengths for $Nb_2^{-/+}$ species are also unknown.

Structural data and nature of electronic states should be cautiously treated at the single-reference framework of computations as these methods are valid for equilibrium geometry only. There is no way to generate proper energy surfaces up to the dissociation limit to figure out whether the dissociated species properly reproducing the atomic state of the niobium atoms. The energy surface crossings in such calculations are also not available and this is one of the guiding features to control the nature of the low-lying states (which can compete with ground state). Since these clusters are mostly observed in gas-phase, the nature of such low-lying states is very important in deciding the proper nature of the ground state.

## 2.2 The Trimers

The electronic state of the triatomic system remains a matter of discussion. Kietzmann et al. [11] and Fournier et al. [48] agreed with each other that the trimer anion $Nb_3^-$ exhibits a triangular $D_{3h}$ shape and a singlet ground state. On the contrary, Majumdar and Balasubramanian [22] found the lowest energy structure of the triatomic anion to be an isosceles triangle ($C_{2v}$) with a high spin ground state $^3A_2$ at the B3LYP level, but a low spin $^1A_1$ state at a MR-CISD level. These authors also concluded that the ground state of the neutral trimer $Nb_3$ is the $^2B_1$ ($C_{2v}$) at both levels. However, this contradicted the earlier DFT results of Goodwin et al. [25] and Kumar et al. [27] which showed the $^2A_2$ ground state for $Nb_3$. Recent DFT/B3LYP calculations by Zhai et al. [55] led to a ground state $^3A_2$ ($C_{2v}$) for $Nb_3^-$ and a distorted $^2A''$ ($C_s$) triangular structure for $Nb_3$.

Nguyen and coworkers identified two quasi-degenerate states $^2A_2$ and $^2B_1$ for $Nb_3$. Both are in fact the two resulting components of Jahn-Teller distortions from the unstable degenerate $^2E''$ ($D_{3h}$) state in two distinct (perpendicular) vibrational modes. While the $^2A_2$ state is characterized as a genuine energy minimum, the other is a first-order saddle-point with an imaginary frequency of $184i$ cm$^{-1}$ and lies $\sim$0.01 eV above (DFT/BPW91). The R/UCCSD(T)/cc-pVTZ-PP calculations using restricted HF and unrestricted CCSD formalism point out that the $^2A_2$ state is 0.11 eV more stable than the $^2B_1$ counterpart. In spite of such a small separation between both states, the results suggest that the $^2A_2$ is the ground state of $Nb_3$.

Concerning $Nb_3^-$, DFT calculations suggested that the most stable form of $Nb_3^-$ is an isosceles triangle ($C_{2v}$) with a $^2A_2$ state [22, 55]. In a $D_{3h}$ form, the corresponding singlet state $^1A_1'$: ... $(e')^4(a_1')^2(e'')^4$ is not subjected to a Jahn–Teller distortion, but this $^1A_1'$ state is located at 0.19 eV higher than the triplet state (BPW91). On the contrary, CCSD(T) single point calculations reverse the state energy ordering in predicting that the singlet state is $\sim$0.6 eV lower than the triplet. In other words, our results concur with those reported in Ref. [23] using MO methods. DFT results thus suggest that the $Nb_3^-$ anion possesses a high symmetry low spin ground state $^1A_1'$ ($D_{3h}$).

For the cation $Nb_3^+$, the most favorable form is an equivalent triangle ($D_{3h}$) with the orbital configuration $^3A_1'$:... $(e')^4(a_1')^2(e'')^2$. Pairing the two electrons in the doubly degenerate orbital e'' reduces its symmetry by Jahn-Teller effect, but this brings no benefit for the singlet state. Actually, the $^1A_1$ state of an isosceles triangle ($C_{2v}$) is $\sim$0.4 eV less favored than the $^3A_1'$ state (BPW91) [34]. This is in line with previous findings by Fowler et al. [28]. They however disagreed with B3LYP results reported in Ref. [22], in that the most stable form of $Nb_3^+$ has a $^3B_1$ state ($C_{2v}$). We found that the $^3B_1$ is an excited state, being 0.18 and 0.40 eV above the $^3A'_1$ state by BPW91 and CCSD(T) calculations, respectively.

Figure 1 displays the vibrational spectra computed at the BPW91/cc-pVTZ-PP level for some lower-lying trimeric forms. While the vibrations of the charged $Nb_3^{+/-}$ species are not experimentally recorded yet, vibrational fundamentals of 227.4 $\pm$ 2.9 and 334.9 $\pm$ 2.8 cm$^{-1}$ were experimentally determined for the neutral $Nb_3$ [56]. As compared to the predicted spectrum of $Nb_3$ ($^2A_2$, $C_{2v}$), experiment missed only the lowest frequency band at around 140 cm$^{-1}$. The spectrum of the $D_{3h}$ cation becomes much simpler, with a doubly degenerate mode at $\sim$240 cm$^{-1}$. The anion in $D_{3h}$ symmetry with the state $^1A_1'$ also contains a doubly degenerate mode at $\sim$225 cm$^{-1}$ (Fig. 1).

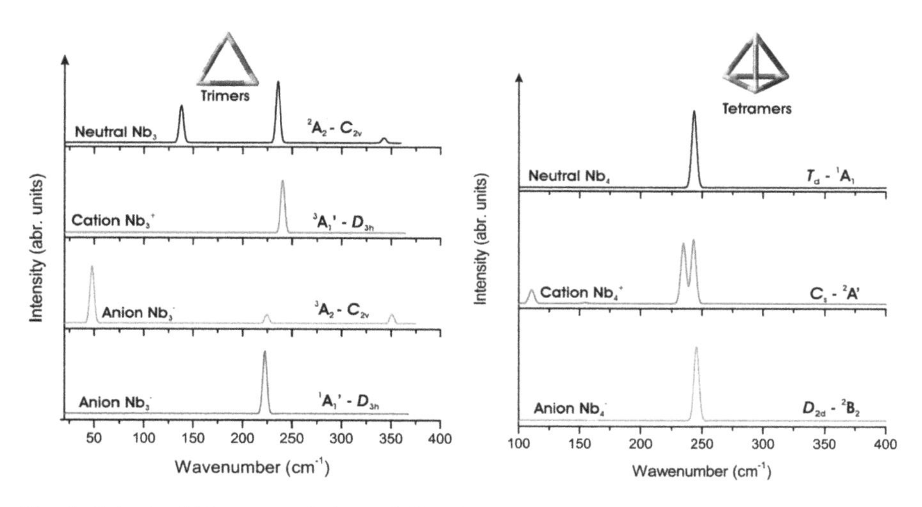


**Fig. 1** Calculated IR spectra of trimer (*left*) and tetramer (*right*) Nb clusters

There are not many experimental evidences to substantiate the predicted ground states of $Nb_3$ cluster ($C_{2v}$ versus $D_{3h}$) and their ionic derivatives. Available experimental IE result was assigned quite accurately through various DFT and MRSDCI calculations, although such experiments are not very sensitive to solve such structural riddle. Actually well-resolved PES could provide answer for such problem in a more convincing way.

## 2.3 *The Tetramers*

In agreement with previous studies [25, 27, 28], a tetrahedral $^1A_1$ ($T_d$) structure is the lowest-energy isomer of $Nb_4$. This neutral can also exist as a rhombus ($D_{2h}$) but both corresponding states $^3A_u$ and $^1A_g$ are 0.73 and 0.83 eV, respectively, higher in energy than the $^1A_1$ (BPW91). The rhombus form is a higher energy local minimum of the cation and anion as well. Their relative energies are listed in Table 2.

For both ions $Nb_4^+$ and $Nb_4^-$, results reported in the literature are not consistent with each other. Because both the HOMO and LUMO of $Nb_4$ are degenerate, attachment or removal of an electron reduces its symmetry due to a Jahn-Teller effect. The structure of $Nb_4^+$ ranges from an ideal to a distorted tetrahedron, which have either a $^2A_1$ ($C_{2v}$) or $^2A'$ ($C_s$) state [22, 28]. Using the BPW91 functional, we found that $Nb_4^+$ is marginally more stable in a lower symmetry form ($C_s$, $^2A'$). The $^2A_1$ ($C_{2v}$) structure is a transition structure with an imaginary frequency of $92i$ cm$^{-1}$. The other component $^2B_2$ of the geometrical distortion corresponding to a $D_{2d}$ structure is also a transition structure with a comparable imaginary frequency. Nevertheless, the three states $^2A'$, $^2B_2$ and $^2A_1$ are nearly identical in terms of

**Table 2** Ground and lower-lying states of $Nb_n^{0/\pm}$ ($n$ = 2–6) and transition energies (TE, eV) computed at the BPW91/cc-pVTZ-PP level

| Cluster[a] | | State | TE | Cluster[a] | | State | TE |
|---|---|---|---|---|---|---|---|
| $Nb_2$ | Linear ($D_{\infty h}$) | $^3\Sigma_g^-$ | 0.00 | $Nb_5$ | **5A** ($C_s$) | $^4A'$ | 0.23 |
| | | $^1\Sigma_g^+$ | 0.37 | | **5B** ($C_{2v}$) | $^2A_1$ | 1.84 |
| $Nb_2^+$ | | $^4\Sigma_g^-$ | 0.00 | $Nb_5^+$ | **5A**$^+$($C_{2v}$) | $^3B_1$ | 0.00 |
| | | $^2\Gamma_g$ | 0.27 | | | $^3A_1$ | 0.001 |
| $Nb_2^-$ | | $^4\Sigma_u^-$ | 0.00 | | | $^1A_1$ | 0.13 |
| | | $^2\Delta_g$ | 0.05 | | | $^1A'$ | 0.18 |
| $Nb_3$ | **3A** ($C_{2v}$) | $^2A_2$ | 0.00 | | **5B**$^+$($C_{2v}$) | $^1A_1$ | 2.02 |
| | | $^2A_2$ | 0.52 | $Nb_5^-$ | **5A**$^-$($D_{3h}$) | $^1A_1'$ | 0.00 |
| $Nb_3^+$ | **3A**$^+$ ($D_{3h}$) | $^3A_1'$ | 0.00 | | **5A**$^-$($C_{2v}$) | $^3A_2$ | 0.18 |
| | **3A**$^+$ ($C_{2v}$) | $^3B_1$ | 0.18 | | **5B**$^-$($C_{2v}$) | $^3B_2$ | 1.69 |
| | | $^1A_1$ | 0.40 | $Nb_6$ | **6B** ($C_{2v}$) | $^3B_2$ | 0.00 |
| $Nb_3^-$ | **3A**$^-$ ($C_{2v}$) | $^3A_2$ | 0.00 | | | $^1A_1$ | 0.13 |
| | **3A**$^-$ ($D_{3h}$) | $^1A_1'$ | 0.19 | | **6B** ($C_2$) | $^1A$ | 0.08 |
| $Nb_4$ | **4A** ($T_d$) | $^1A_1$ | 0.00 | | **6A** ($D_{2h}$) | $^3B_{1u}$ | 0.14 |
| | **4B** ($D_{2h}$) | $^3A_u$ | 0.73 | | | $^1A_g$ | 0.31 |
| | | $^1A_g$ | 0.83 | | **6A** ($D_{4h}$) | $^3A_{1g}$ | 0.24 |
| $Nb_4^+$ | **4A**$^+$ ($C_s$) | $^2A'$ | 0.00 | $Nb_6^+$ | **6B**$^+$ ($C_{2v}$) | $^2B_2$ | 0.00 |
| | **4A**$^+$ ($C_{2v}$) | $^4A_2$ | 0.72 | | | $^4B_1$ | 0.22 |
| | **4B**$^+$($D_{2h}$) | $^2B_{1g}$ | 0.71 | | **6B**$^+$ ($C_2$) | $^4A$ | 0.29 |
| | | $^4A_u$ | 1.32 | | **6A**$^+$ ($D_{2h}$) | $^2B_{3u}$ | 0.04 |
| $Nb_4^-$ | **4A**$^-$($D_{2d}$) | $^2B_2$ | 0.00 | | | $^4B_{3g}$ | 0.10 |
| | **4A**$^-$($C_s$) | $^2A''$ | 0.007 | | | $^2B_{2g}$ | 0.33 |
| | **4A**$^-$($C_{2v}$) | $^4A_2$ | 0.34 | $Nb_6^-$ | **6B**$^-$ ($C_2$) | $^2A$ | 0.00 |
| | **4B**$^-$($D_{2h}$) | $^2B_{1g}$ | 0.28 | | **6B**$^-$ ($C_{2v}$) | $^2A_1$ | 0.01 |
| | | $^4B_{2g}$ | 0.70 | | | $^4A_2$ | 0.13 |
| $Nb_5$ | **5A** ($C_{2v}$) | $^2B_2$ | 0.00 | | **6A**$^-$ ($D_{2h}$) | $^2B_{2g}$ | 0.19 |
| | | $^4B_1$ | 0.39 | | | | |

[a]Structures **nA, nB**… are displayed in figures of the following sections

energy at both BPW91 and CCSD(T) levels. Previous CAS-MCSCF and MRSDCI calculations [23] also suggested that the lowest-lying geometry of $Nb_4^+$ is a $C_s$ pyramid ($^2A'$). The other isomer of $Nb_4^+$ is a rhombus ($D_{2h}$, $^2B_{1g}$) and being 0.71 and 0.81 eV less stable than the ground state by the BPW91 and CCSD(T) methods, respectively. In this context, the $^2A'$ state appears to be the ground state of the tetraatomic cation.

Similarly, the tetraatomic anion is able to appear as a $^2B_2$ ($D_{2d}$) or a $^2A''$ ($C_s$) state of a distorted tetrahedron, with a similar energy content. The gap between both forms is in fact only ~0.01 eV. Another local minimum of $Nb_4^-$ is a rhombus

($D_{2h}$, $^2B_{1g}$) but at 0.28 eV higher. CCSD(T) single point electronic energy calculations using BPW91 geometries confirm the energy ordering but increase this gap to 0.35 eV. It can thus be concluded that the anion $Nb_4^-$ exhibits a $^2B_2$ ground state, but with a highly fluxional structure.

Vibrational spectra for the tetramers are also depicted in Fig. 1. The neutral spectrum is simple with a triply degenerate mode centered at ~245 cm$^{-1}$. The cation contains two peaks in the range 230–240 cm$^{-1}$ and very low intensity peak at around 100 cm$^{-1}$. The anion also has the distinct band around 240 cm$^{-1}$. To the best of our knowledge, no experimental IR data is actually available for the four niobium atoms species.

## 2.4 The Pentamers

DFT calculations concurred that the ground state of $Nb_5$ has a distorted trigonal bipyramid shape with $C_{2v}$ point group and a low multiplicity. In addition, the ground $^2B_2$ state is located below the lowest quartet $^4A'$ state ($C_s$) by 0.23 (BPW91) to 0.53 eV (CCSD(T)). Another $Nb_5$ isomer has a planar W–type shape ($C_{2v}$, $^2A_1$) and is 1.84 eV higher in energy (Table 2).

For the cation, reported results are again not consistent. While wave-function calculations [23] indicated a low spin distorted trigonal pyramid ($C_s$, $^1A'$) to be the most favored isomer, DFT/PBE computations pointed toward a distorted trigonal pyramid as well but with $C_{2v}$ symmetry and a triplet $^3A_1$ state [31].

BPW91 results indicate a distorted trigonal pyramid ($C_s$) with a triplet $^3A'$ state to be the most stable form of $Nb_5^+$. The distorted trigonal pyramid with $C_{2v}$ symmetry and $^3A_1$ state is confirmed to be the transition structure as it has a small imaginary frequency of 70$i$ cm$^{-1}$. The corresponding low spin states $^1A'$ ($C_s$) and $^1A_1$ ($C_{2v}$) are located at 0.18 and 0.13 eV above the $^3A'$ state (BPW91). However, CCSD(T)//BPW91 calculations reverse this energy ordering, in such a way that the $^1A_1$ ($C_{2v}$) state now becomes the lowest-lying state. The other states $^3A'$ and $^1A'$ are calculated at 0.11 and 0.33 eV above the $^1A_1$, respectively. In this context, it can be suggested that $Nb_5^+$ is characterized by a singlet ground state, but with a tiny triplet-singlet gap.

The ground state of the anion $Nb_5^-$ exhibits a high symmetry trigonal bipyramid ($D_{3h}$) singlet state [11], or a distorted trigonal bipyramid with either a low spin state ($C_{2v}$, $^1A_1$) (by B3LYP), or a high spin ground $^3B_1$ state when employing CAS-MCSCF calculations [23]. BPW91 calculations yield the high-symmetry low-spin ($D_{3h}$, $^1A_1'$) as the most stable form of $Nb_5^-$. The orbital configuration related to such state $^1A_1'$: [… $(e')^4(a_1')^2(a_2'')^2(e'')^4$] is stable with respect to Jahn-Teller effect. Subsequent DFT calculations did not locate a $^1A_1$ ($C_{2v}$) state, as all geometry optimizations invariably converge to the $^1A_1'$ ($D_{3h}$) state [34]. The high spin $^3A_2$ ($C_{2v}$) state is an excited state located at 0.18 eV higher. At the CCSD

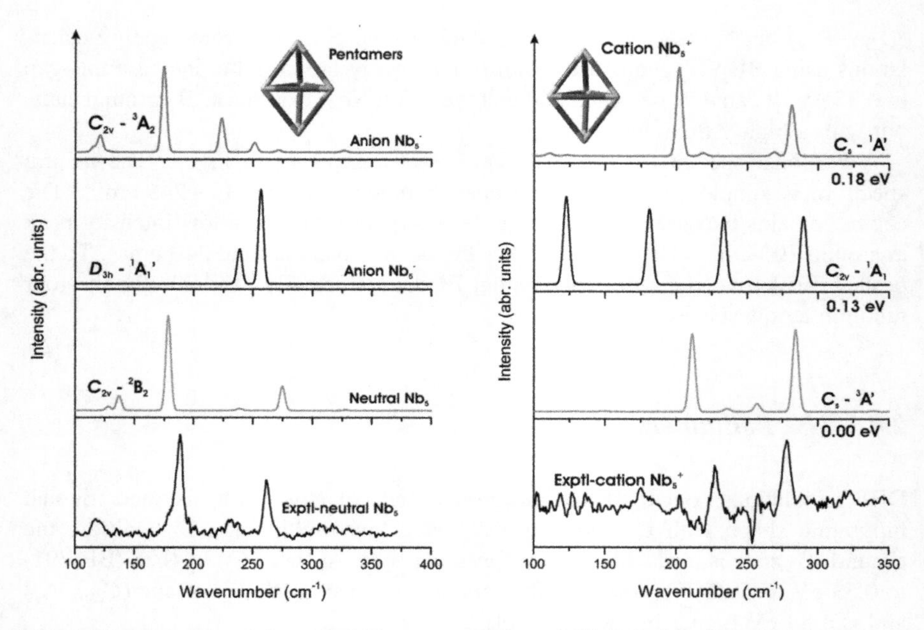


**Fig. 2** Experimental and theoretical IR spectra of $Nb_5$ and $Nb_5^-$ *(left)* and $Nb_5^+$ *(right)*

(T) level, the singlet-triplet separation is significantly reduced, with the $^1A_1'(D_{3h})$ state being only 0.04 eV more stable than the $^3A_2$ ($C_{2v}$) state. These results again suggest that the $Nb_5^-$ anion exhibits a nearly degenerate ground state in both singlet and triplet manifolds.

Figure 2 compares the theoretical and experimental vibrational spectra of the pentamers. Experimental spectra were reported by Fielicke et al. [31]. Calculated vibrational spectrum of the ground state $Nb_5$ ($^2B_2$) covers the range below 300 $cm^{-1}$ with three specific peaks, where the most intense one is centered at $\sim 175$ $cm^{-1}$, which is in line with experimental results. For the purpose of comparison, the calculated spectra of both states of the anion are also plotted in Fig. 2. Vibrational features are thus significantly modified upon electron addition.

The assignment for $Nb_5^+$ becomes more complicated. Although the singlet $^1A_1$ state is lower in energy as compared to the triplet $^3A'$ at the CCSD(T) level, the positions of two intense bands in the calculated spectrum of the triplet state are in better agreement with experimental findings. The $^1A_1$ spectrum covers the range from 100 to 300 $cm^{-1}$, where some bands centered at around 175, 225 and 280 $cm^{-1}$ were also detected experimentally. These facts suggest that we cannot exclude the contribution of the singlet state to the observed spectrum.

## 2.5 The Hexamers

Theoretical studies on the $Nb_6^{0/\pm}$ system were mainly based on DFT calculations, and the identity of the most stable structures remains a matter of debate. For the neutral, Goodwin et al. [25] reported a dimer-capped rhombus to be the lowest energy form. On the contrary, Kumar et al. [27] suggested a distorted triangular prism associated with a singlet state. Recently, Fielicke et al. [31] predicted two isoenergetic isomers including a singlet distorted triangular prism ($C_2$) and a triplet tetragonal-square bipyramid (or distorted octagon, $D_{4h}$).

BPW91 results point toward a dimer-capped $C_{2v}$ rhombus with a high spin $^3B_2$ state as the lowest-energy structure (Fig. 3). The corresponding low spin $^1A_1$ state of such a form is energetically less favored by 0.13 eV. Another $Nb_6$ structure is a distorted $D_{2h}$ octagon. In this shape, the $^3B_{1u}$ state is more stable than the $^1A_g$ counterpart (Fig. 3). Their energies relative to the ground state amount to 0.14 and 0.31 eV, respectively. The high symmetry $D_{4h}$ structure with a triplet state $^3A_{1g}$, which was reported as the lowest-energy form in Ref. [31], turns out to be 0.24 eV above and has a small imaginary frequency of $50i$ cm$^{-1}$.

Let us remind that all results mentioned above are obtained with BPW91/cc–pVTZ–PP computations. CCSD(T)/cc–pVTZ–PP single point calculations however reveal a different energy landscape (Table 3). The distorted high spin octagon ($D_{2h}$, $^3B_{1u}$) becomes now the lowest energy structure, which is, however, quasi

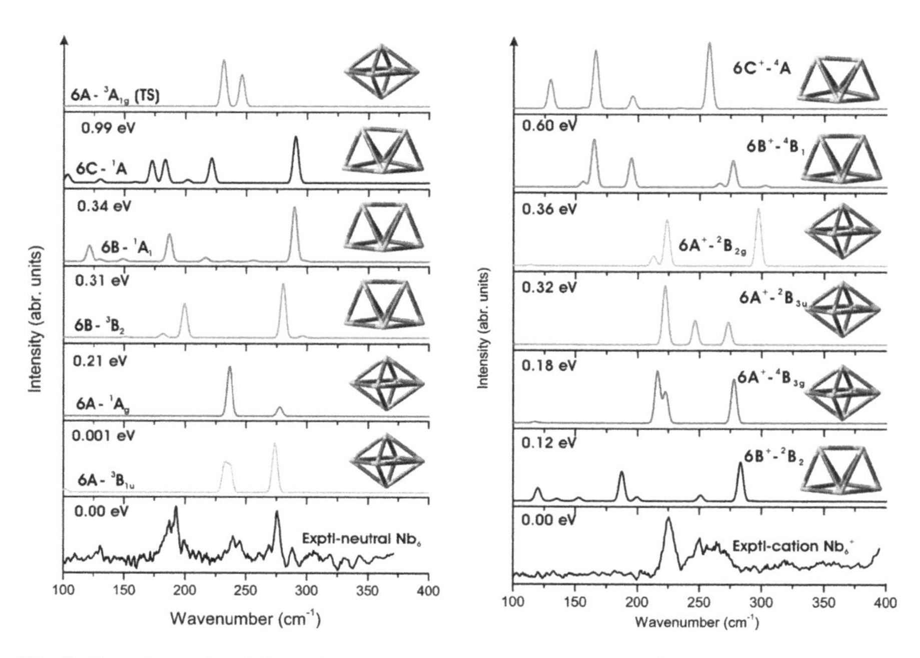


**Fig. 3** Experimental and theoretical IR spectra of $Nb_6$ (*left*) and $Nb_6^+$ (*right*)

**Table 3** Ground and low-lying states of $Nb_6^{0/\pm}$ (CCSD(T)/cc-pVTZ-PP single point calculations at BPW91/cc-pVTZ-PP geometries)

| Cluster | Geometry | Symmetry | State | Relative energy (eV) |
|---|---|---|---|---|
| $Nb_6$ | Distorted octagon (**6A**) | $D_{2h}$ | $^3B_{1u}$ | 0.00 |
| | | | $^1A_g$ | 0.001 |
| | Dimer-capped rhombus (**6B**) | $C_{2v}$ | $^3B_2$ | 0.21 |
| | | | $^1A_1$ | 0.31 |
| | Dimer-capped rhombus (**6B**) | $C_2$ | $^1A$ | 0.34 |
| | Distorted octagon (**6A**) | $D_{4h}$ | $^3A_{1g}$ | 0.99 |
| $Nb_6^+$ | Dimer-capped rhombus (**6B$^+$**) | $C_{2v}$ | $^2B_2$ | 0.00 |
| | | | $^4B_1$ | 0.36 |
| | Distorted octagon (**6A$^+$**) | $D_{2h}$ | $^4B_{3g}$ | 0.12 |
| | | | $^2B_{3u}$ | 0.18 |
| | | | $^2B_{2g}$ | 0.32 |
| | Dimer-capped rhombus (**6B$^+$**) | $C_2$ | $^4A$ | 0.60 |
| $Nb_6^-$ | Distorted octagon (**6A$^-$**) | $D_{2h}$ | $^2B_{2g}$ | 0.00 |
| | Dimer-capped rhombus (**6B$^-$**) | $C_{2v}$ | $^2A_1$ | 0.14 |
| | | | $^4A_2$ | 0.25 |
| | Dimer-capped rhombus (**6B$^-$**) | $C_2$ | $^2A$ | 0.16 |

iso-energetic with the corresponding $^1A_g$ state of this geometry. The dimer-capped rhombus ($C_{2v}$, $^3B_2$) state is now 0.21 eV higher in energy than the ground state. Detailed information on these electronic states is tabulated in Table 3.

Reported results for the $Nb_6^+$ cation again disagreed with each other on its ground state. Local-spin-density calculations [25] identified a doublet dimer-capped rhombus as its lowest-energy form. Subsequent DFT computations [31] assigned a doublet $D_{2h}$ structure as the most stable isomer. BPW91 calculations listed in Table 3 suggest that the lowest energy isomer of $Nb_6^+$ is indeed a dimer-capped rhombus having $C_{2v}$ geometry, and a low spin $^2B_2$ state. The higher spin state $^4B_1$ of this shape is at 0.22 eV above. Nevertheless, a tetragonal bipyramid ($D_{2h}$, $^2B_{3u}$) turns out to be less favored by only 0.04 eV. The isomer bearing a tetragonal bipyramid with a high spin state $^4B_{3g}$ is 0.10 eV higher.

CCSD(T) calculations support the DFT-based observations above, confirming the dimer-capped rhombus structure ($C_{2v}$, $^2B_2$) as the ground electronic state of $Nb_6^+$. The next lower-lying state is the $^4B_{3g}$ of a tetragonal bipyramid ($D_{2h}$), which lies at 0.12 eV higher (see Table 3).

Concerning $Nb_6^-$, BPW91 results point out a distorted triangular prism **6A** ($C_2$, $^2A$) as the lowest-energy isomer (Fig. 3). This is in agreement with previous DFT calculations [11, 48]. The dimer-capped rhombus structure ($C_{2v}$, $^2A_1$) is only 0.01 eV higher but it possesses a tiny imaginary frequency of 50*i* cm$^{-1}$. The anion has also another local minimum, which is a distorted $D_{2h}$ octagon with a $^2B_{2g}$ state and 0.19 eV less stable than the $C_2$ form. However, this energy ordering is actually reversed by CCSD(T)/cc–pVTZ–PP calculations (Table 3). Accordingly, the most stable form of $Nb_6^-$ is now the distorted octagon ($D_{2h}$, $^2B_{2g}$) structure. Relative to the latter, the $^2A$ ($C_2$) and $^2A_1$ ($C_{2v}$) states are calculated at 0.16 and 0.14 eV higher, respectively.

The shapes and IR spectra for some low-energy structures of both $Nb_6$ and $Nb_6^+$ determined by BPW91/cc–pVTZ–PP calculations are shown in Fig. 3, along with the experimental spectra taken from Ref. [31]. For both neutral and cationic species, our assignments are at variance with those proposed by Fielicke et al. [31]. These authors assigned the $C_2$ ($^1A$) structure as the main carrier for the observed spectrum of $Nb_6$. We however found that it likely arises from a superposition of the spectra of several states, namely $^3B_{1u}$, $^1A_g$ and $^3B_2$, rather than from a sole carrier (see Fig. 3).

For $Nb_6^+$, according to Fielicke et al. [31] the $^4B_1$ ($C_{2v}$) structure is responsible for the observed spectrum, even though it is not the most stable form. As discussed above, both BPW91 and CCSD(T) calculations predict the dimer-capped rhombus with a doublet state ($C_{2v}$, $^2B_2$) to be the most stable isomer. However, its calculated IR spectrum does not match experiment well. Instead, the vibrational spectra of the distorted octagon in both low $^2B_{3u}$ and high spin $^4B_{3g}$ states reproduce better the experimental spectra, in particular the low spin structure (Fig. 3). In other words, under experimental conditions described, the lowest-lying isomer is apparently not responsible for the observed IR spectrum, but rather its higher-lying isomers.

## *2.6 The Heptamers*

Two structures **7A** and **7B** are predicted to be the most stable forms of $Nb_7$ (Fig. 4). While the former can be considered as a distorted pentagonal $C_s$ bipyramid, which is substantially distorted from a $D_{5h}$ form, the latter is a capped octahedron (also $C_s$). For the neutral $Nb_7$, the ground state is confirmed to be the $^2A''$ state of **7A** [25, 27]. The corresponding high spin $^4A''$ state of this form is energetically less favorable by 0.48 eV (BPW91/cc-pVDZ-PP). In the second shape **7B**, the $^2A'$ state is also more stable than the corresponding high spin $^4A''$ state. The separation gaps of both doublet and quartet states of **7B,** relative to those of **7A** amount to 0.92 and 1.15 eV, respectively.

At their equilibrium point, the lowest-energy isomer of either the $Nb_7^+$ cation or the $Nb_7^-$ anion is a distorted pentagonal bipyramid having a $C_s$ point group as well. In addition, two states, namely $^1A'$ **7A-1$^+$** and $^3A''$ **7A-3$^+$**, both having the geometrical shape **7A**, were found to be nearly degenerate and compete to be the ground state for $Nb_7^+$. The singlet-triplet gap is calculated to be only 0.07 (BPW91) or 0.01 eV (M06). However, single point CCSD(T) results reverse the state ordering in predicting that the $^1A'$ is $\sim$0.17 eV lower in energy than the $^3A''$ state.

IR spectra of some lower-lying structures of $Nb_7$ and $Nb_7^+$ determined with BPW91/cc–pVDZ–PP calculations are illustrated in Fig. 4, along with experimental spectra taken from Ref. [31]. For the neutral, in agreement with a previous assignment [31], the calculated vibrational spectrum of **7A** ($C_s$, $^2A''$) clearly gives the best match to experiment. The calculated vibrational spectrum for the ground

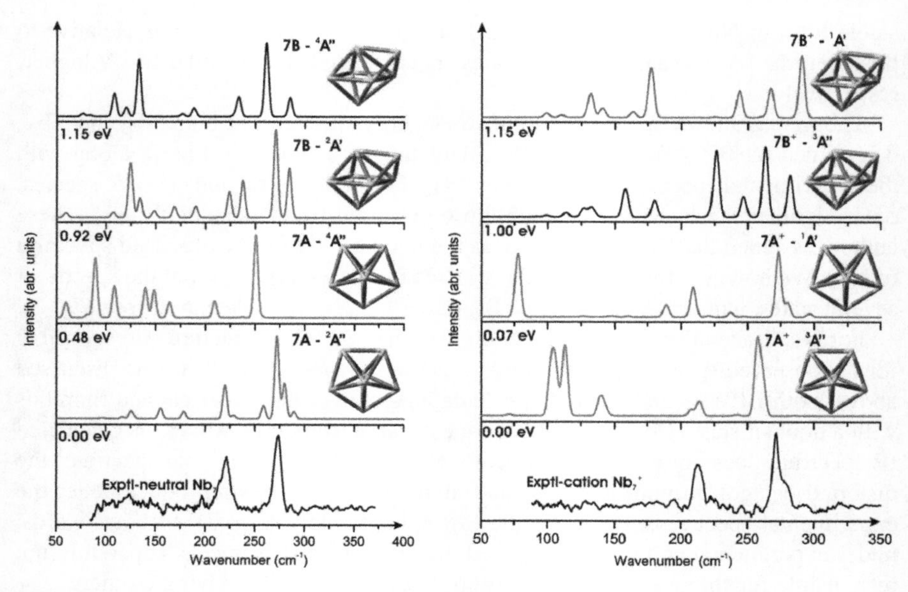


**Fig. 4** Experimental and theoretical IR spectra of $Nb_7$ (*left*) and $Nb_7^+$ (*right*)

state $^2A''$ of $Nb_7$ **7A** covers a range below 300 $cm^{-1}$ with two characteristic peaks centered at around 225 and 275 $cm^{-1}$, which are also experimentally seen. For the cation $Nb_7^+$, comparison of the theoretical IR spectra of the four low-lying isomers with experiment indicates that the singlet **7A-1**$^+$ ($C_s$, $^1A'$) is the main carrier of the observed spectrum (Fig. 4). Thus, along with the energetic results stated above, this agreement on the IR properties allow us to assign that the $Nb_7^+$ cation possesses a closed-shell ground state, but with a small singlet-triplet gap.

## 2.7 The Octamers

Several isomers are located for the octamers including **8A**, **8B**, and **8C** as displayed in Fig. 5. These structures can be viewed as consisting of a six-atom core plus two caps. The lowest-energy $Nb_8$ form is a bicapped distorted octahedron **8A** with a $C_{2v}$ symmetry and a closed-shell $^1A_1$ state. **8B** and **8C** also correspond to capping of a hexagon, with the caps being in either a *trans* ($D_{3d}$) or *cis* ($C_{2v}$) configurations, respectively, but both are of higher energy. **8B** ($^1A_{1g}$) lies 0.33 eV above whereas **8C** ($^1A_1$) is 2.0 eV above **8A** (BPW91 values) (Table 4).

Following electron removal or attachment, the shape **8A** remains in the most stable form **8A**$^+$ for $Nb_8^+$ and **8A**$^-$ for $Nb_8^-$. This is in agreement with earlier studies reported for $Nb_8$ and its singly charged species as well [11, 26]. As both HOMO and LUMO of **8B** ($D_{3d}$) are degenerate, addition or removal of an electron reduces

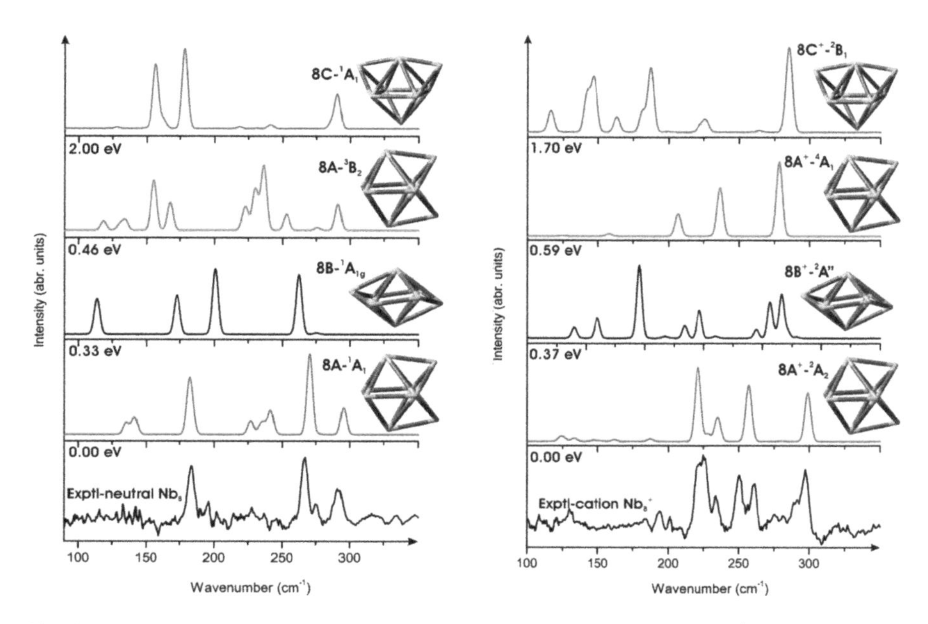


**Fig. 5** Experimental and theoretical IR spectra of $Nb_8$ (neutral, *left*) and $Nb_8^+$ (cation, *right*)

its symmetry because of a Jahn-Teller effect. Therefore, the cation **8B**$^+$ as well as anion **8B**$^-$ prefer a $C_s$ point group. Starting from the third isomer **8C**, which is much less stable than **8A** and **8B**, the two ions **8C**$^+$ (cation, $^2B_1$) and **8C**$^-$ (anion, $^2B_2$) are calculated to lie 1.70 and 1.67 eV higher in energy than the ground states **8A**$^+$ ($Nb_8^+$) and **8A**$^-$ ($Nb_8^-$), respectively.

Experimental far-IR spectra of both $Nb_8$ and $Nb_8^+$ were recorded and reported in Ref. [31]. For the purpose of comparison, we regenerate them in Fig. 5 along with our calculated results. For the neutral, the best match to the experimental IR spectrum obviously comes from that of isomer **8A** ($^1A_1$), which exhibits three intense bands centered at 186, 276 and 300 cm$^{-1}$. Even though both $Nb_8$ and $Nb_8^+$ have a very similar geometry in their ground electronic state, their IR spectra differ much from each other. The IR spectrum of $Nb_8^+$ in the bicapped distorted octahedron **8A**$^+$ ($C_{2v}$, $^2A_2$) contains four typical peaks, all detected experimentally. The higher spin $^4A_1$ states are much higher in energy and their IR spectra do not fit with experiment. These facts suggest that the low-spin states of **8A** and **8A**$^+$ are actually manifested in the IR spectra recorded by Fielicke et al. [31].

## *2.8 The Nonamers*

Gronbeck et al. [26] reported that the distorted bicapped pentagonal bipyramid **9B** is the lowest-energy $Nb_9$ (Fig. 6). Subsequent studies [27, 31] agreed with each

**Table 4** Ground and lower-lying states of $Nb_n^{0/\pm} n = 7–12$ and relative energies (RE, eV, BPW91/cc-pVDZ-PP)

| Cluster | | State | RE | Cluster | | State | RE |
|---|---|---|---|---|---|---|---|
| $Nb_7$ | **7A** ($C_s$) | $^2A''$ | 0.00 | $Nb_9^-$ | **9B**$^-$ ($C_2$) | $^1A$ | 0.03 |
| | | $^4A''$ | 0.48 | | | $^3A$ | 0.11 |
| | **7B** ($C_s$) | $^2A'$ | 0.92 | | **9C**$^-$ ($C_{2v}$) | $^1A_1$ | 0.54 |
| | | $^4A''$ | 1.15 | | | $^3A_2$ | 0.42 |
| $Nb_7^+$ | **7A**$^+$ ($C_s$) | $^3A''$ | 0.00 | | **9D**$^-$ ($C_{2v}$) | $^1A_1$ | 0.28 |
| | | $^1A'$ | 0.07 | | | $^3A_2$ | 0.16 |
| | **7B**$^+$ ($C_s$) | $^3A''$ | 1.00 | $Nb_{10}$ | **10A** ($D_4$) | $^1A_1$ | 0.00 |
| | | $^1A'$ | 1.15 | | **10A** ($D_{4d}$) | $^1A_1$ | 0.45 |
| $Nb_7^-$ | **7A**$^-$ ($C_s$) | $^1A'$ | 0.00 | | | $^3B_1$ | 0.36 |
| | | $^3A'$ | 0.31 | | **10B** ($C_s$) | $^3A'$ | 0.99 |
| | **7B**$^-$ ($C_s$) | $^1A'$ | 0.91 | | **10C** ($D_{2h}$) | $^1A_g$ | 1.66 |
| | | $^3A'$ | 1.10 | $Nb_{10}^+$ | **10A**$^+$ ($D_4$) | $^2A_1$ | 0.00 |
| $Nb_8$ | **8A** ($C_{2v}$) | $^1A_1$ | 0.00 | | **10A**$^+$ ($C_{2v}$) | $^2A_2$ | 0.07 |
| | | $^3B_2$ | 0.46 | | | $^4B_1$ | 0.41 |
| | **8B** ($D_{3d}$) | $^1A_{1g}$ | 0.33 | | **10B**$^+$ ($C_s$) | $^2A'$ | 0.44 |
| | **8C** ($C_{2v}$) | $^1A_1$ | 2.00 | | **10C**$^+$($D_{2h}$) | $^2B_{1u}$ | 1.10 |
| $Nb_8^+$ | **8A**$^+$ ($C_{2v}$) | $^2A_2$ | 0.00 | $Nb_{10}^-$ | **10A**$^-$ ($D_4$) | $^2A_1$ | 0.00 |
| | | $^4A_1$ | 0.59 | | **10A**$^-$ ($C_{2v}$) | $^2A_2$ | 0.01 |
| | **8B**$^+$ ($C_s$) | $^2A''$ | 0.37 | | | $^4B_2$ | 0.71 |
| | **8C**$^+$ ($C_{2v}$) | $^2B_2$ | 1.70 | | **10B**$^-$ ($C_s$) | $^2A'$ | 0.86 |
| $Nb_8^-$ | **8A**$^-$ ($C_{2v}$) | $^2B_1$ | 0.00 | | **10C**$^-$($D_{2h}$) | $^2B_{1g}$ | 1.47 |
| | | $^4B_2$ | 0.83 | $Nb_{11}$ | **11A** ($C_{2v}$) | $^2B_2$ | 0.00 |
| | **8B**$^-$ ($C_s$) | $^2A'$ | 0.39 | | | $^4A_2$ | 0.62 |
| | **8C**$^-$ ($C_{2v}$) | $^2B_2$ | 1.67 | | **11B** ($C_s$) | $^2A''$ | 1.14 |
| $Nb_9$ | **9A** ($C_2$) | $^2A$ | 0.00 | | **11C** ($C_s$) | $^2A''$ | 1.28 |
| | | $^4B$ | 0.40 | $Nb_{11}^+$ | **11A**$^+$ ($C_{2v}$) | $^1A_1$ | 0.00 |
| | **9B** ($C_2$) | $^2A$ | 0.04 | | | $^3A_2$ | 0.43 |
| | | $^4B$ | 0.31 | | **11B**$^+$ ($C_s$) | $^1A'$ | 1.13 |
| | **9C** ($C_{2v}$) | $^2A_1$ | 0.24 | | **11C**$^+$ ($C_s$) | $^1A'$ | 1.45 |
| | | $^4B_1$ | 0.47 | $Nb_{11}^-$ | **11A**$^-$($C_{2v}$) | $^1A_1$ | 0.00 |
| | **9D** ($C_{2v}$) | $^2A_1$ | 0.06 | | | $^3B_2$ | 0.005 |
| | | $^4B_2$ | 0.37 | | **11B**$^-$ ($C_s$) | $^1A'$ | 0.99 |
| $Nb_9^+$ | **9A**$^+$ ($C_2$) | $^3B$ | 0.24 | | **11C**$^-$ ($C_s$) | $^1A'$ | 1.00 |
| | | $^1A$ | 0.31 | $Nb_{12}$ | **12A** ($C_i$) | $^1A_g$ | 0.00 |
| | **9B +** ($C_2$) | $^1A$ | 0.00 | | | $^3A_g$ | 0.04 |
| | | $^3B$ | 0.29 | | **12B** ($C_{2v}$) | $^1A_1$ | 3.36 |
| | **9C**$^+$ ($C_{2v}$) | $^1A_1$ | 0.36 | $Nb_{12}^+$ | **12A**$^+$ ($C_i$) | $^2A_u$ | 0.00 |
| | | $^3B_1$ | 0.47 | | | $^4A_g$ | 0.98 |
| | **9D**$^+$ ($C_{2v}$) | $^1A_1$ | 0.05 | | **12B**$^+$ ($C_{2v}$) | $^2A_1$ | 4.13 |
| | | $^3A_2$ | 0.39 | $Nb_{12}^-$ | **12A**$^-$ ($C_i$) | $^2A_u$ | 0.00 |
| $Nb_9^-$ | **9A**$^-$ ($C_2$) | $^1A$ | 0.00 | | | $^4A_u$ | 0.03 |
| | | $^3A$ | 0.21 | | **12B**$^-$ ($C_{2v}$) | $^2B_1$ | 2.94 |

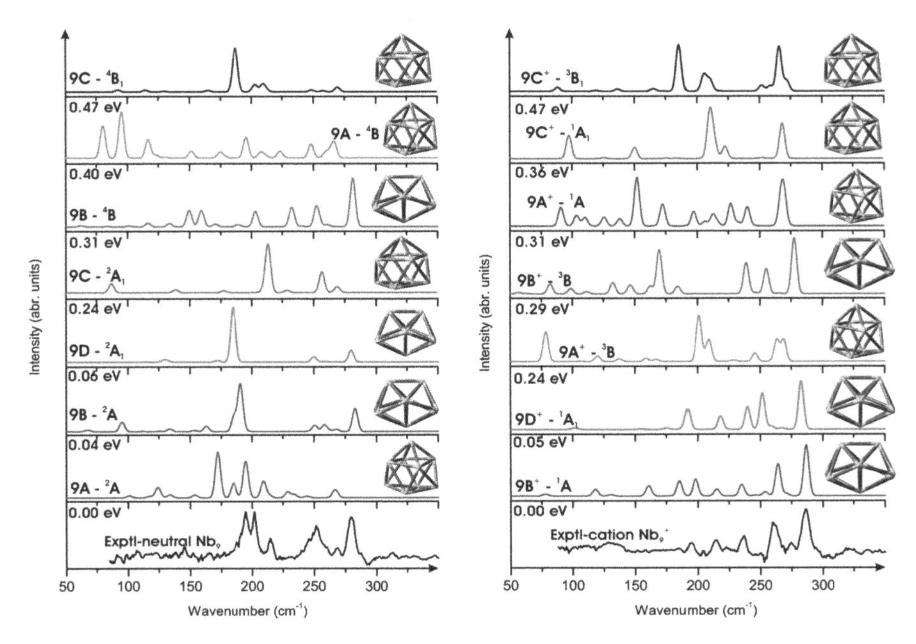


**Fig. 6** Experimental and theoretical IR spectra of $Nb_9$ (neutral, *left*) and $Nb_9^+$ (cation, *right*)

other that the lowest-energy isomer of $Nb_9$ is rather a distorted tricapped trigonal prism **9A**. Nguyen et al. found three nearly isoenergetic structures **9A** ($^2$A), **9B** ($^2$A) and **9D** ($^2A_1$) competing to be the ground state (Fig. 6). **9A** arises from a distortion of a tricapped trigonal prism, and **9B** from a distorted bicapped pentagonal bipyramid with the caps located at two adjacent trigonal faces. **9D** is analogous to **9B** but in higher symmetry ($C_{2v}$). BPW91 calculations yield **9A** as the lowest-energy structure of $Nb_9$ but with two states **9B** ($^2$A) and **9D** ($^2A_1$) being only 0.04 and 0.06 eV above **9A**. Another local minimum is the **9C** form, whose geometry is similar to **9A** but more symmetric ($C_{2v}$) (Fig. 6).

Similarly, structures **9A**$^-$ and **9B**$^-$ turn out to be the lowest and second-lowest isomers of the anion $Nb_9^-$ (for the shape, cf. Fig. 6). The closed-shell $^1$A ($C_2$) states of both forms are also quasi-degenerate and separated by ~0.06 eV (BPW91), in which **9A**$^-$ is the lower state of the two. For the cation $Nb_9^+$, **9B**$^+$ ($C_2$, $^1$A) is confirmed as the lowest-energy isomer, whereas **9D**$^+$ ($C_{2v}$, $^1A_1$) is found to lie marginally close, only from 0.05 (BPW91) to 0.09 eV (M06) above **9B**$^+$ (Fig. 6).

In a further attempt to calibrate the DFT results obtained above using both BPW91 or M06 functionals, CCSD(T)/cc-pVDZ-PP single-point electronic energy calculations were performed for the lower-lying states. Results recorded in Table 5 show a different energy landscape. For the neutral $Nb_9$, **9B** ($C_2$, $^2$A) becomes the lowest-energy structure. The doublet states **9A** ($^2$A) and **9D** ($^2A_1$) are now 0.18 and 0.31 eV higher in energy than **9B**, respectively (Fig. 6).

**Table 5** Relative energies (eV) of several isoenergetic states computed at various levels (using the cc-pVDZ-PP basis set)

| Cluster[a] | | State | Relative energies | | |
|---|---|---|---|---|---|
| | | | BPW91 | M06 | CCSD(T) |
| Cation | **7A**$^+$ | $^3A''$ | 0.00 | 0.00 | 0.17 |
| | | $^1A'$ | 0.07 | 0.01 | 0.00 |
| Neutral | **9A** | $^2A$ | 0.00 | 0.11 | 0.18 |
| | **9B** | $^2A$ | 0.04 | 0.00 | 0.00 |
| | **9D** | $^2A_1$ | 0.06 | 0.08 | 0.31 |
| Cation | **9B**$^+$ | $^1A$ | 0.00 | 0.00 | 0.04 |
| | **9D**$^+$ | $^1A_1$ | 0.05 | 0.09 | 0.00 |
| Anion | **9A**$^-$ | $^1A$ | 0.00 | 0.18 | 1.76 |
| | **9B**$^-$ | $^1A$ | 0.06 | 0.00 | 0.00 |
| Cation | **10A**$^+$ | $^2A_1$ | 0.00 | 0.00 | 0.00 |
| | | $^2A_2$ | 0.07 | 0.27 | 0.17 |
| Anion | **10A**$^-$ | $^2A_1$ | 0.00 | 0.00 | 0.16 |
| | | $^2A_2$ | 0.01 | 0.11 | 0.00 |
| Anion | **11A**$^-$ | $^1A_1$ | 0.00 | 0.00 | 0.00 |
| | | $^3B_2$ | 0.005 | 0.06 | 2.60 |
| Neutral | **12A** | $^1A_g$ | 0.00 | 0.00 | 0.40 |
| | | $^3A_g$ | 0.04 | 0.09 | 0.00 |
| Anion | **12A**$^-$ | $^2A_u$ | 0.00 | 0.00 | 0.00 |
| | | $^4A_u$ | 0.03 | 0.05 | 0.18 |

CCSD(T) calculations are performed based on BPW91 geometries

[a]Shapes of structures are given in figures of the corresponding sections

Concerning the $Nb_9^+$ cation, CCSD(T) calculations pointed out that both states **9B**$^+$ and **9D**$^+$ differ by only 0.04 eV from each other but in favor of **9B**$^+$. Such a small difference could again be changed when a full geometry optimization at the CCSD(T) level could be carried out. Hence, both states **9B**$^+$ and **9D**$^+$ can be regarded as quasi-degenerate states competing for the ground state of $Nb_9^+$ (Fig. 6). For anion $Nb_9^-$, both isomers **9A**$^-$ and **9B**$^-$ in their singlet state are no longer isoenergetic at CCSD(T)/cc-pVDZ-PP level. Accordingly, **9B**$^-$ ($^1A$) is 1.76 eV lower than **9A**$^-$ ($^1A$). Overall, the distorted bicapped pentagonal bipyramid form **9B** emerges as the favored form for nine-niobium systems, irrespective of the charged state.

We now consider the far IR spectra of some low-lying states of $Nb_9$ and $Nb_9^+$ clusters, which are plotted in Fig. 6. For the neutral, comparison of the computed and experimental spectra allows a conclusion to be made that **9A** is not present in the experimental spectrum. In this case, **9B** ($C_2$, $^2A$) appears as the main carrier of the observed spectrum, even though **9D** ($C_{2v}$,$^2A_1$) may contribute to it at a certain extent. Fielicke et al. [31] also suggested the **9B** ($C_2$) structure to be responsible for the observed spectrum albeit this structure is not the most stable form and the intensities do not match well.

For $Nb_9^+$, isomer **9B⁺** ($C_2$) provides us with the best accord with experiment. However, the calculated spectrum of the quasi-degenerate isomer **9D⁺** ($^1A_1$) also agrees well with experimental data. In addition, these states are energetically quasi-degenerate at both DFT and CCSD(T) levels. In this context, we can deduce that the recorded IR spectrum of $Nb_9^+$ most likely arises from vibrations of both isomers, rather than from a sole carrier. Fielicke et al. [31] assigned **9B⁺** as the main carrier for the observed spectrum. As for $Nb_7^-$ and $Nb_8^-$, an experimental far IR spectrum of $Nb_9^-$ is not available yet.

## 2.9 The Decamers

The global minimum of $Nb_{10}$ as well as of their ions $Nb_{10}^+$ and $Nb_{10}^-$ correspond to a distorted bicapped antiprism (structures **10A**, **10A⁺** and **10A⁻**) all with $D_4$ symmetry (Fig. 7). Such a structure with $C_4$ symmetry was confirmed to be the global minimum for both neutral $Nb_{10}$ and cationic $Nb_{10}^+$ [57]. The low spin structure **10A** ($D_4$, $^1A_1$) was found to be strongly favored over all the isomers. Other structures, i.e. **10B** ($^3A'$) and **10C** ($^1A_g$), are respectively 0.99 and 1.66 eV higher. The first excited state, whose energy gap is 0.36 eV, is the high spin $^3B_1$ state of **10A** with $D_{4d}$ symmetry.

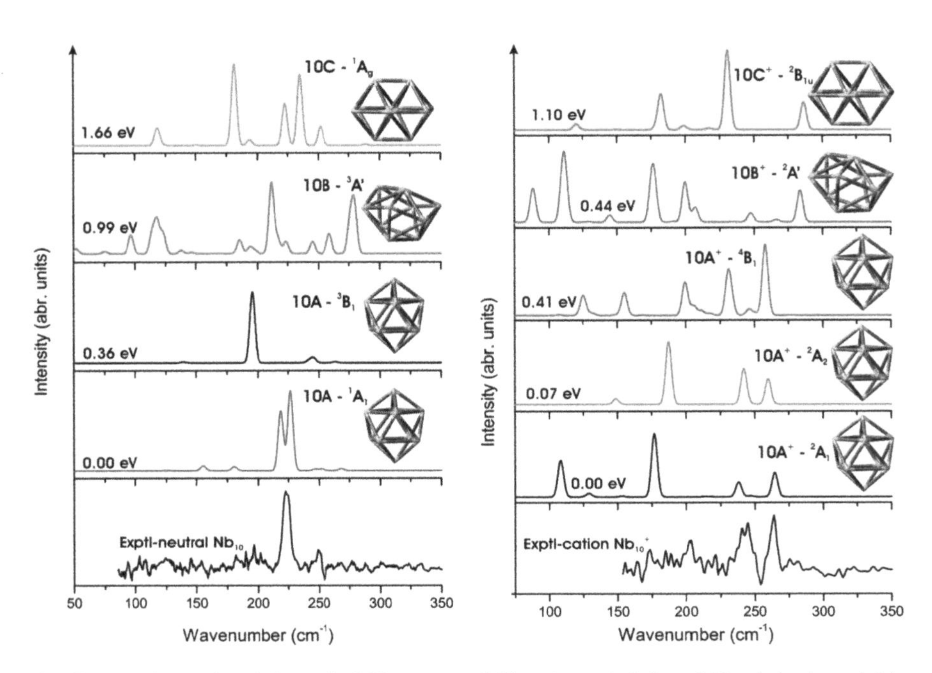


**Fig. 7** Experimental and theoretical IR spectra of $Nb_{10}$ (neutral, *left*) and $Nb_{10}^+$ (cation, *right*)

DFT results indicate that both $Nb_{10}^{+}$**10A**$^{+}$ and $Nb_{10}^{-}$**10A**$^{-}$ ions also prefer the distorted bicapped squared antiprism form with $D_4$ symmetry and a low spin $^2A_1$ state (Fig. 7). Again, at the BPW91 level, other quasi-degenerate states emerge as competing ground state of the ions. For $Nb_{10}^{+}$, the $^2A_2$ ($C_{2v}$) state of **10A**$^{+}$ is computed at only 0.07 eV above the $^2A_1$ state. However, the M06 functional significantly increases this gap to 0.27 eV. CCSD(T) single-point calculations support the latter DFT results in predicting that the $^2A_1$ state is still favored by ~0.17 eV over the $^2A_2$ counterpart.

Similarly for $Nb_{10}^{-}$, both states $^2A_1$ (**10A**$^{-}$, $D_4$) and $^2A_2$ (**10A**$^{-}$, $C_{2v}$) have a comparable energy content, with a separation gap from 0.01 (BPW91) to 0.11 eV (M06). At the CCSD(T) level, this difference is enlarged, but the $^2A_2$ state is now 0.16 eV lower in energy than the $^2A_1$ state. Additional local minima on the potential-energy surface obtained for $Nb_{10}^{+}$ and $Nb_{10}^{-}$ include the forms **10B**$^{+}$ and **10C**$^{+}$ and **10B**$^{-}$ and **10C**$^{-}$ (Fig. 7), but such geometries are much higher than the relevant **10A**$^{+/-}$ counterpart (see Table 5).

The calculated IR spectra for some selected structures of these neutral and cationic systems are also shown in Fig. 7. Removal of one electron leads to a significant change in vibrational motions of the cluster. The predicted spectrum of the neutral **10A** contains two distinct peaks centered at ~220–230 $cm^{-1}$. For the two low-lying states $^3B_1$ and $^1A_1$ ($D_{4d}$), their spectra become simpler with the most intense bands centered at ~200 and ~215 $cm^{-1}$, respectively. On the contrary, the IR spectrum of the cation $Nb_{10}^{+}$ in its ground state becomes much more complicated including the presence of four intense peaks in the range of 100–300 $cm^{-1}$. The quasi-degenerate state $^2A_2$ exhibits three distinct bands in the range of 175–276 $cm^{-1}$. For the anion $Nb_{10}^{-}$ ground state **10A**$^{-}$, we observe two distinct bands appeared at ~215 and ~245 $cm^{-1}$, and the other very low intensity peaks seen in the vicinity.

The experimental spectra of the clusters $Nb_{10}$–$Nb_{12}$ in both neutral and cationic states were recently published by Fielicke and Meijer [58] and we regenerate them here for the purpose of comparison. Vibrational spectra of these clusters in gas-phase have been measured via far-IR photo-dissociation of weakly bound complexes of the clusters with single Ar atoms formed at 80 K. As seen in Fig. 7, a good agreement between experimental and calculated spectra is found for $Nb_{10}$. For the cation $Nb_{10}^{+}$, although the quality of experimental spectra is not exceptionally high, we can detect the existence of the $D_4$ isomer with two distinct bands in the vicinity of 240–270 $cm^{-1}$.

## 2.10 *The Undecamers*

DFT calculations by Kumar et al. [27] revealed that the lowest-energy geometry of neutral $Nb_{11}$ could be described as two pentagonal bipyramids fused at a triangular face. In addition such a shape, **11A** in Fig. 8, is built upon **10A** and turns out to be the energetically optimized configuration for both ions $Nb_{11}^{+}$ and $Nb_{11}^{-}$. The $^2B_2$ state **11A** is the ground state of $Nb_{11}$, whereas a $^1A_1$ state is the ground state for both $Nb_{11}^{+}$ (**11A**$^{+}$) and $Nb_{11}^{-}$ (**11A**$^{-}$). DFT results also predict that the anion has two degenerate low-energy states, namely the $^{1}A_{1}$ and $^{3}B_{2}$ states of structure **11A**$^{-}$. The triplet state is only 0.005 and 0.06 eV higher than the singlet state at the BPW91 and M06 levels, respectively. However, CCSD(T) calculations drastically increase this gap to 0.6 eV in favor of the low spin state.

For $Nb_{11}$ and $Nb_{11}^{+}$, the higher spin states $^{4}A_{2}$ and $^{3}A_{2}$ are respectively located at 0.62 and 0.43 eV higher in energy than the corresponding ground states (BPW91).

The two next lowest-energy isomers are **11B** and **11C** (Fig. 8). Also, both can be considered as arising from addition of one Nb atom to the lowest-energy isomer of $Nb_{10}$ **10A**. However, these geometries are much less stable than **11A**. In fact, for neutral $Nb_{11}$, the doublet $^{2}A''$ states **11B** and **11C** are 1.14 and 1.28 eV above the $^{2}B_{2}$, respectively (BPW91). Concerning the cation $Nb_{11}^{+}$, the singlet $^{1}A'$ state of these shapes **11B**$^{+}$ and **11C**$^{+}$ are 1.13 and 1.45 eV above the corresponding ground state. The relevant energy separations obtained for the anion **11B**$^{-}$ and **11C**$^{-}$ amount to 0.99 and 1.00 eV (BPW91).

Figure 8 shows the vibrational spectra for some low-lying states of the 11-atom clusters in both neutral and cationic states, along with the experimental spectra. The calculated vibrational spectrum of the neutral ground state **11A** ($^{2}B_{2}$) covers the range below 280 $cm^{-1}$ with five specific peaks, where the most intense one is centered at around 270 $cm^{-1}$. As mentioned above, the spectra of these systems are also strongly dependent on the addition or removal of an electron from the neutral.

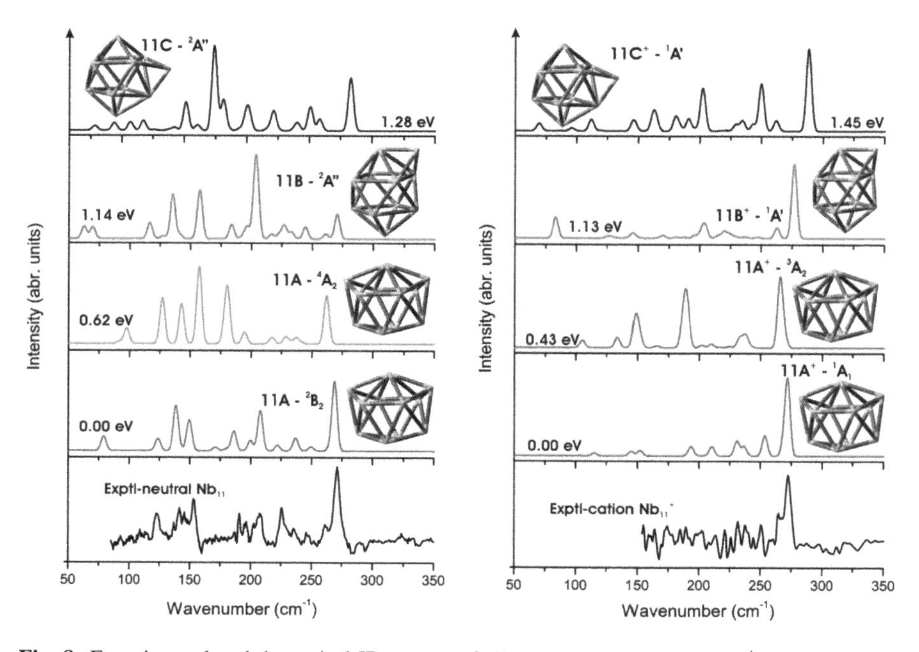


**Fig. 8** Experimental and theoretical IR spectra of $Nb_{11}$ (neutral, *left*) and $Nb_{11}^{+}$ (cation, *right*)

Significantly large changes in band intensities of the cation spectrum were observed. As seen in Fig. 8, the highest intensity peak is at 270 $cm^{-1}$, whereas the intensities of other bands below 250 $cm^{-1}$ are substantially reduced.

For the anion $Nb_{11-1}$**1A**$^-$, at least four distinct absorption bands can be found in the range below 250 $cm^{-1}$, and there exists no band above 250 $cm^{-1}$. We in addition realize that upon electron attachment two intense bands of the neutral at 150 and 138 $cm^{-1}$ are slightly red-shifted by ~10 $cm^{-1}$. The peak at around 128 $cm^{-1}$ becomes mostly visible. Comparison of experimental and calculated spectra shows good agreement for both $Nb_{11}$ and $Nb_{11}^+$.

## 2.11 The Dodecamers

An isomer with an encapsulated Nb atom (cf. structure **12B**, Fig. 9) was predicted to be the most stable form for neutral $Nb_{12}$ [27]. However a distorted icosahedron (cf. structure **12A**) with the $C_i$ symmetry and low spin electronic state was subsequently found to be the global minimum of $Nb_{12}$ and $Nb_{12}^{\pm}$ as well. The $^1A$ state of **12B** is computed to lie around 0.60 eV above the $^1A_g$ of **12A** (BPW91). Furthermore, two states $^1A_g$ and $^3A_g$ of **12A** are evaluated with a tiny energy difference by ~0.04 (BPW91) to 0.09 eV (M06). Nonetheless R/UCCSD(T)/cc-pVDZ-PP calculations point out that the triplet $^3A_g$ state is 0.40 eV lower in

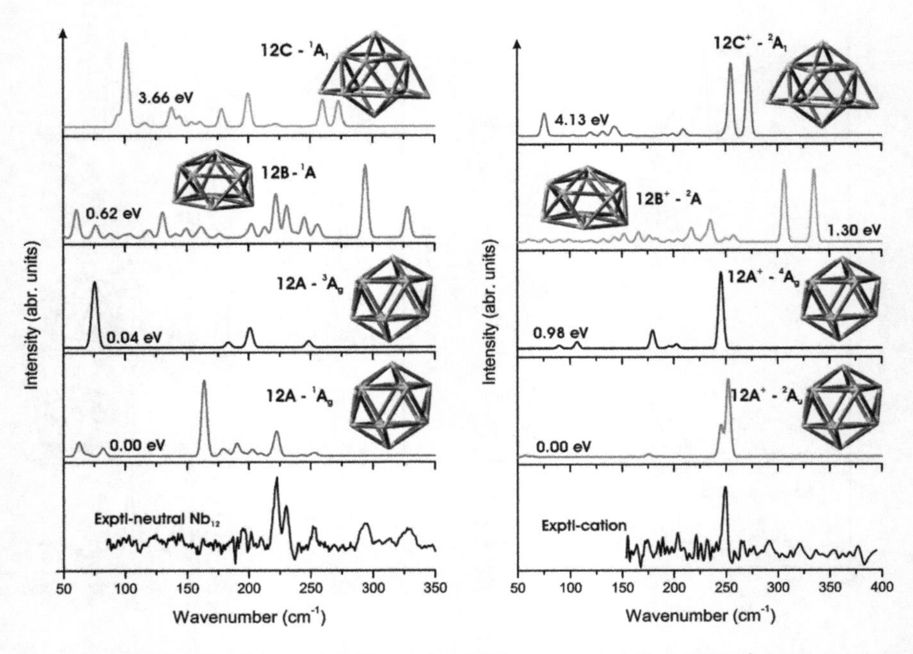


**Fig. 9** Experimental and theoretical IR spectra of $Nb_{12}$ (neutral, *left*) and $Nb_{12}^+$ (cation, *right*)

energy than the $^{1}A_{g}$ counterpart. Moro et al. [59] based on magnetic deflection experiments revealed a total spin of zero for all the even sized Nb clusters.

Again, for $Nb_{12}^{-}$, two lowest-lying states $^{2}A_{u}$ and $^{4}A_{u}$ of structure **12A**$^{-}$ are very close in energy in which the low spin $^{2}A_{u}$ is only 0.03 eV (BPW91) or 0.05 eV (M06) to 0.18 eV (CCSD(T)) more stable than the high spin $^{4}A_{u}$ state. On the contrary, the cation $Nb_{12}^{+}$ (**12A**$^{+}$) clearly prefers a low spin $^{2}A_{u}$ state, as the doublet–quartet ($^{2}A_{u}$–$^{4}A_{g}$) separation gap is markedly enlarged to 0.98 eV (BPW91). The isomer **12B**$^{+}$ is ~1.30 eV higher in energy than isomer **12A**$^{+}$ (BPW91).

Another local minimum **12C,** based on a dimer capping of **10A**, is much less stable than **12A**. In fact, the $^{1}A_{1}$ and $^{2}A_{1}$ states obtained here for $Nb_{12}$ and $Nb_{12}^{+}$, respectively, are much higher energies, namely 3.66 and 4.13 eV, than the respective ground states. For $Nb_{12}^{-}$, the $^{2}B_{1}$ state of this form **12C**$^{-}$ is located at 2.94 eV above the ground $^{2}A_{u}$ state. Such a relatively large energy difference does not suggest the presence of isomer **12C** in experiment (Fig. 9).

The computed IR spectra of some lowest-lying states of $Nb_{12}$ are shown in Fig. 9. The spectrum of the $^{1}A_{g}$ state is rather different with indications for two bands located at around 160 and 220 $cm^{-1}$, and some very low intensity peaks emerged in the vicinity. Again, even though the neutral and its ions have a very similar structural feature, each one owns a typical IR spectrum. The $^{2}A_{u}$ cation ground state **12A**$^{+}$ contains two closely spaced peaks at 245 and 255 $cm^{-1}$ (Fig. 9). In general, the vibrational spectrum of the first excited $^{4}A_{g}$ state is surprisingly similar to that of the $^{2}A_{u}$ ground state, with a small red-shift from the peak around 255–245 $cm^{-1}$. As a result, the $^{4}A_{g}$ state has the most recognizable peak centered at ~245 $cm^{-1}$. There also exists a typical band on the spectroscopic signatures of the $^{2}A_{u}$ anion ground state at ~245 $cm^{-1}$. Besides, (at least) five lower intensity bands in the range below 200 $cm^{-1}$ can be detected. The quasi-degenerate state $^{4}A_{u}$ also exhibits the band at around 250 $cm^{-1}$, but with a very small intensity (Fig. 9).

The computed IR spectrum displayed in Fig. 9 of the lowest-energy isomer **12A** does not match the experimental IR findings. Instead, the predicted spectrum of **12B** is actually in better agreement with experimental band pattern than that of **12A**. The point of interest here is as to whether the most stable form **12A** was present in the IR experiment. As far as calculated results are concerned, results displayed in Fig. 9 tend to suggest that the experimental IR spectrum of $Nb_{12}$ likely arises from vibrations of isomer **12B**, rather than from isomer **12A**. The presence of different structural isomers for $Nb_{12}$ was reported in an earlier study on $N_2$ and $D_2$ reaction kinetics [60]. For $Nb_{12}^{+}$, the isomer **12A**$^{+}$ is actually manifested in the experimentally observed spectrum as the most intense band at ~250 $cm^{-1}$ is clearly visible in both measured and computed spectra.

## 2.12 *The Tridecamers $Nb_{13}$*

Relevant structures and spectra are displayed in Fig. 10. Although an icosahedral shape is found to be the lowest-energy isomer for several 13-atom

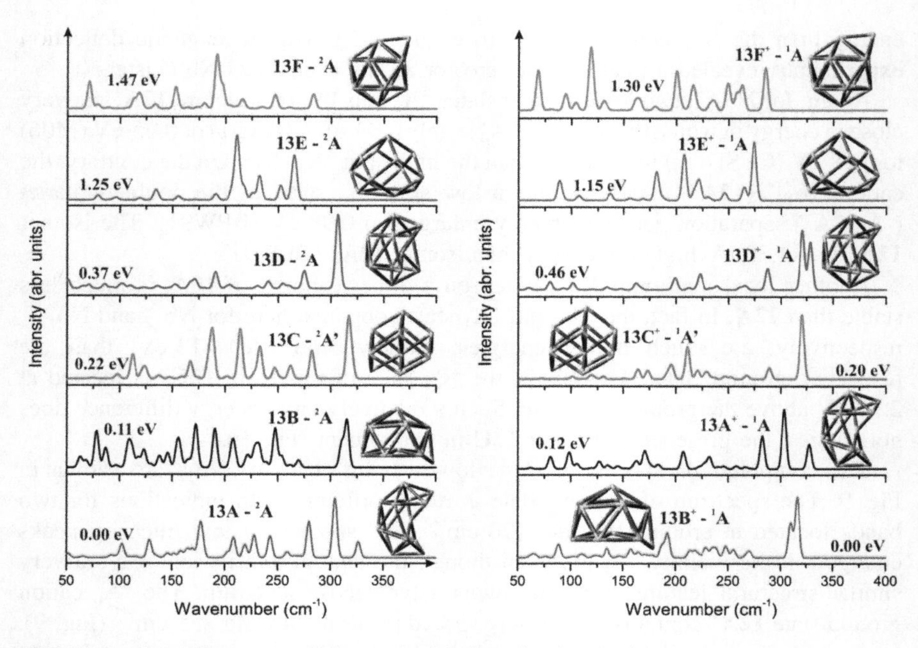


**Fig. 10** Predicted IR spectra of $Nb_{13}$ neutral (*left*) and cation (*right*)

4*d* transition-metal clusters, such as Y, Zr and Pd [61], this form is even not a local minimum for $Nb_{13}$ [27]. A severely distorted icosahedron similar to **13B** was proposed as the lowest-energy form of $Nb_{13}$ [62]. In this study, two isomers **13A** and **13B**, with an energy gap of 0.1 eV, are found to compete for the ground state. Other isomers, namely **13C**, **13D**, **13E** and **13F** (Fig. 10), are located from 0.22 to 1.47 eV above **13A** [62–65].

Following electron detachment or attachment, the forms **13A** and **13B** are also giving rise to the most stable configurations for $Nb_{13}^+$ and $Nb_{13}^-$ as well. However, in these ions, the resulting **13A⁺** is 0.12 eV higher in energy than the **13B⁺**, while the anion **13A⁻** is predicted to be 0.32 eV more stable than the counterpart **13B⁻**. Such an energy difference remains small to allow a definitive assignment for the global minimum to be made.

As shown in Fig. 10, spectroscopic signatures of **13A** contain the highest-energy band centered at $\sim 330\ cm^{-1}$ and the highest-intensity peak at around $300\ cm^{-1}$. The IR spectrum of the iso-energetic isomer **13B** is more complex with the most significant peak at $\sim 320\ cm^{-1}$ and some in the vicinity below $250\ cm^{-1}$. Though the neutral and cation seem to have a similar spectral pattern, there still exist some minor but typical differences between their spectral lines. The spectra of **13A⁺** and **13B⁺** are much smoother than those of **13A** and **13B**. Both isomers **13A⁻** and **13B⁻** are predicted to vibrate in a range below $320\ cm^{-1}$ with the appearance of a high frequency band above $300\ cm^{-1}$ related to the encapsulated atom [62–65].

An experimental spectrum for the neutral $Nb_{13}$ cluster covering the 200–350 $cm^{-1}$ spectral range was reported by Lapoutre et al. [16]. As compared to experiment, the doublet **13A** ($^2A$), which exhibits three intense bands centered at around 260, 310 and 330 $cm^{-1}$ could be present under experimental conditions.

## 2.13 *The Tetradecamers $Nb_{14}$*

A previous study [27] revealed that the global minimum of $Nb_{14}$ can be described as a hexagonal anti-prism with an atom situated at the center and one face capped (cf. **14A** in Fig. 11).

Our results concurred with this, and further demonstrated that such a shape turns out to be the energetically optimized configuration for both $Nb_{14}^+$ and $Nb_{14}^-$ [62–65].

**14A** exhibits higher stability compared to other detected structures of $Nb_{14}$. For example, an incomplete cubic with one encapsulated atom **14B** is computed at 0.54 eV above the ground state $^1A$ of **14A**. Niobium clusters apparently do not favor icosahedral form as a capped icosahedron is not even a local minimum of the 14-atom system.

The calculated vibrational spectrum of **14A** ($^1A$) in the range of 250–350 $cm^{-1}$ contains specific peaks centered at 280 and 320 $cm^{-1}$, that are also experimentally present [16]. As in 13-atom system, the IR spectrum of this species is not much

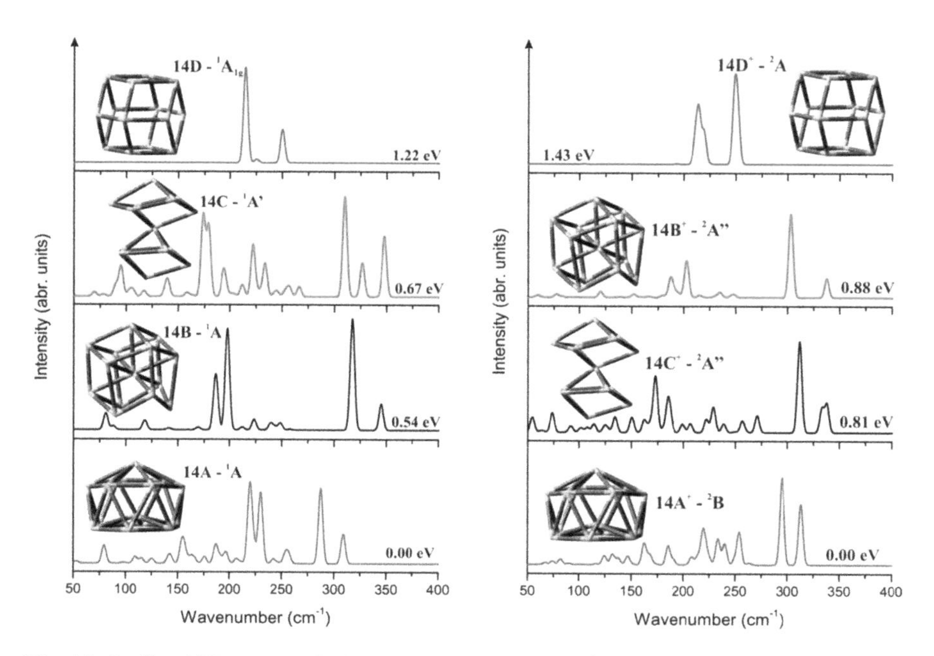


**Fig. 11** Predicted IR spectra of $Nb_{14}$ (neutral, *left*) and $Nb_{14}^+$ (cation, *right*)

changed following removal of an electron from the neutral. We actually observe the similarity in the band pattern of the cation spectrum. However, the intensities of some bands located at ~225 $cm^{-1}$ are now substantially reduced, in addition to a small blue-shift with respect to those of the neutral **14A**. The spectrum of the anion in its ground state **14A$^-$** is slightly distinguishable from that of the neutral **14A**.

## 2.14 The Pentadecamers $Nb_{15}$

Kietzmann et al. [11] predicted that the appearance of a bulk–like body-centered cubic (**bcc**) structure may occur at $Nb_{15}^-$. Besides, upon electron detachment from the anion $Nb_n^-$ the atomic arrangement remains almost unchanged [33–35, 63], so the neutral $Nb_{15}$ is expected to involve a geometric **bcc** shell closing as well. In fact, we find a rhombic dodecahedron (cf. structure **15A** in Fig. 12), to be the most stable form of $Nb_{15}$. Because of a Jahn–Teller distortion, the cluster exists in a low $C_i$ symmetry ($^2A_g$ state) rather than in a perfectly cubic shape ($O_h$). Other isomers include a capped hexagonal antiprism structure **15B**, and an incomplete Frank-Kasper structure **15C**, both have significantly higher energy content (0.77–1.45 eV).

The lowest-energy isomer of the cationic state $Nb_{15}^+$ is also the distorted cube ($C_i$), while that of the anion $Nb_{15}^-$ corresponds to a regular cube ($O_h$). Again, both

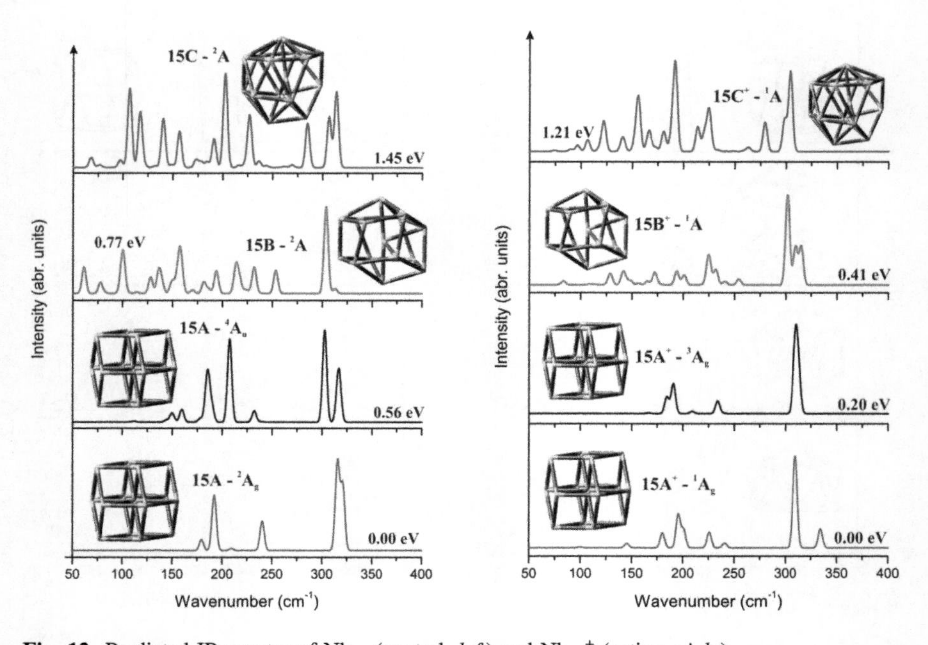


**Fig. 12** Predicted IR spectra of $Nb_{15}$ (neutral, *left*) and $Nb_{15}^+$ (cation, *right*)

**Table 6** Ground and low-lying states of $Nb_n^{0/\pm}$ ($n$ = 13–17) clusters and relative energies (RE, eV) at the BPW91/cc-pVDZ-PP level

| Cluster[a] | | State | RE | Cluster | | State | RE |
|---|---|---|---|---|---|---|---|
| $Nb_{13}$ | **13A** ($C_1$) | $^2A$ | 0.00 | $Nb_{15}$ | **15C** ($C_1$) | $^2A$ | 1.45 |
| | **13B** ($C_1$) | $^2A$ | 0.11 | $Nb_{15}^+$ | **15A⁺** ($C_i$) | $^1A_g$ | 0.00 |
| | **13C** ($C_s$) | $^2A'$ | 0.22 | | | $^3A_g$ | 0.20 |
| | **13D** ($C_1$) | $^2A$ | 0.37 | | **15B⁺** ($C_1$) | $^1A$ | 0.41 |
| | **13E** ($C_1$) | $^2A$ | 1.25 | | **15C⁺** ($C_1$) | $^1A$ | 1.21 |
| | **13F** ($C_1$) | $^2A$ | 1.37 | $Nb_{15}^-$ | **15A⁺** ($C_i$) | $^1A_g$ | 0.00 |
| $Nb_{13}^+$ | **13A⁺** ($C_1$) | $^1A$ | 0.12 | | | $^3A_g$ | 0.35 |
| | **13B⁺** ($C_1$) | $^1A$ | 0.00 | | **15B⁺** ($C_2$) | $^1A$ | 1.25 |
| | **13C⁺** ($C_s$) | $^1A'$ | 0.20 | | **15C⁺** ($C_1$) | $^1A$ | 1.70 |
| | **13D⁺** ($C_1$) | $^1A$ | 0.46 | $Nb_{16}$ | **16A** ($D_3$) | $^1A$ | 0.00 |
| | **13E⁺** ($C_1$) | $^1A$ | 1.15 | | **16B** ($C_1$) | $^1A$ | 0.60 |
| | **13F⁺** ($C_1$) | $^1A$ | 1.30 | | **16C** ($C_1$) | $^1A$ | 1.61 |
| $Nb_{13}^-$ | **13A⁻** ($C_1$) | $^1A$ | 0.00 | | **16D** ($C_3$) | $^1A$ | 2.70 |
| | **13B⁻** ($C_1$) | $^1A$ | 0.34 | $Nb_{16}^+$ | **16A⁺** ($C_1$) | $^2A$ | 0.00 |
| | **13C⁻** ($C_s$) | $^1A'$ | 0.36 | | **16B⁺** ($C_1$) | $^2A$ | 0.26 |
| | **13D⁻** ($C_1$) | $^1A$ | 1.31 | | **16C⁺** ($C_1$) | $^2A$ | 1.68 |
| | **13E⁻** ($C_1$) | $^1A$ | 1.45 | | **16D⁺** ($C_3$) | $^2A$ | 2.90 |
| | **13F⁻** ($C_1$) | $^1A$ | 1.57 | $Nb_{16}^-$ | **16A⁻** ($C_2$) | $^2A$ | 0.00 |
| $Nb_{14}$ | **14A** ($C_{2v}$) | $^1A_1$ | 0.00 | | **16B⁻** ($C_1$) | $^2A$ | 0.46 |
| | **14B** ($C_1$) | $^1A$ | 0.54 | | **16C⁻** ($C_1$) | $^2A$ | 1.53 |
| | **14C** ($C_s$) | $^1A'$ | 0.67 | | **16D⁻** ($C_3$) | $^2A$ | 2.58 |
| | **14D** ($O_h$) | $^1A_{1g}$ | 1.22 | $Nb_{17}$ | **17A** ($C_1$) | $^2A$ | 0.00 |
| $Nb_{14}^+$ | **14A⁺** ($C_2$) | $^2B$ | 0.00 | | **17B** ($C_2$) | $^2A$ | 0.45 |
| | **14B⁺** ($C_s$) | $^2A''$ | 0.88 | | **17C** ($C_1$) | $^2A$ | 1.51 |
| | **14C⁺** ($C_s$) | $^2A''$ | 0.81 | | **17D** ($C_i$) | $^2A_u$ | 1.81 |
| | **14D⁺** ($C_1$) | $^2A$ | 1.43 | $Nb_{17}^+$ | **17A⁺** ($C_1$) | $^1A$ | 0.00 |
| $Nb_{14}^-$ | **14A⁻** ($C_{2v}$) | $^2A_2$ | 0.00 | | **17B⁺** ($C_1$) | $^1A$ | 0.50 |
| | **14B⁻** ($C_1$) | $^2A$ | 0.62 | | **17C⁺** ($C_{3v}$) | $^1A_1$ | 1.60 |
| | **14C⁻** ($C_s$) | $^2A'$ | 0.70 | | **17D⁺** ($C_i$) | $^1A_g$ | 1.69 |
| | **14D⁻** ($O_h$) | $^2B_{2g}$ | 1.62 | $Nb_{17}^-$ | **17A⁻** ($C_1$) | $^1A$ | 0.00 |
| $Nb_{15}$ | **15A** ($C_i$) | $^2A_g$ | 0.00 | | **17B⁻** ($C_1$) | $^1A$ | 0.57 |
| | | $^4A_u$ | 0.56 | | **17C⁻** ($C_1$) | $^1A_1$ | 1.40 |
| | **15B** ($C_2$) | $^2A$ | 0.77 | | **17D⁻** ($C_i$) | $^1A_g$ | 1.95 |

[a]Structures are displayed in figures of the corresponding sections

$Nb_{15}^+$ and $Nb_{15}^-$ ions prefer to exist in a low spin $^1A_g$ and $^1A_{1g}$ states, respectively. Energy separations of some lower-lying states from the ground state are listed in Table 6. The HOMO of the anion **15A⁻** is triply degenerate, in such a way that removal of either one or two electrons to form, respectively, the neutral **15A** and the cation **15A⁺** is accompanied by symmetry reduction from $O_h$ to $C_i$. Overall, the

$Nb_{15}$ clusters can be regarded as the capped cubes, irrespective of their charged state.

Figures 12 shows the IR spectra computed for some lower-lying states of the 15-atom clusters. The calculated spectra of both neutral and cationic ground states **15A** ($^2A_g$) and **15A**$^+$ ($^1A_g$) contain the most intense band between 310–325 $cm^{-1}$, besides several peaks in the vicinity of 175–250 $cm^{-1}$. However, they can still be distinguished from each other. In fact, while the neutral is predicted to vibrate below 325 $cm^{-1}$, a relatively distinct peak around 335 $cm^{-1}$ is seen on the IR pattern of **15A**$^+$. The spectrum of the ground state **15A**$^-$ structure is characterized by four triplets located at around 320, 240, 190 and 180 $cm^{-1}$.

The experimental IR spectra of the neutral $Nb_{15}$ and larger systems up to $Nb_{20}$ were reported [16]. However, it turns out to be rather difficult to make a comparison with our calculated spectra. The measured data in the spectral range of 200–350 $cm^{-1}$ are in fact not in high quality for a clear-cut assignment.

## 2.15 *The Hexadecamers $Nb_{16}$*

According to Kumar and Kawazoe [27], the lowest-energy isomer of $Nb_{16}$ can be described as a hexagonal anti-prism with one side capped by two atoms and the other side capped by a single atom. Our results concurred with this and demonstrated in addition that such a shape **16A** (Fig. 13) is also the most favored configuration not only for $Nb_{16}$ but also for its singly charged states $Nb_{16}^+$ and $Nb_{16}^-$.

Structure **16A** is actually a Frank-Kasper (FK) polyhedron which has 15 vertexes and all triangular faces. Note that a vertex which is the meeting of N faces is called *N-degree vertex*, or vertex having degree of N (5-faces stands for 5 triangular faces). Accordingly, among the vertexes of **16A**, 12 vertexes are the meetings of 5 triangular faces, whereas the 3 remaining vertexes are the meetings of 6 triangular faces. Structure **16A** can thus be viewed as a (12,3) FK polyhedron. However, this 15-vertex FK with three 6° vertices exhibits a $D_3$ point group, rather than a perfect $D_{3h}$ delta-hedron whose faces are all equilateral triangles.

The computed IR spectra of some lower-lying states of $Nb_{16}$ and $Nb_{16}^+$ clusters are shown in Fig. 13. For all states considered, the vibrational fundamentals appear at below 325 $cm^{-1}$. The spectrum of the neutral ground $^1A$ state exhibits an intense peak at 310 $cm^{-1}$, and some lower-intensity ones below 300 $cm^{-1}$. The spectrum of the **16A**$^+$ ($^2A$) is characterized by a doublet at ~305 $cm^{-1}$. In addition, five distinct peaks are present in the range of 100–300 $cm^{-1}$. For the anion, at least four distinct absorption bands can be found in the range below 300 $cm^{-1}$ in the spectrum of **16A**$^-$.

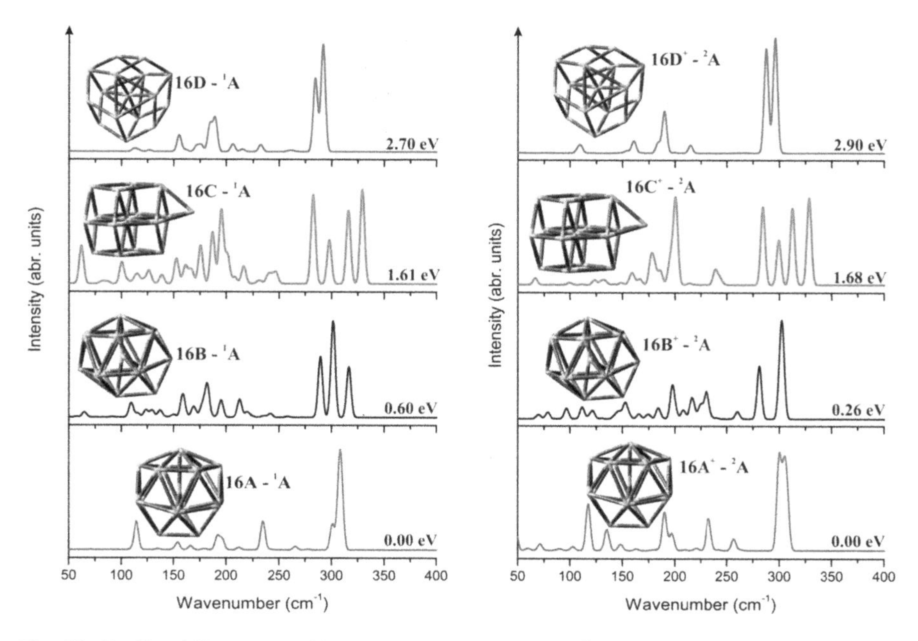


**Fig. 13** Predicted IR spectra of $Nb_{16}$ (neutral, *left*) and $Nb_{16}^{+}$ (cation, *right*)

## 2.16 *The Heptadecamers $Nb_{17}$*

Several possible structures for $Nb_{17}$ were identified that include **17A**, **17B**, **17C** and **17D** (Fig. 14). Of these structures, the distorted (12,4) FK **17A**having $C_1$ symmetry and a low spin electronic state, is found to be the global minimum of the neutral $Nb_{17}$ and also of its ions $Nb_{17}^{+/-}$. In this case of 17-atom systems, a regular FK polyhedron, i.e. a 16-vertex tetracapped truncated tetrahedron ($T_d$ symmetry), is not an energetically preferred structure. An isomer obtained from **16A** by adding one Nb atom (cf. structure **17C**) was predicted to be the most stable form for $Nb_{17}$ [27]. However, **17C** ($^1$A) was later computed to be 1.5 eV above **17A** ($^1$A) [63]. The other isomer **17B,** which is similar to **17A** but more symmetric ($C_2$), is located at 0.45 eV above. The 17-atom species can also appear as a heptagonal anti-prism with each heptagonal face capping one atom (cf. **17D**), but this form is much less stable than **17A** by 1.8 eV.

The IR spectrum of $Nb_{17}^{+}$ is also predicted to exhibit in its ground state a typical peak at $\sim$280 cm$^{-1}$ (Fig. 14). An interesting point here is that they vibrate at a considerably lower energy than the smaller size clusters $Nb_{13}^{0/+}$–$Nb_{16}^{0/+}$. As compared to **17A** ($^2$A), the closed-shell counterpart **17A⁺** ($^1$A) has a somewhat smoother spectrum as it is likely to be more symmetric. Some typical bands also exist as spectroscopic signatures of the $^1$A anion ground state at $\sim$280 cm$^{-1}$, besides several lower intensity bands in the range below 240 cm$^{-1}$.

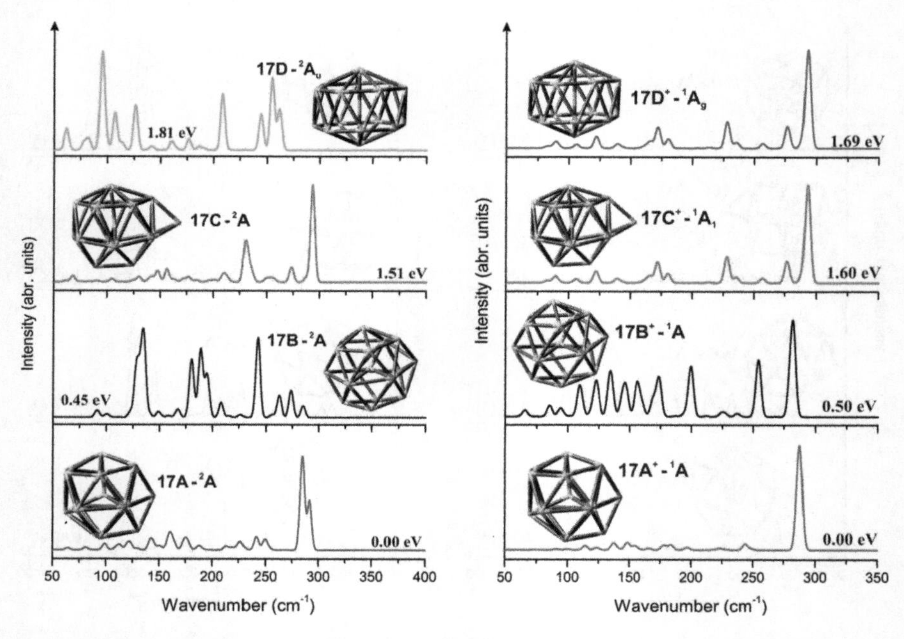


**Fig. 14** Predicted IR spectra of $Nb_{17}$ (neutral, *left*) and $Nb_{17}^+$ (cation, *right*)

## 2.17 *The Octadecamers $Nb_{18}$*

As stated above, the optimal Nb clusters prefers a close-packed growth behavior rather than a successive growth pattern. This also holds true for the 18-atom species. In fact, the isomer obtained from **17A** by adding an extra Nb atom (cf. structure **18C**, Fig. 15) is computed to be around 0.54 eV higher in energy than the more compact **18A**. The latter is also derived from five and six faces, but it is not a FK polyhedron because it has two pairs of adjacent degree six vertices. Only delta-hedra with degree 5 and 6 vertices and with no adjacent degree 6 vertices are considered as FK polyhedral [64]. A two-fold symmetry isomer **18B**, being 0.25 eV higher energy than **18A**, is obtained when one top-side atom is cut from the lowest-energy of the larger size **19A**.

Following electron detachment and attachment, the two shapes **18A** and **18B** are equally giving rise to the energetically optimal configurations **18A⁺** and **18B⁺** for $Nb_{18}^+$ and **18A⁻** and **18B⁻** for $Nb_{18}^-$ (Fig. 15). Both isomers **18B⁺** and **18B⁻** are now found to lie 0.36 and 0.10 eV higher than **18A⁺** and **18A⁻**, respectively. In previous investigations on clusters of Fe, Ca and Sr [65, 66] the global minima for 18-atom systems are related to a double icosahedron by removal of one atom. In the present study, such a form **18D** with $D_{5h}$ symmetry and a low-spin $^1A_1'$ state remains a local minimum, but being $\sim$0.77 eV above the lowest-energy isomer **18A**.

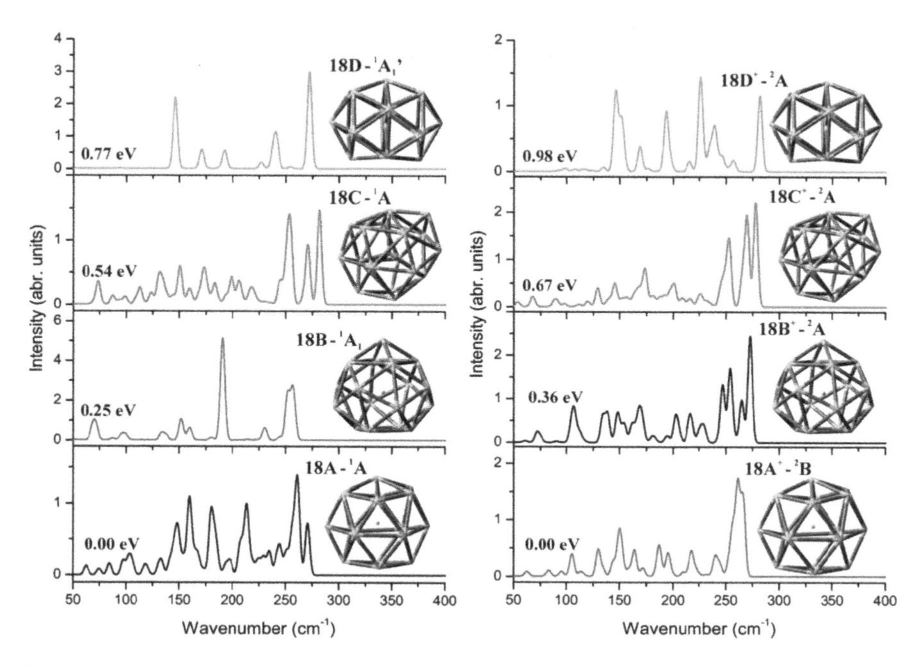


**Fig. 15** Predicted IR spectra of $Nb_{18}$ (neutral, *left*) and $Nb_{18}^+$ (cation, *right*)

The vibrational spectrum of **18A** ($^1$A) contains (at least) six distinct bands located in the range of 150–275 $cm^{-1}$, where the highest one is centered at around 260 $cm^{-1}$ (Fig. 15). The IR spectrum of **18B** ($^1A_1$) is much simpler with the presence of the most intense band at ~190 $cm^{-1}$ and a doublet at ~255 $cm^{-1}$. Removal of one electron from **18A** and **18B** forming **18A**$^+$ and **18B**$^+$ cations is accompanied with a significant change in the vibrational spectrum. While the intensities of bands below 225 $cm^{-1}$ of the **18A** spectrum are significantly reduced, the doublet at ~255 $cm^{-1}$ on **18B** band pattern now splits into several peaks and some substantially shift to higher energy levels. The band patterns of **18A**$^-$ and **18B**$^-$ anions are also quite different from those of the corresponding neutrals and cations.

## 2.18 *The Nonadecamers $Nb_{19}$*

Species containing 19 atoms constitute an extensively studied object in metal clusters, as the icosahedral growth pattern gives rise to a double icosahedrons [67].

However, in the case of $Nb_{19}$, no low-lying isomer is directly derived from a double icosahedron. Instead, it prefers a hexadecahedron-like structure **19A** (Fig. 16), which is built upon **18B**. This lends a further support for the view that an

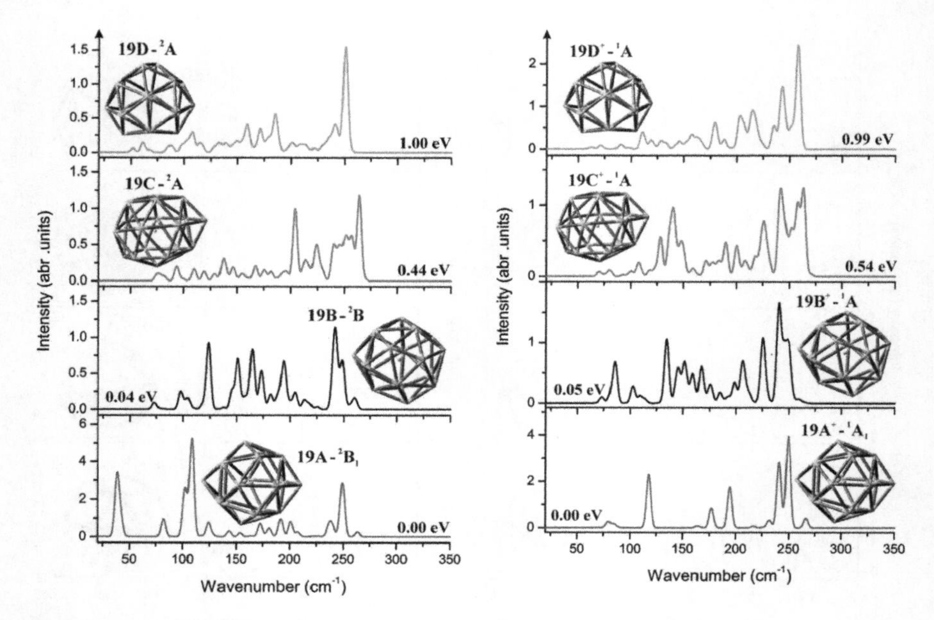


**Fig. 16** Predicted IR spectra of $Nb_{19}$ (neutral, *left*) and $Nb_{19}^+$ (cation, *right*)

icosahedral growth is not favored in niobium clusters. Also particularly stable is a compact structure **19B**, being only 0.04 eV higher in energy than **19A**. In **19B** the numbers of vertices where five and six faces meet, amount to 12 and 6. Another compact isomer, i.e. **19C**, with one degree 4, seven degree 5, and ten degree 6 vertices, is predicted to be 0.44 eV above **19A**. In addition, several local minima can be built upon the fivefold symmetry **18D** of the smaller size by capping one Nb atom randomly. They are computed to be 1.0–2.2 eV less stable than the lowest-energy isomer.

The two lowest-lying structures **19A** and **19B** also turn out to be the energetically favored configurations in both charged states. The state $^1A$ of $\mathbf{19B^+}$ is now only 0.05 eV higher than its counterpart $^1A_1$ of $\mathbf{19A^+}$. Both states are thus basically degenerate. Earlier experiments also observed the presence of different structural isomers for the cationic species $Nb_{19}^+$ [68]. Different structural isomers of the neutral $Nb_{19}$ could exist as well, but this is also dependent on the energy barriers to their interconversion. This issue has not been addressed yet.

Figure 16 displays the vibrational spectra of 19-atom species in both neutral and cationic states. The predicted spectrum of **19A** ($^2B_1$) contains distinct peaks in the vicinity of 110, 200 and 250 $cm^{-1}$, besides the appearance of a very low-frequency band below 50 $cm^{-1}$. The IR spectrum of the quasi-degenerate state **19B** ($^2B$) becomes much more complicated with the presence of over seven typical bands in the range of 125–250 $cm^{-1}$. The vibrational spectrum of $\mathbf{19A^+}$ ($^1A_1$), as in **19A**, is also characterized by recognizable peaks at around 110, 200 and 250 $cm^{-1}$.Their intensities are however somewhat different, and the low-frequency vibration below 50 $cm^{-1}$ is not observed on the band pattern of **19A⁺**.

There also exist some minor but distinguishable differences between the vibrational spectra of **19B** and **19B⁺**. Following the electron detachment to form **19B⁺**, the intensities of bands in the range 150–200 $cm^{-1}$ are substantially reduced, in addition to the emergence of a distinct band at 225 $cm^{-1}$ and a small blue shift from the peak around 125–135 $cm^{-1}$. Let us mention again that an experimental spectrum of $Nb_{19}$ was reported [16] but a comparison is not feasible due to the low quality of the recorded spectrum.

## 2.19 *The Eicosamers $Nb_{20}$*

Recently, Gruene et al. [69] confirmed a regular tetrahedron ($T_d$) as the global minimum for $Au_{20}$. On the contrary, a double icosahedron ($D_{5h}$) with two encapsulated atoms is predicted to be the most stable form of $Ni_{20}$ [70], whereas a face-capped octahedron ($C_{3v}$) is the favored form of $Pd_{20}$ [71].

The cluster $Nb_{20}$ has the most stable form as a polyhedron **20A** (Fig. 17) whose degree five and six vertices are 12 and 7, respectively. Besides, we detect a quasi-isoenergetic form **20B** ($^1A'$) competing for the ground state of $Nb_{20}$ as it is located only 0.07 eV above **20A** ($^1A$). The shape **20B** arises from the fivefold

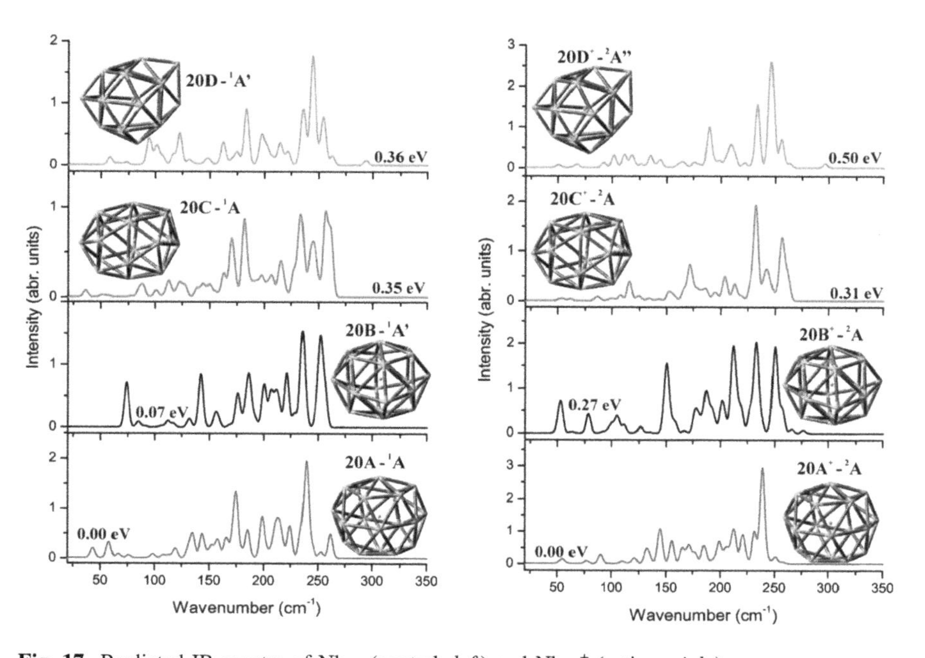


**Fig. 17** Predicted IR spectra of $Nb_{20}$ (neutral, *left*) and $Nb_{20}^+$ (cation, *right*)

symmetry structure **18D** by capping two additional Nb atoms. Both **20A** and **20B** forms are also the lowest and the second-lowest forms of the ions. However, the energy separation between them is enlarged upon ion formation, in which **20B**$^+$ and **20B**$^-$ are now lying 0.27 and 0.13 eV above their respective counterparts **20A**$^+$ and **20A**$^-$.

The vibrational spectra of some lower-lying states of $Nb_{20}$ and $Nb_{20}^+$ are plotted in Fig. 17. The predicted spectrum of **20A** contains two distinct peaks at $\sim$175 and 240 cm$^{-1}$. In addition, the highest-energy band is observed at 265 cm$^{-1}$ and several low-intensity ones in the vicinity of 175–225 cm$^{-1}$. Vibrations of **20B** ($^1$A′) are characterized by two strong peaks at around 235 and 260 cm$^{-1}$, and some lower bands at $\sim$185, 145 and 75 cm$^{-1}$. Regarding the cation **20A**$^+$, it spectrum is similar to that of **20A** that significant peak can be found at $\sim$245 cm$^{-1}$. However, the peak at $\sim$175 cm$^{-1}$ is now significantly weaker. The spectrum of **20B**$^+$ contains three closely spaced peaks at 255, 240 and 220 cm$^{-1}$, and a typical one at $\sim$150 cm$^{-1}$.

# 3 Energetic Properties

Some basic energetics of pure niobium clusters such as electron affinities (EAs) and ionization energies (IEs) have experimentally been determined by several groups [10, 11, 13]. Photoemission studies [11] predicted that even-numbered clusters $Nb_4$, $Nb_8$, $Nb_{10}$, $Nb_{12}$, $Nb_{14}$ and $Nb_{16}$ have a closed-shell singlet state and that the additional electron occupies the corresponding LUMO giving rise to a small peak at low binding energy of the resulting anions. Remarkably large frontier orbital (HOMO–LUMO) energy gaps found for 4-, 8-, 10- and 16-atom neutral species correlate well with experiment. The reactivity of these clusters with small molecules, such as $H_2$, $N_2$ and CO, is abnormally low as compared to other counterparts [14, 15] (Table 7).

As far as we are aware, there are no theoretical data reported for both EAs and IEs of $Nb_n$ larger than $Nb_{12}$. Vertical and adiabatic detachment energies for $Nb_n^{0/-}$ species in the range of size $n = 2$–20 were systematically evaluated [35]. The vertical detachment energies (VDEs) of anions are calculated at the anion geometry, whereas the adiabatic detachment energies (ADEs) refer to the values obtained using the optimized structures of both neutral and anionic forms. Such quantities are also EAs of neutral clusters. Similarly, vertical and adiabatic detachment energies of neutrals, which correspond to the vertical and adiabatic IEs. Table 8 lists, along with available experimental data, the vertical and adiabatic values for both EAs and IEs of the $Nb_n^{0/-}$ clusters considered.

Overall, the calculated VDEs for clusters smaller than $Nb_6^-$ increase monotonously according to the cluster size, except for the minimum at $Nb_4^-$. This can be deduced that removal of one electron from $Nb_4^-$ to form $Nb_4$ is easier than the same process from $Nb_3^-$, $Nb_5^-$ or $Nb_6^-$. The neutral tetramer is somewhat more stabilized than its neighbors, since its electronic structure corresponds to a magic number of 20 valence electrons, according to the phenomenological shell model [72, 73].

**Table 7** Ground and low-lying states of $Nb_n^{0/\pm}$ ($n$ = 18–20) clusters and relative energies (RE. eV) at the BPW91/cc-pVDZ-PP level

| Cluster[a] | | State | RE | Cluster[a] | | State | RE |
|---|---|---|---|---|---|---|---|
| $Nb_{18}$ | **18A** ($C_1$) | $^1A$ | 0.00 | $Nb_{19}^+$ | **19C**$^+$ ($C_1$) | $^1A$ | 0.54 |
| | **18B** ($C_{2v}$) | $^1A_1$ | 0.25 | | **19D**$^+$ ($C_1$) | $^1A$ | 0.99 |
| | **18C** ($C_1$) | $^1A$ | 0.54 | $Nb_{19}^-$ | **19A**$^-$ ($C_{2v}$) | $^1A_1$ | 0.01 |
| | **18D** ($D_{5h}$) | $^1A_1'$ | 0.77 | | **19B**$^-$ ($C_1$) | $^1A$ | 0.00 |
| $Nb_{18}^+$ | **18A**$^+$ ($C_2$) | $^2B$ | 0.00 | | **19C**$^-$ ($C_1$) | $^1A$ | 0.37 |
| | **18B**$^+$ ($C_1$) | $^2A$ | 0.36 | | **19D**$^-$ ($C_1$) | $^1A$ | 0.98 |
| | **18C**$^+$ ($C_1$) | $^2A$ | 0.67 | $Nb_{20}$ | **20A** ($C_1$) | $^1A$ | 0.00 |
| | **18D**$^+$ ($C_1$) | $^2A$ | 0.98 | | **20B** ($C_s$) | $^1A'$ | 0.07 |
| $Nb_{18}^-$ | **18A**$^-$ ($C_1$) | $^2A$ | 0.00 | | **20C** ($C_1$) | $^1A$ | 0.35 |
| | **18B**$^-$ ($C_{2v}$) | $^2A_1$ | 0.10 | | **20D** ($C_s$) | $^1A'$ | 0.36 |
| | **18C**$^-$ ($C_s$) | $^2A$ | 0.42 | $Nb_{20}^+$ | **20A**$^+$ ($C_1$) | $^2A$ | 0.00 |
| | **18D**$^-$ ($C_1$) | $^2A$ | 0.65 | | **20B**$^+$ ($C_1$) | $^2A$ | 0.27 |
| $Nb_{19}$ | **19A** ($C_{2v}$) | $^2B_1$ | 0.00 | | **20C**$^+$ ($C_1$) | $^2A$ | 0.31 |
| | **19B** ($C_2$) | $^2A$ | 0.04 | | **20D**$^+$ ($C_s$) | $^2A''$ | 0.50 |
| | **19C** ($C_s$) | $^2A$ | 0.44 | $Nb_{20}^-$ | **20A**$^-$ ($C_1$) | $^2A$ | 0.00 |
| | **19D** ($C_1$) | $^2A$ | 1.00 | | **20B**$^-$ ($C_s$) | $^2A'$ | 0.13 |
| $Nb_{19}^+$ | **19A**$^+$ ($C_{2v}$) | $^1A_1$ | 0.00 | | **20C**$^-$ ($C_1$) | $^2A$ | 0.35 |
| | **19B**$^+$ ($C_2$) | $^1A$ | 0.05 | | **20D**$^-$ ($C_s$) | $^2A'$ | 0.28 |

[a]Structures are given in figures of the corresponding sections

The VDE value of anion $Nb_2^-$ is also remarkably low, indicating a high stability of the corresponding neutral $Nb_2$.

In general, the computed VDEs are in good agreement with the measured VDE values reported in Ref. [11]. For example, the VDEs of $Nb_{10}^-$ obtained at M06 and BPW91 levels amount to 1.44 and 1.48 eV, respectively, and both compared well to the experimental value of 1.45 eV [11]. Nevertheless, the calculated VDE ($Nb_{12}^-$) of 1.06 (BPW91) or 1.17 eV (M06) appears underestimated as compared to the earlier experimental value of 1.65 eV reported by Kietzmann et al. [11]. In a more recent study on UPE spectra of $Nb_4^-$ to $Nb_{200}^-$, Wrigge et al. [13] however found that the photoelectron spectrum of $Nb_{12}^-$ does exhibit a low intensity peak at a lower binding energy of $\sim 1.0$ eV, which has not been detected in the previous experiment [11]. We, therefore, would suggest that the neutral 12-atom cluster indeed possesses the lowest EA of $\sim 1.0$ eV.

Noticeably, an even-odd alternation of VDEs, as well as of ADEs, with local minima at $n$ = 8, 10, 12, 14 and 16 were found. Such a pattern is not observed for clusters smaller than $Nb_6^-$. For $Nb_n^-$ species with n > 17 the VDEs tends to increase steeply with no clear even/odd alternation. These results concur with the experimental data estimated from the photoelectron spectra of mass-separated $Nb_n^-$ clusters [11].

**Table 8** Ionization energies (IE) of $Nb_n$ and detachment energies (DE) of $Nb_n^-$ computed using the BPW91 functional

| Isomer | Ionization energies (eV) | | | Isomer | Detachment energies (eV) | | |
|---|---|---|---|---|---|---|---|
| | AIE | VIE | Exptl. (± 0.05)[a] | | ADE | VDE | Exptl. (± 0.05)[b] |
| **2A** | 6.26 | 6.29 | 6.20 | **2A**$^-$ | 0.82 | 0.83 | |
| **3A** | 5.72 | 5.86 | 5.81 | **3A**$^-$ | 0.98 | 1.03 | 1.09 |
| **4A** | 5.56 | 5.66 | 5.64 | **4A**$^-$ | 0.83 | 0.86 | 1.10 |
| **5A** | 5.25 | 5.27 | 5.45 | **5A**$^-$ | 1.38 | 1.48 | 1.52 |
| **6A** | 4.96 | 5.22 | 5.38 | **6A**$^-$ | 1.35 | 1.48 | 1.58 |
| **6B** | 4.90 | 5.29 | | **6B**$^-$ | 1.36 | 1.50 | |
| **7A** | 5.13 | 5.34 | 5.35 | **7A**$^-$ | 1.42 | 1.57 | 1.65 |
| **8A** | 5.28 | 5.38 | 5.53 | **8A**$^-$ | 1.31 | 1.33 | 1.45 |
| **9A** | 5.00 | 5.10 | 5.20 | **9A**$^-$ | 1.38 | 1.53 | 1.65 |
| **9B** | 4.77 | 4.81 | 4.92 | **9B**$^-$ | 1.30 | 1.43 | |
| **10A** | 5.29 | 5.41 | 5.48 | **10A**$^-$ | 1.31 | 1.48 | 1.45 |
| **11A** | 4.58 | 4.52 | 4.74 | **11A**$^-$ | 1.25 | 1.30 | 1.65 |
| **12A** | 4.10 | 4.17 | 4.96 | **12A**$^-$ | 0.88 | 1.06 | 1.65 |
| **12B** | 4.73 | 4.80 | | **12B**$^-$ | 1.27 | 1.31 | |
| **13A** | 4.70 | 4.76 | 4.85 | **13A**$^-$ | 1.66 | 1.68 | 1.70 |
| **13B** | 4.47 | 4.50 | | **13B**$^-$ | 1.44 | 1.48 | |
| **14A** | 4.59 | 4.72 | 4.75 | **14A**$^-$ | 1.56 | 1.60 | 1.64 |
| **15A** | 4.60 | 4.65 | 4.53 | **15A**$^-$ | 1.93 | 2.04 | 1.65 |
| **16A** | 4.54 | 4.66 | 4.80 | **16A**$^-$ | 1.48 | 1.54 | 1.62 |
| **17A** | 4.43 | 4.48 | 4.67 | **17A**$^-$ | 1.58 | 1.63 | 1.80 |
| **18A** | 4.36 | 4.41 | 4.65 | **18A**$^-$ | 1.62 | 1.65 | 1.80 |
| **18B** | 4.47 | 4.56 | | **18B**$^-$ | 1.77 | 1.81 | |
| **19A** | 4.37 | 4.40 | 4.63 | **19A**$^-$ | 1.65 | 1.75 | 1.88 |
| **19B** | 4.42 | 4.50 | | **19B**$^-$ | 1.66 | 1.70 | |
| **20A** | 4.43 | 4.45 | 4.62 | **20A**$^-$ | 1.77 | 1.80 | 1.93 |
| **20B** | 4.60 | 4.64 | | **20B**$^-$ | 1.72 | 1.73 | |

Basis sets employed are cc-pVTZ-PP for $n$ = 2–6 and cc-pVDZ-PP for $n$ = 7–20
[a]Taken from Ref. [10]
[b]Taken from Ref. [12]

We also found that, in general, the detachment energies calculated from the lowest-energy structures are in substantially better agreement with experiment than those obtained from higher-energy isomers. For example, the predicted ADEs of **13A**$^-$ and **13B**$^-$ are 1.68 and 1.48 eV, respectively, as compared to the experimental value of 1.70 eV. The nine-niobium cluster appears remarkable as the VDEs of two near-degeneracy isomers **9A** and **9B** are more or less similar. The BPW91 calculated VDEs of **9A**$^-$ and **9B**$^-$ are 1.53 and 1.43 eV, respectively. Using the M06 functional, such quantities are 1.63 and 1.65 eV that can be compared to the experimental value of 1.65 eV. Only in the case of $Nb_{18}^-$, a low-lying structure

**18B**$^-$ yields a closer VDE to experiment (see Table 8), although it is computed to lie ~0.25 eV above the ground state.

The IEs of $Nb_n$ clusters in the size range of $n$ = 2–76 were experimentally determined by Knickelbein and Yan using photoionization efficiency (PIE) spectrometry [10]. Such thermochemical parameters for systems in the size range $n$ = 4–200 were also reported by Wrigge et al. in a more recent paper [13]. For both vertical and adiabatic IEs for $Nb_n$ species with $n$ = 2–20, calculated results, along with experimental data, are listed in Table 8. It appears that the experimental IEs correspond to the vertical values. In general, the predicted results are in good agreement with experiment, with an average deviation of ~0.1 eV.

For systems smaller than $Nb_6$ the IEs decrease monotonously with the increase of clusters size. Several groups also investigated the IEs of small niobium clusters. However, the reported results appear to underestimate the experimental values. For example, CISD calculations [10] yielded the IEs of 4.85, 4.33 and 4.31 eV for $Nb_2$, $Nb_3$ and $Nb_4$, respectively. These are substantially smaller than the measured values given in Table 8. MRSDCI [52] and DFT [25] calculations derived a value of ~5.9 eV for the IE of $Nb_2$ (exptl: 6.2 eV). When performing the MRSDCI + Q calculations, the IE of $Nb_2$ is improved, being 0.1 eV lower than the experimental value.

In the sizes of $n$ = 7–10, an odd-even pattern of VIEs is observed, with local minima at n = 9, 11. For even-numbered clusters $Nb_8$ and $Nb_{10}$ that have closed electron shell and low spin electronic state, calculated VIEs (BPW91) are about 0.1 and 0.3 eV higher than those of $Nb_7$ and $Nb_9$, respectively, that are in line with experimental observation. In contrast, for the pair of $Nb_{11}$ and $Nb_{12}$, the even/odd alternation of VIEs is not clearly identified. According to available experiment, $Nb_{11}$ has a lower VIE (by ~0.2 eV) than $Nb_{12}$. However, calculations for the lowest-energy isomers **11A** and **12A** follow a reversed experimental trend by ~0.3 eV. Overall, the experimental IE value for $Nb_{12}$ is not well evaluated yet. Two measured values of 4.92 and 5.20 eV were reported for reactive and unreactive forms of $Nb_{12}$ [10] as compared to the computed VIE(**12A**) = 4.17 eV and VIE (**12B**) = 4.80 eV.

The computed VIEs for **9A** and **9B** structures amount to 5.10 and 4.81 eV (BPW91), respectively. By employing the M06 functional, such values amount to 5.25 and 4.84 eV. Two IE values of $Nb_9$ cluster have also been recorded experimentally, namely 5.20 and 4.92 eV [10]. This fact supports the generation of a pair of isomers.

It is also worth noting that both $Nb_8$ and $Nb_{10}$ have a particularly large IE and low EA among the Nb clusters considered. This indicates a kinetic stability towards donating or accepting of electrons. In addition, the high global hardness $\eta$ ($\eta \sim$ IE – EA) [74] denotes that the clusters are unwilling to undergo an addition reaction with either an electrophile or a nucleophile. In other words, a "hard" cluster should be less reactive.

For systems in the range of $n$ = 13–20, the best match to experiment is generally related to the lowest-energy structures, except for $Nb_{12}$ and $Nb_{18}$. In the latter case, the calculated VIEs of 4.41 and 4.56 eV for **18A** and **18B**, respectively, can be compared to the experimental value of 4.65 eV. The VIEs computed from the low-lying isomers **19B** and **20B** also coincide better with experiment. In both cases the energy difference between those isomers and their ground state is extremely small, and within the expected accuracy of the DFT methods, they are energetically degenerate. In these cases, the higher energy isomer identified by DFT computations, could legitimately be considered as the observed structure in experiment.

Although a clear odd–even oscillation of IEs for clusters smaller than $Nb_{12}$ exists, such a pattern is not obviously observed for the larger size range $n$ = 13–20. In agreement with experiment, the 13-atom cluster is computed to have the largest IE value. Other local maxima appear at $n$ = 14 and 16. In general, the VIEs computed at the BPW91/cc-pVDZ-PP level compare well to the experiment, but with some exceptions. According to Knickelbein and Yang [10], $Nb_{15}$ has a lower VIE (by ~0.27 eV) than $Nb_{16}$. However, calculations yield comparable values for both sizes. The predicted IE($Nb_{15}$) = 4.65 eV is thus likely overestimated, whereas that of $Nb_{16}$ (4.66 eV) underestimates the measured data. The computed IE ($Nb_{17}$) = 4.48 eV also appears to underestimate the experimental value of 4.67 ± 0.3 eV, even though the deviation still lies within the error margin.

Of the clusters in the range of $n$ = 13–20, the 18-atom cluster appears remarkable as both VIE and VDE values obtained from the higher lying isomer **18B** are closer to the experimental data than those computed from the lowest-energy **18A**. As stated above, it can be expected that **18B** of $Nb_{18}$ is more likely to be present in experiments than **18A**. Such a phenomenon was also observed for $Nb_6$ and $Nb_{12}$. This finding again points out that in many sizes, the global minimum of a cluster cannot be assigned solely on the basis of DFT computed energies.

## 4 Thermodynamic Stabilities

To obtain more quantitative information about the cluster stability, we consider the average binding energy per atom (BE), the second-order difference of energy ($\Delta^2$E), and the stepwise dissociation energy $D_e$. For pure niobium clusters, these parameters are defined by the following equations:

$$BE(Nb_n^x) = [(n-1) \times E(Nb) + E(Nb^x) - E(Nb_n^x)]/n$$
$$\Delta^2 E(Nb_n^x) = E(Nb_{n+1}^x) + E(Nb_{n-1}^x) - 2E(Nb_n^x)$$
$$D_e(Nb_n^x) = E(Nb_{n-1}^x) + E(Nb) - E(Nb_n^x)$$

where $x = 0,\ +1,\ -1$; $E(Nb_n^x)$ is the lowest energy of $Nb_n^x$ cluster.

Theoretical and experimental binding energies per atom for neutral clusters are plotted together in Fig. 18. First, we note that these systems can form clusters with larger sizes, since they are observed to gain energy during the growth. Figure 18 points out that the BEs increase monotonically with respect to the cluster size, with local maxima at $n$ = 8, 10, 12 and 15. The overall tendency is in line with

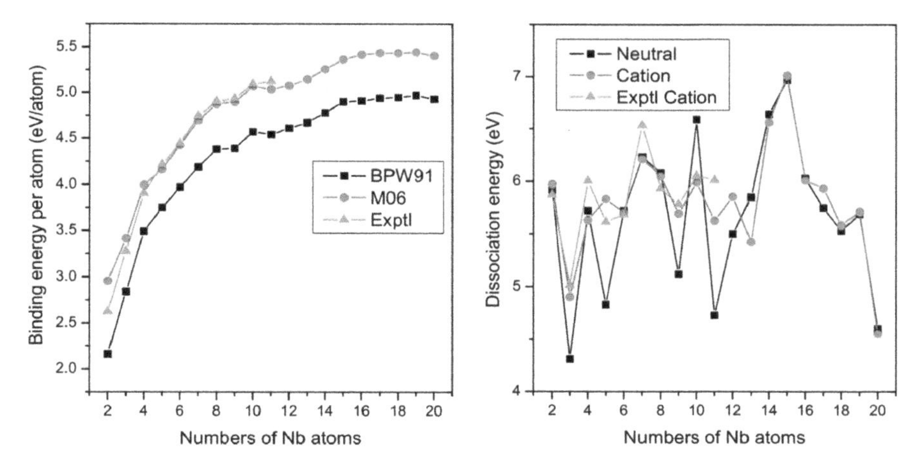


**Fig. 18** Plots of binding energies per atom for the neutral Nb clusters versus cluster size (*left*) and dissociation energies for $Nb_n$ clusters in both neutral and cationic states as a function of cluster sizes (*right*). Experimental values are taken from Ref. [45]

experimental observation. However, the computed BEs are likely underestimated with respect to experimental values.

Calculated BE of $Nb_{10}$ is about 4.60 eV/atom (BPW91) or 5.06 eV/atom (M06) as compared to the experimental value of $5.09 \pm 0.27$ eV/atom. In general, the M06 functional appears to provide better results for this quantity than the BPW91.

The average errors are <0.1 eV for M06 and ~0.5 eV for BPW91. Although the BPW91 functional seems to be more reliable than other functionals in predicting geometries, vibrations and energetics of Nb clusters, it is deficient when computing the binding energy per atom and dissociation energies. For the latter, the hybrid meta-GGA functional M06 describes somewhat better.

The biggest drawback of most GGA approaches is in the evaluation of the atomization energies [75], and as a consequence, quantities involving the energies of free atoms become less relevant. For the sake of comparison, we here report both M06 and BPW91 values for these quantities.

The cations are predicted to have larger BEs than the corresponding neutrals by both functionals (Table 9). Additionally, the BPW91 systematically yields lower BE values than the M06, with the average deviations of ~0.5 and ~0.4 eV for neutral and cationic states, respectively. The BEs of clusters considered are predicted to be in the range of 2.2–5.0 eV/atom by BPW91, as compared to the corresponding values of 3.0–5.4 eV/atom computed using M06. The BEs thus exhibit a gradual growth and tend to reach the maximal value of 5.4 eV/atom for $Nb_{19}$ (M06). However, this quantity is still far away from the cohesive energy of 7.6–7.8 eV/atom of the bulk niobium [76]. While measured values were reported for neutrals smaller than $Nb_{12}$ [45], no experimental data are found for the larger clusters.

**Table 9** Average binding energy per atom (BE, eV), second-order difference of energy ($\Delta^2E$, eV), and dissociation energy ($D_e$, eV) of $Nb_n^{0/+}$clusters ($n$ = 2–20)

| | BE | | $D_e$ | | $\Delta^2E$ | |
|---|---|---|---|---|---|---|
| | BPW91 | M06 | BPW91 | M06 | BPW91 | M06 |
| **n** | *Neutral clusters* | | | | | |
| **2** | 2.16 | 2.95 | 4.32 | 5.92 | 0.14 | 1.61 |
| **3** | 2.84 | 3.41 | 4.18 | 4.31 | −1.28 | −1.42 |
| **4** | 3.49 | 3.99 | 5.46 | 5.72 | 0.68 | 0.89 |
| **5** | 3.75 | 4.16 | 4.78 | 4.83 | −0.31 | −0.88 |
| **6** | 3.97 | 4.42 | 5.09 | 5.72 | −0.39 | −0.46 |
| **7** | 4.19 | 4.69 | 5.84 | 6.23 | 0.18 | 0.14 |
| **8** | 4.38 | 4.87 | 5.67 | 6.08 | 1.15 | 0.96 |
| **9** | 4.39 | 4.89 | 4.52 | 5.12 | −1.62 | −1.47 |
| **10** | 4.57 | 5.06 | 6.14 | 6.59 | 1.87 | 1.87 |
| **11** | 4.54 | 5.03 | 4.27 | 4.73 | −1.16 | −0.77 |
| **12** | 4.61 | 5.07 | 5.43 | 5.50 | 0.53 | 0.28 |
| **13** | 4.67 | 5.14 | 4.93 | 5.85 | −1.22 | −0.68 |
| **14** | 4.78 | 5.25 | 6.15 | 6.64 | −0.41 | −0.33 |
| **15** | 4.90 | 5.36 | 6.56 | 6.97 | 1.38 | 0.94 |
| **16** | 4.91 | 5.41 | 5.20 | 6.03 | −0.20 | 0.28 |
| **17** | 4.94 | 5.43 | 5.40 | 5.75 | 0.17 | 0.22 |
| **18** | 4.95 | 5.43 | 5.22 | 5.53 | −0.06 | −0.17 |
| **19** | 4.97 | 5.44 | 5.28 | 5.69 | 1.13 | 1.10 |
| **20** | 4.93 | 5.40 | 4.15 | 4.60 | | |
| **n** | *Cationic clusters* | | | | | |
| **2** | 2.62 | 2.98 | 5.24 | 5.97 | 0.52 | 1.08 |
| **3** | 3.32 | 3.62 | 4.72 | 4.90 | −0.88 | −0.73 |
| **4** | 3.89 | 4.12 | 5.61 | 5.63 | 0.50 | −0.20 |
| **5** | 4.14 | 4.46 | 5.11 | 5.83 | −0.01 | 0.13 |
| **6** | 4.30 | 4.67 | 5.11 | 5.70 | −0.46 | −0.11 |
| **7** | 4.52 | 4.86 | 5.91 | 6.21 | 0.40 | 0.16 |
| **8** | 4.65 | 5.01 | 5.51 | 6.05 | 0.45 | 0.36 |
| **9** | 4.70 | 5.08 | 5.06 | 5.69 | −0.53 | −0.30 |
| **10** | 4.79 | 5.18 | 5.60 | 5.99 | 0.58 | 0.37 |
| **11** | 4.81 | 5.21 | 5.02 | 5.62 | −0.88 | −0.23 |
| **12** | 4.91 | 5.27 | 5.90 | 5.85 | 0.78 | 0.48 |
| **13** | 4.87 | 5.28 | 4.77 | 5.42 | −1.67 | −1.14 |
| **14** | 4.96 | 5.37 | 6.13 | 6.56 | −0.34 | −0.54 |
| **15** | 5.06 | 5.49 | 6.47 | 7.01 | 1.13 | 1.09 |
| **16** | 5.08 | 5.52 | 5.34 | 6.01 | −0.16 | 0.08 |
| **17** | 5.10 | 5.54 | 5.50 | 5.93 | 0.21 | 0.35 |
| **18** | 5.11 | 5.55 | 5.29 | 5.58 | 0.02 | −0.13 |
| **19** | 5.12 | 5.56 | 5.28 | 5.71 | 1.22 | 1.88 |
| **20** | 5.07 | 5.52 | 4.10 | 4.55 | | |

As shown in Fig. 18 the dissociation energy $D_e$ values as a function of cluster size obey the odd-even staggering in the region of $n = 8$–12. Among these clusters, $Nb_9$ and $Nb_{11}$ are characterized by the lowest $D_e$ values, implying their low thermodynamic stability. On the contrary, $Nb_{15}$ is expected to be the most stable one as it has the largest $D_e$ value. Other noticeable peaks are found at $n = 7$, 8, 10 and 12, indicating that these systems are also relatively stable.

Available experimental dissociation energies of the cationic species are also included in Fig. 18 for comparison. Both DFT methods employed in this analysis follow familiar trends in dissociation energy predictions and tend to underestimate experimental results. The M06 functional reaches again a closer agreement with experiment than the BPW91. For this parameter, the average absolute deviations relative to experiment of BPW91 and M06 methods are ~0.6 and ~0.2 eV, respectively.

The time-of-flight (TOF) mass spectra of Nb clusters recorded by Sakurai et al. [77] show a remarkably high intensity for the systems in the range of $n = 11$–16. Subsequently, the intensity significantly decreases and reaches a minimum at $n = 19$. This observation correlates well with the dissociation energy $D_e$ values tabulated in Table 9. Indeed, predicted values for clusters with $n = 13$–16 are much higher than those of the larger systems. Especially, both $Nb_{15}$ and $Nb_{15}^+$ are expected to be exceptionally stable as they exhibit the especially large $D_e$ values. Of the neutrals considered, $Nb_{15}$ is characterized by the highest $D_e$ value, implying a particular thermodynamic stability even though it has an open-shell electronic structure. These results indicate again the dominance of a *geometric effect*, namely the cluster stability is determined more by the atomic arrangement (shape) than by the electron distribution.

The magic number of $n = 15$ was found for the pure $M_n$ clusters of Fe, Ti, Zr, and Ta, owing to their **bcc** structural unit [77]. In addition, both $Fe_{19}$ and $Ti_{19}$ were also assigned to be magic as they are expected to hold a highly symmetric conformation, i.e. a double icosahedron with two encapsulated atoms. The 19-Nb atom species, on the contrary, does not prefer such a form because, as compared to 3*d* elements, the 4*d* elements have such a larger volume that the fivefold symmetry cage of 17-Nb atoms cannot conveniently encapsulate two Nb atoms. This concurs with the low intensity on the abundance spectra of auto-ionized Nb clusters observed by Sakurai et al. for $Nb_{19}$ [77].

As far as we are aware, no experimental dissociation energies are actually available for clusters containing more than 11 Nb atoms. Based on the agreement between the experiment and our computed values for small systems, it could be suggested to adopt the $D_e$ values predicted by computations using the M06 functional for $Nb_n$ in the range of $n = 13$–20.

Another indicator that can be used to estimate the relative stability of a cluster with respect to its immediate neighbors is the second-order difference of energy ($\Delta^2E$). A cluster with the positive value of $\Delta^2E$ is considered to be more stable, compared to the smaller and larger size neighbors. For systems smaller than $Nb_{12}$, the $\Delta^2E$s clearly exhibit odd-even fluctuations: a cluster with an even number of atoms appears to be much more stable than its neighboring odd-numbered ones.

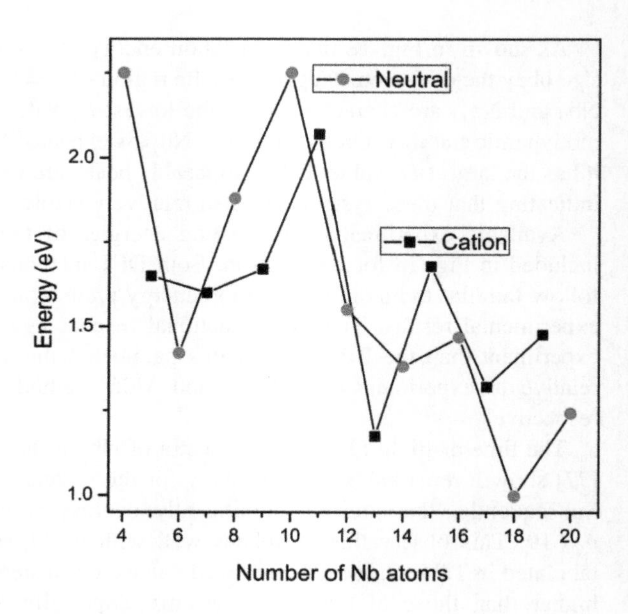


**Fig. 19** Evolution of HOMO–LUMO gaps in several $Nb_n^{0/+}$ clusters

However, such a trend is no longer observed for systems larger than $Nb_{12}$. Of the clusters considered, the 8-, 10- and 15-atom species are found to have particularly large $\Delta^2E$, indicating their higher stabilities. This is again in perfect agreement with the analysis based on dissociation energies given above.

The energy gap between frontier orbitals (HOMO–LUMO) is also used to measure the kinetic stability of chemical species. The graph of the HOMO–LUMO gaps for the lowest-energy structures of neutral and cationic Nb clusters given in Fig. 19. Because such separations of the closed-shell systems and the SOMO–LUMO gaps of the open-shell radicals cannot directly be compared, we only plot the gaps for closed-shell species. As illustrated in Fig. 19, the relatively large HOMO–LUMO gaps of $Nb_8$, $Nb_{10}$ and $Nb_{16}$ correlate well with their low reactivities toward nitrogen, deuterium and hydrogen [14]. Concerning the cations, the large gap obtained for $Nb_{15}^+$ concurs with its high stability discussed above. However, its reactivity with simple molecules in gas phase is comparable to other Nb cationic clusters [78–80]. In general, several groups investigated the reaction of cations $Nb_n^+$ ($n$ up to 30) with simple molecules [17, 78–80] and found that no size is much less reactive than the other; the difference between more reactive and less reactive forms is insignificant. Our theoretical results also support these observations in predicting that no cation is exceptionally inert or particularly stable.

In an early study on clusters of alkali metals [81], it has been revealed that certain sizes composed of a specific number of atoms turn out to be particularly stable and are found to be more abundant than others in the experimental beam. These are called *magic clusters*. In the current work, we found $Nb_2$ and $Nb_4$ to be magic as they hold the number of valence electrons corresponding to the closed shell in the sequence (1S/1P/2S/1D/1F….) [73, 74, 82, 83].

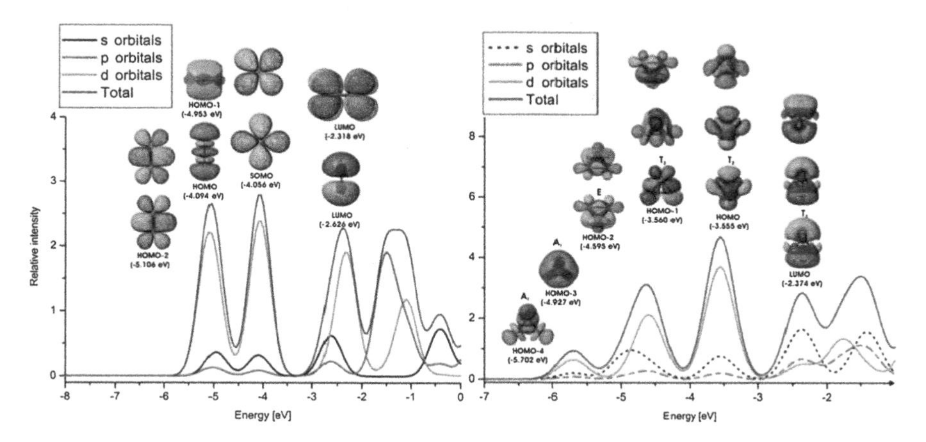


**Fig. 20** Total and partial density of states of $Nb_2$ (*left*) and $Nb_4$ (*right*)

In these clusters, as seen from the densities of states (DOS) graphics plotted in Fig. 20 [84], the valence bands of $Nb_2$ and $Nb_4$ are principally formed by atomic orbitals 4*d* and 5*s* overlaps. Similarly, each niobium atom contributes five electrons in orbitals 4*d* and 5*s* to forming molecular orbitals in the valence bands of $Nb_8$ and $Nb_{10}$.

The particular stability and inertness of $Nb_{10}$ can further be understood on the basis of a three dimensional (3D) aromaticity. The $Nb_{10}$ cluster has a spherical shape containing 50 valence electrons, and it thus is expected to have a spherical aromatic character as its number of valence electrons satisfies the empirical $2(N + 1)^2$ rule [85]. Overall, the spherical aromatic character of $Nb_{10}$ is a parameter to account for its strong magic behavior.

For larger systems, as stated above, the cluster stability appears to be determined more by the atomic arrangement than by the electronic distribution. The $Nb_{15}$ species in both neutral and singly charged states is predicted to be exceptionally stable as they can form a highly symmetric structure.

In order to probe further the high stability of some specific sizes, the shapes and symmetries of MOs generated from interactions between valence orbitals of the encapsulated Nb atoms with the outside niobium frameworks in $Nb_{15}^-$ and $Nb_{19}^-$ systems are displayed in Fig. 21. Such combinations are analogous to those of the homoleptic complexes in which the encapsulated Nb serves the function of central atom whose benefit is a stabilizing effect for the Nb cages. When placed at the center of an $Nb_{14}$ cube, the five *d* orbitals of Nb are no longer degenerate (Fig. 21). Instead they split into two non-equivalent sets of orbitals with $e_{.g.}$ and $t_{2g}$ sub-symmetries of $O_h$ group.

Similar interactions are also observed in larger systems, such as $Nb_{19}^-$. However, in the latter, the five *d* orbitals split into five discrete levels $2xa_1$, $a_2$, $b_1$ and $b_2$ of $C_{2v}$ point group. Furthermore, moving to $Nb_{19}^-$, the $Nb_{18}$ skeleton expands, in such a way that linear combinations of encapsulated Nb valence orbitals with those of the

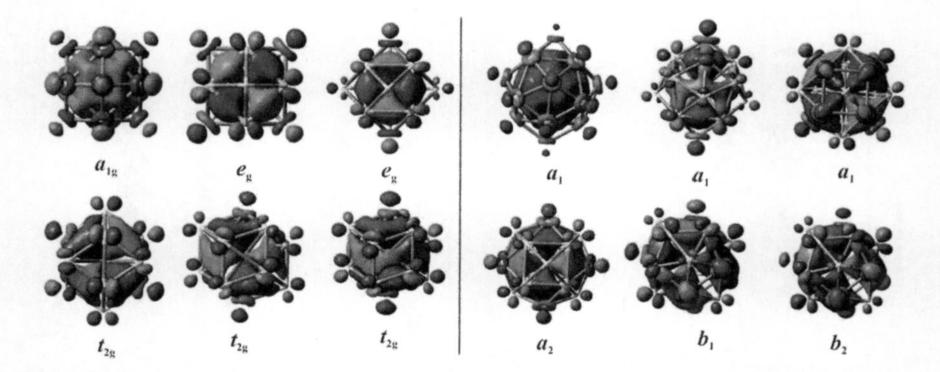


**Fig. 21** Shapes and symmetries of MOs generated from valence orbitals of the central Nb atom in the anions $Nb_{15}^-$ (*left*) and $Nb_{19}^-$ (*right*)

**Table 10** The natural charges (a.u.) on encapsulated Nb atom of $Nb_n$ clusters with $n$ = 13–20 (BPW91/cc-pVDZ-PP)

| Isomers | NBO charges | Isomers | NBO charges |
|---|---|---|---|
| **13A** | −1.3 | **18A** | −1.1 |
| **13B** | −1.0 | **18B** | −1.2 |
| **14A** | −1.0 | **19A** | −1.2 |
| **15A** | −1.5 | **19B** | −1.0 |
| **16A** | −1.3 | **20A** | −1.0 |
| **17A** | −1.2 | **20B** | −1.1 |

$Nb_{18}$ cage become less effective. This fact likely results in a greater stabilization of the pentadecamers, as compared to other species. In addition, we find an electron density flow from the outside Nb atoms to the encapsulated one as the net charges of the latter moiety are consistently negative (see Table 10). The incomplete *d* orbitals of the central atom thus basically play as a global electron acceptor. As expected, the electron transfer in **15A** is much more effective as compared to the similar procedure in other species. Therefore, it can be suggested that the greater the charge involved the more stabilized the cluster is.

## 5 Concluding Remarks

We report a comprehensive review on the structures and related spectroscopic properties of niobium clusters in the range from two to twenty atoms, in three different charged states. We can now draw some general trends for the behaviors of $Nb_n$ clusters:

(i) In some cases, we observed several lower-lying isomers for a certain system that are very close in energy, in such a way that DFT energy computations cannot clearly establish their ground electronic states. The current computed results confirm the existence of structural isomers with comparable energy content for species with $n = 6, 9, 13, 18, 19$, and 20 in both neutral and charged states.

(ii) As in many other transition metal clusters, niobium clusters tend to prefer multi-coordinated geometries. The most stable trimeric form is thus triangular, not linear. Oligomers containing more than four atoms exhibit already nonplanar 3D shape in their most stable structures. In general, $Nb_n$ clusters invariably prefer the lowest possible spin state for their ground state, i.e. closed-shell singlet and doublet states for even- and odd-electron systems, respectively.

(iii) From a growth pattern point of view, we find that the optimal structure of the cluster at a certain size is not simply generated from that of the smaller one by adding an atom randomly. Instead, these clusters prefer a close-packed growth behavior. Systems larger than 12 atoms tend to form a compact structure with one Nb atom encapsulated by a cage constituted from five and six triangles. Unlike many 3*d* transition metals, whose volumes are much smaller, the clusters containing 13 and 19 Nb atoms do not exist as icosahedra and double-icosahedra. A distinct case is $Nb_{15}$ as it bears a slightly distorted solid **bcc** structure.

(iv) The computed vibrational (IR) spectra of the clusters present an interesting observation. While the spectra of smaller systems are found to be strongly dependent on the addition or removal of an electron from the neutral, those of the larger size clusters are likely to be independent on the charged states. The neutrals and their corresponding ions usually have a quite similar IR patterns. This also shows that the electronic effects become less important in medium and large size clusters. The atomic arrangement plays a more significant role than the electronic distribution in determining the stability of larger systems.

(v) Basic energetics including electron affinities, ionization energies, binding energies per atom, and stepwise dissociation energies are presented. In general, energetic properties of small sizes exhibit odd-even oscillations. Comparison with experimental data, both functionals BPW91 and M06 are reliable in predicting the EA and IE values. However, the BPW91 turns out to be deficient in predicting the binding energies per atom and dissociation energies in niobium clusters. The average absolute errors relative to experiment of BPW91 method are $\sim 0.6$ eV for $D_e$ and 0.5 for BE as compared to $\sim 0.2$ eV and $<0.1$ eV using the M06 functional.

(vi) Of the clusters considered, $Nb_2$, $Nb_4$, $Nb_8$ and $Nb_{10}$ are found to be magic as they hold the number of valence electrons corresponding to the closed-shell in the sequence [1S/1P/2S/1D/1F.....] of the phenomenological electron shell model. Owing to the valence electrons that satisfy the empirical $2(N + 1)^2$ rule, $Nb_{10}$ is observed to have a spherically aromatic character, high chemical hardness and large HOMO–LUMO gap. The open-shell $Nb_{15}$ system is also particularly stable as it can form a highly symmetric structure in all charged states. In addition, for species with an encapsulated Nb atom, an electron density flow is operative from the outside skeleton to the central atom, and the greater the charge involved the more stabilized the cluster is.

**Acknowledgments** PVN thanks Can Tho University for financial support. MTN is indebted to the KU Leuven Research Council for continuing support through GOA programs. Jerzy Leszczynski acknowledges the support of NSF CREST (HRD-0833178) and PREM (DMR-1205194) projects. We thank Dr. Andre Fielicke at Technical University of Berlin, Germany, for assistance with experimental IR spectra.

# References

1. Griffith WP, Morris PJT (2003) Notes Rec R Soc Lond 57:299
2. Wollaston WH (1811) Phil Trans R Soc Lond 101:96
3. Rose H (1845) Philos Mag 26:179
4. Blomstrand M, Deville H, Troost L, Hermann R (1866) Fresenius J Anal Chem 5:384
5. Rayner-Canham G, Zheng Z (2008) Found Chem 10:13
6. Peiniger M, Piel H (1985) Nucl Sci 32:3610
7. Patel Zh, Khulka K (2001) Metallurgist 45:477
8. De Heer WA (1993) Rev Mod Phys 65:611
9. Schmid G (2001) Adv Eng Mater 3:737
10. Knickelbein MB, Yang S (1990) J Chem Phys 93(1476):5760
11. Kietzmann H, Morenzin J, Bechthold PS, Ganteför G, Eberhardt W, Yang DS, Hackett PA, Fournier R, Pang T, Chen C (1996) Phys Rev Lett 77:4528
12. Kietzmann H, Morenzin J, Bechthold PS, Ganteför G, Eberhardt W (1998) J Chem Phys 109:2275
13. Wrigge G, Hoffmann MA, Von Issendorff B, Haberland J (2003) Eur Phys J D 24:23
14. Berces A, Hackett PA, Lian L, Mitchell SA, Rayner DM (1998) J Chem Phys 108:5476
15. Holmgren L, Andersson M, Rosen A (1995) Surf Sci 331:231
16. Lapoutre VJF, Haertelt MO, Meijer G, Fielicke A, Bakker JM (2013) J Chem Phys 139:121101
17. Berg C, Schindler T, Niedner-Schatteburg G, Bondybey VE (1995) J Chem Phys 102:4870
18. Vakhtin AB, Sugawara K (1999) Chem Phys Lett 299:553
19. Moro R, Xu X, Yin S, De Heer WA (2003) Science 300:1265
20. Knickelbein MB, Menezes WJC (1992) Phys Rev Lett 69:1046
21. Menezes WJC, Knickelbein MB (1993) J Chem Phys 98:1856
22. Majumdar D, Balasubramanian K (2001) J Chem Phys 115:885
23. Majumdar D, Balasubramanian K (2003) J Chem Phys 119:12866
24. Majumdar D, Balasubramanian K (2004) J Chem Phys 121:4014
25. Goodwin L, Salahub DR (1993) Phys Rev A 47R774
26. Gronbeck H, Rosen A, Andreoni W (1998) Phys Rev A 58:4630
27. Kumar V, Kawazoe Y (2002) Phys Rev B 65:125403
28. Fowler JE, Garcia A, Ugalde JM (1999) Phys Rev A 60:3058
29. Calaminici P, Mejia-Olvera R (2011) J Phys Chem C 115:11891
30. Fielicke A, Ratsch C, Helden GV, Meijer G (2005) J Chem Phys 122:091105
31. Fielicke A, Ratsch C, Helden GV, Meijer G (2007) J Chem Phys 127:234306
32. Haertelt M, Lapoutre VJF, Bakker JM, Redlich B, Harding DJ, Fielicke A, Meijer G (2011) J Phys Chem Lett 2:1720
33. Nhat PV, Ngan VT, Nguyen MT (2010) J Phys Chem C 114:13210
34. Nhat PV, Ngan VT, Tai TB, Nguyen MT (2011) J Phys Chem A 115:3523
35. Nhat PV, Nguyen MT (2012) J Phys Chem A 116:7405
36. Becke AD (1988) Phys Rev A 38:3098
37. Perdew JP, Burke K, Wang Y (1992) Phys Rev B 96:6796
38. Hay PJ, Wadt WR (1985) J Chem Phys 82:299
39. Peterson KA (2003) J Chem Phys 119:11099

40. Peterson KA, Figgen D, Goll E, Stoll H, Dolg M (2003) J Chem Phys. 119:11113
41. Peterson KA, Puzzarini C (2005) Theor Chem Acc 114:283
42. Jug K, Zimmermann B, Calaminizi P, Kçster A (2002) J Chem Phys 116:4497
43. Zhao Y, Truhlar DG (2008) Theor Chem Acc 120:215
44. James AM, Kowalczyk P, Fournier R, Simard B (1993) J Chem Phys 99:8504
45. Hales DA, Lian L, Armentrout PB (1990) Int J Mass Spectrom Ion Proc 102:269
46. Aydin M, Lombardi JR (2004) Int J Mass Spectrom 235:91
47. James AM, Kowalczyk P, Langlois E, Campbell E, Ogawa A, Simard B (1994) J Chem Phys 101:4485
48. Fournier R, Pang T, Chen C (1998) Phys Rev A 57:3683
49. Moskovits M, Limm W (1986) Ultramicroscopy 83:20
50. Loh SK, Lian L, Armentrout PB (1989) J Am Chem Soc 111:3167
51. Lombardi JR, Davis B (2002) Chem Rev 102:2431
52. Balasubramanian K, Zhu XL (2001) J Chem Phys 114:10375
53. Schultz NE, Zhao Y, Truhlar DG (2005) J Phys Chem A 109:4388
54. Aydin M, Lombardi JR (2009) J Phys Chem A 113:2809
55. Zhai HJ, Wang B, Huang X, Wang LS (2009) J Phys Chem A 113:3866
56. Wang H, Craig R, Haouari H, Liu Y, Lombardi JR, Lindsay DM (1996) J Chem Phys 105:5355
57. Walsh TR (2006) J Chem Phys 124:204317
58. Fielicke A, Meijer G (2011) J Phys Chem A 115:7869
59. Moro R, Yin S, Xu X, De Heer WA (2004) Phys Rev Lett 93:086803
60. Hamrick Y, Taylor S, Lemire GW, Fu ZW, Shui JC, Morse MD (1988) J Chem Phys 88:4095
61. Sun Y, Fournier R, Zhang M (2009) Phys Rev A 79:043202
62. Kumar V (2006) Comput Mater Sci 35:375
63. Nhat PV, Ngan VT, Tai TB, Nguyen MT (2011) J Phys Chem A 115:3523
64. King RB (1995) Inorg Chim Acta 235:111
65. Besley NA, Johnston RL, Stace AJ, Uppenbrink J (1995) Theor Chim Acta 341:75
66. Hearn JE, Johnston RL (1997) J Chem Phys 107:4674
67. Lloyd LD, Johnston RL (2000) J Chem Soc, Dalton Trans 3:307
68. Elkind JL, Weiss FD, Alford JM, Laaksonen RT, Smalley RE (1988) J Chem Phys 88:5215
69. Gruene P, Rayner DM, Redlich B, Van Der Meer AFG, Lyon JT, Meijer G, Fielicke A (2008) Science 321:674
70. Song W, Lu WC, Zang QJ, Wang CZ, Ho KM (2012) Int J Quant Chem 112:1717
71. Zhang H, Tian D, Zhao D (2008) J Chem Phys 129:114302
72. Janssens E, Neukermans S, Lievens P (2004) Curr Opin Solid State Mater Sci 8:185
73. Höltzl T, Veszpremi T, Lievens P, Nguyen MT (2011) In: Chattarij PK (Ed) Aromaticity and metal clusters. CRC Press, Boca Rota, FL, USA, Chapter 14:271
74. Geerlings P, De Proft F, Langenaeker W (2003) Chem Rev 103:1793
75. Rappoport D, Crawford NRM, Furche F, Burke K (2009) In: Solomon EI, King RB, Scott RA (ed) Computational inorganic and bioinorganic chemistry. Wiley, Chichester, USA
76. Gronbeck H, Rosen A (1996) Phys Rev B 54:0163
77. Sakurai M, Watanabe K, Sumiyama K, Suzuki K (1999) J Chem Phys 111:235
78. Berg C, Schindler T, Kantlehner M, Schatteburg GN, Bondybey VE (2000) Chem Phys 262:143
79. Pfeffer B, Jaberg S, Schatteburg GN (2009) J Chem Phys 131:194305
80. Wu Q, Yang S (1999) Int J Mass Spectrom 184:57
81. Knight WD, Clemenger K, de Heer WA, Saunders WA, Chou MY, Cohen ML (1984) Phys Rev Lett 52:2141
82. de Heer WA, Knight WD, Chou MY, Cohen ML (1987) Solid State Phys. 40:93
83. Alonso JA, March NH (1989) Electrons in metals and alloys. Academic Press, London
84. Tenderholt A (2005) PyMOlyze-2.0. Stanford University, Stanford
85. Hirsch A, Chen Z, Jiao H (2000) Angew Chem Int Ed 39:3915

# Submersion Kinetics of Ionized Impurities into Helium Droplets by Ring-Polymer Molecular Dynamics Simulations

F. Calvo

**Abstract** Alkali dopants interact with helium droplets very differently depending on their size and ionization state, leading to non-wetting behavior for small neutral impurities or to more homogeneous embedding in the case of ionized or large neutral systems. In the present contribution we examine by means of atomistic computer modeling the equilibrium state and out-of-equilibrium submersion kinetics of sodium atoms and dimers in helium clusters containing between 55 and 560 atoms in the normal fluid state, after ionization typically produced by appropriate laser excitation. Our modeling relies on the path-integral molecular dynamics framework, using simple but realistic pair potentials to describe all interactions. Ring-polymer molecular dynamics trajectories shed light onto the various stages of the submersion process, namely initial shell formation around the impurity followed by the slower sinking of this 'snowball' to the droplet center and accompanied by the evaporation of several helium atoms in the process. Characteristic times are evaluated as a function of impurity and cluster sizes.

**Keywords** Helium nanodroplets · Alkali impurities · Snowball formation · Solvent rearrangement · Atomistic modeling · Path-integral molecular dynamics

## 1 Introduction

Helium nanodroplets are a fascinating cryogenic medium of high chemical inertia, allowing high-resolution studies of the spectroscopy of molecular compounds ranging from atomic impurities [1–6] to clusters of fullerenes [7]. As a natural model for cold environments, helium droplets also provide a laboratory for reactivity and dynamics under extreme conditions such as those met in the interstellar

F. Calvo (✉)
University of Grenoble-Alpes and CNRS, LIPHy, Grenoble, France
e-mail: florent.calvo@univ-grenoble-alpes.fr



M.T. Nguyen and B. Kiran (eds.), *Clusters*, Challenges and Advances in Computational Chemistry and Physics 23,
DOI 10.1007/978-3-319-48918-6_4

medium [8]. Although droplets formed of $^3He$ or $^4He$ are naturally valuable for themselves, especially to address phenomena such as superfluidity at the nanoscale [9], their significance in chemical physics mainly arises when they are doped with a compound of interest difficult to obtain, stabilize or simply cool down in vacuum. Being chemically inert, helium atoms tend to bind very weakly to the impurities, but usually this interaction remains stronger than helium to itself owing to huge zero-point effects that considerably damp helium binding.

However, in some cases the electronic structure and small size of the dopant molecule give rise to interactions with helium that are even lower than helium to itself [10]. Such a situation occurs for small neutral alkali impurities, which have attracted a lot of attention from theoreticians and experimentalists who notably investigated the dynamics and relaxation following electronic excitation [11–18]. The importance of interactions has also been confirmed on the unexpected presence of high-spin states in such dopants [19–22], which despite being intrinsically higher in energy are formed preferentially under helium environment due to the lower energy released by the association process [6, 19]. Larger alkali clusters are expected to lie in their electronic ground state [21] and should be embedded in the droplet owing to favorable dispersion interactions with the surrounding medium [23], only separated by a vacuum bubble originating from the Pauli repulsion between the metal electron cloud and the helium fluid. For sodium, theoretical models [23] and experiments [24] have recently concurred to show that this size-induced surface to interior transition takes place near $n = 21$ atoms in the case of sodium [24].

Ionic impurities display quite different chemistry and physics, because the strong polarization interaction readily makes individual atoms much more attracted by helium [25]. Not only do charged impurities favor interior solvation, but the stronger binding tends to localize the helium atoms at least in the vicinity of the first solvation shell, giving rise to so-called "snowballs" in which helium behaves essentially classically or rigidlike but is surrounded by an outer fluid part [26–29], or simply to bubbles if the ions do not have a closed electronic shell as in the case of alkaline earth metals [30]. The snowball phenomenon caused by electrostriction was originally detected in bulk liquid helium [31], and has since been experimentally observed indirectly in mass spectrometry experiments on helium droplets [32]. In between neutral and ionic dopants, it is also worth mentioning the very unusual properties of high Rydberg states of some atoms such as rubidium, extending beyond the droplet boundaries and thus allowing the contribution of this dielectric medium on the excited states to be evaluated [33].

The specific dynamics of alkali dopants under helium environments and following electronic excitation has been studied in various situations covering exciplex formation and fluorescence relaxation [12, 34–37] or charge redistribution [38] as well as desorption [16, 17, 39–41] or diffusion [42] mechanisms. The processes at play subsequently to ionization triggered by photoabsorption [40, 43–45] or electron impact [46] have also received some attention, although generally for weakly bound atomic or molecular dopants. One interesting exception is the recent work by Buchta and coworkers who performed extreme UV excitations of helium droplets doped with rare gas or alkali atoms [47]. In this energy range corresponding to so-called Penning ionization, and similarly to the basic process at play in electron impact ionization, the helium droplet is ionized and the charge is transferred to the impurity over times that could be measured [47]. The solvation dynamics of rubidium atoms upon photoionization has been indirectly addressed experimentally by von Vangerow and coworkers [40] through pump-probe photoexcitation-photoionization spectroscopy, the delay between the two laser impulsions needed to desolvate the excited atom without producing a fully solvated cation being of the order of 1 ps.

Following their ionization, larger dopants may by simply ejected [6, 48] or even dissociate and exhibit different patterns when embedded in helium droplets with respect to the isolated case [46, 49, 50]. In particular, the possibility that the surrounding solvent may hinder fragmentation or promote specific channels, suggested already in the 1930s by Franck and Rabinowitch as the so-called "cage effect" [51], has been debated in the case of neon clusters [49, 50].

The multiple ionization of metal clusters in helium droplets has been investigated by pump-probe spectroscopy [52], the subsequent relaxation showing several neutral or ionic fragments travelling away from the droplet and possibly carrying with them snowballs of helium atoms. At even stronger excitation regimes, Coulomb explosion has been observed [53] and shown to be decelerated due to the presence of helium. The metal clusters addressed in these studies were sufficiently large to be naturally embedded at the center of the droplet prior to their ionization. Small neutral alkali dopants, which lie on the surface, should thus display rather different phenomenology upon ionization.

From the theoretical perspective, the equilibrium properties of helium clusters doped with neutral or ionic impurities are relatively well documented, especially in the case of sodium [15, 26–30, 37]. However, the kinetics following some electronic (or vibrational) excitation have received much less attention, simply because these processes involve electronic and nuclear degrees of freedom to be all treated on a quantum mechanical footing. Instead of searching for an excessively difficult solution of the Schrödinger equation, some groups have occasionally followed simplified treatments, generally based on semiclassical or even coarse-grained descriptions of the droplet medium and devoting most of the computational effort to the impurity. Semiclassical approaches have been used e.g. to describe nuclear motion of helium following the ionization-induced fragmentation of rare-gas clusters [50], or to model the electronic response to ultrashort pulses [53]. Density-functional theory, which has provided a wealth of information about the global structure of doped helium clusters [14], has also recently been combined with Bohmian dynamics [54] to model the desorption of excited metal impurities [16]. Very recently, this approach has been used to model the dynamics of snowball formation upon ionization, which was shown to proceed over typically 1 ps [55].

While all these efforts have brought invaluable insight into the response of the impurity to the excitation in presence of the helium solvent, the response of helium itself has not been scrutinized to the same extent, except in the aforementioned work by Leal and coworkers [55] but indirectly through continuum densities. This investigation was also limited to impurities no larger than a single atom. It is the purpose of the present contribution to examine this issue in more details, taking as an archetypal example the case of the post-ionization dynamics of small sodium impurities sitting initially on the side of the helium droplet. As the interactions suddenly change, the impurities should sink into the weakly bound cluster. The submersion process of alkali atoms has been studied experimentally by Ernst and coworkers [56], who were able to adjust the laser excitation in order to trigger ionization or simply desorption. These authors also showed that the ionization efficiency is influenced by the neighboring helium droplet [57]. However, so far the detailed dynamics of the newly formed ion has not been studied until very recently [40, 55].

The questions raised by these processes are fourfold. (i) Is the submersion process a one-step mechanism, or does the snowball form first and remains on the periphery of the droplet before sinking? (ii) How fast is the submersion process? (iii) Do molecular impurities submerse differently, given that they initially carry some excess kinetic energy owing to internal conversion? (iv) How many helium atoms are evaporated in the process, depending on the size of the initial droplet? The ring-polymer molecular dynamics (RPMD) framework [58] was chosen to tackle these issues and simulate the quantum dynamics of helium clusters doped by atomic or molecular sodium and undergoing sudden ionization. This approach is based on the path-integral formalism of Feynmann and Hibbs [59], adapted to the computation of quantum statistical properties at thermal equilibrium several decades ago [60, 61] but especially suited here to provide approximate quantum time correlation functions [58, 62].

Although the RPMD approach is conceptually powerful, it is computationally demanding due to the need to perform statistical numbers of trajectories. This constraint has lead us to develop dedicated potential energy surfaces in the pairwise approximation, but constructed on accurate ab initio reference data. In addition, we have neglected all exchange effects and the bosonic nature of the $^4$He atoms, therefore studying helium at a temperature of 1 K relevant to the normal fluid state. Our results suggest that the formation of the snowball indeed precedes the submersion by several picoseconds depending on the droplet size, and that this difference is a manifestation of the viscous hydrodynamics. The submersion process also carries some heat and this kinetic energy is converted into the evaporation of several helium atoms. In the case of the diatomic molecule, ionization is even more dramatic because of the additional heat released by internal conversion to the impurity first, and then to the host droplet.

In the next section, we describe the methods implemented to investigate and quantify the submersion kinetics of sodium impurities in $^4$He clusters, and in particular the protocol to simulate neutral and ionic complexes at and out of equilibrium. We also present the potential energy surfaces used in the modeling and the electronic structure calculations performed to calibrate these potentials. The results are presented and discussed in Sect. 3, before some concluding remarks and perspectives close the chapter in Sect. 4.

## 2 Methods

The methodology chosen in the present work [61, 63] provides a convenient framework to describe many-body quantum systems that is rigorous at equilibrium [63] and which benefits from the recent development of some dynamical extensions allowing time-dependent properties to be evaluated as well [58, 64]. The ability of the RPMD method to address dynamical problems has been demonstrated in various works dealing with spectroscopy [65–67] or transport [68, 69] properties and even reaction dynamics [70, 71]. In addition, Althorpe and coworkers [72, 73] recently showed that the RPMD transition state theory is closely related to the semiclassical instanton approximation in the deep tunnelling regime, further validating the method for modeling the quantum dynamics of many-body molecular systems.

Throughout this section we denote by $\mathbf{R}$ the set of Cartesian coordinates of the $Na_x^{(+)}He_n$ system, with $x = 0$–$2$ and $n$ ranging from 55 to 561, and by $V(\mathbf{R})$ the potential energy in the ground electronic state, explicited below in Sect. 2.3. The individual positions and atomic masses are denoted as $\mathbf{r}_i$ and $m_i$, respectively.

### 2.1 Path-Integral and Ring-Polymer Molecular Dynamics

It is not our purpose to present the path-integral MD methods in details, as recent reviews are available on this subject [74, 75]. We use the RPMD approach in its thermostatted and microcanonical versions to address equilibrium and kinetic properties, respectively. Common to these two versions is the replacement of classical pointlike particles by a number of $P$ monomers that form a polymer necklace. These monomers act as imaginary time slices along the thermal path in the path-integral theory [63], and can be seen as a practical realization of Trotter discretization of the exact continuous path. These monomers interact successively through effective harmonic bonds in such a way that the dynamics of the entire system is ruled by the effective potential energy

$$V_{\mathrm{eff}}(\{\mathbf{R}_\alpha\}) = \frac{1}{P}\sum_{\alpha=1}^{P} V(\mathbf{R}_\alpha) + \sum_{\alpha=1}^{P}\sum_{i\in\mathrm{atoms}} \frac{m_i P}{2\beta^2\hbar^2}\left\|\mathbf{r}_{\alpha,i} - \mathbf{r}_{\alpha,i+1}\right\|^2, \quad (1)$$

where $\beta = 1/k_{\mathrm{B}}T$ with $k_{\mathrm{B}}$ and $\hbar$ the Boltzmann and reduced Planck constants, respectively. In Eq. (1) $\mathbf{r}_{\alpha,i}$ denotes the position of particle $i$ of replica $\alpha$ with the cyclic condition $\mathbf{r}_{P+1,i} = \mathbf{r}_{1,i}$. In conventional (thermostatted) path-integral molecular dynamics, each of the $P \times (n+x)$ particles with position $\mathbf{r}_{\alpha,i}$ is associated with a momentum $\mathbf{p}_{\alpha,i}$ and a mass $m_{\alpha,i}$ that can be freely adjusted in order to improve sampling efficiency. In the specific approach of ring-polymer MD, all masses are set to the physical values: $m_{\alpha,i} = m_i$ for all $\alpha = 1,\ldots,P$.

In practice, all variables are transformed into normal modes in order to decouple the harmonic part of the Hamiltonian, which in turns enables the use the reference system propagation algorithm (RESPA) [76] to accelerate the propagation of the trajectory. At equilibrium, it is necessary to ensure that all polymer variables are properly thermostatted at the temperature of interest, and in this purpose the particles were all coupled to individual Nosé-Hoover variables. Massively thermostatted PIMD simulations were thus performed for pure and doped helium clusters with neutral and ionic, atomic and diatomic impurities in order to characterize their equilibrium properties.

The submersion kinetics of impurities upon ionization was addressed by selecting randomly an initial phase space configuration sampled along the trajectories for the neutral dopant, ionizing it suddenly and evolving the equations of motion for the ion instead of the neutral and, also importantly, without thermostats. Here we assume that ionization was the consequence of a sufficiently intense laser excitation, the emission of the electron being transparent to nuclear motion. In the case of the diatomic impurity, care has to be taken that the ionized molecule carries some kinetic energy resulting from internal conversion of the electronic excitation energy into vibrational energy on the potential energy surface of the cation. A realistic excitation is provided by three 627 nm photons [77], close to one main absorption band of $Na_2$ (Ref. [78]) and leaving the dimer in its $^2\Sigma_g^+$ ground electronic state but with an excess energy of about 1 eV. We model this excitation assuming immediate internal conversion, the excess energy being described by an equivalent kinetic energy along the molecular axis.

All path-integral simulations were carried out at $T = 1$ K, using a Trotter discretization into $P = 64$ beads and a time step of 1 fs. At equilibrium, the simulations also employed a soft-wall container in order to prevent evaporation especially in the case of the barely bound neutral sodium impurities:

$$V_{\rm wall}(\mathbf{R}) = \sum_{i \in \rm atoms} \frac{1}{4}\kappa(r_i - R_{\rm wall})^4, \tag{2}$$

where $r_i$ is the distance of atom $i$ to the center of the (laboratory) reference frame, $R_{\rm wall}$ is the radius of the container, and $\kappa = 10^{-4}$ eV/Å$^4$. The container radii were taken from the starting geometries, adding 15 Å in order to allow for sufficient expansion upon quantum delocalization. This container was not used during the submersion simulations in order not to artificially alter the energy relaxation mechanisms away from equilibrium. The equilibrium trajectories were propagated for 1 ns, 500 phase space configurations being periodically saved along the way for further propagation without thermostats.

## 2.2 Observables

Some natural properties to consider before addressing the submersion dynamics are at equilibrium. In view of the different chemical interactions between neutral and ionic dopants, it seems relevant to determine the extent to which some global properties could be affected by the presence and nature of the dopant.

As far as geometric quantities are concerned, the radial distribution functions of both helium and sodium atoms relative to the center of mass of the reference frame should provide the obvious information needed to characterize the relative positions of the impurities. Another property considered in the present work is the viscosity, which we expect to differ in presence of a strongly binding impurity such as alkali cations. We have evaluated the shear viscosity coefficient $\eta_S$ of the various helium clusters from the RPMD trajectories, using the appropriate Green-Kubo relation [79] which involves the time autocorrelation function of the off-diagonal elements of the microscopic stress tensor $\sigma_{\gamma\nu}$, namely [79]

$$\eta_S = \frac{1}{\mathscr{V} k_B T} \int_0^\infty \langle \sigma_{\gamma\nu}(t) \sigma_{\gamma\nu}(0) \rangle dt, \qquad (3)$$

$\mathscr{V}$ denoting the volume of the system, $\gamma$ and $\nu \neq \gamma$ two arbitrary components among the three Cartesian variables, and the average $\langle \cdot \rangle$ being taken over initial conditions. The volume was calculated by integrating the radial distribution functions, and the stress tensor determined from its definition [80, 81],

$$\sigma_{\gamma\nu}(t) = \sum_{i \in \text{atoms}} \frac{p_i^{(\gamma)} p^{(\nu)}}{m_i} + \frac{1}{2} \sum_{i,j \neq i \in \text{atoms}} \left[ r_i^{(\gamma)} - r_j^{(\gamma)} \right] \times \left[ f_i^{(\nu)} - f_j^{(\nu)} \right], \qquad (4)$$

where $r_i^{(\gamma)}$ and $p_i^{(\gamma)}$ designate the $\gamma$ component of the position and linear momentum of particle $i$, $f_i^{(\gamma)}$ being the corresponding physical force exerted on it. The above expressions are valid in the classical limit, the extension to RPMD is straightforward and requires replacing the classical expressions by the arithmetic average over the $P$ monomers [58].

When simulating the submersion kinetics, various local geometric properties were also calculated. The coordination number, or average number of helium atoms around the impurity, was evaluated based on simple distance criteria, one helium being declared as coordinated to a given sodium atom if its distance does not exceed a certain threshold $r_{\max} = 5$ Å, a value chosen to yield an approximate first solvation shell of the $Na^+$ ion equal to 12 and corresponding to icosahedral completion. For diatomic impurities, a global coordination was evaluated by counting the number of helium atoms close enough to either of the two alkali atoms. The distance of the impurity to the center of mass was also monitored in an attempt to probe the progressive submersion of the impurity as it becomes ionized, but this quantity turned out to be sometimes deceptive, because due to evaporation events and the conservation of linear momentum in the process the remaining large fragment connected to the impurity drifts away from the center of mass of the laboratory frame (vide infra). It was thus necessary to determine the spatial repartition of fragments on-the-fly, and recalculate the distance of the impurity but with respect to the center of mass of the main fragment to which the impurity belongs. Here fragments are defined accordingly with the Stillinger criterion [82], two atoms belonging to a same fragment if they can be successively connected by particles lying at most 10 Å from each other (this is a very unrestrictive definition). From this spatial repartition at the end of the trajectory, the number of evaporated atoms and the final cluster size were both accumulated. The simulations of submersion kinetics used 500 independent trajectories initiated from the thermostatted trajectories.

## 2.3 Potential Energy Curves

Although approximate, the RPMD method is computationally intensive due to the replication of the classical system into $P$ monomers. While path-integral methods have been used with explicit descriptions of electronic structure in the past [66, 83], those approaches are still far from routine to evaluate accurate kinetics and accumulate statistics for large systems. Considering the relatively simple chemistries of $Na_x^{(+)}He_n$ clusters for $x \leq 2$, we have followed the simplest possible model for the potential energy with the pairwise approximation:

$$V(\mathbf{R}) = \sum_{i,j \in \text{atoms}, i \neq j} V_{ij}(r_{ij}), \tag{5}$$

where $r_{ij}$ denotes the distance between particles at sites $i$ and $j$ and $V_{ij}$ the pair energy between those particles. For the present systems five pair potentials are required, describing respectively $He_2$, $Na_2$, $Na_2^+$, NaHe and $Na^+He$. Computational efficiency was further optimized by adopting very simple but chemically suitable forms for these potentials, namely of the Lennard-Jones (LJ) and Morse (M) forms possibly completed by a polarization contribution:

$$\begin{aligned} V_{\text{LJn}}(r) &= 4\varepsilon\left[\left(\frac{\sigma}{r}\right)^{12} - \left(\frac{\sigma}{r}\right)^{6}\right] \\ V_{\text{LJq}}(r) &= V_{\text{LJn}}(r) - \frac{\alpha_{\text{He}} q^2}{2r^4} \\ V_{\text{M}}(r) &= De^{-\rho(r-r_0)}[e^{-\rho(r-r_0)} - 2] \end{aligned} \tag{6}$$

The LJ form between neutral atoms (LJn) was adopted for $He_2$ and NaHe, while the Morse form is more appropriate for the two alkali dimers $Na_2$ and $Na_2^+$.

The interaction between helium and charged sodium ions, denoted as LJq, also accounts for the polarization of the helium atom with atomic polarizability $\alpha_{He} = 5.08$ Å$^3$ due to the charge $q$ taken as +1 for $Na^+$ and +1/2 on each atom of $Na_2^+$.

The various parameters of these potentials were fitted to reproduce available data for $He_2$, or potential energy curves obtained from dedicated *ab initio* all-electron coupled-cluster calculations at the level of single, double, and perturbative triple excitations and the aug-ccpv5z basis set. These parameters are given in Table 1, and the corresponding potential energy curves are represented in Fig. 1 for the four molecules involving sodium. As judged from this figure, the *ab initio* energy curves are very satisfactorily represented by the above forms, which confers our computational approach a first-principles character. Our calculated energy curves also agree with available results for $Na_2$ ($X^1\Sigma_g$ state, Ref. [84]) and $Na_2^+$ ($X^2\Sigma_g$ state, Ref. [85]). The even lower binding energy of neutral sodium to helium with respect to helium to itself (by a sevenfold factor) is in good quantitative agreement with recent CASSCF calculations for this $X^2\Sigma$ ground electronic state [37]. In view of these results, the neutral impurities can indeed be expected to lie on the outer

**Table 1** Values of the different parameters of the pairwise interactions fitted to reproduce reference potential energy curves

| Molecule | Type | Distance parameter $\sigma$, $r_0$ (Å) | Energy parameter $\varepsilon$, $D$ (cm $^{-1}$) | Range $\rho$ (Å $^{-1}$) |
|---|---|---|---|---|
| $He_2$ | LJn | 2.64 | 7.471 | |
| NaHe | LJn | 5.99 | 1.095 | |
| $Na^+He$ | LJq | 2.78 | 4.982 | |
| $Na_2$ | M | 3.07 | 5915 | 0.917 |
| $Na_2^+$ | M | 3.66 | 7945 | 0.575 |

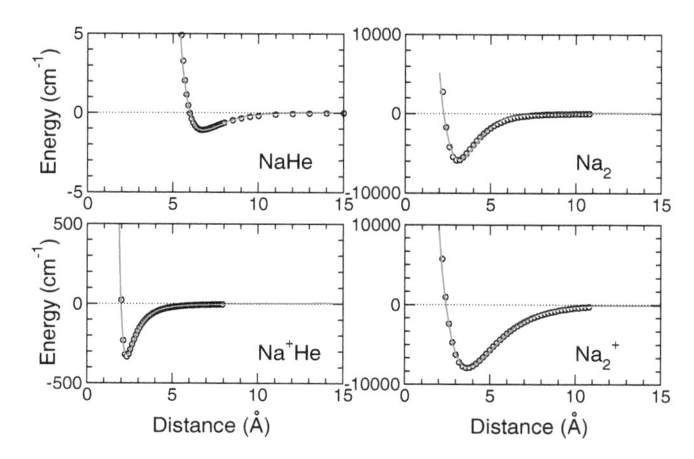


**Fig. 1** Potential energy curves calculated for the ground electronic states of NaHe, $Na^+He$, $Na_2$ and $Na_2^+$ at the CCSD(T) level of theory (*black circles*), and fitted forms based on various expressions given in the text (*red lines*)

surface of the helium clusters, as shown experimentally [11–17, 40]. In contrast, ionized dopants should definitely submerge into the droplet due to their stronger binding, by more than two orders of magnitude with respect to the neutral species.

# 3 Results and Discussion

## 3.1 Equilibrium Properties

The radial distribution functions are the most natural quantity to consider to characterize the overall structure of $^4$He clusters doped with impurities. We show in Fig. 2 these properties for the specific examples of $Na_2He_{559}$ and $Na_2^+He_{559}$.

The neutral cluster exhibits a rather flat helium density near the center, as typical for a fluid, and an impurity clearly localized on the surface, in agreement with aforementioned energetic arguments. The rather broad distribution found for the sodium dimer further indicates that the molecule undergoes some significant motion at the cluster surface, and inspection of the PIMD trajectories reveals some ongoing rotational motion. The same behavior is found for all helium cluster sizes. In the case of the neutral atomic impurity, similar radial densities are found with the alkali sitting on a dimple at the surface of the droplet.

The ionized cluster shows drastically different radial densities, with the impurity localized near the center and clear shells of helium around it extending over the entire volume. The sodium peak is much narrower than in the neutral cluster, which suggests stronger localization. The presence of several shells around the cation was also expected as a manifestation of snowball formation. From these equilibrium

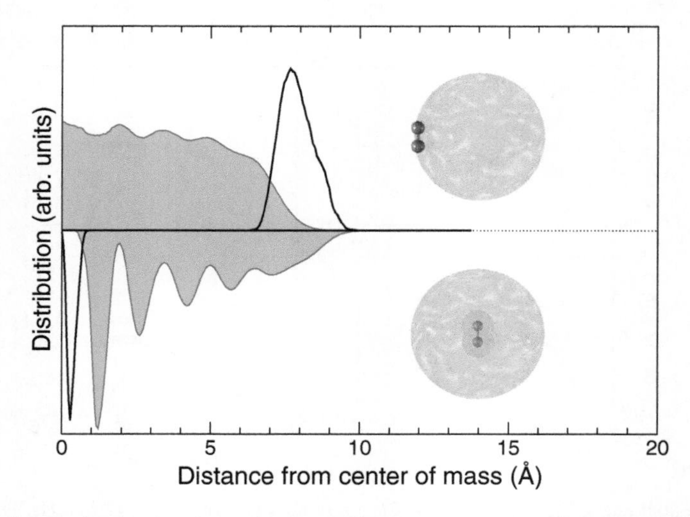


**Fig. 2** Radial distribution functions of helium (*filled blue lines*) and sodium (*black lines*) atoms in $Na_2He_{559}$ (*upper panel*) and $Na_2^+He_{559}$ (*lower panel*)

results, the preference of neutral (charged) impurities for outer (interior) locations is clearly established, which legitimates our potential energy surfaces and the general modeling. Using the radial distribution functions, a global density can be evaluated as 0.19 g/L, and the volumes of the clusters are found approximately as 1800, 4600, 10500, and 19300 Å$^3$ for the 55-, 147-, 309-, and 561-atom systems, respectively. Those volumes are necessary ingredients for the determination of absolute viscosity coefficients through the Green-Kubo formulas, Eq. (3).

From the equilibrium PIMD simulations performed at 1 K, ring-polymer MD trajectories were carried out and propagated for 50 ps each and the time variations of the stress tensor were recorded. The time autocorrelation function of this quantity, averaged over the three directions *xy*, *xz* and *yz*, is represented in Fig. 3 for the three clusters $He_{561}$, $NaHe_{560}$ and $Na^+He_{560}$. These autocorrelation functions quickly decay beyond a few picoseconds, and show very comparable values for the two neutral systems, indicating the limited influence of the external impurity. The corresponding function for the ionic impurity reaches significantly higher values and decays a bit faster. Integrating these quantities eventually yields shear viscosity coefficients that are represented in the inset of Fig. 3 as a function of cluster size and impurity. The values obtained for $\eta_S$ are of the same magnitude as measurements of viscosity coefficients of helium in the normal fluid phase [86].

The case of the pure helium clusters is easier to discuss first. As clusters grow, their viscosity increases and reaches about 15 $\mu$P at $n = 561$. Finite size effects are monotonic, which is the anticipated result for liquidlike systems with no defined geometric or shell structure. That the viscosity is lower in smaller clusters is also expected, owing to the greater proportion of exterior atoms at the liquid-vapor

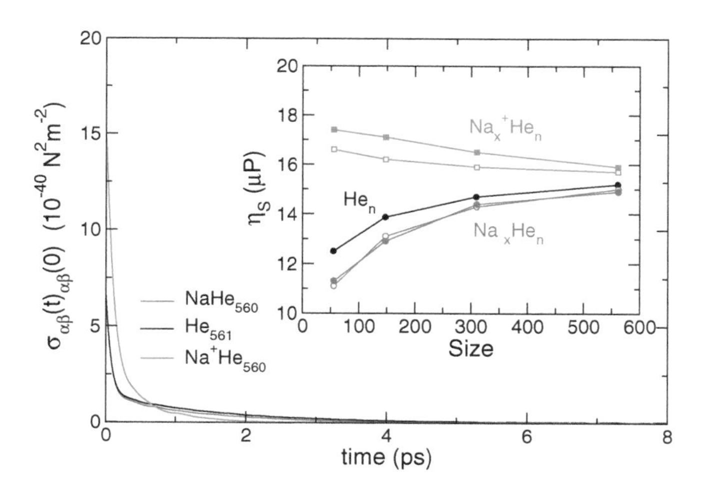


**Fig. 3** Off-diagonal stress tensor time autocorrelation function of $He_{561}$, $NaHe_{560}$ and $Na^+He_{560}$. The *inset* shows the variations of the shear viscosity coefficient $\eta_S$ as a function of cluster size for pure and doped helium clusters. All empty symbols refer to diatomic impurities, whereas full symbols refer to atomic impurities or no impurity at all

interface. Replacing one or two neutral helium atoms by sodium, the viscosity of doped neutral clusters is even lower especially at small sizes, with no major effect of the impurity being a molecule or an atom. This lower viscosity originates from the lower binding of the impurity, which tends to stay on the cluster surface and often detach from it at 1 K, before being bound again after reflection on the soft wall container. Ionic impurities tend to generally increase the viscosity coefficient especially in small clusters. This result is consistent with the rigidlike behavior of a proportion of atoms in the first solvation shell around the impurity, and with the increasing proportion of such atoms as size decreases. For those charged impurities the molecular dopant leads to a slightly smaller viscosity, which we interpret as the consequence of the particularly stable icosahedral snowball shell around the $Na^+$ dopant. In contrast, the solvation shell of the $Na_2{}^+$ molecule is not as fully localized as the magic icosahedron and produces a slightly less viscous fluid.

The viscosity coefficients converge rather slowly to a common bulk value, and in absence of any data for larger droplets it is not obvious to extract scaling laws based on the present simulation data. We have also tried to evaluate the bulk viscosity coefficient $\eta_B$ from the diagonal stress tensor time autocorrelation function, and generally found $\eta_S/\eta_B$ ratios of the order 2–3, in contrast with classical fluids where this ratio does not exceed 1. However, this is consistent with the results of Yonetani and Kinugawa [87] who performed centroid molecular dynamics simulations [62] of para-$H_2$ and found ratios close to 5 for this quantum system.

### 3.2 Submersion Kinetics

Having validated the path-integral methodology for treating $^4$He clusters doped with alkali impurities in the normal fluid equilibrium conditions, we turn to the out-of-equilibrium situation of a neutral dopant undergoing sudden ionization at time $t = 0$. For the molecular impurity, this ionization is accompanied by some heating of the resulting molecule due to internal conversion, the excitation being converted in purely vibrational (kinetic) energy along the molecular axis.

Figure 4 shows the time evolution of the coordination number around the atomic and molecular impurities, averaged over 500 independent trajectories, and for the four cluster sizes. In all cases, the coordination number increases very rapidly as the impurity is pulled toward the droplet, and increases further but more slowly along the subsequent tens of picoseconds. The first stage represents the initial fast coating around the impurity, and the time scale of a typical picosecond is consistent with the results of van Vongerow and coworkers [40], but also more directly with the time-density functional calculations from the Pi group [55].

The subsequent slower events are associated with the more progressive relaxation to the new equilibrium state corresponding to the transition to isotropic solvation and immersion of the snowball to the center of the droplet. For atomic impurities in small clusters solvation appears as a single-stage process, with the

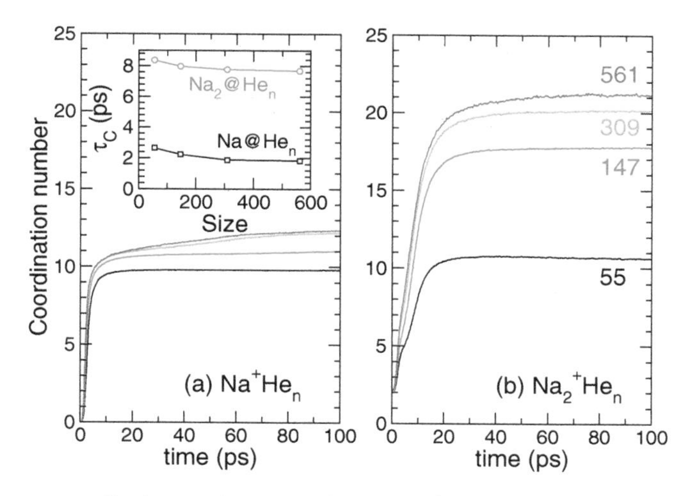


**Fig. 4** Average coordination number of **a** $Na^+$ and **b** $Na_2^+$ as a function of time after ionization, for impurities embedded in $^4He$ clusters of total sizes 55, 147, 309 and 561 atoms. The *inset* in the *left panel* shows the variations of the characteristic solvation time $\tau_C$ for the two impurities as a function of cluster size

snowball not fully reaching the expected icosahedral limit of 12. As seen below, this is a consequence of significant evaporation occurring as the ion is pulled to the center, thereby producing excessively large rearrangements in the solvent structure concomitant with several atoms being thermally released in the process. In larger clusters, the slower growth of the solvation shell in the second stage is understood as resulting from the sinking of the partially solvated impurity toward the droplet center. As this sinking process goes on, the solvation shell rearranges and can occasionally accept one or two more atoms, eventually filling the icosahedral shell.

The submersion process upon ionization of sodium dimers proceeds in a similar way but the variations of the coordination number are markedly slower at short times, which we interpret as the longer time needed for the more numerous helium atoms to initially cover the $Na_2^+$ molecule. In the smallest 55-atom cluster the solvation shell barely exceeds the shell obtained for the atomic impurity. This is again a consequence of the many evaporations that compete with solvent rearrangement. The snowball size grows in larger droplets but is not yet converged in the 561-atom cluster, suggesting that its full completion to at least 21 atoms requires more than 100 ps.

From these plots some characteristic solvation time scales can be estimated from the time where the variations of the coordination number mark an inflection, which approximately also represents half the time needed to reach the second stage where the filling of the shell becomes much slower. These solvation times $\tau_C$ are represented in Fig. 4 as a function of cluster size, for the atomic and molecular impurities. Although the time scales are of the same magnitude for the two types of impurities and for all cluster sizes, the solvation times are about four times longer for the alkali dimer. It would be interesting to determine how this factor further increases in larger impurities, although no simple law is expected due to the spreading of the charge on the various atoms, making the individual helium-metal interactions globally decrease with increasing impurity size.

For both atomic and molecular impurities the solvation times exhibit some minor decrease with increasing host cluster size, the convergence to large droplets being reached already at a few hundred atoms. That the solvation time is slightly longer in smaller helium clusters is interpreted as the consequence of a relatively stronger perturbation caused by ionization, which heats the system up, leading to evaporations that take the system further away from equilibrium and slows down the stabilization of the snowball.

Figure 5 shows the instantaneous distance between the impurity (or its center of mass) and the center of the laboratory frame, and its distance to the main cluster center of mass, as a function of time after ionization and in the representative cases of $Na^+He_{146}$ and $Na_2{}^+He_{145}$. These results, which were again averaged over 500 trajectories, illustrate another important aspect of the submersion kinetics, namely the sinking of the snowball to the center of the helium droplet. The two distances initially coincide, but soon diverge from each other after a few picoseconds. They exhibit some fast decrease at short times, concomitantly with the fast solvation that pulls the ionic impurity and the immediate helium neighbors closer together to form the snowball. However, at longer times the distance to the laboratory center of frame reaches a minimum and steadily increases after this moment. In contrast, the distance to the main cluster keeps decreasing but at smaller rates. The steady increase in the distance to the laboratory frame is a signature that the impurity and the fragment it belongs to drift away from each other, a recoil phenomenon caused by the motion of evaporating atoms. The greater slope in the case of the diatomic impurity indicates a stronger recoil motion, consistently with the greater amount of energy dissipated by evaporation by this heated molecule with respect to the atomic dopant.

These variations of the distance to the laboratory center of mass do not give direct insight into the sinking process, hence it is necessary to consider the largest fragment attached to the impurity instead. By monitoring this fragment on-the-fly, the impurity is found to reach the central region over a time of a few tens of picoseconds. Here we arbitrarily define the submersion time $\tau_S$ as the duration needed for the impurity to become close to its droplet center of mass by less than 2 Å. Averaging over trajectories and repeating the evaluation for the two types of impurities yields values for $\tau_S$ that are represented in the inset of Fig. 5 as a function of initial cluster size before ionization. Generally the submersion times are much longer than the solvation times and also increase significantly with cluster size. However this is easily interpreted by noticing that the time needed to travel across the fluid droplet should scale linearly with the diameter, that is as $n^{1/3}$. The limited number of sizes covered in the present study precludes from asserting whether this scaling law is already satisfied for such small clusters. Yet it is remarkable that the molecular dopant sinks faster than the atomic ion. We

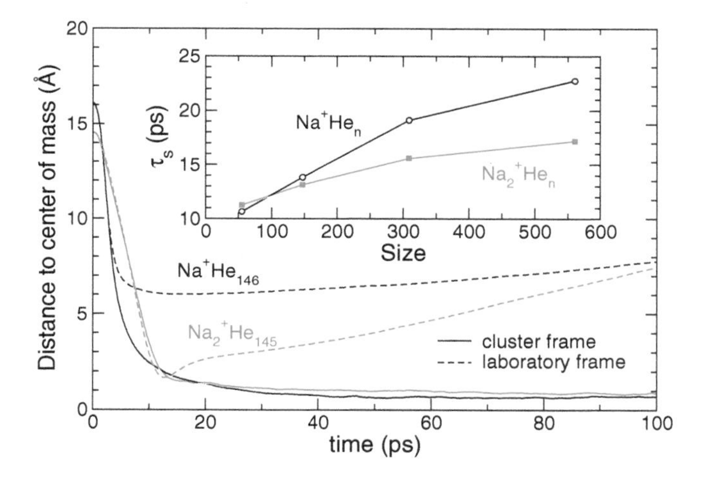


**Fig. 5** Average distance between the sodium impurity and the laboratory frame as a function of time after ionization for $Na^+He_{146}$ and $Na_2^+He_{145}$ (*solid lines*), and average distance to the main cluster center of frame (*dashed lines*). The *inset* shows the variations of the submersion time $\tau_S$ with increasing total cluster size for both impurities

conjecture that this result reflects the highly stable nature of the icosahedral shell solvating $Na^+$, whereas $Na_2^+$ is surrounded by a larger but not as stable shell of helium atoms. In order to clarify this issue it would definitely be valuable to extend the present investigation to larger impurities, starting of course with trimers.

The previous results provide indirect evidence for the release of atoms by evaporation through the recoil motion of the main droplet, however the number of atoms having effectively dissociated 100 ps after ionization can be precisely enumerated, and the resulting histograms have been represented in Fig. 6 for all clusters considered. For both atomic and molecular dopants, the number of evaporations decreases significantly with increasing initial cluster size. This is of course expected, as larger clusters are able to accommodate better to a localized and approximately fixed excitation. However, even in very large clusters it is likely that the localized excitation would produce some dissociation events in the vicinity of the impurity due to the sudden change in interaction forces. Once the new equilibrium has been established, the excess energy would be dissipated by statistical evaporation but over characteristic times that exceed the nanosecond regime even barely touched upon here. The excess vibrational energy carried by the ionized molecule also conveys into further evaporations that can be estimated as 10–20 more than for the ionized atom, depending on host cluster size.

Trying now to extrapolate the present results to larger droplets, the two-step mechanism identified from the ring-polymer molecular dynamics simulations should remain, but with somewhat different kinetics. The initial solvation process takes only a few picoseconds to complete, and this time is probably realistic even in the case of a superfluid droplet, because superfluidity should precisely be

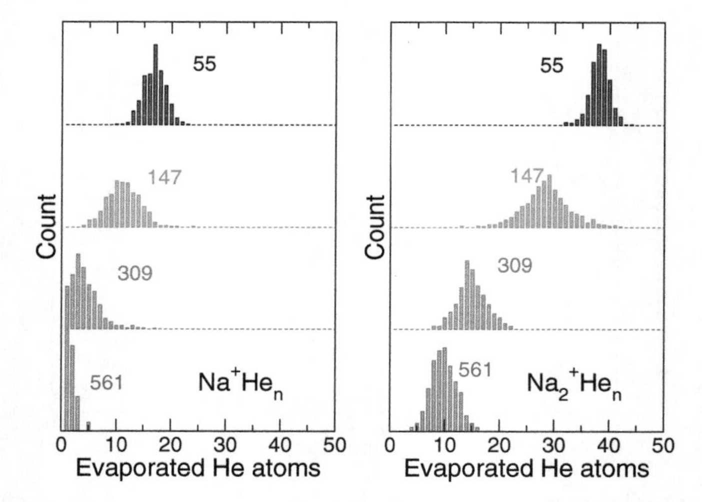


**Fig. 6** Distributions of the numbers of evaporated helium atoms 100 ps after ionization of sodium atoms or dimers, for different total cluster sizes

suppressed upon ionization, at least in the immediate vicinity of the impurity. While the submersion or sinking process is expected to remain as well, its kinetics is of course bound to the time needed to travel accross the entire radius of the droplet. In very large droplets, it is not clear whether the snowball would drift steadily to the center, or whether this motion would couple to some thermal diffusion that could slow it down. In the superfluid state the situation could be simpler, with thermal diffusion being attenuated and the snowball finding its way to the new equilibrium state more rapidly. Unfortunately, the methodology employed here is not practically suited to account for indistinguishability of particles [88] and coarse-grained approaches such as density-functional theory [4] focusing on the impurity and treating its dynamics, using e.g. Bohmian dynamics [16] could shed more light into the importance of superfluidity on the submersion process.

## 4 Concluding Remarks

Although helium droplets naturally constitute an ideal inert environment to isolate chemical compounds, they also provide a unique medium sensitive to the weakest interactions. Alkali dopants, which prefer to reside on the surface of helium droplets when they are neutral, undergo major changes in their potential energy surface when they are ionized. While conventional electronic excitation below the ionization threshold leads to the spontaneous desorption of the impurity [16, 18, 40, 56], ionization deeply modifies the chemistry of the system, which leads to a reorganization of the solvent. The present work was aimed to model the kinetics of those processes using a computational framework suitable to address the dynamics of many-body systems, taking into account the quantum mechanical nature of the nuclei through the path-integral formalism. Employing simple but chemically accurate pair potentials fitted to reproduce CCSD(T) quantum chemical calculations, the equilibrium properties of $^4He_n$ clusters in the range of 55–561 atoms and doped with neutral or cationic sodium impurities were addressed using thermostatted PIMD simulations, focusing on the normal fluid state at 1 K. Besides the general confirmation that neutral impurities reside outside the droplet, and that ionic impurities are surrounded by "snowballs" or shells of localized atoms, we have evaluated the shear viscosities of these droplets and shown that they were rather sensitive to the presence of the dopant and its nature.

The out-of-equilibrium processes occurring upon sudden ionization and, in the case of the dimer, assuming fast internal conversion of electronic excitation, were simulated also using ring-polymer MD, but without any thermostatting procedure. The submersion process was generally found to proceed stepwise, with the initial solvation of the impurity by most of the snowball occurring over a rather short time scale of a couple of picoseconds. This fast increase in the coordination number due to the sudden change in the interactions is manifested by a rapid pulling of the ion to the droplet and the motion of several helium atoms to form the solvation shell, under time scales of about 1 ps in rather good agreement with recent time-dependent density functional calculations [55]. The snowball thus formed subsequently sinks into the droplet but over a characteristic time closer to tens of picoseconds and with a greater dependence on the droplet size.

Several issues are naturally raised by the present work. As aforementioned, the importance of superfluidity of the helium host remains to be determined as it could notably contribute to accelerate the submersion process. The size of the dopant molecule was also explored only in the limit of atoms and diatoms, and in larger clusters the effects of ionization could be quite different. In particular, clusters exceeding several tens of atoms could be already large enough to spontaneously submerge into the droplet due to favorable collective van der Waals attraction forces [23]. While those theoretical predictions have recently received indirect experimental support [24], the role of the droplet size was not discussed, and it is rather clear that a sufficiently large helium cluster is needed in order that those dispersion interactions among the helium atoms or between the metal and rare-gas atoms balance each other. The methodology used here could be extended to tackle this issue at the atomistic level and confirm the size of 21 atoms at which this interior to surface transition was found to take place in the case of sodium [23, 24]. Following recent simulations by Leal and coworkers [89], the RPMD approach would also be suitable to shed light on other detailed processes involving alkali impurities interacting with helium droplets, such as their capture and the response of the solvent as a function of collision parameters.

## References


1. Goyal S, Schutt DL, Scoles G (1992) Phys Rev Lett 69:933
2. Kwon Y, Huang P, Patel MV, Blume D, Whaley KB (2000) J Chem Phys 113:6469
3. Toennies JP, Vilesov AF (2004) Angew Chem Int Ed 43:2522
4. Barranco M, Guardiola R, Hernández S, Mayol R, Navarro J, Pi M (2006) J Low Temp Phys 142:1
5. Choi MY, Douberly GE, Falconer TM, Lewis WK, Lindsay CM, Merrit JM, Stiles PL, Miller RE (2006) Int Rev Phys Chem 25:15
6. Tiggesbäumker J, Stienkemeier F (2007) Phys Chem Chem Phys 9:4748
7. Echt O, Kaiser A, Zöttl S, Mauracher A, Denifl S, Scheier P (2013) ChemPlusChem 78:910
8. Stienkemeier F, Lehmann KK (2006) J Phys B 39:R127
9. Hartmann M, Miller RE, Toennies JP, Vilesov A (1995) Phys Rev Lett 75:1556
10. Bellet D, Breckenridge WH (2002) Chem Rev 102:1595
11. Stienkemeier F, Higgins J, Callegari C, Kanorsky SI, Ernst WE, Scoles G (1996) Z Phys D: At Mol Clusters 38, 253
12. Reho J, Callegari C, Higgins J, Ernst WE, Lehmann KK, Scoles G (1997) Faraday Discuss 108:161
13. Bünermann O, Droppelmann G, Hernando A, Mayol R, Stienkemeier F (2007) J Phys Chem A 111:12684
14. Hernando A, Barranco M, Mayol R, Pi M, Ancilotto F, Bünermann O, Stienkemeier F (2010) J Low Temp Phys 158:105
15. Nakayama A, Yamashita K (2001) J Chem Phys 114:780
16. Hernando A, Barranco M, Pi M, Loginov E, Langlet M, Drabbels M (2012) Phys Chem Chem Phys 14:3996
17. Pifrader A, Allard O, Auböck G, Callegari C, Ernst WE, Huber R, Ancilotto F (2010) J Chem Phys 133:164502
18. von Vangerow J, Sieg A, Stienkemeier F, Mudrich M, Leal A, Mateo D, Fernando A, Barranco M, Pi M (2014) J Phys Chem A 118:6604
19. Higgins J, Callegari C, Reho J, Stienkemeier F, Ernst WE, Lehmann KK, Gutowski M, Scoles G (1998) J Phys Chem A 102:4952
20. Reho JH, Higgins J, Nooijen M, Lehmann KK, Scoles G, Gutowski M (2001) J Chem Phys 115:10265
21. Vongehr S, Kresin VV (2003) J Chem Phys 119:11124
22. Bünermann O, Stienkemeier F (2011) Eur Phys J D 61:645
23. Stark C, Kresin VV (2010) Phys Rev B 81:085401
24. An der Lan L, Bartl P, Leidlmair C, Schöbel H, Jochum R, Denifl S, Märk TD, Ellis AM, Scheier P (2011) J Chem Phys 135:044309
25. Grandinetti F (2004) Int J Mass Spectrom 237:243
26. Rossi M, Verona M, Galli DE, Reatto L (2004) Phys Rev B 69:212510
27. Di Paola C, Sebastianelli F, Bodo E, Baccarelli I, Gianturco FA, Yurtsever E (2005) J Chem Theory Comput 1:1045
28. Marinetti F, Coccia E, Bodo E, Gianturco FA, Yurtsever E, Yurtsever M, Yildirim E (2007) Theor Chem Acc 118:53
29. Issaoui N, Abdessalem K, Ghalla H, Yaghmour SJ, Calvo F, Oujia B (2014) J Chem Phys 141:174316
30. Galli DE, Ceperley DM, Reatto L (2011) J Phys Chem A 115:7300
31. Atkins KR (1959) Phys Rev 116:1339
32. An der Lan L, Bartl P, Leidlmair C, Jochum R, Denifl S, Echt O, Scheier P (2012) Chem Eur J 18:4411
33. Lackner F, Krois G, Koch M, Ernst WE (2012) J Phys Chem Lett 3:1404
34. Reho J, Higgins J, Lehmann KK, Scoles G (2000) J Chem Phys 113:9694
35. Pacheco AB, Thorndyke B, Reyes A, Micha DA (2007) J Chem Phys 127:244504

36. Schlesinger M, Mudrich M, Stienkemeier F, Strunz WT (2010) Chem Phys Lett 490:245
37. Dell'Angelo D, Guillon G, Viel A (2012) J Chem Phys 136:114308
38. Pentlehner D, Riechers R, Vdovin A, Pötzl GM, Slenczka A (2011) J Phys Chem A 115:7034
39. Loginov E, Drabbels M (2014) J Phys Chem A 118:2738
40. von Vangerow J, John O, Stienkemeier F, Mudrich M (2015) J Chem Phys 143:034302
41. Sleg A, von Vangerow J, Stienkemeier F, Dulieu O, Mudrich M (2016) J Phys Chem A 120:7641
42. Mateo D, Hernando A, Barranco M, Loginov E, Drabbels M, Pi M (2013) Phys Chem Chem Phys 15:18388
43. Zhang X, Drabbels M (2012) J Chem Phys 137:051102
44. Loginov E, Rossi D, Drabbels M (2005) Phys Rev Lett 95:163401
45. Kim JH, Peterka DS, Wang CC, Neumark DM (2006) J Chem Phys 124:214301
46. Callicoatt BE, Mar DD, Apkarian VA, Janda KC (1996) J Chem Phys 105:7872
47. Buchta D, Krishnan SR, Brauer NB, Drabbels M, O'Keeffe P, Devetta M, Di Fraia M, Callegari C, Richter R, Coreno M, Prince KC, Stienkemeier F, Moshammer R, Mudrich M (2013) J Phys Chem A 117:4394
48. Lindebner F, Kautsch A, Koch M, Ernst WE (2014) Int J Mass Spectrom 365–366:255
49. Ruchti T, Callicoatt BE, Janda KC (2000) Phys Chem Chem Phys 2:4075
50. Bonhommeau D, Lewerenz M, Halberstadt N (2008) J Chem Phys 128:054302
51. Franck J, Rabinowitch E (1934) Trans Faraday Soc 30:120
52. Döppner T, Diederich T, Göde S, Przystawik A, Tiggesbäumker J, Meiwes-Broer K-H (2007) J Chem Phys 126:244513
53. Döppner T, Fennel T, Diederich T, Tiggesbäumker J, Meiwes-Broer K-H (2005) Phys Rev Lett 94:13401
54. Wyatt RE (1999) J Chem Phys 111:4406
55. Leal A, Mateo D, Hernando A, Pi M, Barranco M, Ponti A, Cargnoni F, Drabbels M (2014) Phys Rev B 90:224518
56. Theisen M, Lackner F, Ernest WE (2011) J Chem Phys 135:074306
57. Theisen M, Lackner F, Krois G, Ernest WE (2011) J Phys Chem Lett 2:2778
58. Craig IR, Manolopoulos DE (2004) J Chem Phys 121:3368
59. Feynman RP, Hibbs A (1965) Quantum mechanics and path integrals. McGraw-Hill, New York
60. Doll JD, Coalson RD, Freeman DL (1985) Phys Rev Lett 55:1
61. Berne BJ, Thirumalai D (1986) Annu Rev Phys Chem 37:401
62. Cao J, Voth GA (1993) J Chem Phys 124:154103
63. Ceperley DM, Mitas L (1996) Adv Chem Phys 93:1
64. Braams BJ, Manolopoulos DE (2006) J Chem Phys 125:124105
65. Habershon S, Fanourgakis GS, Manolopoulos DE (2009) J Chem Phys 129:074501
66. Shiga M, Nakayama A (2008) Chem Phys Lett 451:175
67. Calvo F, Parneix P, Van-Oanh N-T (2010) J Chem Phys 132:124308
68. Habershon S, Markland TE, Manolopoulos DE (2009) J Chem Phys 131:024501
69. Calvo F, Costa D (2010) J Chem Theory Comput 6:508
70. Collepardo-Guevara R, Suleimanov YV, Manolopoulos DE (2009) J Chem Phys 130:174713
71. Menzeleev AR, Bell F, Miller TF III (2014) J Chem Phys 140:064103
72. Richardson JO, Althorpe SC (2009) J Chem Phys 131:214106
73. Hele TJH, Althorpe SC (2013) J Chem Phys 138:084108
74. Pérez A, Tuckerman ME, Müser MH (2009) J Chem Phys 130:184105
75. Habershon S, Manolopoulos DE, Markland TE, Miller TF III (2013) Ann Rev Phys Chem 64:387
76. Tuckerman ME, Berne BJ, Martyna GJ (1992) J Chem Phys 97:1990
77. Engel V, Baumert T, Meier Ch, Gerber G (1993) Z Phys D At Mol Clusters 28, 37
78. Fredrickson WR, Watson WW (1927) Phys Rev 30:429
79. Kubo R, Toda M, Hashitsume N (1992) Statistical physics II. Springer, Berlin
80. Zwanzig R (1965) Annu Rev Phys Chem 16:67
81. McQuarrie DA (1976) Statistical mechanics. Harper and Row, New York

82. Stillinger FH (1963) J Chem Phys 38:1486
83. Kaczmarek A, Shiga M, Marx D (2009) J Phys Chem 113:1985
84. Matsunaga N, Zavitstas AA (2004) J Chem Phys 120:5624
85. Cerjan CJ, Docken KK, Dalgarno A (1976) Chem Phys Lett 38:401
86. Borese van Groenou A, Poll JD, Delsing AMG, Gorter CJ (1956) Physica 22:905
87. Yonetani Y, Kinugawa K (2003) J Chem Phys 119:9651
88. Miura S, Okazaki S (2000) J Chem Phys 112:10116
89. Leal A, Mateo D, Hernando A, Pi M, Barranco M (2014) Phys Chem Chem Phys 16:23206

# Structure, Stability and Electron Counting Rules in Transition Metal Encapsulated Silicon and Germanium Clusters

**Prasenjit Sen**

**Abstract** Transition metal (TM) doped silicon and germanium clusters have been studied widely over the last two-three of decades. The initial motivation was to understand metal-semiconductor interfaces, and to couple magnetism in the TM atoms with the semiconducting properties of silicon. However, later on, the focus shifted to production of stable, TM encapsulated silicon or germanium cage clusters, possibly having magnetic moment. The most fundamental question the experiments threw up is, what are the most stable clusters in these series? And, how to understand their stability? Like other branches of physics and chemistry, simple electron counting rules, such as the 18-electron rule of inorganic and organometallic chemistry, and the idea of shell filling in a spherical electron gas, were invoked to explain stability of some of the clusters. However, different clusters were found to be most stable under different experimental conditions, making it somewhat unclear whose stability one has to explain. Faced with such challenges, a large number of experimental and theoretical studies were devoted to elucidating these issues. Quite early on, some authors (Sen and Mitas, e.g.) argued that electron counting rules may not explain all the observations in TM–Silicon clusters, though some of the observations can ostensibly be explained by such rules. The most recent works from Bandyopadhyaya and Sen, and Khanna and co-workers show that electron counting rules only have limited applicability to these clusters. A detail review of the literature discussing these issues is presented, and the still open questions are pointed out.

**Keywords** Transition metal-silicon clusters · Transition metal-germanium clusters · Cage clusters · Magic clusters · Ground state structure · 18-electron rule · Shell model

P. Sen (✉)
Harish-Chandra Research Institute, Chhatnag Road, Jhunsi, Allahabad 211019, India
e-mail: prasen@hri.res.in



M.T. Nguyen and B. Kiran (eds.), *Clusters*, Challenges and Advances in Computational Chemistry and Physics 23,
DOI 10.1007/978-3-319-48918-6_5

# 1 Introduction

Two seminal papers by Beck [1, 2] in the late 1980s may be taken as the starting point for the studies of doped metal–silicon clusters, though Beck's original motivation for studying these clusters was quite different from the motivation driving the later works post-1990. Silicon has been, and still remains, the most widely used material in electronics industry. It was known in the 1980s that the chemistry at the metal-semiconductor interfaces had a direct influence on the electrical properties of such contacts. Hence, understanding the metal–Si interface was crucial. However, the chemical processes at metal–Si interfaces, or the reaction products at these interfaces were still unknown [3, 4]. Two very different mechanisms were proposed, one by Tu [5] and the other by Hiraki [6], to explain how a metal film could react with a silicon substrate to form an alloy, or silicide. However, there was no direct experimental or theoretical proof of any of these. One of the technical difficulties was the large number of atoms involved at the interface without the simplifying aspect of symmetry of the bulk phases. The surface sensitive experimental techniques also produced properties that were statistical averages over many possible local geometries, and not any information about the local chemistry of metal–Si bonding.

In order to gain knowledge about metal–silicon bonding and chemistry, Beck produced and studied small binary clusters composed of a single transition metal (TM) atom of three different types, Cr, Mo and W with Si. Beck's experiments showed that incorporation of a single TM atom could stabilize Si clusters in the size range in which bare Si clusters were unstable under the experimental conditions. We will discuss the issue of stability of metal–Si clusters in detail later. What is important here is that the field of metal–Si clusters received renewed attention after the discovery of carbon fullerenes. Carbon formed stable fullerene cage structures which interacted weakly, and so could form stable molecular solids. This gave rise to the idea that if one could identify such stable, weakly interacting Si cage clusters, those could be used to assemble nano-scale materials whose electronic and optical properties could be tuned by choosing the right building blocks.

Search for small Si clusters as building blocks started in the 1990s. But it was soon realized that there were several difficulties in this approach. Firstly, elemental Si clusters have dangling bonds which make them highly reactive [7], and hence unsuitable as building blocks for cluster-assembled materials. Secondly, while C can have stable *sp*, $sp^2$ or $sp^3$ hybridization, Si only prefers *sp* hybridization [8, 9]. This rules out stable cage structures. In fact, ion mobility experiments revealed that small Si clusters form non-planar prolate structures rather than cage structures [10, 11]. These structures were also unlike bulk fragments [12].

In the meantime, Jackson and Nellemoe [13], using first-principles electronic structure calculations, showed that a $Si_{20}$ cage could be stabilized by encapsulating a Zr atom inside it. This generated new interest in Beck's experiments. Encapsulating a metal atom was seen as a possible way of generating stable Si cage clusters. It should be mentioned that at this time there was no evidence to suggest that the stable TM–Si clusters in Beck's experiments had the TM atoms encapsulated inside a Si cage.

The next breakthrough in the field of TM–Si clusters came through the experiments of Hiura et al. [14]. These authors produced $TMSi_n H_x^+$ cluster ions for TM = Hf, Ta, W, Re and Ir. Depending on the TM atom, at certain sizes, completely dehydrogenated clusters were formed indicating stability of those TM–Si clusters. In particular, $WSi_{12}$ was found to be a stable cluster. Through supporting ab initio calculations, these authors proposed that a W encapsulating $Si_{12}$ hexagonal prism (HP) cage is the lowest energy structure for this cluster. This work started a flurry of activities on metal–Si clusters in general, and TM–Si clusters in particular. Structural, electronic and magnetic properties of these clusters were studied both experimentally and theoretically. Attempts were made to rationalize the observed stability of TM–Si clusters at particular sizes in terms of various electron counting rules prevalent in chemistry, and in the field of metal clusters. These ideas generated a fascinating debate in the literature. Along with TM–Si clusters, TM-encapsulating Ge clusters have also been studied. Attempts have been made to understand their properties within the same theoretical framework as the TM–Si clusters.

In this chapter we will review the most interesting aspects of the structural and electronic properties of binary clusters composed of a single TM atom and a number of Si or Ge atoms. We will mostly focus on the 3d elements Sc–Ni, 4d elements Y–Ru, and the 5d elements Hf–Ir. Since experiments have been performed on Lu–doped clusters, it will come into our discussion briefly. We will particularly focus on the various electron counting rules that have been invoked to understand the relative stabilities of the TM encapsulated Si and Ge cage clusters at different sizes, i.e., different number of Si or Ge atoms. We will also try to analyze how well these rules explain the observed phenomena.

# 2 Transition Metal–Silicon Binary Clusters

## *2.1 Structure*

Like any other atomic cluster, knowledge of the structure of TM–Si clusters is crucial in understanding their properties. Many theoretical and experimental studies have attempted to find the structures of these clusters. As is well known, there are no direct experimental probes for structures of small atomic clusters. A synergy of theory and experiment is required. Theoretically, the most interesting and difficult question is what is the lowest energy structure of a given cluster. Experimentally, however, the most relevant question would be which isomers are produced in greatest abundance. While these are interesting and challenging questions, with our focus on TM encapsulating Si cage clusters, we will start with a different question here: what is the smallest Si cage in which a TM atom gets encapsulated? This question has been addressed both experimentally and theoretically, and the answer, of course, depends on the TM atom involved. We will first review the experimental efforts at determining whether the TM atom is attached exohedrally or endohedrally to the Si cluster. After that we will discuss the theoretical calculations trying to find the structures of TM–Si clusters. Although experiments and theory should go together, and they do in some of the works, there are also some differences between the two, as there are differences between different calculations. That is why we prefer to present them separately.

**Experimental Efforts**

Kaya and Nakajima group [15–17] studied adsorption reactivity of the TM–Si clusters towards $H_2O$ to determine the position of the metal atom. Water vapor was introduced as a reactant in the flow tube reactor (FTR) coupled to their dual laser vaporization cluster source by mixing it with the helium (He) carrier gas. A time of flight (TOF) mass spectrometer was used to measure the abundance of the clusters before and after the reactions. This way, the relative adsorption reactivity of the clusters could be measured.

Bare Si clusters are much less reactive towards $H_2O$ than metal atoms. This was demonstrated by Ohara et al. [18]. It was found that the abundance of bare Si clusters did not change after exposure to $H_2O$, but the abundance of Tb–Si clusters changed significantly after such exposure. Hence adsorption reactivity of a $TMSi_n$ cluster depends sensitively on the location of the metal atom. At small sizes, when the metal atom takes an exohedral position, the reactivity is high. It usually decreases with increasing number of Si atoms. Eventually a threshold size is reached beyond which the clusters are not reactive. Relative adsorption reactivity at a particular cluster size is defined as—$\ln(I^f/I^0)$, where $I^0$ and $I^f$ are the intensities of the cluster in the mass spectrum before and after the reaction. Nakajima and co-workers studied structures of a number of 3d, 4d and 5d TM atom-doped Si clusters in the neutral, and singly charged cation and anion states using this chemical probe method [17]. These include Sc, Ti, V, Y, Zr, Nb, Lu, Hf and Ta. Plots of relative adsorption reactivity of these clusters with varying number of Si atoms are shown in Figs. 1 and 2. The threshold size for each TM atom is indicated on the plots. Two features concerning the threshold sizes become clear from these plots: (i) For any particular TM atom the threshold size depends on the charge state of the cluster. The threshold size decreases from cation to neutral to the anion charge state. This can be seen from the sequence of panels (a)–(c); (d)–(f); and (g)–(i) in Fig. 1. Lu and Ta are exceptions to this rule. (ii) Threshold size also depends on the element in the same period. Within the same period, and the same charge state of the cluster, the threshold size decreases with increasing atomic number of the TM atom. This can be seen from the progression in the panels (a)–(d)–(g); (b)–(e)–(h) and (c)–(f)–(i) in Fig. 1. Within the same group and same charge state there is generally a slight increase in the threshold size with the atomic number. Thus, the threshold size increases from Ti to Zr and Hf in the neutral clusters, and from V to Nb and Ta in the cations as seen in the sequence of panels (d)–(e)–(f); and (g)–(h)–(i) in Fig. 2. In the anion clusters however, there is an increase from Sc to Y, but then there is a slight decrease in case of Lu. These findings

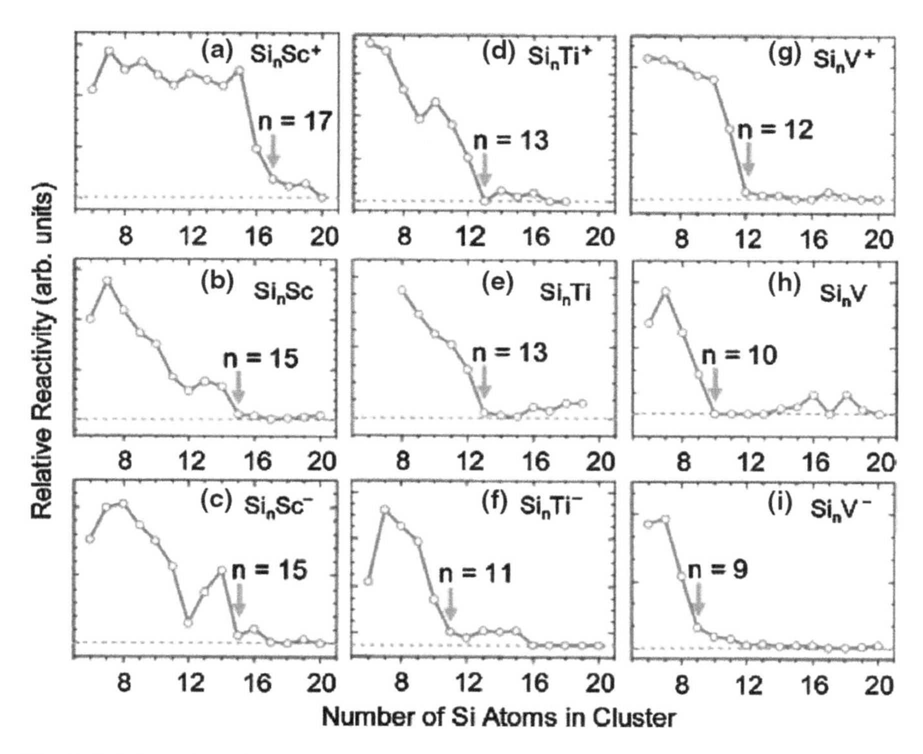


**Fig. 1** Relative adsorption reactivity of Sc-, Ti- and V-Si clusters towards $H_2O$ with varying size. The threshold size at which reactivity is completely lost is indicated in each panel. Reprinted with permission from J Phys. Chem. A 2007, 111, 42. © 2007 American Chemical Society

are consistent with the fact that the metallic bond radii of the TM atoms considered here decrease across a period with increasing atomic number. The metallic radii increase in a group from 3d to the 4d elements, with a slight decrease for the 5d elements in some cases. Another observation in support of the interpretation of the loss of reactivity due to complete encapsulation of the TM atom is that the reactivity is recovered at the threshold sizes and beyond when a second metal atom is present. Since a Si cage at the threshold size or just bigger can encapsulate only one metal atom, the second metal atom has to be located on the surface of the cluster, leading to enhanced reactivity. This is shown by the filled symbols in the panels in Fig. 2. We have listed the threshold sizes for the above TM atoms for different charge states of the clusters as reported in Ref. [17] in Table 1 for an easy reference. The metallic bond radii for the TM atoms are also given.

In their photodissociation experiments of metal–Si clusters, Jaeger et al. [19] found that $CrSi_7^+$ clusters primarily lost the metal atom but $CrSi_{15}^+$ and $CrSi_{16}^+$ lost Si atoms. This can be understood if the Cr atom is encapsulated in a Si cage at the

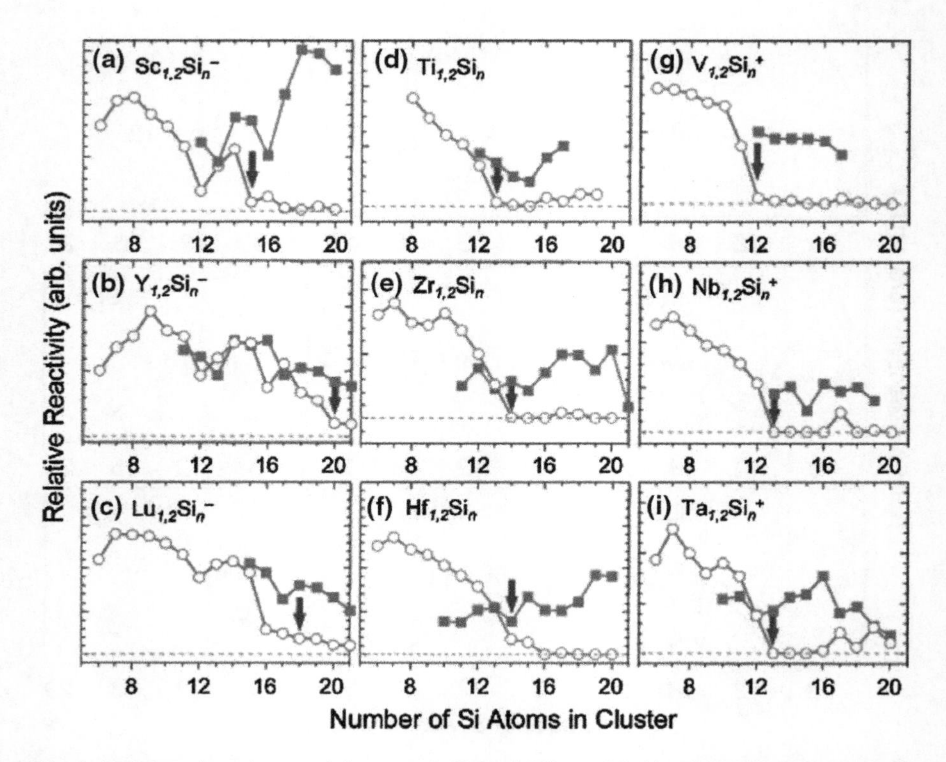


**Fig. 2** Relative adsorption reactivity towards $H_2O$ of different 3d, 4d and 5d TM–Si clusters with varying size. Open symbols are for clusters with a single TM atom, and filled symbols are for those with two TM atoms. The threshold size at which reactivity is completely lost for single TM–Si clusters is indicated in each panel. Reprinted with permission from J Phys. Chem. A 2007, 111, 42. © 2007 American Chemical Society

**Table 1** The metallic bond radii and threshold size for loss of adsorption reactivity of $TMSi_n$ clusters towards $H_2O$ in three different charge states for nine different TM atoms belonging to the 3d, 4d and 5d series

| TM atom | Metallic radius (Å) | Cation | Neutral | Anion |
|---|---|---|---|---|
| Sc | 1.63 | 17 | 15 | 15 |
| Ti | 1.45 | 13 | 13 | 11 |
| V | 1.31 | 12 | 10 | 9 |
| Y | 1.78 | 21 | 20 | 20 |
| Zr | 1.59 | 15 | 14 | 12 |
| Nb | 1.43 | 13 | 12 | 11 |
| Lu | 1.72 | 21 | 16 | 18 |
| Hf | 1.56 | 14 | 14 | 12 |
| Ta | 1.43 | 13 | 10 | 11 |

larger sizes. However, these authors could not determine the threshold size as only these three sizes were produced in enough abundance in their cluster source to perform dissociation experiments.

Lievens and co-workers used noble gas (NG) physisorption as a probe for endoherdally doped Si clusters [20]. The basic idea is as follows. Bare Si, and TM–Si clusters of different sizes were produced in a dual-target dual-laser vaporization source. Helium was used as a carrier gas with the source. When about 1 % of Ar was added to the He gas, the cation clusters formed complexes with one or two Ar atoms. The mass spectrum contained $Si_n^+$, $TMSi_n^+$, $(TM)_2Si_n^+$ clusters and $(TM)_{1,2}Si_n^+Ar_{1,2}$ complexes. Notably, no $Si_n^+{\cdot}Ar_{1,2}$ complexes were seen. This suggested that Ar atoms physisorb only with the metal atoms and not with the Si atoms. More polarizable Xe atoms, however, formed complexes with Si atoms. This fact was exploited by these authors in their infrared multiple photon dissociation (IR-MPD) experiments to identify low energy structures of TM–Si clusters. This is discussed later. In the present work [20], they measured the fraction of $(TM)_{1,2}Si_n^+{\cdot}Ar_{1,2}$ complexes at different sizes. This fraction fell sharply at a specific size for a given cluster depending on the TM atom. This threshold was taken to indicate the size at which the metal atom(s) was/were completely encapsulated by the Si cage. Experimental results from Ref. [20] is shown in Fig. 3. The fact that the threshold size is larger for clusters containing two metal atoms is consistent with the interpretation that it corresponds to their complete encapsulation inside the cage. From these measurements, the threshold size for Ti, V, Cr and Co were found to be 13, 12, 9 and 8 respectively. Perfect agreement with the findings of Koyasu et al. [17] for $TiSi_n^+$ and $VSi_n$ clusters discussed above gives confidence in both the results.

Lievens's group [21–24] extended the NG physisorption technique to the IR-MPD of clusters which, in conjunction with density functional theory (DFT) calculations, could yield information about structure of the cluster isomers produced in experiments. In IR-MPD experiments, the beam of cluster-NG complexes is overlapped with an intense infrared laser beam propagating in the opposite direction. The frequency of the infrared laser is varied over a large range. If the cluster happens to have an infrared active vibrational mode at a particular frequency, it can absorb one or more photons. The subsequent vibrational heating of the cluster can lead to detachment of the weakly bound NG atom. IR-MPD depletion spectrum (difference in intensity of cluster-NG complex before and after the laser pulse) is measured, from which absorption cross section of the IR radiation

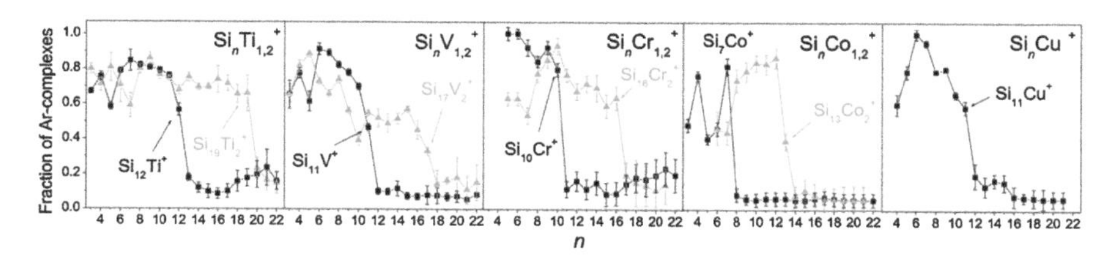


**Fig. 3** Fraction of argon complexes formed with different $TMSi_n$ clusters with varying n. Complexes for clusters with one or two TM atoms are studied. Reprinted figure with permission from E. Janssens, P. Gruene, G. Meijer, L. Wöste, P. Lievens and A. Fielicke, Phys. Rev. Let. 99, 063401 2007. © 2007 by the American Physical Society

at different frequencies can be extracted. This in effect gives the vibrational spectrum of the cluster. Vibrational spectra for a number of low energy isomers are calculated using DFT techniques, and the one reproducing the experimental spectrum most accurately is thought to have been produced in the cluster beam.

Using this technique Gruene et al. [21] found that $VSi_n^+$ clusters for n = 6–8 have a V atom attached exohedrally. In a subsequent work Claes et al. [22] found that a V atom is encapsulated inside a cage for n = 12–16. These observations are consistent with the chemical probe experiments. They were also able to identify the lowest energy isomers in most cases. The details can be found in the original papers. Ngan et al. [23] used the same method to identify the low energy structures for $MnSi_n^+$ clusters. They showed that for n = 6–10 the Mn atom is endohedrally attached. For n = 12–14 and 16, the Mn atom is encapsulated in a cage. They could not identify an isomer for $MnSi_{11}^+$ that explained the measured vibrational spectrum, and could not measure the spectrum for $MnSi_{15}^+$ due to its mass coincidence with $Mn_2Si_{13}^+$. Nevertheless, it is confirmed that $MnSi_{12}^+$ onwards, the Mn atom is encapsulated. A very similar study was reported by Claes et al. [24] which confirmed that for n = 6–9, V and Mn atoms are attached exohedrally in $VSi_n^+$ and $MnSi_n^+$ clusters.

Although the IR-MPD technique has been used for structure assignment of $TMSi_n$ clusters extensively, there are limitations that it suffers from. Firstly, it is based on the assumption that the structures and vibrational spectra of the clusters do not change even after absorption of a NG atom. Structure assignment crucially depends on this assumption. Secondly, neutral TM–Si clusters do not form complexes with NG atoms as efficiently as the charged clusters. Hence, it is difficult to use this method for structure assignment of neutral clusters. One requires further ionization of the cluster for mass analysis, adding to the complexity of the experimental process. To avoid these, Lievens's group employed infrared-ultra violet two color ionization (IR-UV2CI) process to assign structures to neutral $CoSi_n$ clusters [25]. In these experiments, the cluster beam was exposed to a counter-propagating tunable IR laser before being ionized by a 7.87 eV UV laser. The ionized clusters were mass analyzed by a reflectron TOF mass spectrometer. Exposure to IR laser prior to the ionizing UV laser increases the internal energy of the clusters, and increases their ionization efficiency. For example, in absence of the IR laser, signals for $Si_{10}$ and $CoSi_{10}$ clusters in the mass spectra were rather weak. But exposure to 430 $cm^{-1}$ IR laser increased their signals dramatically, indicating that these clusters have IR active vibrational modes at this wave number. The measured vibrational spectra were compared with the calculated spectra from DFT. It was concluded that $CoSi_n$ clusters for n = 10–12 (these are the sizes for which IR-UV2CI spectra could be recorded) have the Co atom encapsulated in a Si cage in all the isomers that were likely to have been produced in the experiments. Measured IR-UV2CI spectra and the calculated IR spectra of a few low-energy isomers of $CoSi_n$ clusters for n = 10–12 are shown in Fig. 4 for illustration.

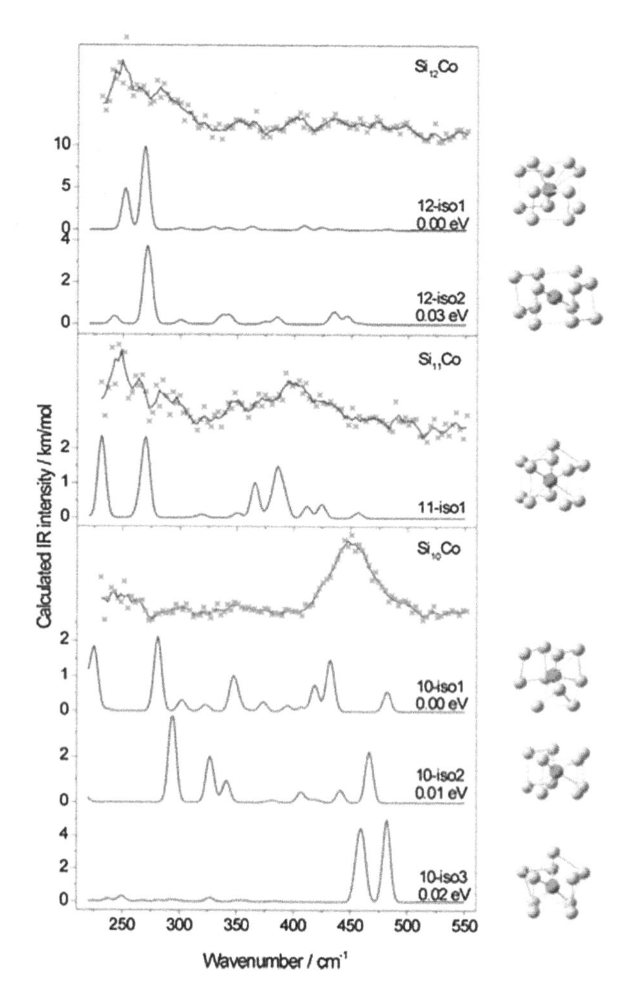


**Fig. 4** Measured IR-UV2CI spectra and the calculated IR spectra of a few low-energy isomers of $CoSi_n$ clusters (n = 10–12). Reprinted with permission from J Phys. Chem. A 118, 8198 2014. © 2014 by the American Physical Society

## Theoretical search for structures

There are a number of theoretical works on $TMSi_n$ clusters that have studied their structures. However, most of these studies are confined either to small range of sizes, or to a few TM atoms. There are very few studies exploring low energy structures of TM–Si clusters for many different TM atoms and a wide range of size. Also, very few studies have employed techniques for a global search for the lowest energy structures. Two works we came across that have used global search techniques are by Ma et al. [26] for Co–Si clusters, and by Li et al. [25] in synergy with their IR-UV2CI experiments, also for Co–Si clusters. Given this state of affairs, and our focus on TM encapsulated Si cage clusters, here we take a look at what is known theoretically about the structure of larger $TMSi_n$ clusters, in particular, the smallest size at which a TM atom is encapsulated in a Si cage for the first time. For clarity, we will present results on the 3d, and 4d and 5d TM encapsulated Si cage clusters in different subsections.

### 3d TM encapsulated $Si_n$ clusters

Guo et al. [27] and Gueorguiev et al. [28, 29] performed some of the most extensive studies for the structures of TM–Si clusters. Gueorguiev et al. [28, 29] studied structures of $TMSi_n$ clusters for n = 1–14 for a total of twenty-four 3d, 4d and 5d TM atoms. Guo et al. addressed the question of the smallest metal encapsulated cage structure for all the 3d TM atoms. The smallest Si cage encapsulating a Sc atom was found to be $Si_{14}$. Ti, V and Cr atoms were encapsulated by a $Si_{12}$ cage. Mn was encapsulated by a $Si_{11}$ cage, while the smallest cage encapsulating Fe, Co and Ni atoms was $Si_{10}$. Gueorguiev et al. reported a TM encapsulating cage structure for $TMSi_{10}$ clusters for TM = Ti, V, Fe, Co and Ni. Rest of the 3d TM atoms occupy endohedral positions starting at n = 12. In $FeSi_{10}$ and $CoSi_{10}$ clusters the TM atoms occupy a central position inside a Si cage which can be viewed as a two-layer structure: one quadrilateral, and one pentagon, with the pentagon capped by the last Si atom. This structure is different from the ones reported by Guo et al. Guo et al. also reported a two-layer Si cage for $CoSi_{10}$, but in this the quadrilateral face was capped rather than the pentagonal face. Moreover, for $FeSi_{10}$ they found a perfectly symmetric ($D_{5h}$) Si pentagonal prism encapsulating the Fe atom. Gueorguiev et al. [28] claims that the cage structures they found for $FeSi_{10}$ and $CoSi_{10}$ are energetically not very favorable. Therefore it seems that these are not the lowest energy structures they found. In fact, they do not make any clear statement about which are the lowest energy structures in their calculations. All the TM atoms were encapsulated by the time the cage size increased to $Si_{12}$ [28, 29]. In fact, they only reported a HP cage at this size, with distortions in some cases. For example, they found small Jahn-Teller distortions of the HP cage from a perfect $D_{6h}$ symmetry in $TiSi_{12}$, $VSi_{12}$ and $NiSi_{12}$.

The result for Sc found by Guo et al. does not match the experimental result obtained by the chemical probe method [17]. Moreover, both Sen and Mitas [30], and Reveles and Khanna [31] studied the $ScSi_{12}$ cluster in which the Sc is encapsulated in a hexagonal prism (HP) cage. Reveles et al. reported considerable distortion of the HP cage from a perfectly symmetric $D_{6h}$ symmetry. Sen et al. found a doublet spin state to be the lowest in energy while Reveles et al. found a quartet state. The reason for this difference is not clear. However, one may note that Sen et al. used the B3LYP functional for exchange-correlation energy in their calculations while Reveles et al. used the PBE functional. It is also not clear if this structure is the ground state in Sen et al. and Reveles et al's calculations. Sen et al. did not consider any other structure. Reveles et al. did consider other structures suggested previously in the literature, but how extensive that search was is not clear. No global search was performed for the lowest energy structure in any of these works.

For Ti also, Sen and Mitas [30], and Reveles and Khanna [31] studied structures in which the Ti atom is encapsulated in a $Si_{12}$ HP cage. Reveles et al. found a larger distortion of the cage in case of Ti compared to Sc. They also reported a singlet spin state for this cluster. Sen et al. found a triplet state to be lower in energy for $TiSi_{12}$ at all levels of theory. Most importantly, accurate quantum Monte Carlo (QMC) calculations produced a rather significant energy difference (0.7 eV) between the lowest energy triplet state and the singlet. Kawamura et al. [32] also studied growth behavior of $TiSi_n$ clusters. Interestingly, they found a basket-like structure to be most stable at n = 12, and found Ti-encapsulated structures for n = 13 and higher. What is intriguing is that they found a distorted HP cage structure to be 0.273 eV higher than the basket-like ground state at n = 12. Again, striking differences between different DFT calculations show that either the approximations involved in these calculations do not allow for an unambiguous identification of the lowest energy structure, or that in absence of a global search, the lowest energy structure has not been identified in some or all of these calculations.

In agreement with Guo et al., Andriotis et al. [33] found that a V atom is encapsulated by a slightly distorted HP $Si_{12}$ cage. Their attempt to encapsulate V in a $Si_{10}$ cage with either $D_{5d}$ or $D_{5h}$ symmetry was unsuccessful. Sen et al. [30] and Reveles et al. [31] also studied $VSi_{12}$ clusters with slightly distorted HP cage structure. This result of Andriotis et al. is in obvious disagreement with those of Gueorguiev et al. [29]. Khanna et al. [34], Sen et al. [30], Reveles et al. [31] and Abreu et al. [35] also found a Cr atom to be encapsulated by a HP $Si_{12}$ cage. In fact, Reveles et al. [31] reported that this cluster had a perfectly symmetric $D_{6h}$ structure. Kawamura et al. [36] also studied $CrSi_n$ clusters over a size range n = 8–17 and found that $Si_{12}$ is the smallest cage which can encapsulate a Cr atom. They also found that $CrSi_{12}$ has a HP cage structure. Li et al. [37] studied Mn doped Si clusters and found that the Mn atom gets encapsulated in the Si cage at n = 11. This is in agreement with Guo et al. [27].

Apart from Guo et al. [27] and Gueorguiev et al. [28], Khanna et al. [38] studied $FeSi_n$ clusters. They also found that $FeSi_{10}$ is the smallest cluster in which a Fe atom is encapsulated in a Si cage. Although the Fe atom occupies an interior position in $FeSi_9$, its encapsulation in a Si cage does not seem to be quite complete at this size (Fig. 1 in Ref. [38]). The ground state structure Khanna et al. reported is very similar to the structure reported by Gueorguiev et al. [28]. Sen et al. [30] had also studied a $FeSi_{10}$ cluster in a pentagonal pyramid structure with the Fe atom encapsulated between two $Si_5$ pentagons. Although they did not explore any other structures, the Fe-encapsulated pentagonal pyramid was found to be a stable one, in agreement with the other calculations. Ma et al. [26] also found a Fe-encapsulated cage structure starting with $FeSi_{10}$, and the lowest energy structure they obtained is a pentagonal prism of Si atoms with $D_{5h}$ symmetry. In the most recent work, Chauhan et al. [39] also found that the smallest cage cluster encapsulating a Fe atom is $Si_{10}$. However, the ground state structure they obtained is similar to the one reported in Ref. [38], and it is not a pentagonal prism.

As in Refs. [27] and [28], Wang et al. [40] also found that a Co atom gets encapsulated in a Si cage for the first time in $CoSi_{10}$. In an earlier work [26] these authors had claimed that $CoSi_9$ has a cage-like structure. However, in Ref. [40] they showed that a basket-like structure is lower in energy for $CoSi_9$. As for Ni, Ren et al. [41] and later on Li et al. [42] claimed that an endohedral structure becomes most favorable $NiSi_8$ onwards. This is in disagreement with Refs. [27] and [28] and also the results of Wang et al. [43]. Reference [43] found that a Ni atom gets encapsulated in a Si cage at $NiSi_9$.

**4d and 5d TM encapsulated $Si_n$ clusters**

The question of the smallest metal encapsulated Si cage structure has not been addressed so systematically for the 4d and 5d TM atoms. We will discuss what is available in the literature. From the works of Kawamura et al. [32] and Gueorguiev et al. [29] it is found that by n = 12 both the group-IV atoms Zr and Hf are encapsulated in a $Si_{12}$ cage. The cage is distorted from a perfect $D_{6h}$ symmetry in case of Zr. Gueorguiev et al. [29] also claim that Hf takes an endohedral position even in $Si_{10}$. But Zr takes an exohedral position at this size. Uchida et al. [44] also found a distorted HP cage for $HfSi_{12}$. Sen and Mitas [30] considered a structure in which Hf was encapsulated inside a HP $Si_{12}$ cage. Therefore group-IV elements are definitely encapsulated by n = 12. The group-V element Nb is also found to occupy an endohedral position in $NbSi_{10}$. The 5d group-V atom Ta, however, takes an exohedral position in $TaSi_{10}$. It is encapsulated in a perfectly symmetric $D_{6h}$ HP cage in $TaSi_{12}$. Uchida et al. [44] also found a similar structure for this cluster.

Group-VI atoms Mo and W have attracted more attention as they are expected to satisfy the 18-electron rule when encapsulated in a $Si_{12}$ cage. This was largely motivated by the work of Hiura et al. [14]. All the works [28, 30, 32, 44–46] agree that they are encapsulated inside a $Si_{12}$ cage. In their recent work, Abreu et al. [47] find that the W atom is partially encapsulated in a $Si_{11}$ cage, and it gets completely encapsulated in $WSi_{12}$. Likewise, all the remaining TM–Si clusters have the TM atoms (Tc–Pd in the 4d series and Re–Pt in 5d series) encapsulated by n = 12 [28, 29, 44]. Out of these, Pd gets encapsulated at n = 10, but Pt takes an exohedral position at this size [29]. Tc, Ru, Rh, Re, Os and Ir take endohedral positions at n = 12, and take exohedral positions below this size [28].

## 2.2 *Relative Stability of $TMSi_n$ Clusters*

One of the fundamental questions while studying a particular series of clusters is their relative stability with changing size and composition. This question has also been addressed in the context of TM–Si clusters. The most relevant question is, given a particular TM atom, which $TMSi_n$ cluster has the greatest stability? Then, what is this 'magic number' for different TM atoms? Experimentally, these questions are addressed by measuring the mass abundance spectra of these clusters. However, only identifying the stable clusters is not enough. One would also like to understand of the origin of their stability. Particularly relevant points are: whether the observed stabilities are due to geometric close packing, or some electronic shell closing effects. On these points, both experiments and theory play equally important roles.

We will discuss the present state of our knowledge regarding these issues. First, we will discuss the experimental results regarding stability of TM–Si clusters, and then will go on to discuss the possible origins of this stability, and our theoretical understanding of it.

**Experimental Scenario**

As mentioned earlier, Beck was the first to produce TM–Si clusters in experiments. In his first experiment [1] Beck used a laser vaporization supersonic expansion source in which the carrier He gas was seeded with a small fraction of the respective metal carbonyl. He found that for Cr, Mo and W, the mass spectra were dominated by the $TMSi_{15}^+$ and $TMSi_{16}^+$ clusters. Under the experimental conditions, no pure Si clusters at these sizes were observed. Clearly, incorporation of TM atoms enhanced stability of the Si clusters. It was speculated that either (i) metal atoms acted as 'seed' around which a compact atomic shell corresponding to a specific number of Si atoms formed (15 and 16 in these cases), and this led to enhanced stability of these clusters, or (ii) the $TMSi_{15}^+$ and $TMSi_{16}^+$ clusters were the most stable fragments in the photo-fragmentation of larger clusters during photoionization process.

In a subsequent work [2] Beck produced $TMSi_n$ clusters for TM = Cr, Mo and W using three different variations of the laser vaporization setup. In the first, metal carbonyls were introduced along with the He carrier gas as in the previous experiment. In the second method, metal atoms were deposited on the Si target in the laser vaporization source in a fine grid pattern. The laser spot was large enough to overlap with both silicon and metal atoms. Thus both elements were vaporized simultaneously, and they reacted in the vapor phase to produce the mixed clusters. In the third method, the metal was used as the target and Si was introduced into the reaction zone in the form of silane ($SiH_4$) seeded into the carrier gas. Interestingly, in all these situations, the mass spectra looked strikingly similar, and were dominated by the $TMSi_{15}^+$ and $TMSi_{16}^+$ clusters when high laser fluence was used to ionize the neutral clusters.

Hiura et al's [14] experiment is a landmark in the field of binary metal–Si clusters. In their setup, metal vapor was generated by resistively heating a metal wire. It was subsequently ionized by electron irradiation. The ions were trapped in an external quadrupole static ion trap (EQSIT), and were then made to react with silane gas. This process, in general, produced $TMSi_nH_x^+$ clusters. For example, in case of W, a sequential growth from $W^+$ up to $WSi_{12}H_x^+$ was seen. Any such cluster for $n > 12$ was rarely seen for W. A similar behavior was found for all other TM atoms. That is, reactivity with $SiH_4$ almost stopped when n reached a specific value m depending on the TM atom. m was found to be 14 for Hf, 13 for Ta, 12 for W, 11 for Re, and 9 for Ir. What is even more interesting is that the clusters tended to lose all hydrogen atoms at $n = m$. Thus $TMSi_m^+$ clusters were highly abundant. This also means that the $TMSi_m$ clusters have enough stability to compensate for the entire lack of H atoms. Figure 5 from Ref. [14] shows the relative abundance of $TMSi_n^+$ clusters for various TM atoms as a function of n.

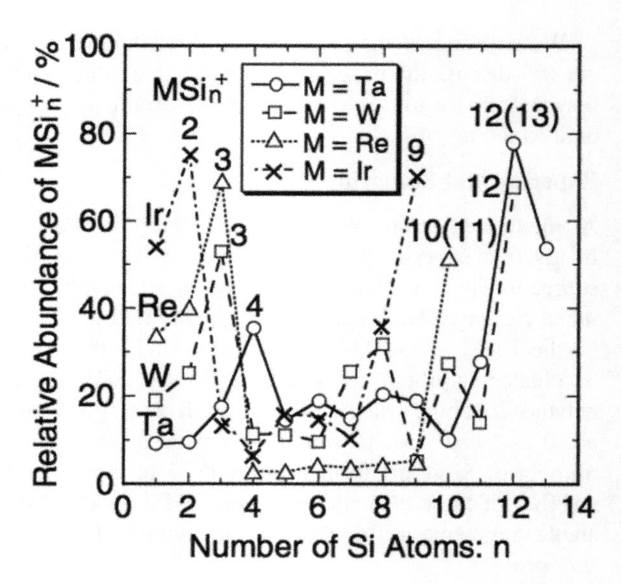


**Fig. 5** Fraction of $TMSi_n^+$ among $TMSi_nH_x^+$ clusters with varying n. Reprinted figure with permission from H. Hiura, T. Miyazaki and T. Kanayama, Phys. Rev. Let. 86, 1733 2001. © 2001 by the American Physical Society

Two features of these observations are important. Firstly, no $TMSi_{15}^+$ or $TMSi_{16}^+$ clusters were observed. This is in contrast to Beck's results. Secondly, the sum of m and the atomic number of the TM atom is always 86, the atomic number of the rare gas atom Rn. Hiura et al. speculated that different method of producing TM–Si clusters could be the reason behind the difference between their observations and those of Beck. However, as we noted earlier, Beck found $TMSi_{15}^+$ and $TMSi_{16}^+$ clusters dominating the mass spectra even when TM–Si clusters were produced through reaction of metal ions and $SiH_4$. The second observation gave rise to the idea that $TMSi_m$ clusters are particularly stable because of the so called 18-electron rule of inorganic and organometallic chemistry. If one assumes that each Si atom donates one electron to the central metal atom, the W atom in $WSi_{12}$ would have 18 outermost valence electrons leading to a closed electronic shell. The same argument holds for $TMSi_m$ clusters for TM = Hf, Ta, Re and Ir. This argument holds strictly for the neutral clusters. But the metal atoms were ionized positively when the reaction with $SiH_4$ started. The authors argued that once they were completely encapsulated the by Si atoms, the metal atoms would be 'almost neutral' as the 'positive charges would be widespread on the whole cluster'. This simple idea was explored theoretically in a number of subsequent studies. We will discuss these later.

After Hiura et al's experiments, a series of experiments on TM–Si clusters came from the groups of Kaya and Nakajima [15–18], and Lievens [20, 48]. We have already mentioned these in the context of structure determination of the clusters. Ohara et al. [15] measured the mass abundance spectra for $TiSi_n$, $MoSi_n$, $HfSi_n$ and $WSi_n$ cluster anions. Strikingly, all the spectra were dominated by the $TMSi_{15}$ and $TMSi_{16}$ clusters. This is in clear contrast to the observations of Hiura et al. [14].

Ohara et al. attempted to explain this by the difference in the cluster production method in the two experiments. However, as we have mentioned before, this cannot explain the observed difference. Ohara et al. also attempted to explain the observed stability of $TMSi_{16}^-$ clusters based on their measurements of the electronic structure of these clusters. From their PES experiments they found that the $TMSi_n$ clusters have a local minimum in the electron affinity (EA) at n = 16 as a function of n. EA, as usual, is defined as the energy gained in adding an electron to a neutral cluster. A cluster having a filled electronic shell gains little energy on addition of an extra electron, and has a small EA. Therefore, it was argued, that the observed stability of $TMSi_{16}^-$ clusters cannot be due to electronic effects. Geometry of the clusters was claimed to be the reason for this stability. Enhanced stability of $TiSi_{16}^-$ and $HfSi_{16}^-$ clusters is in complete contrast to another set of experiments [16] by the same group that we discuss next.

Koyasu et al. [16] studied Sc–Si, Ti–Si and V–Si clusters. In the mass spectrum of neutral $TiSi_n$ clusters, the highest abundance was found for $TiSi_{16}$. $ScSi_n$ and $VSi_n$ clusters also had the greatest abundance at n = 16 but in the anionic and cationic states respectively. The mass spectra reported in Ref. [16] are shown in Fig. 6. This analysis was further extended to include the 4d and 5d TM atoms belonging to the same groups as Sc, Ti and V, i.e., Y, Lu, Zr, Hf, Nb and Ta [17]. Zr and Hf, being tetravalent like Ti, produced the highest abundance of neutral $ZrSi_{16}$ and $HfSi_{16}$. Similar to $VSi_{16}^+$, $NbSi_{16}^+$ and $TaSi_{16}^+$ cation clusters also had the highest abundance. Abundance of $LuSi_{16}^-$ compared to its neighbors was not very marked. The number of $YSi_{16}^-$ clusters was, in fact, slightly less than those of its neighbors: $YSi_{15}^-$ and $YSi_{17}^-$. Enhanced stability of $ScSi_{16}$ in the anionic state, group-IV TM–$Si_{16}$ clusters in the neutral state, and the group-V TM–$Si_{16}$ clusters in the cationic state was explained in terms of a cooperative effect of a compact geometry and a filled 20 electron shell. Firstly, $ScSi_{16}^-$, $TiSi_{16}$ and $VSi_{16}^+$ all have compact FK polyhedron structures [49, 50]. Then, in $TiSi_{16}$ for example, the tetravalent Ti atom forms a closed electronic shell of 20 electrons with one electron

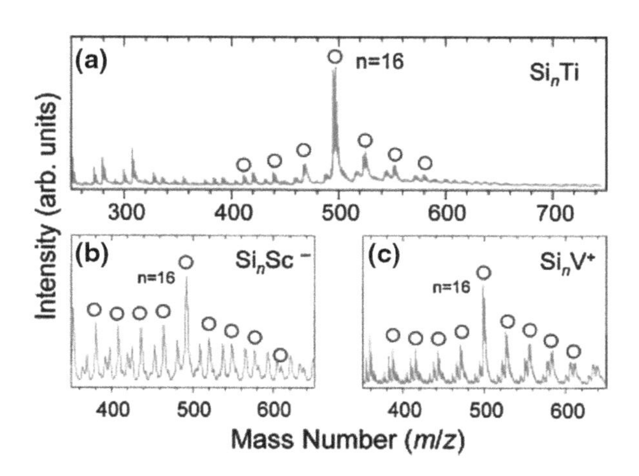


**Fig. 6** Mass spectra of **a** $ScSi_n^-$, **b** $TiSi_n$ and **c** $VSi_n$ clusters. Reprinted with permission from J Am. Chem. Soc. 2005, 127, 4998. © 2005 American Chemical Society

from each of the Si atoms. This idea is borrowed from metal clusters where it has been shown that within the approximation of a confined spherical electron gas, valence electrons occupy one particle levels that appear in bunches, and can be labelled as $1S^21P^61D^{10}$ … [51, 52]. 2, 8, 18, 20, … electrons appear as shell filling numbers, and clusters with these numbers of electrons have enhanced stability. Since Sc and V have one valence electron less and more respectively than Ti, $ScSi_{16}$ and $VSi_{16}$ form 20 electron closed shells in the anionic and the cationic states respectively. The highly symmetric FK polyhedron cage leads to high degree of degeneracy of the electronic levels which is necessary for a pronounced shell filling effect. Absence of marked peaks in the abundance of $YSi_{16}^-$ and $LuSi_{16}^-$ has also been explained in terms of their structures. Since Y and Lu atoms are larger then the Sc atom, they do not fit the Si 16 cage very well. Indeed these two atoms are encapsulated in a cage cluster at larger sizes [17, 53]. It is possible that absence of TM encapsulated cage structures, or at least presence of other isomers, is the reason for the absence of marked peaks.

This conjecture of electronic shell closing leading to enhanced stability of these clusters was confirmed by their photoelectron (PES) spectra. Experimental anion PES spectra of $ScSi_{16}^-$, $TiSi_{16}^-$ and $VSi_{16}^-$ reported by Koyasu et al. [16] are

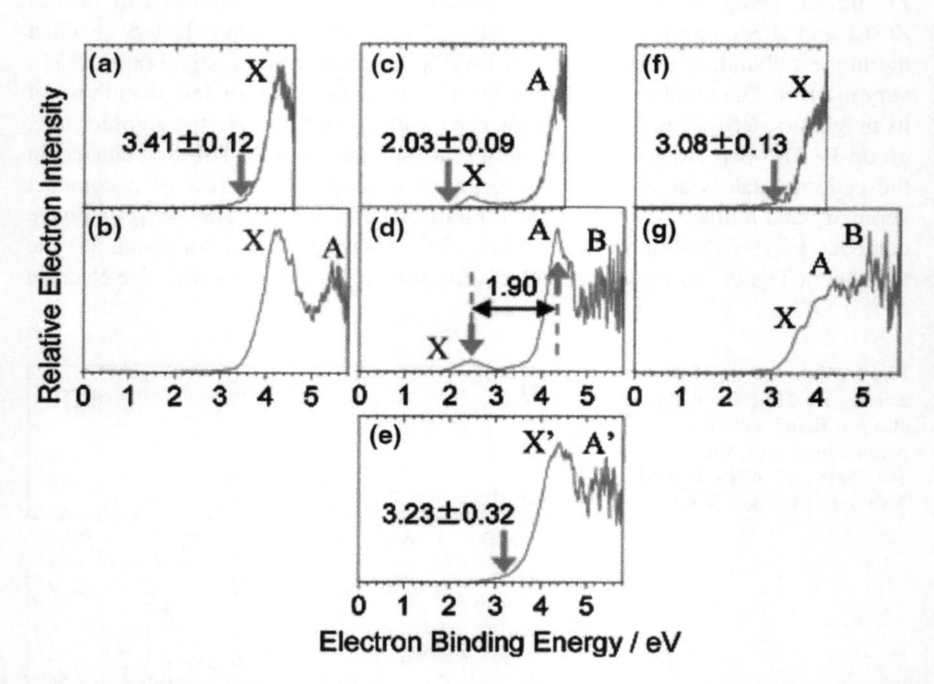


**Fig. 7** Photoelectron spectra of **a** and **b** $ScSi_{16}^-$, **c** and **d** $TiSi_{16}^-$, **f** and **g** $Vsi_{16}^-$ and **e** $TiSi_{16}F^-$ clusters. **a**, **c** and **f** measured at 4.66 eV, and the remaining four spectra measured at 5.82 eV. Reprinted with permission from J Am. Chem. Soc. 2005, 127, 4998. © 2005 American Chemical Society

shown in Fig. 7. PES spectra of $ScSi_{16}^-$ and $VSi_{16}^-$ are distinctly different from that of $TiSi_{16}^-$. The spectra of $TiSi_{16}^-$ at both the photon energies (4.66 and 5.82 eV) have a low energy peak marked X in panels C and D in Fig. 7. This is followed by a large energy gap and more discrete transitions at higher electron binding energies. The spectra of $ScSi_{16}^-$ and $VSi_{16}^-$ do not have this low energy feature. Koyasu et al. assume that the low energy feature X is due to a singly occupied molecular orbital (MO). If that is the case then this spectral feature suggests that neutral $TiSi_{16}$ has a closed electronic shell with a large HOMO-LUMO gap. Indeed the measured HOMO-LUMO gap is 1.9 eV. The conjecture that the feature X comes from a singly occupied MO is supported by the fact that $TiSi_{16}^-$ clusters reacted with $F_2$ to produce $TiSi_{16}F^-$ molecules. The feature X is not see in the PES spectrum of $TiSi_{16}F^-$ while the rest of the spectral features remain intact (Fig. 7e). This is understood as the electron from the singly occupied MO being donated to the F atom to form an ionic complex.

$ZrSi_{16}^-$ and $HfSi_{16}^-$ have PES spectra strikingly similar to that of $TiSi_{16}^-$. Namely, they all have the low energy feature X followed by a large gap. The measured HOMO-LUMO gap of $ZrSi_{16}$ and $VSi_{16}$ are also large, 1.36 and 1.37 eV respectively. Likewise, the spectra of $YSi_{16}^-$ and $LuSi_{16}^-$ are similar to that of $ScSi_{16}^-$, and the spectra of $NbSi_{16}^+$ and $TaSi_{16}^+$ are similar to that of $VSi_{16}^+$. Just as $TiSi_{16}^-$ forms a stable ionic complex with F, $VSi_{16}$, being one electron more than a filled shell configuration, forms a stable complex $VSi_{16}F$. $VSi_{16}^+$, on the other hand, is non-reactive towards $F_2$ because of its filled shell electronic structure. That the feature X is due to a singly occupied MO got further support from the measured PES spectra of the group-III, -IV and -V TM doped $Si_n$ clusters over the entire size range n = 6–20 [53]. The threshold energy for the PES had a maximum at n = 16 for the group-III elements due to the fact that these anions form closed electronic shells. It was, on the other hand, a minimum for the group-IV elements due to the emergence of the low energy feature. And it is only at size 16 that the low energy feature is found. Moreover, the importance of a symmetric structure, and a cooperative effect of the atomic and electronic structures leading to enhanced stability of these clusters was established when it was sown that the calculated PES spectrum of $TiS_{16}$ in the (Jahn-Teller distorted) Frank-Kasper (FK) polyhedron matches the experimental spectrum very well [53].

Neukermans et al. [48] studied Cr and Mn doped Si clusters produced in a dual target laser vaporization source. Neutral clusters produced in the source were ionized by a 6.4 or a 7.9 eV laser to produce cations whose mass spectra were measured. Mass spectrum of $CrSi_n^+$ clusters after irradiation by a 6.4 eV laser is dominated by $CrSi_{15}^+$ and $CrSi_{16}^+$ with much less abundance at n = 14, 17 and 18. Mass abundance of $MnSi_n^+$ clusters is also dominated by $MnSi_{15}^+$ and $MnSi_{16}^+$. In addition, $MnSi_{17}^+$ and $MnSi_{18}^+$ were also produced abundantly. These authors speculated that the observed stability of $CrSi_{15}^+$, $CrSi_{16}^+$, $MnSi_{15}^+$ and $MnSi_{16}^+$ is due to their metal encapsulated cage structure. Electronic structure of the clusters were not invoked by these authors.

The discussions so far show that different sets of experiments give very different results for the relative abundance of various TM–Si clusters. Beck's original experiments [1, 2] found $TMSi_{15}^+$ and $TMSi_{16}^+$ clusters to be the most abundant ones for group-VI TM atoms. Hiura et al. [14], on the other hand, found that the size n at which the most stable $TMSi_n$ occurs depends on the TM atom. In particular, they found $WSi_{12}$ to be exceptionally stable. As reported by Abreu et al. [35], most recent experiments find $WSi_{14}$ to be less reactive compared to $WSi_{12}$. Kaya and Nakajima group, in their earlier work [15], had found anionic $TMSi_n^-$ clusters for group-IV and -VI TM elements to be most abundant at n = 15 and 16. In their later works they showed that the higher abundance at n = 16 for different TM atoms occurs in different charge states. In all these experiments TM–Si binary clusters are produced in laser vaporization sources. These sources produce clusters in cold conditions. The cluster size distribution depends on the conditions of the source. Sufficiently long residence time in the formation channel, and a proper equilibrium between cluster growth and evaporation are required for the mass spectrum to properly reflect the enhanced stability of certain cluster species [48]. One may ask if these conditions are reached in all the experiments in view of the differing results they obtained. Indeed, neutral $TiSi_{16}$ clusters were selectively produced after fine-tuning the source conditions such as the laser fluences, and the flow of the He carrier gas [16]. Therefore, how much of these observations represent the relative thermodynamic stability of these clusters, and how much is determined by the formation kinetics is still an open question.

## Theoretical Studies

A number of theoretical studies have been undertaken to identify $TMSi_n$ clusters with particularly high stability, to explain the experimental mass spectra, and to understand the origin of stability in these clusters. As we will soon discuss, three different frameworks have been invoked to explain the observed and measured enhanced stability of specific $TMSi_n$ clusters. These are (i) compact structure of metal encapsulated Si cage clusters and strong TM–Si bonding; (ii) 18-electron rule as suggested by the experiments of Hiura et al; (iii) shell fillings of the electron gas formed inside the Si cage around the metal atom. Theoretically, several descriptors have been used to identify stable TM–Si clusters. These are: (i) Binding energy per atom (BE) of the cluster; (ii) Embedding energy (EE) of the TM inside the Si cage. This is the energy gain in putting the TM atom inside the Si cage; (iii) First order energy difference $\Delta_1(n)$ that measures the energy gain in forming the $TMSi_n$ cluster by adding a Si atom to the $TMSi_{n-1}$ cluster; (iv) Second order energy difference $\Delta_2$ (n) which is the difference between $\Delta_1(n)$ and $\Delta_1(n + 1)$; (v) The energy difference between the highest occupied molecular orbital and the lowest unoccupied molecular orbital (HOMO-LUMO gap). While the first four quantities measure thermodynamic stability of the clusters in various ways, the HOMO-LUMO gap measures the so-called kinetic stability, i.e., the stability of a cluster in chemical reactions. The larger the HOMO-LUMO gap the less reactive a cluster is. These quantities are defined as follows.

$$BE = (E(TM) + nE(Si) - E(TMSi_n)) / (n + 1), \quad (1)$$

$$EE = E(TM) + E(Si_n) - E(TMSi_n), \quad (2)$$

$$\Delta_1(n) = E(Si) + E(TMSi_{n-1}) - E(TMSi_n), \quad (3)$$

$$\Delta_2(n) = E(TMSi_{n+1}) + E(TMSi_{n-1}) - 2E(TMSi_n). \quad (4)$$

E's are the total energies of the respective systems. Physical significance of BE, EE and $\Delta_1$ are rather straightforward, or has already been stated. $\Delta_2$ requires some elaboration. It is the difference between the energy gained in going from size n − 1 to n, and that in going from n to n + 1. If the latter happens to be smaller, there will be more clusters of size n being formed from size n − 1 than clusters of size n + 1 being formed from size n. That is, there will be more clusters of size n. A positive $\Delta_2$, therefore, indicates an enhanced stability with respect to the two neighboring sizes. The same quantities have sometimes been denoted by different symbols in the literature. But we will use the symbols defined above. In addition, vertical and adiabatic ionization potentials (VIP and AIP) of neutral clusters, and the vertical detachment energy (VDE) and the adiabatic detachment energy (ADE) of the anion clusters have sometimes been calculated to check their stability. VIP is the energy required to detach an electron from a neutral cluster without changing its structure. AIP is the energy difference between the ground states of a neutral cluster and its cation. Clusters with filled electronic shells tend to have high VIP and/or AIP, while those with a single electron in excess of a filled shell have low VIP/AIP. VDE is the energy difference between anion and neutral charge states of a cluster both at the anion structure. ADE is the energy difference between the anion and neutral ground state structures. ADE is also the adiabatic electron affinity (AEA) of the corresponding neutral. If a neutral cluster has a filled electronic shell, it has a small AEA. Clusters having one electron short of a filled shell usually have large AEA's. These ideas are borrowed from the field of metal clusters [51, 52].

We will now discuss specific theoretical results on the stability of TM–Si clusters. First, we will discuss those works that did not invoke electron counting rules to explain these stabilities, and then we will take up applicability of the electron counting rules in these clusters.

### Not invoking electron counting rules

One of the first theoretical works studying the stability of TM–Si clusters was performed by Kumar et al. [49]. This study was motivated by the works of Jackson et al. [13] and Hiura et al. [14]. These authors used the so-called shell-shrinkage method to find the most stable TM–Si cage clusters for Ti, Zr, Hf, Fe, Ru and Os. This method worked as follows. When they optimized the structure of $ZrSi_{20}$ reported by Jackson et al. [13], the cage shrunk, and one of the Si atoms stuck out. When this atom was removed and the structure was re-optimized, a $ZrSi_{17}$ cage was found with two Si atoms sticking out. Removal of these two atoms led to a $ZrSi_{16}$ cage with one capping Si atom. Removal of this last Si atom, and re-optimization produced a compact, symmetric fullerene(f)-like cage of $ZrSi_{16}$. For $TiSi_{16}$ a FK polyhedron isomer was found to be 0.781 eV lower than the f-like isomer. This FK isomer has a very large HOMO-LUMO gap of 2.358 eV. For Hf, the FK, f-like and a capped hexagonal antiprism structures are nearly degenerate. For smaller TM atoms such as Fe, Ru and Os, stable $TMSi_{14}$ cages were found. A cube-derived structure was found to be the lowest in energy for Fe. For Ru and Os the f-like and the cube-derived structures were nearly degenerate.

All these clusters have large HOMO-LUMO gaps in excess of 1 eV. They also have large BEs. The details can be found in Ref. [49]. What is most interesting is the explanation offered for the large stability of the $TiSi_{16}$ cage cluster. It turned out that both the FK and f-like $Si_{16}$ cages had four unpaired electrons, and moments of 4 $\mu_B$. Therefore they could both share four electrons, and formed strong bonds with the tetravalent Ti atom. In fact, a comparison of the charge densities of the bare metal atom and the empty Si cage cluster with that of the metal encapsulated cluster showed depletion of charge on the Si–Si bonds in $Si_{16}$. Similarly, charge density depleted at the centre of the faces in $Si_{14}$ after metal encapsulation. In addition, there was accumulation of charge in the TM–Si bonds. Thus stability of these clusters could be explained by formation of strong TM–Si bonds without invoking any electron counting rules.

Similar analyses were performed for the group-VI TM elements Cr, Mo and W by Kumar et al. [54]. A Cr atom, being smaller than Ti, does not fit a $Si_{16}$ cage very well. Optimization of the f-like $CrSi_{16}$ structure leads to a $CrSi_{15}$ structure capped by a Si atom. Removal of this capping atom and re-optimization leads to a stable f–$CrSi_{15}$ cage. The lowest energy structure for $CrSi_{15}$ was, however, derived from the capping of the cubic structure found for $FeSi_{14}$. For Mo and W also, this cube derived structure turned out to have the lowest energy. All the three clusters have large HOMO-LUMO gaps in excess of 1.5 eV. The largest HOMO-LUMO gaps were, however, obtained for the $TMSi_{14}$ clusters. But they had smaller BEs compared to the corresponding $TMSi_{15}$ clusters. This was cited as the reason that clusters at size 14 were not seen as magic in the experiments of Beck et al. [1, 2]. For $TMSi_{16}$ clusters with Mo and W, structures with the metal atoms coordinated to all the 16 Si atoms were found to be stable, unlike Cr. But the lowest energy structure turned out to be a capped $Si_{15}$ f-cage. Based on all these observations it was argued that a 15 member metal encapsulated cage structure is the most stable one for the group-VI elements. Almost equal abundance of $TMSi_{16}$ clusters in the mass spectra could be due to the capped isomers at this size. The lowest energy structures of these clusters obtained by Kumar and co-workers can be found in the original papers [49, 54].

A number of other studies [26, 28, 29, 32, 36, 40, 42, 43, 46] have also addressed the question of relative stability of various TM–Si clusters without necessarily invoking electron counting rules. Kawamura et al. [32, 36] studied structure and properties of Cr, Ti, Zr and Hf doped Si clusters using two different functionals within DFT. For the Cr doped clusters the BE vs number of Si atoms plot was rather featureless, except that it became flat after $n = 12$ and 15, and decreased slightly after $n = 16$. $\Delta_2$ values were positive at $n = 10, 12, 14, 15$, and 16 within a pure GGA functional, indicating stability at these sizes. The hybrid

B3PW91 functional however, gives a negative value of $\Delta_2$ at n = 15. The B3PW91 functional produces the largest gap at n = 12 (2.939 eV) indicating its kinetic stability. The GGA functional, on the other hand, gives a rather small gap at n = 12 (0.847 eV), and produces a maximum (1.382 eV) at n = 11. This shows that DFT based calculations are limited by the approximations involved. These authors also studied charged clusters. For the cation clusters, $\Delta_2$ becomes larger for n = 13 than for n = 12 within the B3PW91 functional. n = 15 turns out to have a negative value of $\Delta_2$ although it was seen abundantly in the mass spectrum. For the anions, n = 12 and 14 have positive values of $\Delta_2$, while n = 13 and 15 have negative values.

In a similar study on Ti-, Zr- and Hf-doped Si clusters [32], using the PW91 GGA functional, these authors again found that there were no special features in the BE plot of the $TiSi_n$ clusters for n = 8–16. $\Delta_2$ was found to be positive for n = 11 and 13. Anion clusters had positive $\Delta_2$ at n = 12 and 15. Cation clusters had positive $\Delta_2$ at n = 12. BE for the Hf-doped clusters also increased monotonically with size. For Zr, on the other hand, BE showed a dip from n = 12 to 13 indicating a higher stability of $ZrSi_{12}$. $\Delta_2$ shows higher stability at n = 11 and 12 for Zr, and at n = 11 and 14 for Hf. As was mentioned earlier, $TiSi_{16}$ has a rather large HOMO-LUMO gap. $ZrSi_{16}$ actually has a larger gap (2.448 eV), and HOMO-LUMO gap of $HfSi_{16}$ is nearly equal to that of $TiSi_{16}$ (2.352 eV).

Lu and Nagase [46] studied various TM doped Si clusters over a wide size range. Most importantly, they found that both the BE and the EE of W, Zr, Os and Pt doped Si clusters had the highest values at n = 16. In addition there were local peaks at n = 8 and 12 for W, at n = 14 for Zr and Pt, and n = 12 for Os. These findings were cited as a unified explanation for all the diverse observations in different experiments. Since TM–Si clusters were produced by reaction of metal and Si vapors, it was argued that a complete reaction would produce $TMSi_{16}$ clusters. An incomplete reaction could, however, produce locally stable clusters. In case of W, for example, an incomplete reaction could stop at $WSi_{12}$. This argument means that it is the kinetics of formation that determines the relative abundance of various sizes in a given experiment.

Gueorguiev et al. [28, 29] found that for most of the 3d, 4d and 5d TM atoms, n = 12 turned out to be particularly stable among the sizes at which TM encapsulated clusters are formed. This was indicated by peaks in the BE and $\Delta_2$ with changing n. The only exception to this seems to be Ti, Hf and Ta. Hf, in fact, has a local minimum in $\Delta_2$ at n = 12. The HOMO-LUMO gap for many of these clusters do not have a maximum at n = 12. HP structure for all the TM atoms at n = 12 is argued to be the reason for the observed similarity in their energetics at this size.

Ma et al. [26] and Wang et al. [40] studied Co–Si clusters. Ma et al. found a peak in $\Delta_2$ at n = 10. $\Delta_1$, on the other hand showed a peak at n = 9. Since they did not plot $\Delta_2$ beyond n = 12 it is not clear if there is a peak at this size. The HOMO-LUMO gap has peaks at 10 and 12. Since two different descriptors indicate two different sizes to have enhanced stability, it is important to understand what they really mean in the context of the experiments. $\Delta_1$ measures the energy gain in attaching a Si atom to a cluster of the previous size. This would be the relevant criterion when cluster production proceeds by addition of a single Si atom at a time.

This is actually the situation in laser vaporization sources in which clusters are produced in cold conditions. $\Delta_2$ measures the relative stability of a particular size compared to a size higher and a size lower. This criterion is relevant when the final step in cluster production involves both atom aggregation and dissociation, for example in a supersonic nozzle source. Therefore, relevance of $\Delta_2$ to laser vaporization sources is doubtful, although it has been used as a stability descriptor by many authors. Wang et al. [40] found peaks in BE, $\Delta_2$ and HOMO-LUMO gaps at n = 10 and 12 amongst the larger-size clusters indicating their enhanced stability.

Wang et al. [43] and Li et al. [42] studied Ni doped Si clusters. Wang et al. found large $\Delta_2$ and HOMO-LUMO gaps at sizes 10 and 12 among the larger clusters. BE showed a monotonic increase with size with small peaks at 10 and 12. EE showed a large prominent peak at n = 12. n = 10 had a minimum in EE. Li et al. also found a monotonic increase in BE with size for Ni–Si clusters. $\Delta_1$ showed prominent peaks at sizes 10, 12 and 14. $\Delta_2$ and HOMO-LUMO gap also had peaks at these sizes. EE had the most dominant peak at n = 12 with smaller peaks at 10, 14 and 16. This marked difference in the behavior of EE at n = 10 in the two studies requires a closer look. It turns out that Wang et al. have a Ni encapsulated cage structure of $C_1$ symmetry as the lowest energy structure for $NiSi_{10}$. Li et al., on the other hand found a $C_{3v}$ structure to have the lowest energy. Since Li et al. employed a global search based on genetic algorithm, one can put more faith in their calculations. Whether the difference in the theoretical methods used in the two calculations also played any role is difficult to tell.

**Invoking electron counting rules**

Perhaps the most interesting aspect of TM–Si clusters is how well electron counting rules explain the relative stability of these clusters. This was motivated by the suggestion in Hiura et al's work [14] that the observed non-reactivity of particular $TMSi_n$ clusters towards $SiH_4$ can be explained by the empirical 18-electron rule of inorganic and organometallic chemistry. We have already discussed this in some detail. This idea led to a number of theoretical studies, most notably by Sen and Mitas [30], Khanna and co-workers [31, 34, 35, 38, 39, 47], Uchida et al. [44], and Guo et al. [27]. Reveles et al. [50] also attempted to explain the observed stability of $ScSi_{16}^-$, $TiSi_{16}$, $VSi_{16}^+$ clusters using electron counting rule as applied to a spherical electron gas. We will now discuss the most fascinating aspects of these ideas.

In order to explore how generally the 18-electron rule can explain the stability of the $TMSi_n$ clusters, Sen et al. [30] studied all the 3d, and a few 4d and 5d TM doped $Si_{12}$ clusters. To further explore the applicability of the 18-electron rule, they also studied the $FeSi_{10}$ and $ReSi_{11}$ clusters that are expected to obey this rule from a nominal valence electron count. Nominally, group-VI TM atoms Cr, Mo and W, with six valence electrons each, are expected to obey the 18-electron rule with 12 Si atoms. Re with seven valence electrons is expected to obey it in a $ReSi_{11}$ cluster, and since Fe has eight valence electrons, $FeSi_{10}$ is also expected to obey this rule and show enhanced stability. Sen et al. [30] calculated the EE's and the

HOMO-LUMO gaps for all these clusters. Among all the TM's, $WSi_{12}$ has the largest and a remarkably large EE (8.44 eV). But within the 3d series, it was V rather than Cr that had the largest EE (4.46 and 2.99 eV respectively). Both $VSi_{12}$ and $CrSi_{12}$ had quite large HOMO-LUMO gaps, however. $ReSi_{12}$ had a larger EE (7.44 eV) than $ReSi_{11}$ (6.34 eV) in contradiction to the expectations from the 18-electron rule. EE for $FeSi_{10}$ was also rather small, 1.07 eV. It was argued that the 18-electron rule is just one of the aspects determining stability of these clusters, and is not of general applicability. One of the criticisms of this work could be that a thorough structure search was not performed. Only HP cage structures were considered for $TMSi_{12}$ clusters. This may be adequate as other authors also found HP cages at this size. But only a pentagonal prism at size 10, or a cage structure composed of one pentagon and one hexagon at size 11 may not have been adequate.

Khanna et al. [34, 38] addressed the same question from a slightly different point of view. They essentiality focused on the quantity $\Delta_1$ for $CrSi_n$ and $FeSi_n$ clusters. Among the $CrSi_n$ clusters for n = 11–14, $\Delta_1$ was the largest at n = 12. They also found that $Si_{12}$ was the only pure Si cluster that gained more energy by bonding to a Cr atom than to another Si atom. These were taken as indicators to the fact that $CrSi_{12}$ has special stability. This was claimed to be due to the 18-electron rule. Appearance of a uniform charge density around the Cr atom inside the Si cage, and a charge transfer of 2.6e from the Si cage to Cr were taken as further evidences that a filled shell of electrons is formed around the metal atom. It should also be noted that $CrSi_{12}$ has neither the largest BE, nor the largest HOMO-LUMO gap among the clusters studied. This is of importance in the context of the most recent calculations on the $CrSi_n$ clusters over a larger size range by Khanna's Group. We will come to this shortly.

In an early work, studying $FeSi_9$, $FeSi_{10}$ and $FeSi_{11}$ clusters, Khanna et al. [38] claimed that $FeSi_{10}$ is more stable than the other two as it has the largest $\Delta_1$. It was also found to have the largest BE, and the largest HOMO-LUMO gap. It also happens to have the largest VIP. However, it does not have the largest EE. This was claimed to be due to 'magic' nature of $Si_{10}$ among the pure Si clusters. Among the pure clusters, $Si_{10}$ has a smaller $\Delta_1$ (3.14 eV) than $Si_9$ (5.12 eV). In fact, in the experimental mass spectrum of pure Si clusters, the peak for $Si_{10}^+$ is significantly lager than those of its neighbors [55]. It also has a large HOMO-LUMO gap of 2.08 eV making it non-reactive. Formation of $FeSi_{10}$ is thus likely to proceed by a single Si addition to $FeSi_9$, and not by incorporation of Fe in $Si_{10}$. In any case, the largest stability of $FeSi_{10}$, as found in this work, was in accordance with the 18-electron rule. But a very recent work shows that this is not the complete picture. This is discussed later.

In view of Sen and Mitas's assertion [30] that the 18-electron rule does not always apply to the TM–Si clusters, Reveles and Khanna studied the 3d TM doped $Si_{12}$ cage clusters again [31]. Since Sen and Mitas had found the largest EE for $VSi_{12}$ among the 3d TM atoms, Reveles and Khanna focussed on this quantity. In addition to calculating EE using Eq. 2, they calculated this quantity by enforcing the Wigmer-Witmer (WW) spin conservation rule also. The essential idea of this rule is that the total spin has to be conserved when a TM atom is encapsulated in a Si cage. Therefore EE, while the WW rule is enforced, is defined as

$$EE^{WW} = E(Si_{12}) + E^{M}(TM) - E^{M}(TMSi_{12}) \quad (5)$$

M denotes the spin multiplicity of the $TMSi_n$ cluster. In this equation, instead of taking the TM atom necessarily in its ground state, it is taken to have the same spin multiplicity as the $TMSi_n$ cluster so that the reactants and the products have the same total spin. In addition to the neutral clusters, Reveles and Khanna calculated EE and $EE^{WW}$ for the anion clusters as well. They also calculated the AEA and the VIP of the clusters. All these clusters were found to have HP cage structures with distortions that were large towards both ends of the 3d series, but were small near the middle. $CrSi_{12}$ had a perfectly symmetric $D_{6h}$ structure. The EE and $EE^{WW}$ calculated by these authors are reproduced in Fig. 8. EE for the neutral clusters has the largest values for V and Fe, and not for Cr. For the anions it is largest for V. $EE^{WW}$, on the other hand, has the largest value for Cr. It was claimed that spin

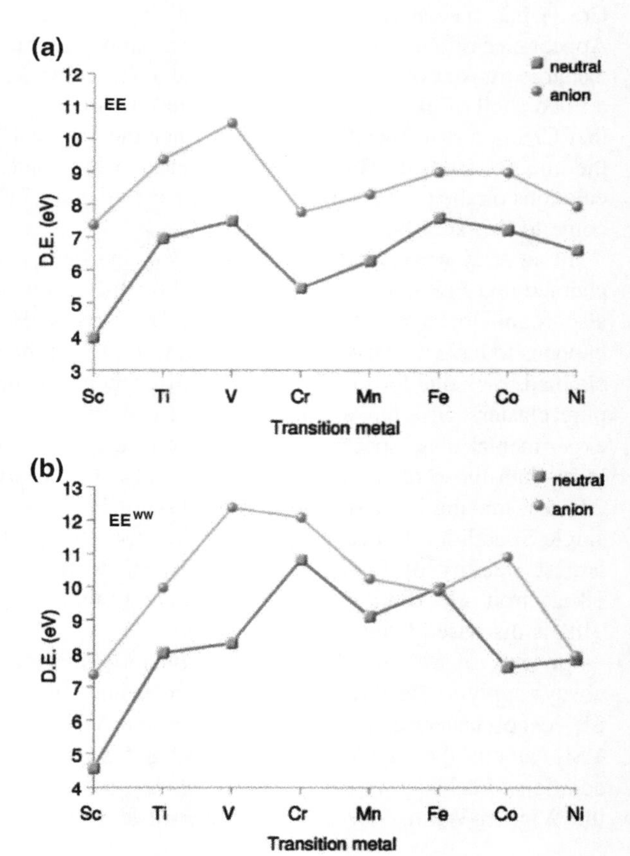


**Fig. 8** Calculated **a** EE and **b** $EE^{WW}$ of 3d TM encapsulated $Si_{12}$ cage clusters. Reprinted figure with permission from J. U. Reveles and S. N. Khanna, Phys. Rev. B 72, 165413 (2005). © 2005 by the American Physical Society

conservation must be enforced while calculating EE to get the right picture about relative stability at different sizes. Once that is done, $CrSi_{12}$ turns out to have the greatest stability in accordance to the 18-electron rule. However, in view of the recent work [35] by Khanna and co-workers showing that $CrSi_{14}$ obeys the 18-electron rule, this argument becomes questionable. In this context, it is also worth noting that Koyasu et al. [53] did not find any evidence of an electronic shell (18-electron) closure in their PES spectra of the group-V TM–Si cluster anions. In particular, the threshold of the PES was smooth around size 12, and no maxima in the threshold was found for either $VSi_{12}^-$ or $NbSi_{12}^-$ or $TaSi_{12}^-$.

A second peak in $EE^{WW}$ is found for Fe which is expected to be a 20 electron cluster [31]. This was claimed as a proof of a filled shell electronic structure of this cluster. As a further proof of the validity of the electron counting rules in these clusters it was cited that among the anion clusters VSi $^-$ has the largest $EE^{WW}$, $CrSi_{12}$ has the highest VIP, and $CoSi_{12}$, being a 21 electron cluster, one more then a filled shell, has a low VIP. In addition, $VSi_{12}$ neutral has a large AEA as it gains a large amount of energy by going to an 18 electron filled shell in the anionic state. It must, however, be noted that $VSi_{12}^-$ has the highest EE among anions even when spin conservation was not taken into account. Also the peak expected for the 20-electron cluster $MnSi_{12}^-$ was absent, and $CoSi_{12}^-$ which is expected to be a 22 electron cluster, has a peak in $EE^{WW}$.

In another interesting work, Reveles and Khanna [50] tried to understand the observed mass spectra of the $ScSi_n^-$, $TiSi_n$ and $VSi_n^+$ clusters as measured by Koyasu et al. [16]. Remember, Koyasu et al. found all these three clusters to be most abundant at n = 16. Apart from finding the structures of the neutral and charged Sc, Ti and V doped Si n clusters for n = 15–17, Reveles and Khanna calculated $BE^{WW}$ and $EE^{WW}$. The WW spin conservation rule was taken care of while calculating both $BE^{WW}$ and $EE^{WW}$. Another interesting aspect these authors presented is the bond critical point (BCP) analysis of these clusters [56]. This has important bearing on the valence electron count. $ScSi_{16}^-$, $TiSi_{16}$ and $VSi_{16}^+$, which are expected to be 20 electron clusters, are found to have symmetric FK polyhedron structures. Other clusters in this series, that deviate from the 20 electron count, have distortions from a perfect FK polyhedron. $BE^{WW}$ for the $ScSi_{16}$, $TiSi_{16}$ and $VSi_{16}$ clusters were plotted for different charge states. The Sc doped clusters had the largest $BE^{WW}$ in the anionic state. The Ti doped clusters were found to have the largest $BE^{WW}$ in the neutral state, and the V doped clusters had the largest $BE^{WW}$ in the cationic state. The $EE^{WW}$ was plotted for the $ScSi_n^-$, $TiSi_n$ and $VSi_n^+$ clusters, and they all had the largest value at n = 16. These results undoubtedly show that $ScSi_{16}^-$, $TiSi_{16}$ and $VSi_{16}$ are the most stable clusters in the anion, neutral and cation series respectively. These nicely explained the experiments of Koyasu et al. [16]. If such a straightforward reasoning was used generally, one would expect $ScS_{17}$, $TiSi_{17}^+$ and $VSi_{15}$ to also have 20 electrons and to be stable. But none of these clusters were seen as magic in the experiments. To explain this, Reveles and Khanna proposed that a Si atom can be considered to donate one electron to the metal atom only if it is bonded to the latter. To check which Si atoms the TM is bonded to, they analyzed the BCPs of these clusters. Existence of a BCP between the TM atom and a Si atom indicates that they are bonded. In the $ScS_{16}^{-}$, $TiSi_{16}$ and $VSi_{16}^{+}$ clusters, the TM atoms are bonded to all the Si atoms. On the other hand, in $ScSi_{17}$ the Sc atom is bonded to only 12 Si atoms leading to a valence electron count of 15. Similarly, in $TiSi_{17}^{+}$ and $VSi_{15}$ the TM atoms are bonded to 10 and 12 Si atoms respectively leading to valence electron counts of 13 and 17. These ideas were later extended to TM–Ge clusters and that is the topic of Sect. 3.

Guo et al. [27] also studied properties of different 3d TM encapsulated Si cage clusters. $VSi_{12}$ had the largest BE and EE among the $TMSi_{12}$ clusters, in agreement with Sen and Mitas. In fact, these were the largest BE and EE among all the cage clusters. $NiSi_{10}$ had the largest HOMO-LUMO gap and VIP among all the clusters. Among the $TMSi_{12}$ clusters, surprisingly, $TiSi_{12}$ had the largest HOMO-LUMO gap and VIP. Considering their results, and those of Refs. [30, 31, 44], Guo et al. also concluded that shell filling rule (whether 18 or 20 electrons) is just one of the aspects determining stability of these clusters.

We alluded to some more recent work on TM–Si clusters from Khanna's group. Let us now discuss these in detail. Cr–Si clusters only up to size 14 were studied in Ref. [34], and therefore the authors could not comment on the enhanced stability or otherwise of $CrSi_{14}$. In a very recent work [35], Khanna and co-workers have extended this study to include clusters from $CrSi_6$ to $CrSi_{16}$. Quite interestingly, $CrSi_{14}$ turns out to have the largest BE, and the largest HOMO-LUMO gap. The EE increases monotonically with size but shows a larger than trend value at size 14. Incidentally, $CrSi_{14}$ also has the largest AIP value. $CrSi_{12}$ has the second largest BE and AIP values. The HOMO-LUMO gap at size 12 is quite small. Various quantities calculated by Abreu et al. [35] for these clusters are shown in Fig. 9. These data clearly establish the highest stability of $CrSi_{14}$ in this series of clusters. The central question, of course, is why $CrSi_{14}$ happens to be more stable than $CrSi_{12}$ when the latter should nominally obey the 18-electron rule.

An understanding of this requires a close look at both the atomic and electronic structures of these two clusters. The lowest energy structures of $CrSi_{12}$ and $CrSi_{14}$ reported by Abreu et al. are shown in Fig. 10. $CrSi_{12}$ has an oblate HP structure. This means that the extent of the cluster in the hexagonal planes is larger than the distance between the planes. The oblate structure causes a crystal field-like splitting of the 3d orbitals of the Cr atom. Since $CrSi_{12}$ is a singlet, the 54 valence electrons are arranged in 27 spin paired MO's that can be designated as follows:

$$|1S^2 \mid 1P^4 \mid 1P^2\, 1D^8 \mid 1F^8\, 1D^2\, 2S^2 \mid 1F^4\, 2P^6\, 2D^{12}\, 1G^4 \mid\mid 1D^2|.$$

Here | demarcates orbitals of nearly same energy, and || separates the occupied and the unoccupied orbitals. MO energy diagram, and the isosurfaces of some of the frontier orbitals of this cluster as reported in Ref. [35] are reproduced in Fig. 11. A $2D^{12}$ electronic configuration does not appear in a spherical electron gas. Here it appears because of covalent bonding in the system. Therefore the electronic structure of this cluster cannot be explained in terms of an electron gas model. Through a more detailed analysis of the electronic structure of the cluster it was found that the Cr atom had an electronic configuration $4s^2 4p^6 3d^8$. The $3d_z{}^2$ orbital

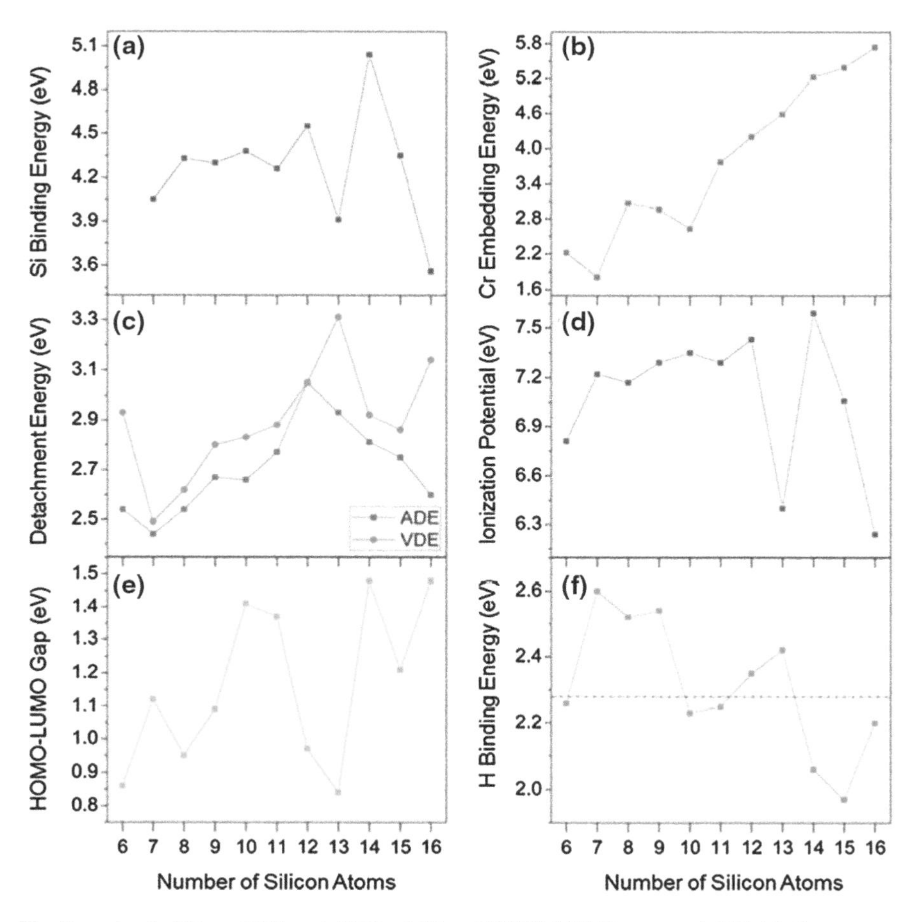


**Fig. 9** **a** $\Delta_1$, **b** EE, **c** ADE and VDE, **d** IP, **e** HOMO-LUMO gap and **f** H binding energy calculated for the $CrSi_n$ clusters with varying n. Reprinted with permission from J Phys. Chem. Lett. 2014, 5, 3492. c 2014 American Chemical Society

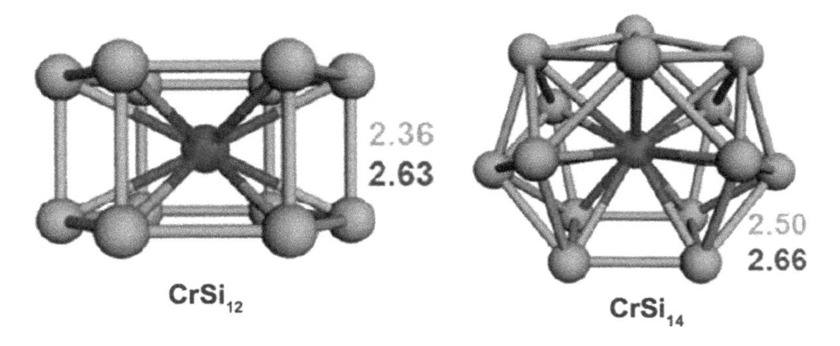


**Fig. 10** Lowest energy structures of the $CrSi_{12}$ and $CrSi_{14}$ clusters. Reprinted with permission from J Phys. Chem. Lett. 2014, 5, 3492. © 2014 American Chemical Society

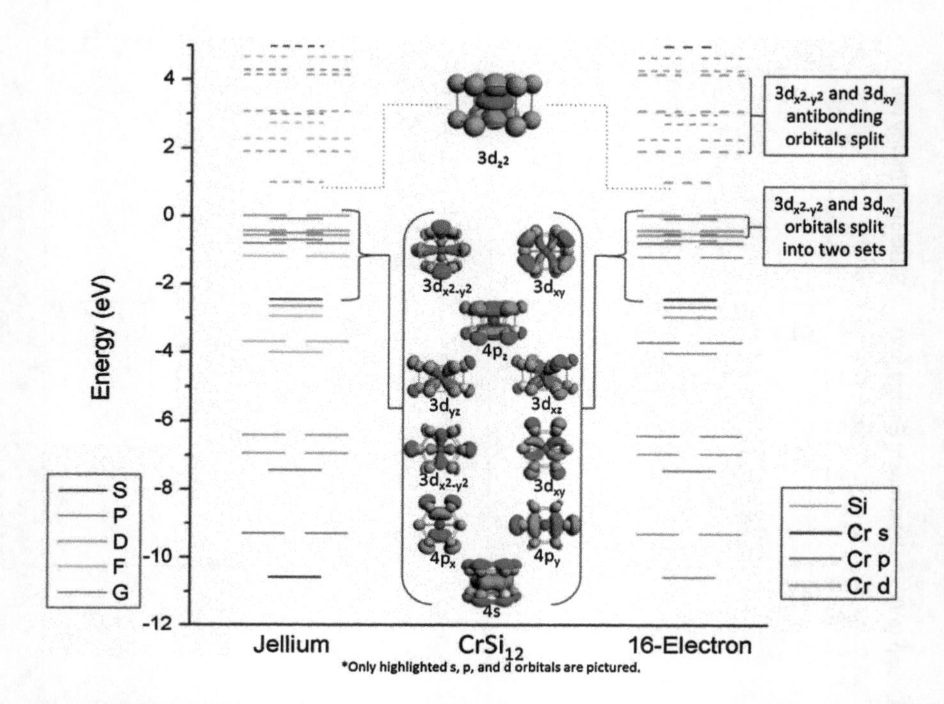


**Fig. 11** Molecular orbital energy level diagrams and isosurfaces some frontier orbitals of the $CrSi_{12}$ cluster. Reprinted with permission from J Phys. Chem. Lett. 2014, 5, 3492. © 2014 American Chemical Society

of the Cr atom was unoccupied as it was split to a higher energy in the crystal-field of the oblate structure of the cluster. This orbital, in fact, forms the LUMO of the cluster. An increased stability at 18 electrons requires a $ns^2(n-1)d^{10}np^6$ electronic configuration on the TM atom. Since the $3d_z{}^2$ orbital remains unoccupied, it is clear that the $CrSi_{12}$ cluster does not obey the 18-electron rule.

The structure of $CrSi_{14}$ is more spherical with no major crystal field splitting. Through a similar analysis of the MOs of this cluster the authors show that all the 3d atomic orbitals of Cr are occupied, and its electronic configuration is $4s^2 3d^{10} 4p^6$. The Cr atom can be said to have 18 effective valence electrons. The MO energy level diagram for $CrSi_{14}$ is reproduced in Fig. 12. It is important to note that naive electron counting can often fail. One has to study the energetics and electronic structure of a series of clusters in detail in order to identify the stable ones, and to understand the origins of their stability. We will find more such examples as we go along. For now, these findings have implications for the W–Si clusters also. Abreu et al. [35] reports a large HOMO-LUMO gap (1.42 eV) for $WSi_{12}$ due to a crystal field splitting of the 5d orbitals of W similar to $CrSi_{12}$, but finds a larger gap (1.80 eV) in $WSi_{14}$. $WSi_{14}$ is thus expected to be less reactive than $WSi_{12}$. Indeed, $WSi_{12}$ reacts with oxygen, while $WSi_{14}$ is resilient to such a reaction (Ref. 53 in ref.

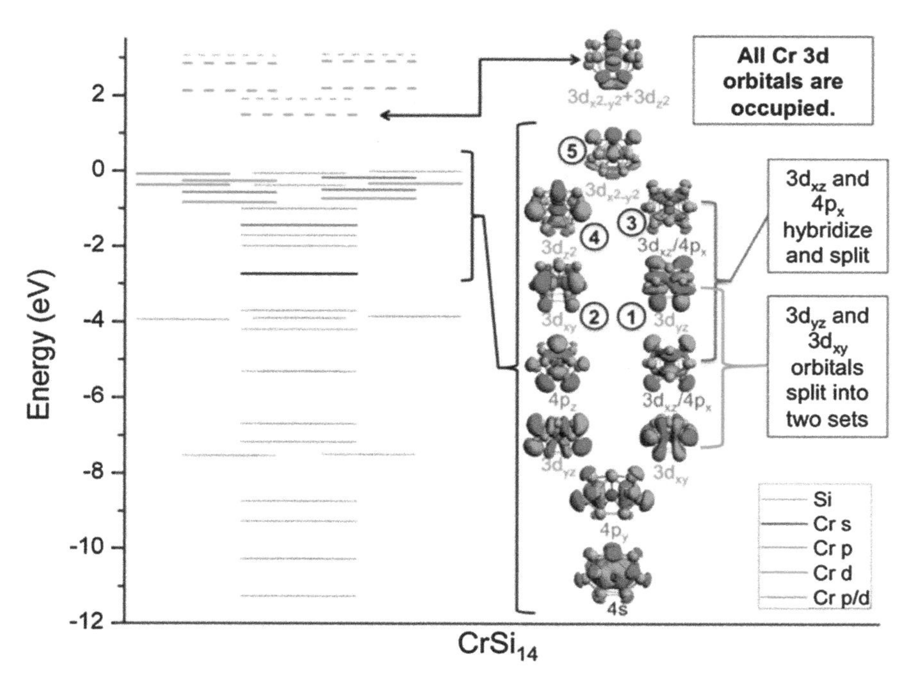


**Fig. 12** Molecular orbital energy level diagrams and isosurfaces some frontier orbitals of the $CrSi_{14}$ cluster. Reprinted with permission from J Phys. Chem. Lett. 2014, 5, 3492. © 2005 American Chemical Society

[35]). Recent theoretical work [47] also suggests that $WSi_{14}$ is more stable than $WSi_{12}$ by all measures.

These findings on Cr–Si clusters made Khanna and co-workers et al. re-visit Fe–Si clusters. In a recent study [39], Chauhan et al. studied $FeSi_n$ clusters over the size range n = 6–16. Just as in case of the Cr-doped clusters, $FeSi_{14}$ turned out to be the most stable cluster. It has a local peak in HOMO-LUMO gap, the largest $\Delta_1$, large EE, peak in AIP, and the lowest ADE and VDE. In addition to $FeSi_{14}$, $FeSi_9$ and $FeSi_{11}$ show enhanced stability in that they have local peaks in HOMO-LUMO gap, $\Delta_1$, EE, AIP, VIP and have low ADE and VDE. Indeed, $FeSi_{11}$ has the largest HOMO-LUMO gap and the lowest ADE. $FeSi_9$ has the second largest $\Delta_1$, and the largest AIP and VIP. In contrast to the previous study, $FeSi_{10}$ or $FeSi_{12}$ does not show any signatures of enhanced stability. $FeSi_{12}$ has the lowest HOMO-LUMO gap, the second lowest EE and the lowest AIP and VIP. So the applicability of 18-electron rule or shell models becomes questionable as $FeSi_{14}$ is a 22-electron cluster. These authors address the question of filling of the 3d orbitals on the Fe atom with changing cluster size. It turns out that the 3d orbitals are completely filled in $FeSi_9$ and all higher sizes. Therefore, the contention that Fe acquires a $4s^2 3d^{10} 3p^6$ configuration only at $FeSi_{12}$ is obviously not correct. In order to understand the observed stability better, Chauhan et al. calculated the number of Fe–Si and Si–Si bonds through an analysis of the BCPs of the electron charge density. In $FeSi_9$ and $FeSi_{14}$, all the Si atoms are bonded to the encapsulated Fe atoms, $FeSi_{11}$ also has a large number of Fe–Si bonds. In contrast, only 8 of the Si atoms are bonded to the Fe atom in $FeSi_{10}$. Nearest neighbor Si–Si bonds increases with size, but has local peaks at n = 9 and 11, and has a particularly large value at n = 14. All these together indicate that stability in these clusters has contributions from Si–Si and Fe–Si bonds. A compact symmetric structure, greater number of Si–Si and Fe–Si bonds, all contribute to the high stability of $FeSi_{14}$. Simple electron counting rules seem to have limited applicability.

Uchida et al. [44] addressed the question of electron counting rules in the 5d TM encapsulated $Si_{12}$ clusters. The most important point of their paper is that EE has the largest value for W irrespective of whether the WW spin conservation rule is imposed or not. This is claimed to indicate that the 18-electron rule is more clearly applicable to the case of 5d TM–Si clusters than the 3d TM–Si clusters. It was also reported that the 5d elements Hf–Os get encapsulated in a HP cage, and Ir–Au get encapsulated in a cage with four pentagonal faces (FPF). Specifically for $OsSi_{12}$, which is expected to be a 20 electron cluster, both HP and FPF cages were favored. Noting the example of $FeSi_{12}$, which was found to have a HP structure in Reveles et al's work [31], Uchida et al. claimed that stable HP structures can be found for both 18 and 20 electron clusters, but stable FPF structures can be found only for 20 electron clusters. This has to do with the symmetry of these structures rather than the exact atomic arrangements. The sequence of electronic levels in a spherical electron gas is $1S^2 1P^6 1D^{10} 2S^2 \cdots$. Although the FPF structure has a lower symmetry ($D_{2d}$) than the HP structure ($D_{6h}$), it is more spherical. This spherical shape lowers the energy of the 2S level. The energy difference between the 1D and 2S levels decreases so that shell filling at 20 electrons becomes more prominent than that at 18 electrons.

The new findings on Cr–Si and Fe–Si clusters prompted Khanna and co-workers to re-examine electron counting rules in W–Si clusters. Abreu et al. [47] found $WSi_{14}$ to be the most stable cluster in the $WSi_n$ series for n = 6–16, as already mentioned. In fact, it is the most stable cluster by all measures of stability, except for EE. $WSi_{12}$ shows signatures of electronic stability as it has a large HOMO-LUMO gap, large VIP and low ADE and VDE. $WSi_{13}$ turns out to be particularly less stable compared to $WSi_{12}$ and $WSi_{14}$. Analyses similar to the Cr–Si clusters show that the W 5d orbitals are not completely filled in $WSi_{12}$. The $d_z{}^2$ orbital is pushed to a higher energy due to crystal-field effect. But all the d orbitals are filled in $WSi_{14}$. It was argued that enhanced stability of $WSi_{14}$ originates from a greater number of Si–Si and W–Si bonds rather than from electron counting rules, a situation similar to $FeSi_{14}$.

Abreu et al. [47] also addressed the question why Beck found $TMSi_{15}{}^+$ and $TMSi_{16}{}^+$ as stable clusters for TM = Cr, Mo and W, while Hiura et al. [14] found $WSi_{12}$ as most stable. It is argued that in silane based experiments, $SiH_4$ activation takes place on the TM atom. Therefore, production of larger clusters will proceed as long as the TM atom is not completely encapsulated by the Si cage. The W atom is remains exposed till n = 11, but in $WSi_{12}$ (and $WSi_{12}{}^+$) the W atom is completely encapsulated. Therefore, reaction with silane stops at this size. Moreover, since $WSi_{13}$ is particularly less stable, growth to higher sizes is hindered. On the other hand, if the clusters grow by incorporation of TM atoms in already formed Si clusters, there is no reason for the growth to stop at $WSi_{12}$. These authors also show that hydrogen adsorption energy on $WSi_{12}$ and $WSi_{14}$ are lower than the BE of a $H_2$ molecules. On clusters smaller than $WSi_{11}$ opposite is the case. Therefore, formation of hydrogenated clusters is unlikely for $WSi_{12}$ and beyond, in agreement with Hiura et al's [14] findings. This, however, does not answer the question why Beck found $TMSi_{15}^+$ and $TMSi_{16}^+$ to be stable in all three different ways of cluster production, including using silane as a source of Si atoms.

# 3 Transition Metal–Germanium Binary Clusters

There have been fewer studies of TM encapsulated Ge clusters as compared to TM encapsulated Si clusters. This is perhaps consistent with the relative importance of these two materials in the electronic industry. In all the studies that have been undertaken on the TM–Ge clusters, the most important question still is the relative stability of different clusters and an understanding of their stability. As a necessary precondition to all such studies, structure of TM–Ge clusters have also been explored experimentally (in one case) and theoretically. When it comes to understanding relative stability of various TM–Ge clusters, electron counting rules have again been invoked. In this section we will focus on this aspect of the TM–Ge clusters. We will keep the discussion short, and will focus on some of the key experimental and theoretical works only.

## *3.1 Experimental Studies*

One of the first experimental studies exploring stability of TM–Ge clusters was performed by Zhang et al. [57]. In the mass spectra of Co–Ge clusters, a strong signal was observed for $CoGe_{10}^-$. Geometric and electronic shell closure was given as the reason for its stability. $CoGe_{10}^-$ happens to have 20 valence electrons. Neukermans et al. [48] measured the mass spectra of the $CrGe_n^+$ and $MnGe_n^+$ clusters. Among the Cr-doped clusters, $CrGe_{15}^+$ and $CrGe_{16}^+$ had the highest abundance. Among the Mn-doped clusters also $MnGe_{15}^+$ had the highest abundance although $MnGe_{14}^+$ was also produced in sufficient numbers. The enhanced stability of these clusters relative to their neighbors was explained by their stable, compact cage structures.

Nakajima's group had studied Ti-, Zr- and Hf-doped Ge clusters but only at size 16 [58]. In a more recent work they have studied a number of group-III, -IV and -V TM doped Ge clusters [59]. In particular, they studied Sc, Y, Lu, Tb, Ti, Zr, Hf, V, Nb and Ta doped clusters over the size range 5-22 in different cases. They explored the relative stability of these clusters as revealed by their mass spectra. Among the group-III TM–Ge cluster anions, $YGe_{16}^-$ and $LuGe_{16}^-$ had marked peaks in their mass spectra. The peak for $ScGe_{16}^-$ was not as sharp. These are in clear contrast to the $TMSi_{16}^{-}$ clusters as found by the same group [17]. Among the group-IV TM–Ge neutral clusters, $TiGe_{16}$, $ZrGe_{16}$ and $HfGe_{16}$ clearly had the highest abundance. Among the group-V TM–Ge cations, $VGe_{16}^{+}$, $NbGe_{16}^{+}$ and $TaGe_{16}^{+}$ had the highest abundance. The last two results are similar to what were observed for the doped Si clusters. The mass spectra reported by these authors are shown in Fig. 13.

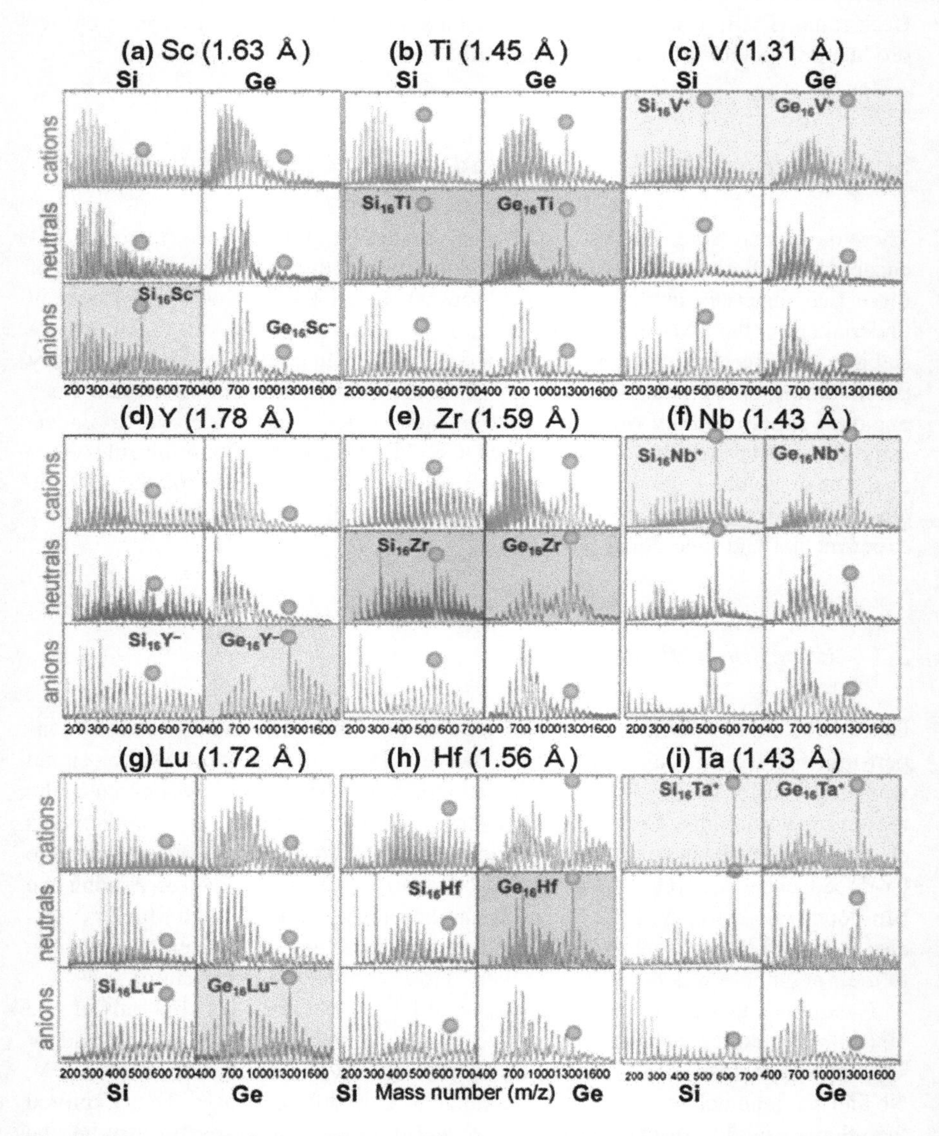


**Fig. 13** Mass spectra of group-III, group-IV and group-V TM-Ge clusters in anion, neutral or cation charge states as indicated. Reprinted with permission from Phys. Chem. Chem. Phys. 2012, 14, 9403 with permission of the PCCP Owner Societies

Greater stability of group-III TM–Ge anions, group-IV TM–Ge neutrals and group-V TM–Ge cations at size 16 is again claimed to be due to 20 electron filled shells. Firstly, the group-III TM–$Ge_{16}$ anions, group-IV TM–$Ge_{16}$ neutrals and the group-V TM–$Ge_{16}$ cations all have symmetric FK structures. Secondly, Y and Lu atoms, which are too large to be encapsulated by a $Si_{16}$ cage, better fits the larger cavity of $Ge_{16}$. Sc, on the other had fitted the $Si_{16}$ cage well but is too small for the $Ge_{16}$ cage. This explains the observed lack of sharpness in the mass spectrum of $ScGe_{16}^{-}$. Therefore, as in the case of TM–Si clusters, a cooperative effect between the atomic and electronic structures is responsible for the observed stability of these clusters.

This picture was supported by their measured anion PES spectra. Atobe et al. [59] measured PES spectra of these clusters at all sizes. For their obvious importance, we reproduce the spectra of the $TMGe_{16}$ clusters only in Fig. 14. The PES spectra of group-III TM–Ge cluster anions have maxima in the threshold energy due to their filled shell electronic configuration. The group-IV TM–Ge cluster anions had a minimum in the threshold energy at size 16 due to emergence of the low energy feature (panels (e)–(g) in Fig. 14). Since the low energy feature appears only at size 16, it was argued that this comes from a singly occupied MO. This feature indicates large HOMO-LUMO gaps for the neutral group-IV TM–$Ge_{16}$ clusters. Indeed the measured gaps of the $TiGe_{16}$, $ZrGe_{16}$ and $HfGe_{16}$ clusters are all between 1.8–1.9 eV. Note that the HOMO-LUMO gaps of the Zr and Hf doped $Ge_{16}$ clusters are much larger than the corresponding $TMSi_{16}$ clusters. This again goes on to show that geometric factor plays a major role in the stability of these clusters. Since the Zr and Hf atoms are bigger than Ti, they fit the larger cavity of the $Ge_{16}$ cage better, leading to stronger bondings and larger gaps. Another interesting and important difference between the TM doped Si and Ge clusters shows up in the anion PES spectra of the group-V TM–Ge clusters. In the doped Si clusters, the PES threshold energy had a maximum at size 14 or 15 for the group-V elements. But for the doped Ge clusters, there is a minimum in the threshold energy at size 16. This is also ascribed to geometric effects. Since the group-V atoms are smaller than the group-III and group-IV atoms, they lead to distortions in the $Ge_{16}$ cage.

Another aspect of these TM–Ge clusters studied in Ref. [59] is the question of the smallest size at which the TM atoms get encapsulated inside a Ge cage. Adsorption reactivity towards $H_2O$ in a FTR is used as a chemical probe method as in case of Si clusters. In the group-III TM–Ge cluster anions, the threshold sizes for the loss of adsorption reactivity are found to be 13, 16 and 16 for Sc, Y and Lu respectively. These sizes are smaller than the corresponding threshold sizes for the Si clusters. This is consistent with the fact that the Ge cage is bigger than the Si cage. For the neutral group-IV TM–Ge clusters, the threshold sizes turned out to be 12, 13 and 13 for Ti, Zr and Hf respectively. Finally, for the group-V TM–Ge cluster cations, these sizes were 11, 12 and 12 for V, Nb and Ta respectively. In these cases also the threshold sizes are smaller than the corresponding sizes for the Si clusters.

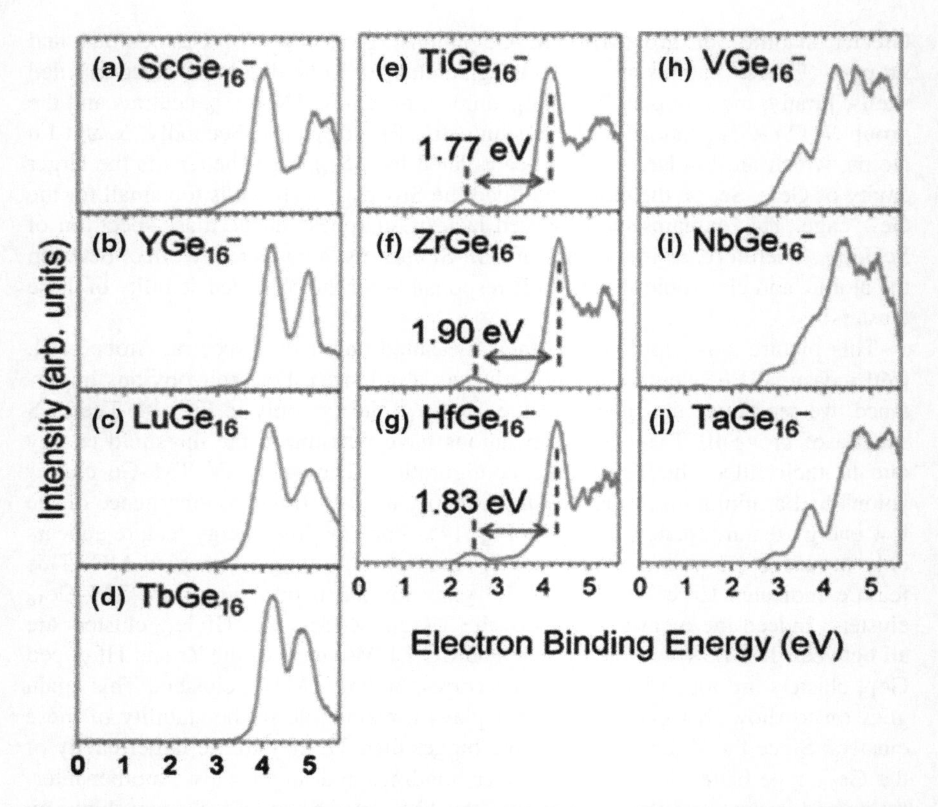


**Fig. 14** Anion photoelectron spectra of Sc-, Ti- and V-doped $Ge_{16}$ clusters. Reprinted with permission from Phys. Chem. Chem. Phys. 2012, 14, 9403 with permission of the PCCP Owner Societies

We also came across a very recent set of PES measurements by Deng et al. [60]. Although the anion PES spectra of $VGe_n$ clusters for n = 3–12 have been measured, the relative stability at different sizes have not been explored. In the same study, however, structures of neutral and anionic V–Ge clusters have been studied through DFT methods.

## *3.2 Theoretical Studies*

One of the first theoretical studies on TM–Ge clusters was performed by Kumar et al. [61]. In this work they used DFT methods to study structures of a number of TM doped Ge clusters at sizes 14, 15 and 16. We will focus on size 16 in view of the enhanced stability of certain TM–Ge clusters at this size. For all the group-IV TM atoms, the FK structure becomes most favorable. Not only that, for all these atoms, the f-like structure transforms to the FK structure implying that the former is not even a stable isomer. Thus the growth pattern of the TM–Ge clusters is different from that of the TM–Si clusters. All of the group-IV $TMGe_{16}$ clusters have large EEs in excess of 10 eV, and large HOMO-LUMO gaps close to 2 eV. The Ti, Zr and Hf encapsulated $Ge_{16}$ clusters have gaps of 1.790, 1.955 and 1.979 eV respectively. Their EEs are 10.608, 11.674 and 11.774 eV respectively. $CrGe_{16}$, on the other hand, has much lower gap and EE of 0.463 and 6.196 eV. Partial quenching of the large magnetic moment on the Cr atom is claimed to be responsible for its low EE. However, it is also possible that its smaller size compared to the group-IV atoms leads to less than optimal bonding with the $Ge_{16}$ cage resulting in low EE. Covalent bonding between the Ge cage states and the valence orbitals of the group-IV atoms is claimed to be the reason for enhanced stability of these clusters. Occupation of all the bonding orbitals also leads to large HOMO-LUMO gap. A few other studies have explored structures of TM–Ge clusters. Kapila et al. [62] studied $CrGe_n$ in the size range n = 1–13. They found that Cr takes exohedral positions in all these clusters. Bandyopadhyaya and Sen [63] studied Ni doped Ge clusters and found that $Ge_9$ is the smallest cage completely encapsulating a Ni atom.

A very interesting set of calculations came from Sen's Group in which they attempted to study the stability of TM–Ge clusters in terms of electron counting rules. In the first of these [63], these authors studied $NiGe_n$ clusters in the size range n = 1–20. They studied the relative stability of these clusters by plotting their EE, $\Delta_2$ and HOMO-LUMO gap with changing n. Along with the neutrals, they studied the cation clusters also. EE, $\Delta_2$ and HOMO-LUMO gap, all had the highest values at n = 10 among the neutral clusters, and at n = 11 among the cations. Since a Ni atom has 10 valence electrons, this indicated that clusters with 20 valence electrons have enhanced stability. If the enhanced stability of the 20-electron $NiGe_{10}$ cluster is due to a filled shell configuration of an electron gas inside the cage, then there should be a sharp drop in IP as the next Ge atom is added. Indeed there was a peak in IP at n = 10 and a sharp drop at n = 11. Sharp drop in IP from n = 10–11 was the strongest indication that the assumption of an electron gas inside the Ge cage is a good model for Ni–Ge cage clusters. Wang and Han [64] had also studied Ni–Ge clusters and had found $NiGe_{10}$ to be the most stable. But they did not invoke electron counting rules to explain its stability.

It is interesting that the second work from Sen's group [65] had predicted many of the observations in Atobe et al's [59] experiments. They had studied the stability of Sc, Ti and V doped Ge clusters in the size range 14-20. $ScGe_{16}^-$ was found to have the highest $EE^{WW}$ among all the Sc encapsulated clusters. Among all the Ti encapsulated clusters, $TiGe_{15}$ had the highest $EE^{WW}$, but those of $TiGe_{15}^+$ and $TiSi_{16}$ were very close. $VGe_{16}^+$ and $VGe_{18}^+$ had the highest $EE^{WW}$s among the V encapsulated clusters. $ScGe_{16}^-$, $TiSi_{16}$ and $VGe_{16}^+$ had by far the highest HOMO-LUMO gaps among the respective TM encapsulated clusters. These facts establish enhanced stability of these three clusters. The calculated $EE^{WW}$s and HOMO-LUMO gaps of all the TM encapsulated Ge clusters studied in this work are reproduced in Fig. 15. To further check their stability, EE $^{WW}$s of the Sc, Ti and V

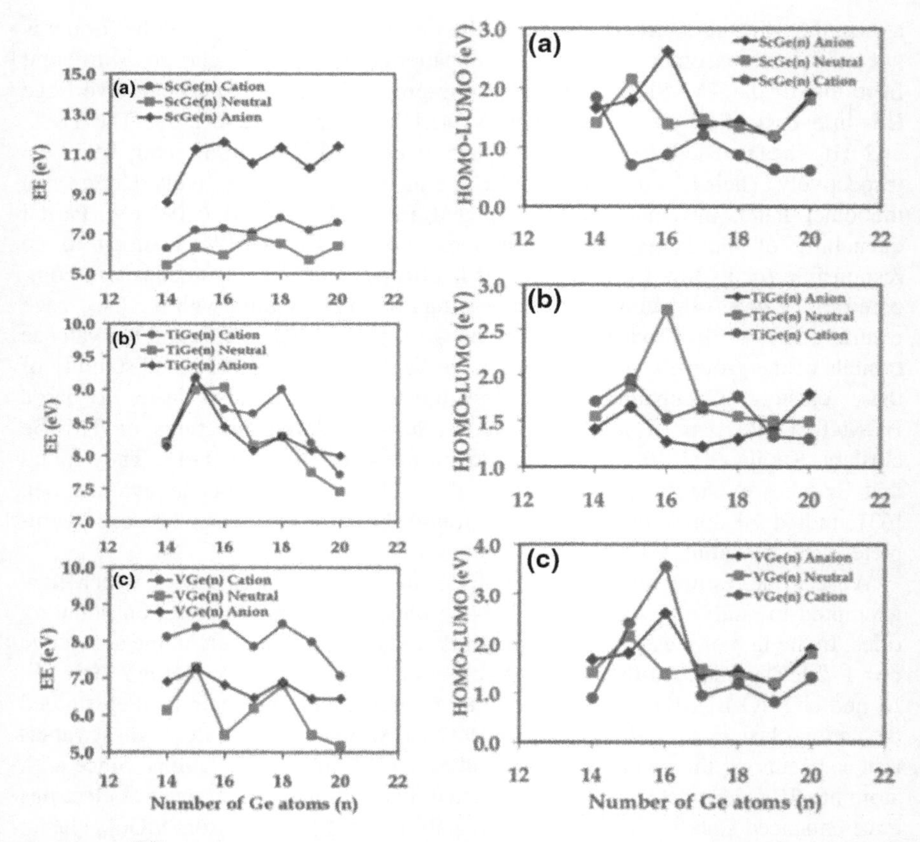


**Fig. 15** $EE^{WW}$s and HOMO-LUMO gaps of **a** Sc, **b** Ti, and **c** V encapsulating $Ge_n$ clusters. $EE^{WW}$s and gaps of neutral and singly ionized anion and cation clusters are shown. Reprinted with permission from J Phys. Chem. A 2010, 114, 12986. © 2010 American Chemical Society

encapsulated $Ge_{16}$ clusters were calculated and plotted in three different charge states: singly charged cation and anion, and neutral. It turned out, $ScGe_{16}$ had the highest $EE^{WW}$ in the anion state, $TiGe_{16}$ had the highest $EE^{WW}$ in the neutral state while $VGe_{16}$ had the highest $EE^{WW}$ in the cation state.

In this work the authors tried to have a deeper understanding of whether these observed stabilities really originated due to filled electronic shells, and at a more basic level, whether an electron gas really formed inside the Ge cages. To address these questions they analyzed the critical points of the molecular electrostatic potential (MEP) scalar field in these clusters. Other scalar fields can also be used, but there are certain advantages to using the MEP [65]. As is well known, the (3, +3) critical points of this scalar field identify points of local minima, i.e., they locate electron rich regions in the cluster or molecule. The (3, −1) critical points (the BCPs), on the other hand, indicate bonds between pairs of atoms. Reveles and

Khanna [50] had only analyzed the BCPs indicated by the (3, −1) critical points in TM–Si clusters. The (3, +3) and the (3, −1) critical points for some of the clusters calculated in Ref. [65] are shown in Fig. 16. Focusing on the panels (a), (c) and (e) we find that the TM atoms are bonded to all the 16 Ge atoms in each of these clusters. Assuming that each of the Ge atoms donates one electron to the central TM atom, these clusters can justifiably thought of as 20 electron clusters. The panels (b), (d) and (f) show that in all these clusters there is a large number of (3, +3) critical points inside the cage almost uniformly surrounding the TM atoms indicating a uniform electron-rich region. This was the first direct evidence of formation of a uniform electron-rich region due to delocalized electrons inside the cage.

While the above picture explains the stability of the $ScGe_{16}^{-}$, $TiSi_{16}$ and $VGe_{16}^{+}$ clusters, a naive application of electronic counting rules tells us that some other clusters should also have 20 electrons, and hence should have enhanced stability. These include $ScGe_{17}$ and $ScGe_{18}^{+}$ among the Sc encapsulated clusters, $TiGe_{15}^{-}$ and $TiGe_{17}^{+}$ among Ti encapsulated clusters, and $VGe_{15}$ among the V encapsulated

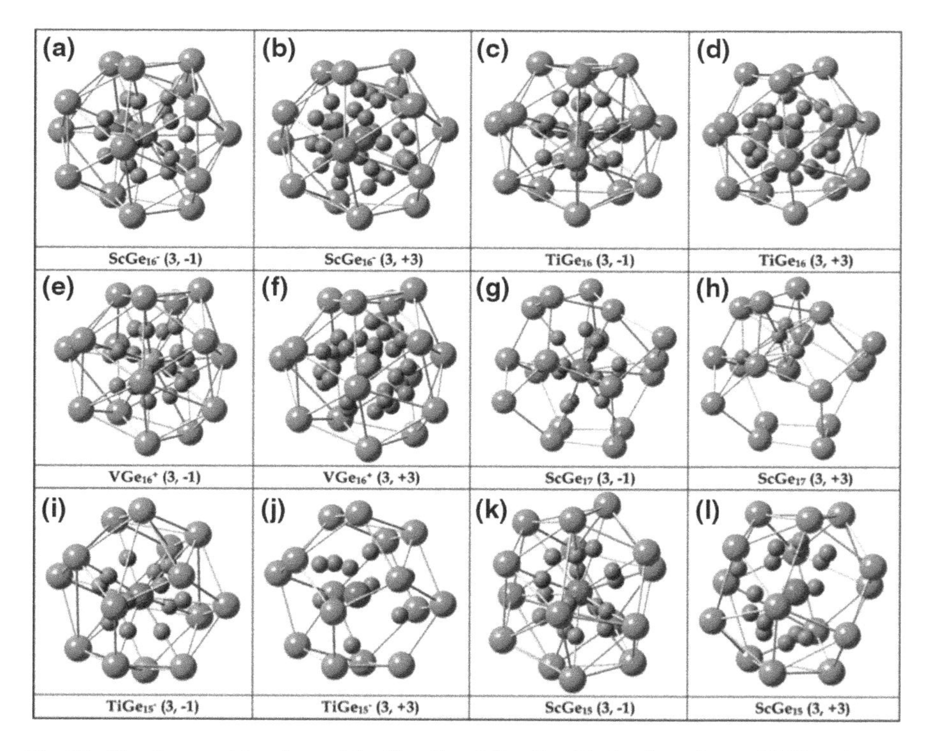


**Fig. 16** Structures and locations of the (3, −1) and (3, +3) critical points of some $TMGe_n$ clusters. *Purple* and *pink* (*larger*) *balls* represent the Ge and TM (at the center of each cage) atoms. *Red* and *green* (*smaller*) *balls* represent the (3, −1) and (3, +3) critical points respectively, as marked below each panel. Reprinted with permission from J Phys. Chem. A 2010, 114, 12986. © 2010 American Chemical Society

clusters. $ScGe_{17}$ and $ScGe_{18}^+$ do have the highest EE among the Sc encapsulated neutral and cation clusters respectively. However, they do not have peaks in their HOMO-LUMO gaps. It was argued that these clusters attain stability due to geometric effects. Indeed an analysis of the (3, −1) critical points show that the Sc is bonded to 13 and 11 Ge atoms in $ScGe_{17}$ and $ScGe_{18}^+$ respectively. Analysis of the (3, +3) critical points also shows that there is no formation of uniform electronic gas in these clusters. Critical points for the $ScGe_{17}$ cluster are shown in the panels (g) and (h) in Fig. 16. $TiGe_{15}^-$ shows peaks in EE and HOMO-LUMO gap. $TiGe_{17}^+$ however, does not show peaks in any of these quantities. If $TiGe_{15}^-$ is indeed a filled shell cluster, one would expect $TiGe_{15}$ to have a large EA. EAs and IPs all these clusters, as calculated in Ref. [65] are shown in Fig. 17. Note that what was called VDE in Ref. [65] is really the vertical IP of the neutral clusters. $TiGe_{15}$ indeed has a large EA which drops sharply at $TiGe_{16}$. The critical points of this cluster are shown in panels (i) and (j) in Fig. 16. The (3, −1) critical points show that Ti is actually bonded to 13 Ge atoms, suggesting that $TiGe_{15}^-$ is really an 18 electron cluster. But an electron-rich region is formed only on one side of the Ti atom as indicated by the locations of the (3, +3) critical points. In $TiGe_{17}^+$ the Ti atom is bonded to only 10 Ge atoms leading to an electronic count of 13. $VGe_{15}$ also shows peaks in EE and HOMO-LUMO gap indicating its enhanced stability due to a filled shell electronic configuration. If $VGe_{15}$ is a filled shell cluster, the IP should drop at $VGe_{16}$. As panel (b) of Fig. 17 shows, IP is the minimum for $VGe_{16}$. An analysis of the critical points shows that the V atom is really bonded to all the Ge atoms, so that the cluster has 20 valence electrons. However, the (3, +3) critical points are located only on one side of the V atom as in the case of $TiGe_{15}^-$.

Among the possible 18 electron clusters, $ScGe_{15}$, $ScGe_{16}^+$ and $TiGe_{15}^+$. $ScGe_{15}$ have peaks in EE and HOMO-LUMO gap. Critical points of MEP in this cluster are shown in the panels (k) and (l) in Fig. 16. The Sc atom is bonded to all the 15 Ge atoms making this an 18 electron cluster. There are also a large number of (3, +3) critical points located nearly uniformly all around the Sc atom, indicating formation of a uniform electron-rich region. IP shows a sharp drop at $ScGe_{16}$ indicating a filled shell nature of $ScGe_{15}$. $ScGe_{16}^+$ does not have a peak in either EE or HOMO-LUMO gap. An analysis of the critical points shows that the Sc atom is bonded to only 10 Ge atoms, and there is no electron-rich region inside the cage. $TiGe_{15}^+$ shows peaks in EE and HOMO-LUMO gap. The Ti atom is bonded to 14 of the Ge atoms making the valence electron count 17 for this cluster. But surprisingly an electron-rich region forms on one side of the Ti atom.

There are other clusters in these series which show major or minor peaks in EE. Of these $TiGe_{18}^+$ also has a peak in the HOMO-LUMO gap. Origin of these stabilities has been explained by either geometric or electronic effects but no evidence of formation of a uniform electron gas is found. This prompted the authors to argue that one should not try to explain all the observed stability in TM-semiconductor clusters through simple minded electron counting rules. The real situation in these clusters are far more complex.

Another interesting idea that has been forwarded to explain the observed stability of $TMSi_{16}$ and $TMGe_{16}$ clusters with Group-IV TM atoms is the aromaticity of the

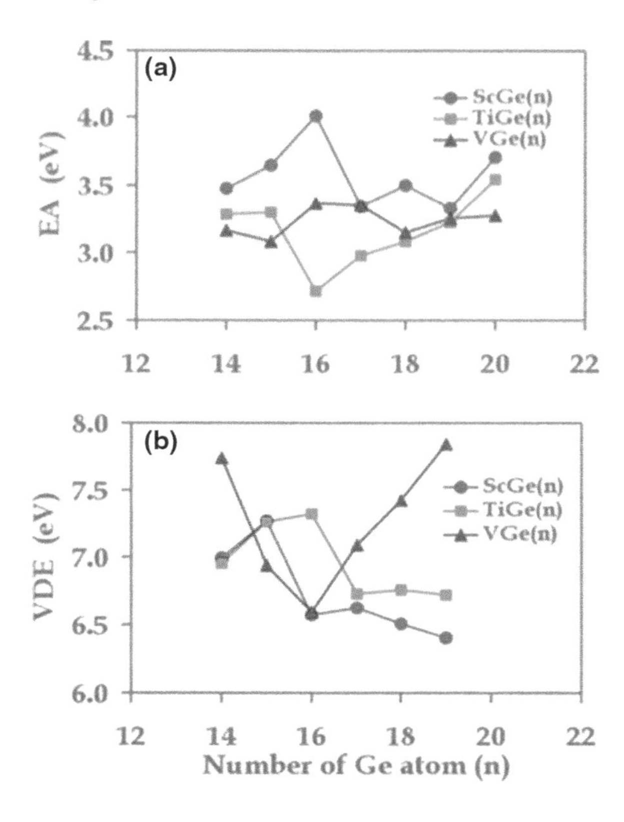


**Fig. 17** EA's **a** and IP's **b** of Sc, Ti, and V encapsulating $Ge_n$ clusters. Reprinted with permission from J Phys. Chem. A 2010, 114, 12986. © 2010 American Chemical Society

$Si_{16}^{4-}$ and $Ge_{16}^{4-}$ cages [58]. Aromaticity is a widely used but poorly defined concept in chemistry [66]. Anyway, one of the most widely used measures of aromaticity is the nucleus independent chemical shift (NICS). Negative values of NICS indicate aromatic character of a molecule or cluster. Furuse et al. [58] calculated the NICS values at the center of the FK $Si_{16}^{4-}$ and $Ge_{16}^{4-}$ cages. Both of these cages turned out to have negative NICS values with the $Si_{16}^{4-}$ cage being more aromatic than $Ge_{16}^{4-}$. As aromatic molecules tend to have higher HOMO-LUMO gaps, the large gaps found in these clusters can also be explained by the aromatic character of the cages. It must be pointed out that there is a fundamental difference between this way of looking at these clusters, and the picture of electron gas inside the cages. If aromaticity is really responsible for stability of these clusters, then the TM atoms have to donate all the four valence electrons to the Si or Ge cage. The electron gas, on the other hand, requires that the cage donates electrons to the metal atoms. The latter picture is more consistent with the electronic structure calculations that clearly show charge transfer to the metal atoms [67].

## 4 Concluding Remarks

More than two decades of research on binary TM–Si and TM–Ge clusters has helped us understand many aspects of these interesting systems. Even then a lot remains unclear. The basic fact one would like to be clear about is which are the clusters that are more stable than others. These can be used as the building blocks for cluster assembled materials. Answer to this fundamental question remains sketchy. Beck's original experiments found $TMSi_{15}^+$ and $TMSi_{16}^+$ clusters (TM = Cr, Mo and W) to be most stable. The same results were found by Ohara et al. [15] for the group-IV TM atoms. Hiura et al. [14], on the other hand, found that the most stable cluster size depends on the TM atom involved. As in all other branches of physics and chemistry, attempts have been made to rationalize these observations with simple rules. The 18-electron rule has been invoked to explain the stability of $WSi_{12}$, and based on the same rule, $CrSi_{12}$ was claimed to have enhanced stability. 20 electron filled shell configuration, along with highly symmetric FK structure has been cited as the reason for the stability of the Sc-, Ti- and V- encapsulated $Si_{16}$ and $Ge_{16}$ cage clusters in the anion, neutral and cation states respectively. As we have discussed in detail above, the simple electron counting rules cannot explain all the observed stability of TM–Si and TM–Ge clusters. Some clusters that are nominally expected to be stable do not show such behavior. Bandyopadhyay and Sen's calculations [65] bring out many of the complexities of these clusters. In light of the recent theoretical findings that $TMSi_{14}$ rather than $TMSi_{12}$ or $TMSi_{10}$ clusters (TM = Cr, Fe and W) are more stable [35, 39, 47], one is faced with the predicament that simple electron counting rules do not apply to these systems. Moreover, still unanswered is the question why Beck [1, 2] and Hiura et al. [14] obtained different sizes as stable clusters. Abreu et al. [47] answered it only partially. Therefore, we would conclude by saying that a lot more of experimental and theoretical studies are needed to develop a complete understanding of these clusters. We hope such studies will come soon.

**Acknowledgments** I am thankful to Akansha Singh for all her help in organizing the material and to Arpita Sen for her help in typing and editing.

## References

1. Beck SM (1987) J Chem Phys 87:4233
2. Beck SM (1989) J Chem Phys 90:6306
3. Brillson LJ (1982) Surf Sci Rep 2:123
4. Rubloff G (1983) Surf Sci 132:268
5. Tu KN (1975) Appl Phys Lett 27:221
6. Hiraki A (1986) Surf Sci 74:168
7. Röthlisberger U, Andreoni W, Parrinello M (1994) Phys Rev Lett 72:665
8. Broyer M, Pellarin M, Baguenard B, Lermé J, Vialle JL, Melinon P, Tuaillon J, Dupuis V, Prevel B (1996) Cluster assembled materials. Zurich-Utikon, p 27

9. Zdetsid A (2007) Phys Rev B 75:085409
10. Jarrold MF (1085) Science 1991:252
11. Hudgins RR, Imai M, Jarrold MF, Dugourd PJ (1999) J Chem Phys 111:7865
12. Ho K-M, Shvartsburg AA, Pan B, Lu Z-Y, Wang C-Z, Wacker JG, Fye JL, Jarrold MF (1998) Nature 392:582
13. Jackson K, Nellemoe B (1996) Chem Phys Lett 254:249
14. Hiura H, Miyazaki T, Kanayama T (2001) Phys Rev Lett 86:1733
15. Ohara M, Koyasu K, Nakajima A, Kaya K (2003) Chem Phys Lett 371:490
16. Koyasu K, Akutsu M, Mitsui M, Nakajima A (2005) JACS 127:4998
17. Koyasu K, Atobe J, Furuse S (2008) Nakajima. A J Chem Phys 129:214301
18. Ohara M, Miyajima K, Pramann A, Nakajima A, Kaya K (2002) J Phys Chem A 106:3702
19. Jaeger JB, Jaeger TD, Duncan MA (2006) J Phys Chem A 110:9310
20. Janssens E, Gruene P, Meijer G, W ̈oste L, Lievens P, Fielicke A (2007) Phys Rev Lett 99:063401
21. Gruene P, Fielicke A, Meijer G, Janssens E, Ngan VT, Nguyen MT, Lievens P (2008) Chem Phys Chem 9:703
22. Claes P, Janssens E, Ngan VT, Gruene P, Lyon JT, Harding DJ, Fielicke A, Nguyen MT, Lievens P (2011) Phys Rev Lett 107:173401
23. Ngan VT, Janssens E, Claes P, Lyon JT, Fielicke A, Nguyen MT, Lievens P (2012) Chem Eur J 18:15788
24. Claes P, Ngan VT, Haertelt M, Lyon JT, Fielicke A, Nguyen MT, Lievens P, Janssens E (2013) J Chem Phys 138:194301
25. Li Y, Tam NM, Claes P, Woodham AP, Lyon JT, Ngan VT, Nguyen MT, Lievens P, Fielicke A, Janssens E (2014) J Phys Chem A 118:8198
26. Ma L, Zhao J, Wang J, Lu Q, Zhu L, Wang G (2005) Chem Phys Lett 411:279
27. Guo L-J, Zhao G-F, Gu Y-Z, Liu X, Zeng Z (2008) Phys Rev B 77:195417
28. Gueorguiev GK, Pacheco JM (2003) J Chem Phys 119:10313
29. Gueorguiev GK, Pacheco JM, Stafström S, Hultman L (2006) Thin Solid Films 515:1192
30. Sen P, Mitas L (2003) Phys Rev B 68:155404
31. Reveles JU, Khanna SN (2005) Phys Rev B 72:165413
32. Kawamura H, Kumar V, Kawazoe Y (2005) Phys Rev B 71:075423
33. Andriotis AN, Mpourmpakis G, Froudakis GE, Menon M (2002) New J Phys 4:78
34. Khanna SN, Rao BK, Jena P (2002) Phys Rev Lett 89:016803
35. Abreu MB, Reber AC, Khanna SN (2014) J Phys Chem Lett 5:3492
36. Kawamura H, Kumar V, Kawazoe Y (2004) Phys Rev B 70:245433
37. Li J-R, Wang G-H, Yao C-H, Mu Y-W, Wan J-G, Han M (2009) J Chem Phys 130:164514
38. Khanna SN, Rao BK, Jena P, Nayak SK (2003) Chem Phys Lett 373:433
39. Chauhan V, Abreu MB, Reber AC (2015) khanna. S N Phys Chem Chem Phys. doi:10.1039/C5CP01386K
40. Wang J, Zhao J, Ma L, Wang B, Wang G (2007) Phys Lett A 367:335
41. Ren Z-Y, Li F, Guo P, Han J-GJ (2005) Mol Struct THEOCHEM 718:165
42. Li J-R, Yao C-H, Mu Y-W, Wan J-G, Han MJ (2009) Mol Struct THEOCHEM 916:139
43. Wang J, Ma Q-M, Xie Z, Liu Y, Li Y-C (2007) Phys Rev B 76:035406
44. Uchida N, Miyazaki T, Kanayama T (2006) Phys Rev B 74:205427
45. Hagelberg F, Xiao C, Lester WA (2003) Phys Rev B 67:035426
46. Lu J, Nagase S (2003) Phys Rev Lett 90:115506
47. Abreu MB, Reber AC (2015) khanna. S N J Chem Phys 143:074310
48. Neukermans S, Wang X, Veldeman N, Janssens E, Silverans RE, Lievens P (2006) Int J Mass Spectrometry 252:145
49. Kumar V, Kawazoe Y (2001) Phys Rev Lett 87:045503
50. Reveles JU, Khanna SN (2006) Phys Rev B 74:035435
51. de Heer WA (1993) Rev Mod Phys 65:611
52. Sen P (2010) In aromaticity and metal clusters. In: Chattaraj PK (ed) CRC Press, Taylor & Francis Group

53. Koyasu K, Atobe J, Akutsu M, Mitsui M, Nakajima A (2007) J Phys Chem A 111:42
54. Kumar V, Kawazoe Y (2002) Phys Rev B 65:073404
55. Kuk Y, Jarrold MF, Silverman PJ, Bower JE (1989) Brown. W L Phys Rev B 39:1168(R)
56. Bader RFW (1994) Atoms in molecules: a quantum theory. Oxford University Press, Oxford
57. Zhang X, Li G, Gao Z (2001) Rapid Comm Mass Spect 15:1573
58. Furuse S, Koyasu K, Atobe J, Nakajima A (2008) J Chem Phys 129:064311
59. Atobe J, Koyasu K, Furuse S, Nakajima A (2012) Phys Chem Chem Phys 14:9403
60. Deng X-J, Kong X-Y, Xu H-G, Xu X-L, Feng G, Zheng W-J (2015) J Phys Chem C 119:11048
61. Kumar V, Kawazoe Y (2002) Phys Rev Lett 88:235504
62. Kapila N, Garg I, Jindal V, Sharma H (2012) J Magn Magn Mat 324:2885
63. Bandyopadhyay D, Sen P (1835) J Phys Chem A 2010:114
64. Wang J, Han J-G (2006) J Phys Chem B 110:7820
65. Bandyopadhyay D, Kaur P, Sen P (2010) J Phys Chem A 114:12986
66. Chattaraj PK (2010) Aromaticity and metal clusters. CRC Press, Taylor & Francis Group
67. Wang J, Han J-G (2006) J Phys Chem A 110:12670

# Transition Metal Doped Boron Clusters: Structure and Bonding of $B_nM_2$ Cycles and Tubes

**Hung Tan Pham and Minh Tho Nguyen**

**Abstract** This chapter consists of a review on the geometric, electronic structure and chemical bonding in a number of small boron clusters doped by two transition metal elements. First-row transition metals introduce not only new class of boron clusters but also particular growth patterns. In the bimetallic cyclic motif, the two metals are vertically coordinated to the planar $B_n$ strings along the symmetry axis. The same M–M axis persist in double ring tubular forms. A bimetallic configuration model has been used to rationalize the electronic structure and stability for both bimetallic cyclic and tubular boron clusters. The anti-bonding $\pi^*$ and $\delta^*$ MOs of dimeric metals enjoy stabilizing interactions with the $B_7$, $B_8$ strings and $B_{14}$ double ring, thus inducing an enhanced stability for the doped cluster. Formation of bimetallic tube requires the occupancy of the molecular orbital configuration of $[(\sigma_{4s})^2 (\pi)^4 (\pi^*)^4 (\delta)^4 (\sigma_{3d})^2 (\delta^*)^4]$. At least 20 electrons are thus needed to populate the electron shell. However, there is no fixed electron count, but this rather depends on the nature of the metallic dopants.

**Keywords** Boron clusters · Transition metal doped boron clusters · Doubly metal doped boron clusters · Bimetallic configuration model

## 1 Introduction

The element of boron exists in polyforms, such as the α-, β-, γ-rhombohedral and α-tetragonal forms [1]. Only four pure elemental phases have been synthesized [2–4]. The boron nanotube (BNT) was also synthesized, and investigation on its work function subsequently indicated that the conductivity of BNT is close to that of the carbon nanotube (CNT) [5–7]. Motivated by the well-known polyforms of carbon,

H.T. Pham
Institute for Computational Science and Technology (ICST), Ho Chi Minh City, Vietnam

M.T. Nguyen (✉)
Department of Chemistry, KU Leuven, Celestijnenlaan 200F, 3001 Leuven, Belgium
e-mail: minh.nguyen@kuleuven.be



M.T. Nguyen and B. Kiran (eds.), *Clusters*, Challenges and Advances in Computational Chemistry and Physics 23,
DOI 10.1007/978-3-319-48918-6_6

much efforts have been devoted to the investigations of structural features and chemical bonding of boron clusters. The richness of carbon fullerenes gives a strong motivation for studies of boron fullerenes including the bucky-balls, volley-balls, core-shell stuffed boron fullerenes and many other forms [8–11]. Studies on boron sheets [12–18] were stimulated by results of two-dimensional (2D) sheets of carbon including graphene. The boron element possesses inherently small radius and electron deficiency, and thereby a high capacity to form bonds with itself and various elements. Boranes [19] present as an example for the rich and diverse chemistry of boron. Many fundamental bonding concepts were established through investigation of this class of compounds.

Studies of atomic clusters induce a profound impact not only on the understanding of structure and bonding, but also on the rational design of novel materials with tailored chemical properties [20]. As a cluster, the $B_{12}$-icosahedral structure was shown to be less stable than a planar isomer, which is the global minimum of $B_{12}$. Four different type of structures have so far been determined for pure boron clusters involving the planar (or quasi-planar), tubular (double ring, triple ring and multiple ring), bowl shaped and fullerene-like or cage forms.

The geometry of a boron cluster is basically controlled by two factors including its size and charge state. The planar or quasi-planar structures are the most favorable for neutral B clusters whose size is smaller than 20 atoms [21], except for $B_{14}$. The double ring (DR) form, the simplest tubular structure, consists of two $B_m$ strings often connected in anti-prism fashion, emerges as the ground state structure of even sizes $B_{2m}$ in the range of 2m = 20–26 [22]. For the neutral state, both $B_{14}$ and $B_{27}$ clusters have special characteristics. $B_{14}$ has a fullerene-type [23], whereas $B_{27}$ appears to be a triple ring tube [24], resulting from a superposition of three $B_9$ strings in anti-prism motif. Regarding the cationic state, three-dimensional (3D) structure is already found for the size $B_{17}^+$ [25]. On the contrary, the planar and quasi-planar forms are global minima structures for anionic boron clusters up to the size of $B_{27}^-$ [24, 26, 27]. The dianion $B_{20}^{2-}$ is also planar and exhibits a disk aromatic character [28].

In view of the complexity of geometries of pure boron clusters, in particular of the growth motif, theoretical models have been introduced to emphasize relationships between geometry, bonding and thermodynamic stability. The planar shape of small boron clusters tends to be a frequent target for application of the classical (4N +2) Hückel rule for their π electrons [26]. However, many clusters such as $B_{18}$, $B_{19}^-$ and $B_{20}^{2-}$ [26] do not follow the classical Hückel rule. Classical concepts are thus no longer helpful to rationalize all aspects of the electronic structure of boron clusters.

Let us mention a few typical cases. Based on the model of a particle moving in a circular box, the concept of disk-aromaticity was proposed, which turns out to be more general than the classical (4N +2) rule for planar or quasi-planar clusters [28]. As stated above, the dianion $B_{20}^{2-}$ follows the electron count of a disk aromaticity. The bowl-shaped boron clusters also belong to the class of disk-aromatic species. A hollow cylinder model (HCM) was proposed to rationalize the particular shape of tubular boron clusters. The HCM reproduces quite well the MO patterns of tubular clusters, and suggests a new type of tubular aromaticity [24, 29].

The boron element has a remarkable capacity to bind to various elements including transition metals. Extensive combined theoretical and experimental investigations revealed that the doping of one transition metal atom into small boron clusters leads to formation of the $B_nM$ cyclic structures, in which the M atom is usually located at the center of a $B_n$ ring [26]. This hyper-valent structure has even been observed in some pure boron clusters involving $B_7^-$, $B_8$, $B_8^{2-}$ and $B_9^-$ [26]. The central atom M tends to connect the $B_n$ ring through both delocalized π and σ bonds, formed by interaction between *d*-AO(M) and *sp*-AO(B). The B–B bonds of the $B_n$ ring are rather simple 2 centre–2 electron bonds. The cyclic structure often features a double aromaticity in which the number of occupied electrons on delocalized π and σ MOs satisfy the (4N +2) electron count.

A way of searching for new $B_nM$ cyclic structures is based on the electron requirement to form electronically stable and doubly aromatic clusters, in which electrons participate in n (2c–2e) B–B peripheral σ bonds, and two sets of aromatic delocalized MOs (6σ and 6π electrons). Thus, the electron count can be written as (2n + 12). Such a design principle was successfully applied to two rings $B_nM$ with n = 8 and 9. Several boron rings doped by other main group elements were theoretically investigated, but none of them was found to be a global minimum on the corresponding potential energy surface [30, 31]. The central atom M should form delocalized π and σ bonds with the $B_n$ ring, but when a main group element having a larger electronegativity, it favors connection with B atoms through 2c–2e bonds. Geometrical factors also play an important role in determining global minima. High symmetry cycles $B_nM$ were not observed for the sizes smaller than 8 ($n < 8$), and for larger clusters $n > 10$ B atoms, they become leaf-like, pyramid-like and umbrella-like structures [32].

Studies on the sizes $n \geq 10$ of doped $B_n$ clusters are rather limited. The Fe dopant can remarkably stabilize $B_n$ with n = 14, 16, 18 and 20, in fullerene and tubular forms [33]. The $B_{14}Fe$ and $B_{16}Fe$ are thus stable tubes, whereas $B_{18}Fe$ and $B_{20}Fe$ are stable fullerenes. Their formation and high thermodynamic stability suggest the use of transition metal dopants to induce different growth paths leading to larger cages, fullerenes and tubes of boron.

According to our literature survey, investigations on boron clusters multiply doped by transition metal atoms are even more scarce, even though some were experimentally observed in bipyramid-shape of organometallic compounds in solid state. A series of bimetallic Re complexes of boron were synthesized and structurally characterized to have a triple-decker $[(Cp^*Re)_2B_nX_n]$ with n = 5–10, X = H or Cl and Cp* = $Me_5C_5$, in which the $B_nX_n$ ring are coordinated by two Re atoms. The $B_5X_5$ and $B_6X_6$ units were found to be perfectly planar pentagon and hexagon motifs [34, 35]. A planar $B_6$ ring was found as a structural motif in the solid $Ti_7Rh_4Ir_2B_8$ [36], in which the $B_6$ unit is sandwiched by two Ti atoms.

However, the $B_6$ core appears to be too small to form a cycle with the presence of two metals. Similarly, the $B_{12}$ cluster is not large enough to cover a metallic dimer within a 2 × $B_6$ DR tube. With Fe and Co as dopants, the larger $B_7Co_2$, $B_7Fe_2$ and $B_7CoFe$ clusters have global minima in a B-cyclic motif, in which a perfectly planar $B_7$ is coordinated with two metallic atoms vertically placed along the $C_7$ axis. These $B_{14}M_2$ clusters turn out to exhibit a $2 \times B_7$ double ring tubular shape in which one metal atom is encapsulated by the $B_{14}$ tube and the other is located at an exposed position [37].

In this chapter, we review some important aspects of the structures of boron clusters doped by one and two metallic atoms. The effects of first-row transition metal atoms on the structural evolution, the growth mechanisms are established for both singly and doubly metallic doped boron clusters. Chemical bonding features are rationalized using the simple electron counts and shell models. Thermodynamic stability of cyclic bimetal doped boron clusters are rationalized in terms of orbital interactions of the metal–metal dimer MOs with the eigenstates of the disk-configuration, whereas the hollow cylinder model (HCM) is used for the bimetallic tubular structures.

## 2 Structural Trend of Pure Boron Clusters

For an understanding of the modifications induced by dopants, a brief overview of structural characteristics and growth pattern of the pure boron clusters appears necessary. For the sizes $B_3$–$B_{30}$, the structural landscape has clearly been identified for the cationic, neutral and anionic states. The structural landscape of small all-boron clusters can now be established and displayed in Fig. 1 [22]. $B_n$ clusters with n = 2–19 prefer the quasi-planar or planar form, excluding $B_9$ and $B_{14}$. Within the series of $B_{20}$–$B_{26}$, the DR tube, which results from a superposition of two $B_m$ strings in anti-prism motif, turns out to be the ground state of $B_{20}$, $B_{22}$, $B_{24}$ and $B_{26}$ clusters, whereas the open-shell $B_n$ (with n = 2m + 1) adopt a quasi-planar motif. Remarkably, $B_{27}$ emerges as a triple ring (TR) tube which is formed by joining three $B_9$ strings in an anti-prism fashion. On the contrary, $B_{28}$ tends to favor a quasi-planar shape.

A new structural motif is observed for the larger sizes, $B_n \geq 29$. A theoretical discovery of a novel structural motif was for the $B_{30}$ neutral cluster, which exhibits a bowl-shaped form. The highly symmetrical $B_{30}$ bowl has a pentagonal hole and satisfies the motif of $B_n@B_{2n}@B_{3n}$ with n = 5 and a disk aromaticity [38]. The $B_{32}$ [39] and $B_{36}$ [40] clusters were also classified as bowled structure. Thus, the bowl $B_{32}$ and $B_{36}$ are found to possess pentagonal and hexagonal holes, respectively. Other known structures such as $B_{29}$ and $B_{35}$ can be considered as defect species of these stable structures.

The shape of $B_{35}$ is similar to that of $B_{36}$, with one B atom being removed. $B_{29}$ can be regarded as one large fragment of $B_{36}$. Recent investigations indicated that all boron fullerenes appear at the sizes of n = 38, 39 [41]. However, it has been shown that for the size $B_{38}$, the fullerene B38A and quasi-planar B38B displayed in Fig. 1 are nearly degenerate on the energy landscape, in such a way that $B_{38}$ can be regarded as a critical size for structural transition from quasi-2D bowl to 3D fullerene form [42].

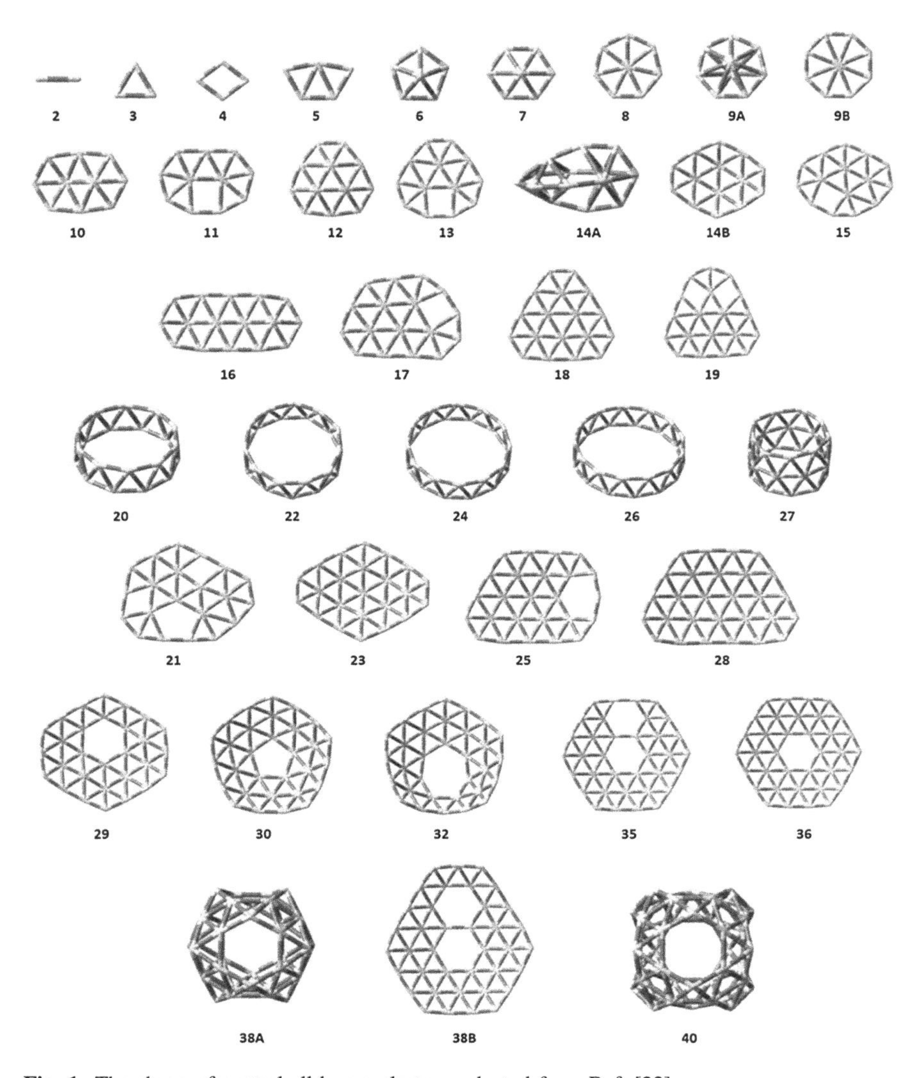


**Fig. 1** The shape of neutral all-boron clusters, adapted from Ref. [22]

For fullerenes, several extensive investigations and careful predictions were reported [43–47] on their geometrical and electronic structure, and bonding properties, from there, the $B_{40}$ emerges as a boron fullerene.

The neutral $B_{42}$ is stable in a triple ring (TR) tube resulting from a superposition of three $B_{14}$ strings in an anti-prism pattern. This tubular form is nearly degenerate with a cage-like structure containing octagonal holes. The latter form was located as the lowest-lying isomer for the cation $B_{42}^{+}$ [48].

The neutral $B_{44}$ cluster has a chiral cage-like structure containing two hexagonal, two heptagonal and two nonagonal holes [49]. Similarly, the neutral $B_{46}$ has a cage-like structure containing two hexagonal, two heptagonal and two decagonal holes [50]. The presence of octagonal, nonagonal and decagonal holes was a remarkable finding since it not only provided us with new bonding motif, but also marked a breakthrough in structural characteristics of elemental clusters. Each of these cages is composed of both delocalized σ and π electron systems that make it aromatic and thermodynamically stable.

## 3 $B_nM$ Clusters with M = Sc and Ti: Formation of Symmetrical Cycles

For both Ti and Sc dopants, the geometries of singly doped clusters indicate that they can replace a B atom of a $B_{n+1}$ cluster, or add to a $B_n$ cluster to release a $B_nM$ cluster [51, 52]. Figure 2 depicts the growth mechanism of the singly doped $B_nM$ clusters with M = Sc and Ti. Within the sizes of n = 2–6, each $B_nM$ is formed upon substitution of a B atom of a $B_{n+1}$ cluster, excluding $B_4Ti$ which results in an addition of Ti to $B_4$. $B_7M$ and $B_8M$ are preferentially formed along the addition pathway in which M is capped to the face of the corresponding $B_7$ and $B_8$ cycles.

Of the small bimetallic $B_nSc_2$ with n = 1–10, DFT computations [53] suggested that $B_6Sc_2$, $B_7Sc_2$ and $B_8Sc_2$ exhibit symmetrical cyclic motif in their most stable forms. Each planar boron string is coordinated by two Sc atoms vertically located along the $C_n$ axis, at two opposite sides with respect to the $B_n$ plan. Such a bimetallic cycle was also observed with two Ti dopants in $B_6Ti_2$, $B_7Ti_2$ and $B_8Ti_2$ having the same size [54]. Our extensive search on the mixed Ti and Sc bi-doping indicates that the mixed $B_6TiSc$, $B_7TiSc$ and $B_8TiSc$ clusters also feature bimetal cycles [55]. Thus, compared to the corresponding singly doped $B_nM$, the presence of an additional M atom, M = Ti and Sc, tends to stabilize the symmetrical cyclic structure at some special sizes, in which a planar boron string is multiply coordinated.

On the basis of the geometrical feature of $B_nM_2$, the growth sequence of bimetal-doped boron clusters can be established and illustrated in Fig. 3. For the mixed $B_2TiSc$, a competition for the ground state between both 3D and 2D structures. $B_3TiSc$ ends up in a 3D structure which is formed by an addition of a new B atom to $B_2TiSc$, and three B atoms are put around the TiSc axis. Similarly, $B_4TiSc$ and $B_5TiSc$ result from an addition of a B atom into $B_3TiSc$ and $B_4TiSc$, respectively. Finally, the cyclic structure is emerged at the $B_6TiSc$, $B_7TiSc$ and $B_8TiSc$ sizes. The presence of two metallic atoms tends thus to form 3D structure, whereas bare boron clusters prefer planar or quasi-planar shape for clusters having the same number of atoms. In this situation, the two metals create a vertical M–M axis which is suitable for interacting with B atoms arranged in string. The $B_6$, $B_7$ and $B_8$ strings are apparently large enough to keep the cyclic structure table,

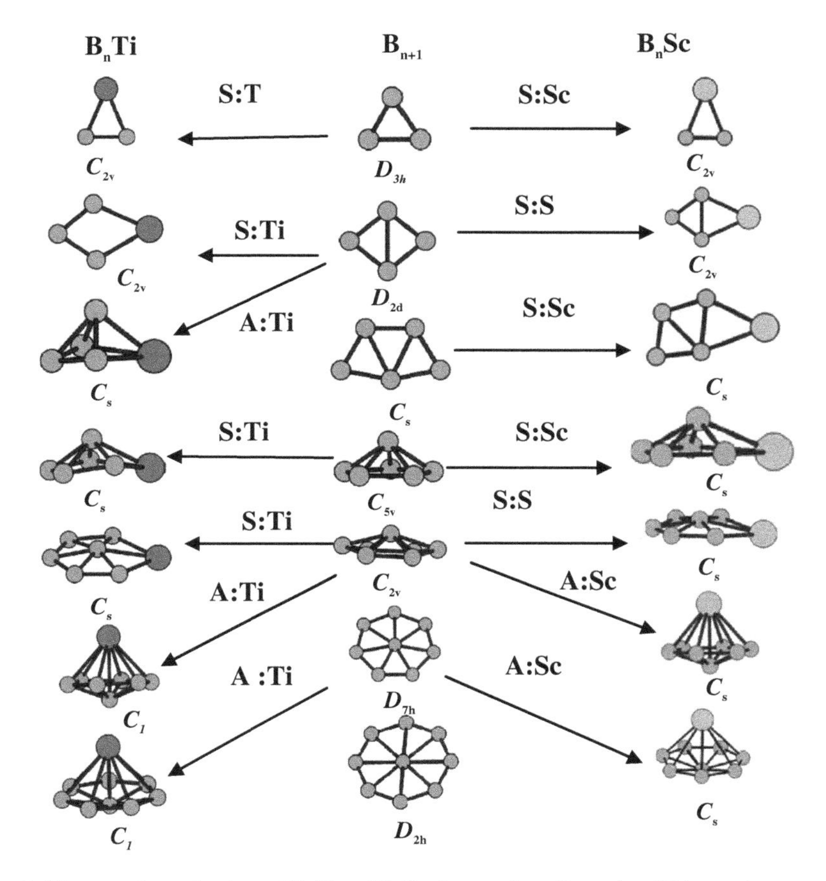


**Fig. 2** The growth mechanism of $B_nTi$ and $B_nSc$ clusters from $B_{n+1}$. **A**: addition pathway and **S**: substitution pathway. Larger ball = metal, small ball = boron

whereas the smaller strings $B_3$–$B_5$ are not suitable, by geometrical reason, for formation of cyclic motif.

# 4 The Effects of Iron and Cobalt Dopants

## *4.1 The Effect of Fe*

The global minima of $B_nFe$ with n = 1–10 were determined [56]. The Fe dopant is capped on a face of each $B_n$ counterpart resulting in 3D structures from $B_2Fe$ to $B_5Fe$, whereas $B_6Fe$ exhibits a quasi-planar structure formed by replacement of one B by Fe in the $B_7$ global minimum. The $B_7Fe$, $B_8Fe$ and $B_9Fe$ clusters do not favor

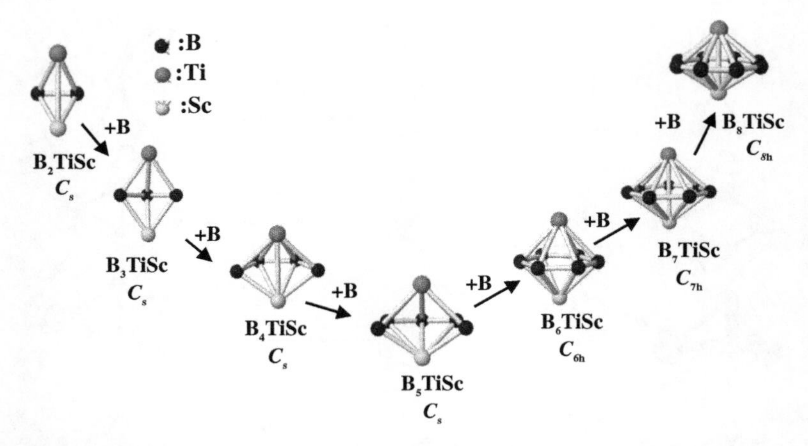


**Fig. 3** Formation of bimetal cyclic motif of structure in $B_nTiSc$

symmetrical cyclic motif, but rather an addition of Fe dopant to the global minima of $B@B_6$, $B@B_7$ and $B@B_8$, respectively, is preferred. The 3D structure is again found in $B_{10}Fe$, in which Fe occupies a high coordination position. However, DFT calculations indicated that the anions $B_8Fe^-$ and $B_9Fe^-$ are stable in cyclic structure [27].

Extensive search on the larger sizes [33] emphasized that both $B_{14}Fe$ and $B_{16}Fe$ exhibit a double ring tubular form in which Fe is encapsulated by two $B_7$ and $B_8$ strings, respectively, connected in an anti-prism fashion. The fullerene-like structure in which Fe is located at the central position of a $B_n$ counterpart, is identified as the most stable isomer for $B_{18}Fe$ and $B_{20}Fe$ (Fig. 4).

For the bare $B_{14}$ cluster, its neutral state exists as a fullerene-like structure, whereas the double ring form only appears in its dicationic state. While both $B_{16}$ and $B_{18}$ exhibit planar structure, incorporation of Fe leads to a tubular form in both resulting $B_{14}Fe$ and $B_{16}Fe$. Such a double ring tube is observed from the size of $B_{20}$ for bare cluster. While $B_{38}$ is a critical size for a transition between bowl and fullerene forms, a fullerene-like structure already appears at the size of 18 and 20 B atoms when a Fe dopant is inserted. Subsequently, Fe exerts a strong effect on the growth sequence of boron clusters in favoring 3D forms, like double ring or fullerene, at smaller sizes as compared with bare boron clusters. Figure 3 illustrates a new growth pattern induced by the presence of a Fe atom: from the cyclic $Fe@B_8^-$ and $Fe@B_9^-$, through the tubular $B_{14}Fe$ and $B_{16}Fe$ to the fullerene $B_{18}Fe$ and $B_{20}Fe$ structures.

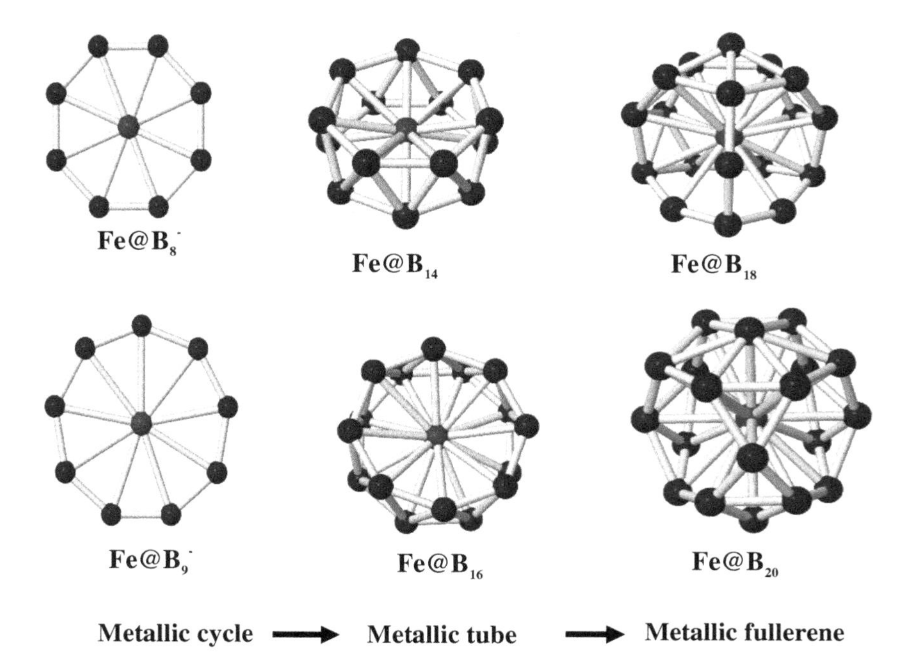


**Fig. 4** The growth pattern of some singly iron doped boron clusters

## 4.2 $B_nM_2$ and $B_{2n}M_2$ with $M_2 = Co_2$, $Fe_2$ and CoFe

### 4.2.1 Geometrical Aspects

Figure 5 displays the shape, point group of $B_nM$ clusters with n = 6, 7, 12 and 14 and M = Co and Fe. The results on $B_6M_2$ and $B_7M_2$, $B_{12}M_2$ and $B_{14}M_2$ are depicted in Figs. 5 and 6 respectively, in which the structures are denoted by **nM$_2$.y** where **n** is the cluster size, namely 6, 7, 12 and 14, **M$_2$** = Co2, Fe2 and CoFe stand for two dopants, and finally **y** = 1 and 2 labels the isomers with increasing energy ordering.

$B_6M_2$ Clusters

**$B_6Co_2$**. The high spin **6Co2.1** ($C_s$, $^5A''$) and **6Co2.2** ($D_{6h}$, $^5A_{2u}$) are competitive for the ground state, with a small energy gap of only 1 kcal/mol between them. **6Co2.1** is formed by adding a Co atom onto the $B_6Co$ structure (Fig. 6). **6Co2.2** is actually generated by replacement of a B atom on the axis of the $B_7Co$ global minimum, which also has a bi-pyramidal shape [57].

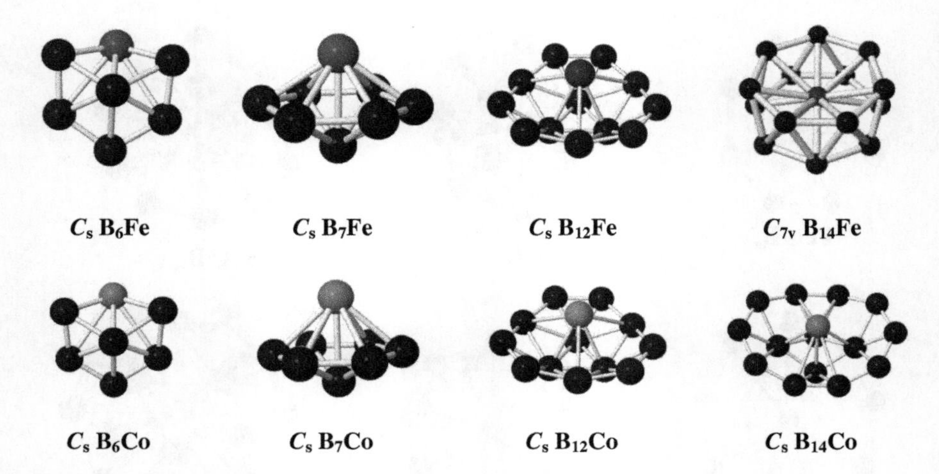


**Fig. 5** Shape of the lowest-lying $B_nM$ isomers with n = 6, 7, 12 and 14 and M = Co, Fe

**$B_6Fe_2$**. The triplet **6Fe2.1**($C_s$, $^3A''$), a structure having a similar shape with the global minimum **6Co2.1** of $B_6Co_2$, is identified as the lowest-energy structure. However, the corresponding quintet state **6Fe2.1** ($C_s$, $^5A''$) is only at 2 kcal/mol higher in energy, thus both spin states are likely nearly degenerate as the ground state for $B_6Fe_2$ (Fig. 6).

**$B_6CoFe$**. The sextet bipyramid **6CoFe.1** ($C_2$, $^6A$), a cyclic derivative in which Co and Fe are caped on two opposite sides of the $B_6$ plane, is located as the ground state. Due to the asymmetry of the dopants, the resulting by-pyramid is slightly distorted along the molecular axis (Fig. 6).

### $B_7M_2$: Stabilized Cyclic Isomers

**$B_7Co_2$**. The quartet heptagonal bi-pyramid **7Co2.1** ($D_{7h}$, $^4A''_2$) is identified as the ground state for $B_7Co_2$ (Fig. 6). This is again found in a beautiful seven-membered cyclic shape, in which two Co atoms coordinated in two opposite sides of the planar $B_7$ ring to form the highest point group of $D_{7h}$.

**$B_7Fe_2$**. Similarly, the heptagonal bi-pyramid **7Fe2.1** ($D_{7h}$, $^6A''_2$), possessing a sextet spin state, turns out to be the most stable isomer (Fig. 6). **7Fe2.1** is another bimetallic cyclic boron structure, in which the dimer $Fe_2$ is symmetrically placed along the central $C_7$ axis of the planar $B_7$ ring.

**$B_7CoFe$**. The high spin and high symmetry heptagonal bi-pyramid **7CoFe.1** ($C_{7v}$, $^5A''_2$) remains the lowest-lying structure which is thus a new bimetallic cyclic motif containing two different metals.

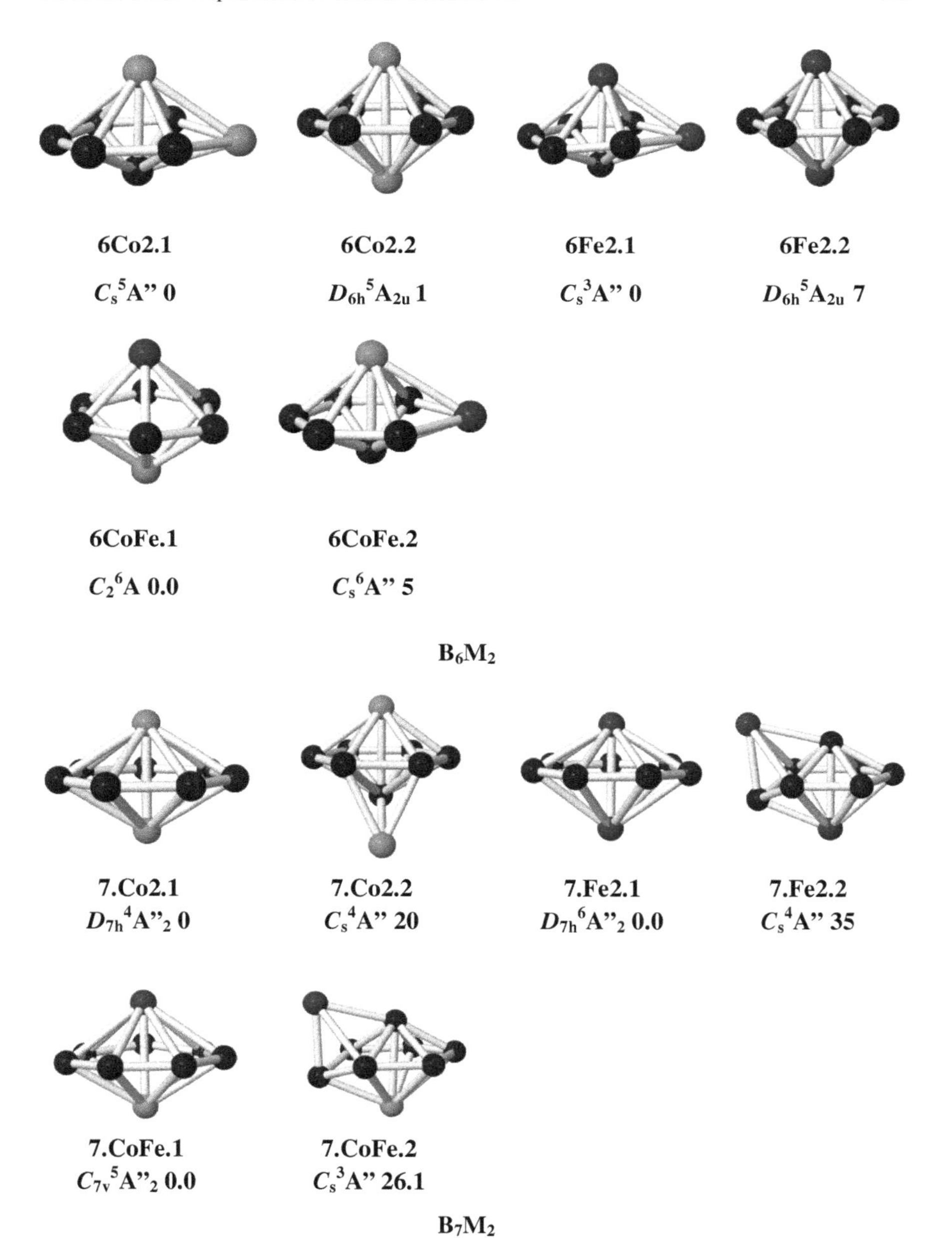


**Fig. 6** The two lowest-energy isomers of $B_6M_2$ and $B_7M_2$, with $M_2 = Co_2$, $Fe_2$ and CoFe. (TPSSh/6-311 + G(d) : Co; : Fe and : B

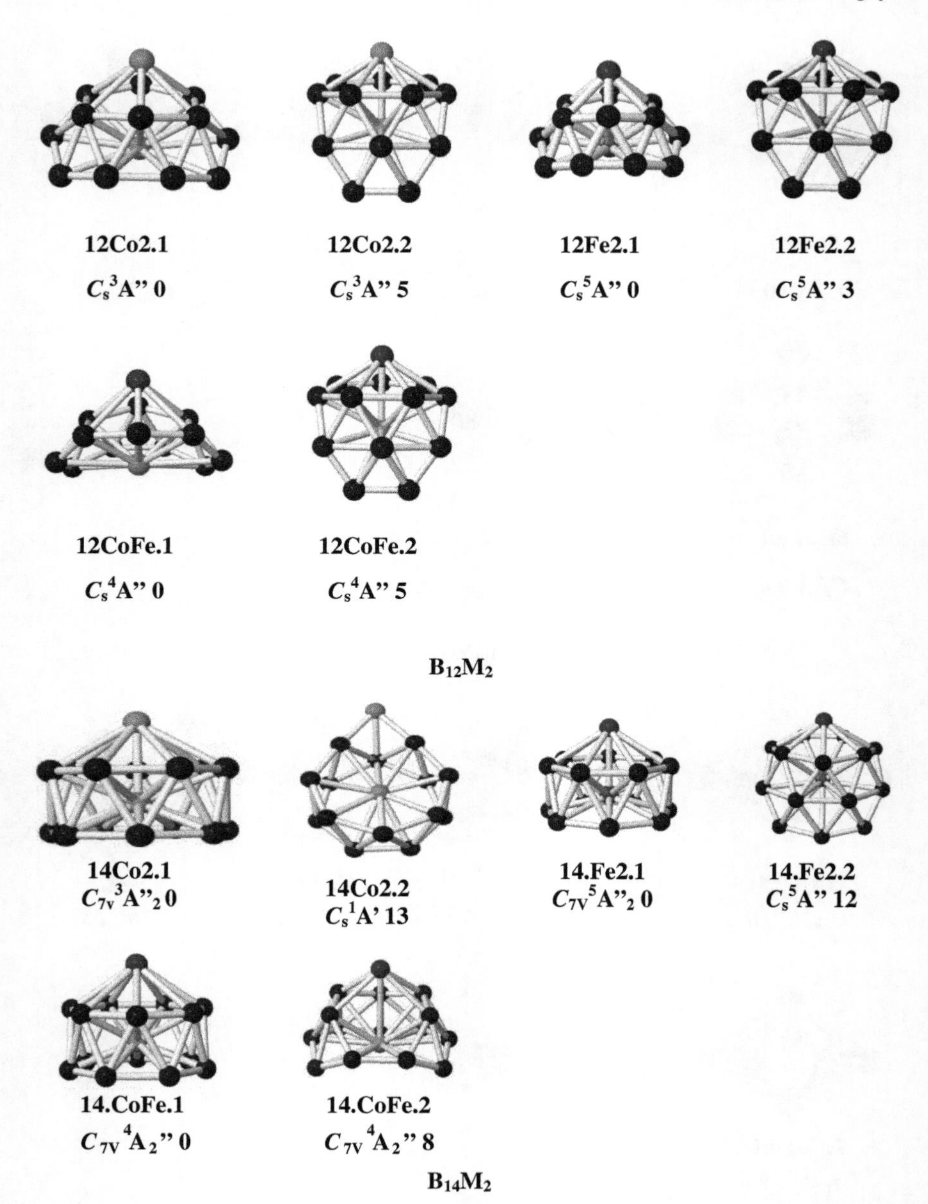


**Fig. 7** The two lowest energy isomers of $B_{12}M_2$ and $B_{14}M_2$, with $M_2 = Co_2$, $Fe_2$ and CoFe (TPSSh/6-311 + G(d)). : Co; : Fe and : B

## Bimetallic Doped $B_{12}M_2$

We now consider the $B_{2n}$ clusters having the sizes of 2n = 12 and 14. The main purpose in selecting these specific sizes is to find out as to whether a tube superposing either two $B_6$ strings or two $B_7$ strings could be formed and stabilized (Fig. 7).

**$B_{12}Co_2$**. The triplet **12Co2.1** ($C_s$, $^3A''$), which exhibits a strong distortion of a double ring tube surrounding the first Co atom, and the second Co atom is caped outside on a hexagonal face, is the most stable structure (Fig. 7). For this shape, the quintet state **12Co2.1** ($C_s$, $^5A''$) is quite close to the triplet state, with a small energy separation of 6 kcal/mol.

**$B_{12}Fe_2$**. A competition for the ground state also occurs for the di-iron doped clusters. Both quintet states **12Fe2.1** ($C_s$, $^5A''$) and **12Fe2.2** ($C_s$, $^5A''$) are in fact equally stable with a small energy separation of ~3 kcal/mol (Fig. 7).

**$B_{12}CoFe$**. Extensive search for the structures of the mixed $B_{12}CoFe$ converges to the lowest-energy **12CoFe.1** ($C_s$ $^4A''$) which possesses a similar shape as **12Co2.1** and **12Fe2.1** but with a quartet state (Fig. 7). The most characteristic feature of **12CoFe.1** is that the iron atom is caped outside on the hexagon, and the cobalt is inserted inside the tube surrounded by 12 B atoms.

## Bimetal Doped $B_{14}M_2$

**$B_{14}Co_2$**. DFT calculations indicated the high-symmetric triplet ground state **14Co2.1** ($C_{7v}$, $^3A''_2$) is the lowest energy structure of **$B_{14}Co_2$** (Fig. 7). This structure is composed of two seven-membered strings connected with each other in an anti-prism fashion. One Co atom is inserted at the central region, while the other is exposed outside the $B_{14}$ double ring. A double ring form exists in the di-cation $B_{14}^{2+}$ and was considered as the smallest boron double ring tube [58]. The neutral $B_{14}$ was found to be stable in a fullerene-type form [59]. Our extensive calculations revealed that the doping by two Co atoms tends to stabilize more the double ring tube than the fullerene counterpart.

**$B_{14}Fe_2$**. Both high spin states of **14Fe2.1**, having a similar shape to **14Co2.1**, are competitive for the ground state. The quintet **14Fe2.1** ($C_{7V}$, $^5A''_2$) and the triplet **14Fe2.1** ($C_{7v}$, $^3A''_2$) structures possess comparable energy content, and differ from each other by only ~2 kcal/mol. In contrast, the closed-shell singlet **14Fe2.1** ($C_s$, $^1A'$) is high in energy (56 kcal/mol).

**$B_{14}CoFe$**. The high symmetry quartet **14CoFe.1** ($C_{7V}$ $^4A''_2$) is once more identified as the ground structure. This double ring tube is characterized by the occupation of Co at the central position of the (2 × 7) boron tube, and the exohedral capping of Fe on a $B_7$ face. **14CoFe.1** can be viewed as a particular tubular structure in which both metallic nuclei form the $C_7$ axis and are half encompassed by a (2 × 7) B tube.

### 4.2.2 Geometrical Requirements of Bimetal Cyclic and Tubular Structures

The **6M2.1** structures described in Fig. 6 can be formed by adding a metal on the hexagonal face of the corresponding singly doped $B_6M$, which exhibits a hexagonal shape but with a B atom located at the central position [56]. The ground state of $B_7M$ cluster has a hexagonal bi-pyramid containing an M-B $C_6$ axis. Replacement of an axial B atom by an M atom releases the **6M2.2** motif. As for a prediction, the $B_8M$ cluster is expected to be stable in a heptagonal bi-pyramid with a M-B $C_7$ axis, which is similar to the **7M2.1** structure in which an axial M atom is replaced by a B atom. Accordingly, both minima **7M2.1** and **6M2.2** have a similar structural motif and are produced by a similar growth sequence. However, while **6M2.2** is competing with **6M2.1** for the global minimum status, **7M2.1** is the clear-cut ground state. In other words, the $B_6$ ring is not favored for formation of a bimetallic cycle. For the latter purpose, a $B_7$ ring appears more readily to satisfy the geometrical requirement.

For bimetal doped $B_{12}M_2$ clusters, **12M2.1** results from a strong distortion of a $(2 \times 6)$ $B_{12}$ tube with one metal caped on the hexagonal face, and but no tubular-type structure could be retained. In **12M2.2**, the competing isomer against **12M2.1** for the ground state, the seven B atoms make a heptagonal face, and the five remaining B atoms cannot introduce the second ring but instead undergo a distortion giving rise to another hexagonal ring. No tubular structure can thus be formed. In view of the emergence of a heptagonal ring in **12M2.2**, the $B_{14}$ system becomes significantly stabilized upon superposition of the second heptagon, and the most remarkable consequence is a production of a bimetallic double ring tube **14M2.1** (Fig. 7).

**Fig. 8** Optimized structures of $B_{14}Co$–Co–$CoB_{14}$, $B_{14}Fe$–Fe–$FeB_{14}$, $B_{14}Fe$–Co–$FeB_{14}$ and $B_{14}Co$–Fe–$CoB_{14}$ (TPSSh/6-311 + G(d))

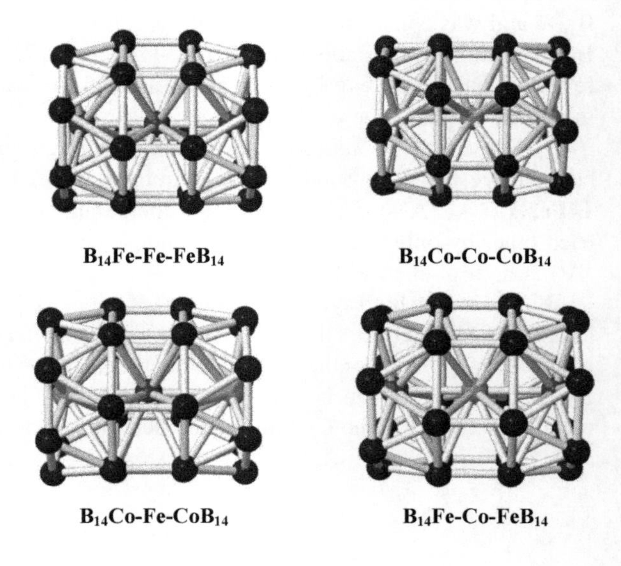

#### 4.2.3 Formation of Tubes from Bimetallic Tubular Units

In view of the tubular shape of the stable **14M2.1**, it is tempting to suggest that the exohedral metal atom can serve as a linker which connects another $[B_{14}M]$ unit, with M = Co and Fe, to form the doped boron tubes. Figure 8 depicts the optimized shapes of dimeric structures produced by combination of two bimetallic tubular $[B_{14}M]$ units. The final configuration of the equilibrium structure remains in a tubular shape. The transition metal atoms form the axis of the tube, whereas the merging of two $B_{14}$ units ends up in a quadruple ring having the shape of a slightly distorted heptagonal tube. The persistence of an external metal atom following fusion of $[B_{14}M_2]$ units apparently suggests it as another linker allowing a continuing formation of a boron tube (Fig. 8).

## 5 Bimetallic Cyclic Clusters $B_nM_2$ with N = 7 and 8

### 5.1 Geometries of $B_7M_2$ and $B_8M_2$

Systematic investigations on $B_nM_2$ clusters described above with M = Ti, Sc illustrate that $B_7M_2$ and $B_8M_2$ exhibit a cyclic ground state. In an attempt to search for more bimetallic cyclic boron clusters, we systematically considered the doubly doped $B_7$ and $B_8$ clusters, with *3d* transition metal atoms as the dopants. Geometries of $B_7M_2$ and $B_8M_2$ clusters with M ranging from Sc to Cu using DFT method (at the TPSSh/cc-pVTZ level) are displayed in Figs. 9 and 10, respectively. The bimetallic cyclic motif appears as a general tendency for $B_7M_2$ while some metals favor cyclic structure with $B_8$ string.

Studies on small $B_nM$ clusters with M = Cr, Mn, Fe, Co and Ni, and n = 1–7 have pointed out that the 3D structure in which the metal atom is capped on the hexagonal face of $B@B_6$ structure serves as the most stable structure of $B_7M$ clusters, excluding the Ni dopant which keep a planar form [57]. Similarly, with Ti and Sc dopants, the clusters are generated by adding them to a hexagonal face of the global minimum $B_7$ structure. The presence of second metal dopant, in both neutral and anionic states, tends to stabilize the $B_7$ cyclic form further, which is extremely unstable without dopant, and subsequently generate the bimetallic cyclic structure. The structural characteristic for $B_7M_2$ clusters show that two transition metals are coordinated oppositely with respect to the $B_7$ string, giving rise to the highest symmetry ($D_{7h}$) structure (Fig. 9). Overall, formation of bimetallic cyclic motif of structure becomes a general trend of the first-row transition metals in combining with the $B_7$ unit.

In contrast to $B_7M_2^{0/-}$, the existence of bimetallic cycles in the $B_8M_2^{0/2-}$ species rather depends on the type of transition metal and the charged state. As given in Fig. 10, $B_8Cu_2$ does not adopt a cyclic shape, but its most stable isomer results from an addition of a Cu atom onto the heptagonal face of $B@B_7$ cluster, and the second

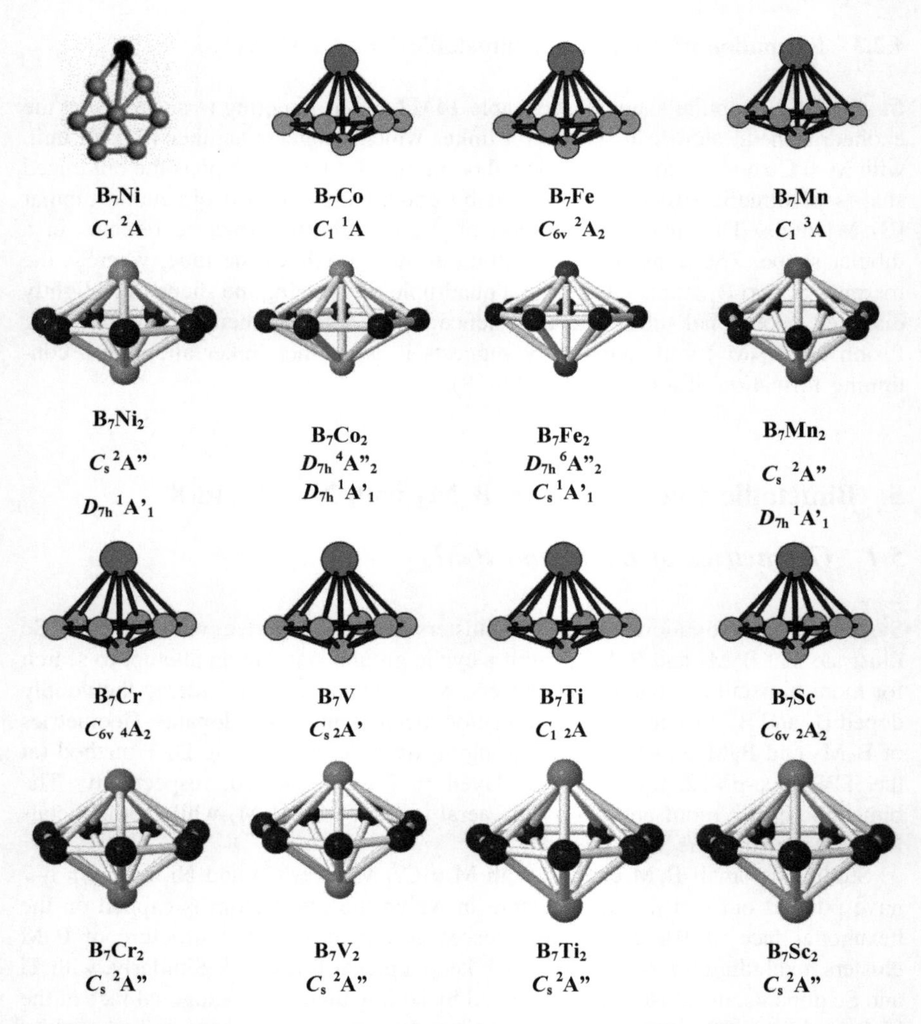


**Fig. 9** Shape of singly and doubly first-row transition metal doped $B_7M$ and $B_7M_2$ clusters. For $B_nM_2$, the upper banal is neutral and lower banal is anion state. Shape of $B_7M$ is obtained from Ref. [57]. $B_7V$, $B_7Cu$ and all $B_7M_2$ clusters were optimized using TPSSh/cc-pVTZ computations

Cu is added to a B–B edge. The Ni, Co and Fe dopants again form bimetallic cycles with eight B atoms for the neutral clusters. Within this motif, the planar $B_8$ cycle is coordinated by two metals vertically located along the main axis and in opposite positions with respect to the ring. On the contrary, both $B_8Co_2^{2-}$ and $B_8Fe_2{}^{2-}$ are not stable in cyclic form, but have a fish-like shape, whereas $B_8Ni_2^{2-}$ is a distorted cycle.

$B_8M_2$ clusters with M = Mn, Cr, V and Ti do not favor a cyclic motif in the neutral state, whereas the di-anion $B_8M_2{}^{2-}$ are stable in cyclic structure (Fig. 10).

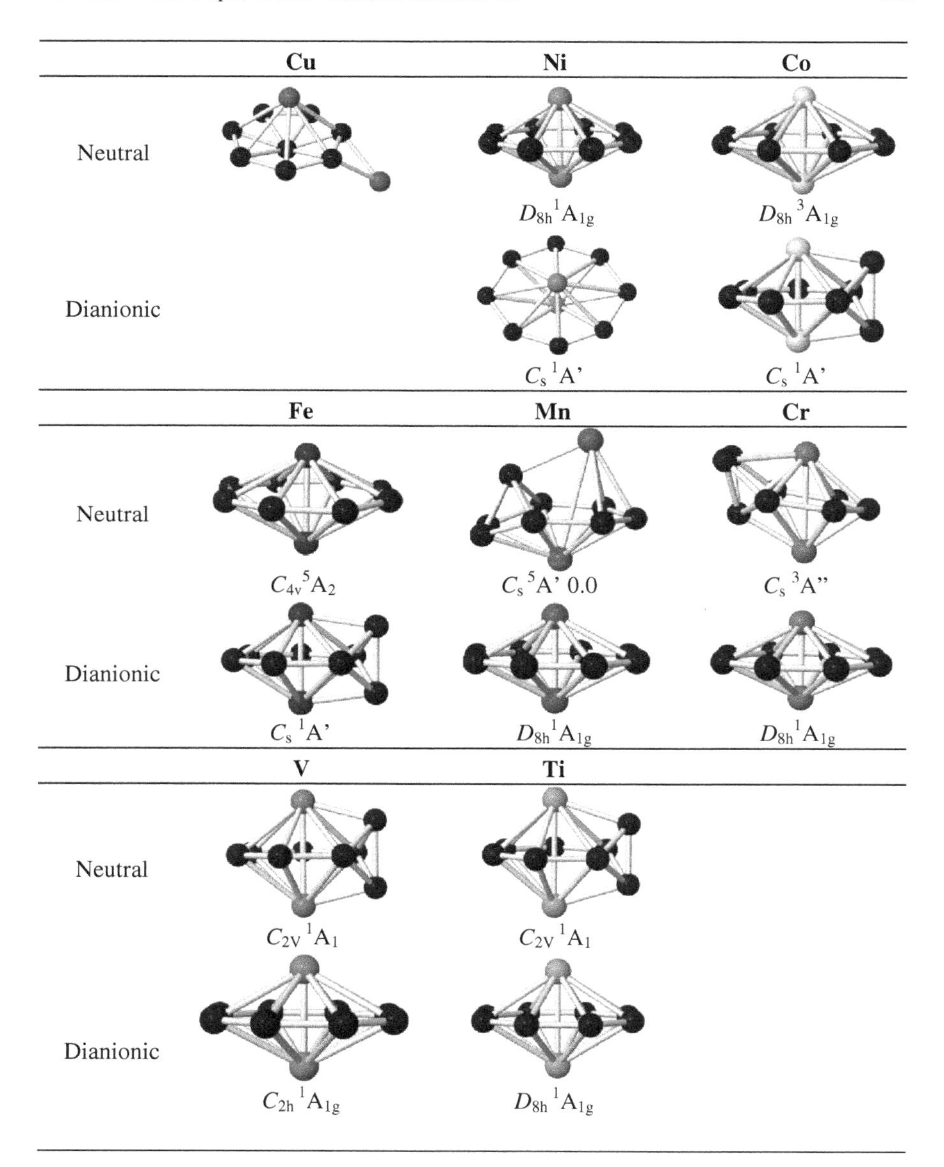


**Fig. 10** Shape of $B_8M_2^{0/2-}$ clusters with M = Sc – Cu (TPSSh/cc-pVTZ computations)

Accordingly, the charge state has a significant influence on the formation of bimetal cyclic structure, beside the nature of the metal involved.

Table 1 lists the bond length (Å) and Wiberg bond order (WBI) of the M–M connection in both bimetallic cycles and free metallic dimers. The WBI values of bare $M_2$ dimers are calculated for their electronic ground state using experimental

**Table 1** The bond length (Å) and Wiberg bond order (WBI) of the M–M connection in bimetallic cycles and free metallic dimers

| Species | Parameter | Ni | Co | Fe | Mn | Cr | V | Ti |
|---|---|---|---|---|---|---|---|---|
| $M_2$ | d(M–M) | 2.15 | (2.56) | 2.02 | 3.4 | 1.68 | 1.78 | 1.94 |
| | WBI | 0.8 | 1.3 | 1.3 | 1.0 | 6.0 | 1.0 | 0.8 |
| $B_7M_2$<br>$B7M_2^-$ | d(M–M) | 2.24<br>2.30 | 2.29<br>2.14 | 2.40<br>2.21 | 2.73<br>2.25 | 2.70<br>2.75 | 2.70<br>2.38 | 2.80<br>2.75 |
| | WBI | 0.2<br>0.3 | 0.3<br>0.8 | 0.3<br>1.2 | 0.2<br>1.3 | 1.4<br>1.9 | 0.3<br>1.9 | 0.2<br>1.0 |
| $B_8M_2$<br>$B_8M_2^{2-}$ | d(M–M) | 2.27<br>2.23 | 2.17<br>2.38 | 2.21<br>2.29 | 3.04<br>2.08 | 3.10<br>1.93 | 2.77<br>2.14 | 2.92<br>2.49 |
| | WBI | 0.3<br>0.3 | 0.3<br>0.6 | 0.3<br>1.2 | 0.1<br>1.3 | 0.3<br>1.6 | 0.2<br>1.6 | 0.6<br>1.8 |

The *lower lines* are values of $B_7M_2^-$ and $B_8M_2^{2-}$ clusters

bond lengths [60, 61]. The bond lengths of M–M connections in bimetallic boron cycles are significantly longer than those of free $M_2$ dimers in most cases (Table 1).

According to the usual convention of WBI, and the changes in bond length, the M–M bonds in $B_7M_2^{0/-}$ and $B_8M_2^{0/2-}$ cycles are consistently weaker than those of free $M_2$ dimers. Formation of bimetallic cycles with $B_7$ and $B_8$ strings thus induces a reduction on the strength of $M_2$ dimers. Such an effect is also recovered in $B_8M_2$ with M = Co, Fe and Ni. The WBI values of $Fe_2$, $Co_2$ and $Ni_2$ amount to ~0.3 in $B_8Fe_2$, $B_8Co_2$ and $B_8Ni_2$ structures, whereas they are identified for free diatomic species as 0.8, 1.3 and 1.3, respectively. The WBI values given in Table 1 for M–M connection of other $B_8M_2$ clusters (M = Sc, Ti, V, Cr and Mn), that do not adopt cyclic structure, are, as expected, significantly smaller than values of corresponding free metallic dimers, whereas their di-anionic state, which exhibits cyclic form, have remarkably greater values. Similarly, the WBI values calculated for anionic $B_7M_2$ boron cycles are much greater than those of the corresponding neutrals (Table 1). In the same vein, the Co–Co distance in $Co_2$ amounts to ~2.6 Å in its ground state $^5\Sigma_g^+$ [60] which is much shorter than that of Co–Co connections in $B_7Co_2^{0/-}$ and $B_8Co_2^{0/2-}$. Interaction of $M_2$ with $B_7$ and $B_8$ strings reduces the strength of the M–M bonds and these connections are weak bond in neutral state, while in the negative charge state the M–M bond strengths are enhanced as indicated by the WBI values.

## 5.2 Chemical Bonding of Bimetallic Cycles

As for an understanding of the chemical bonding and stability features of bimetallic cycles, it is useful to consider a bond model which raises from *d–d* interactions [62]. As for a typical example, Figs. 11 and 12 illustrate the orbital interaction diagrams

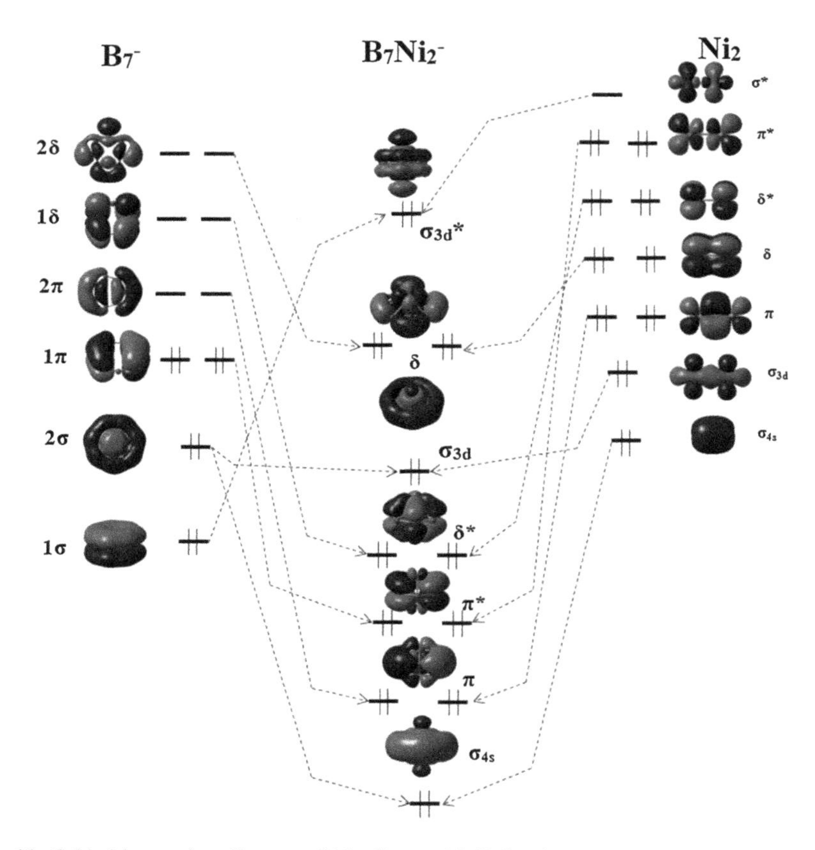


**Fig. 11** Orbital interaction diagram of $Ni_2$ dimer with $B_7^-$ string

of the Ni–Ni bond with $B_7^-$ and $B_8$ strings, respectively. It is obvious that results obtained for $Ni_2$ can be generalized for other metals.

The nickel atom has an electron configuration of $[Ar](3d)^8\ (4s)^2$, and consequently, the $Ni_2$ dimer has enough electrons to fully occupy an electronic configuration of $[\ldots\sigma_{4s}^2\ \sigma_{3d}^2\ \pi^4\ \delta^4\ \delta{*}^4\ \pi{*}^4]$. We note that there are two σ MOs in which $\sigma_{4s}$ is resulted by overlap of 4s AOs, and $\sigma_{3d}$ is a combination of 3*d* AOs. The anti-bonding $\delta^*$, $\pi^*$ and $\sigma^*$ MOs are filled, and as a result, the strength of metal–metal bond is reduced. For both $B_7^-$ and $B_8$ string moieties, the delocalized MO pattern appears to satisfy a disk aromatic framework, which is constructed from a model of a particle in a circular box [63]. Within this model, the boundary conditions are characterized by two quantum numbers, namely the radial quantum number n and the rotational quantum number m. The radial quantum number has value of n = 1, 2, 3,... whereas the rotational quantum number m = 0, ±1, ±2, ... which correspond to m = σ, π, δ, .... A state with non-zero value of m will be doubly degenerated. Thus the lowest eigenstates in ascending energy are 1σ, 1π,

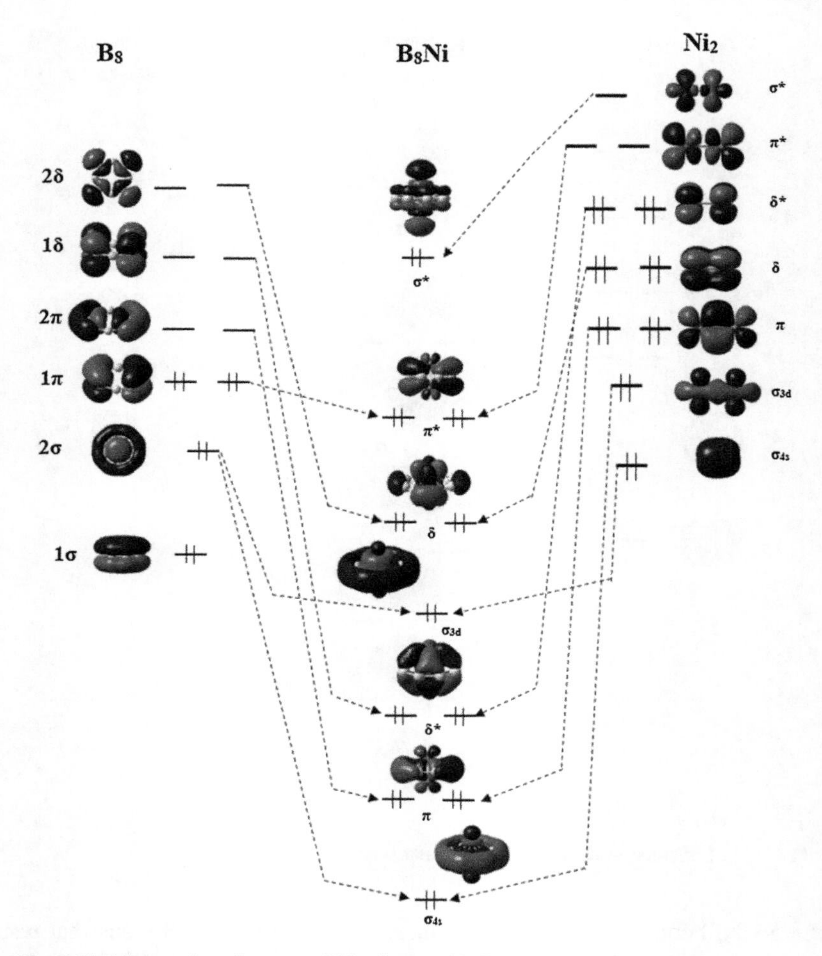


**Fig. 12** Orbital interaction diagram of $Ni_2$ dimer with $B_8$ strings

1δ… etc. A full occupation of degenerate eigenstates which correspond to 2, 6, 12, 16, 20,… electrons leads to a disk-aromaticity.

The fully occupied anti-bonding $\sigma_{3d}^*$ MOs turns out to have a stable interaction with the 1σ level of the disk aromatic configuration, and generate the $\sigma_{3d}^*$ MO. The $\pi^*$ MOs of $Ni_2$ dimer whose occupation significantly reduces the strength of the Ni–Ni bond, are stabilized upon interaction with the doubly degenerate 1π levels of either $B_7^-$ or $B_8$, and thereby produces the $\pi^*$ levels. Interestingly, the $\delta^*$-MOs of $Ni_2$ enjoy a stabilizing overlap with the vacant 1δ level and give raise to the doubly degenerate δ* levels as given in Figs. 11 and 12. Interaction of the bonding $\sigma_{4s}$ and $\sigma_{3d}$ MOs with the 2σ level ends up in the creation of the two σ sub-shells involving $\sigma_{4s}$ and $\sigma_{3d}$ MOs for $B_7Ni_2$ and $B_8Ni_2$. The π bonding MOs of $Ni_2$ have an enhanced overlap with the 2π level of either $B_7^-$ or $B_8$ disk, whereas the 2δ-MOs of $B_7^-$ and $B_8$ string undergo a

**Table 2** Electronic configuration and number of electrons of bimetallic boron cycles

| | Electron configuration | Number of electrons |
|---|---|---|
| $B_7Sc_2^-$ | $(\sigma_{4s})^2(\pi^*)^4(\pi)^4\ (\delta^*)^2(\sigma_{3d})^0(\delta)^0(\sigma_{3d}^*)^0$ | 12 |
| $B_7Ti_2^-$ | $(\sigma_{4s})^2(\pi)^4(\pi^*)^4(\delta^*)^4(\sigma_{3d})^0(\delta)^0(\sigma_{3d}^*)^0$ | 14 |
| $B_7V_2^-$ | $(\sigma_{4s})^2(\pi)^4(\pi^*)^4(\delta^*)^4(\sigma_{3d})^2(\delta)^0(\sigma_{3d}^*)^0$ | 16 |
| $B_7Cr_2^-$ | $(\sigma_{4s})^2(\pi)^4(\pi^*)^4(\delta^*)^4(\sigma_{3d})^2(\delta)^2(\sigma_{3d}^*)^0$ | 18 |
| $B_7Mn_2^-$ | $(\sigma_{4s})^2(\pi)^4(\pi^*)^4(\delta^*)^4(\sigma_{3d})^2(\delta)^4(\sigma_{3d}^*)^0$ | 20 |
| $B_7Fe_2^-$ | $(\sigma_{4s})^2(\pi)^4(\sigma_{3d})^2(\delta^*)^4\ (\pi^*)^4(\delta)^2(\sigma_{3d}^*)^0$ | 18 |
| $B_7Co_2^-$ | $(\sigma_{4s})^2(\pi)^4(\delta^*)^4(\sigma_{3d})^2(\pi^*)^4(\delta)^4(\sigma_{3d}^*)^0$ | 20 |
| $B_7Ni_2^-$ | $(\sigma_{4s})^2(\pi)^4(\pi^*)^4(\delta^*)^4(\sigma_{3d})^2(\delta)^4(\sigma_{3d}^*)^2$ | 22 |
| $B_8Sc_2^{2-}$ | $(\sigma_{4s})^2(\pi)^4(\pi^*)^4(\delta^*)^4(\sigma_{3d})^0(\delta)^0(\sigma_{3d}^*)^0$ | 14 |
| $B_8Ti_2^{2-}$ | $(\sigma_{4s})^2(\pi)^4(\pi^*)^4(\delta^*)^4(\sigma_{3d})^2(\delta)^0(\sigma_{3d}^*)^0$ | 16 |
| $B_8V_2^{2-}$ | $(\sigma_{4s})^2(\pi)^4(\pi^*)^4\ (\sigma_{3d})^2(\delta^*)^4(\delta)^2(\sigma_{3d}^*)^0$ | 18 |
| $B_8Cr_2^{2-}$ | $(\sigma_{4s})^2(\pi)^4(\pi^*)^4\ (\sigma_{3d})^2(\delta^*)^4(\delta)^4(\sigma_{3d}^*)^0$ | 20 |
| $B_8Mn_2^{2-}$ | $(\sigma_{4s})^2(\pi)^4(\pi^*)^4\ (\sigma_{3d})^2(\delta^*)^4(\delta)^4(\sigma_{3d}^*)^0$ | 20 |
| $B_8Ni_2^{2-}$ | $(\sigma_{4s})^2(\pi)^4(\pi^*)^4\ (\sigma_{3d})^2(\delta^*)^4(\delta)^4(\sigma_{3d}^*)^2$ | 22 |
| $B_8Ni_2$ | $(\sigma_{4s})^2(\pi)^4(\delta^*)^4\ (\sigma_{3d})^2\ (\delta)^4(\pi^*)^4\ (\sigma_{3d}^*)^2$ | 22 |

stabilizing interaction with the δ-MOs of $Ni_2$ yielding the δ levels. As a result, this analysis of orbital interaction reveals the electronic configuration of $B_7Ni_2^-$ and $B_8Ni_2$ cyclic structure as $[\ldots(\sigma_{4s})^2\ (\pi)^4\ (\delta^*)^4\ (\sigma_{3d})^2\ (\delta)^4\ (\pi^*)^4\ (\sigma_{3d}^*)^2]$ which is fully occupied by 22 electrons.

The electronic configurations and corresponding electron numbers of some bimetallic cyclic boron clusters are listed in Table 2. Orbital interactions suggest that the high thermodynamic stability of bimetallic cycles is a consequence of two actions: on the one hand, the empty levels of $B_7^-$ and $B_8$ involving 2π, 1δ and 2δ are occupied. On the other hand, bonding and anti-bonding MOs of free $M_2$ are involved in stabilizing interactions with different levels of the disk $B_7^-$ or $B_8$ which enhance the stability of the resulting clusters. Subsequently, the $B_7M_2^{0/-}$ and $B_8M_2^{0/2-}$ clusters, with M being a first-row transition metal, are stabilized in the bimetallic cyclic motif.

## 5.3 *Aromatic Feature of Bimetallic Boron Cyclic Structures*

The aromatic character is analyzed using the ring current concept, which is the magnetic response of the total electron density. In combination with the ipsocentric model [64–66], the current density maps of π and σ electrons are produced with respect to the $B_n$ plane. Orbital contributions to the total ring current are also calculated, and, as a result, the participation of each MO to the aromatic character of whole cluster can be established. Figures 13 and 14 display the π, σ and total (involving both π and σ orbitals) ring current maps for some bimetallic boron

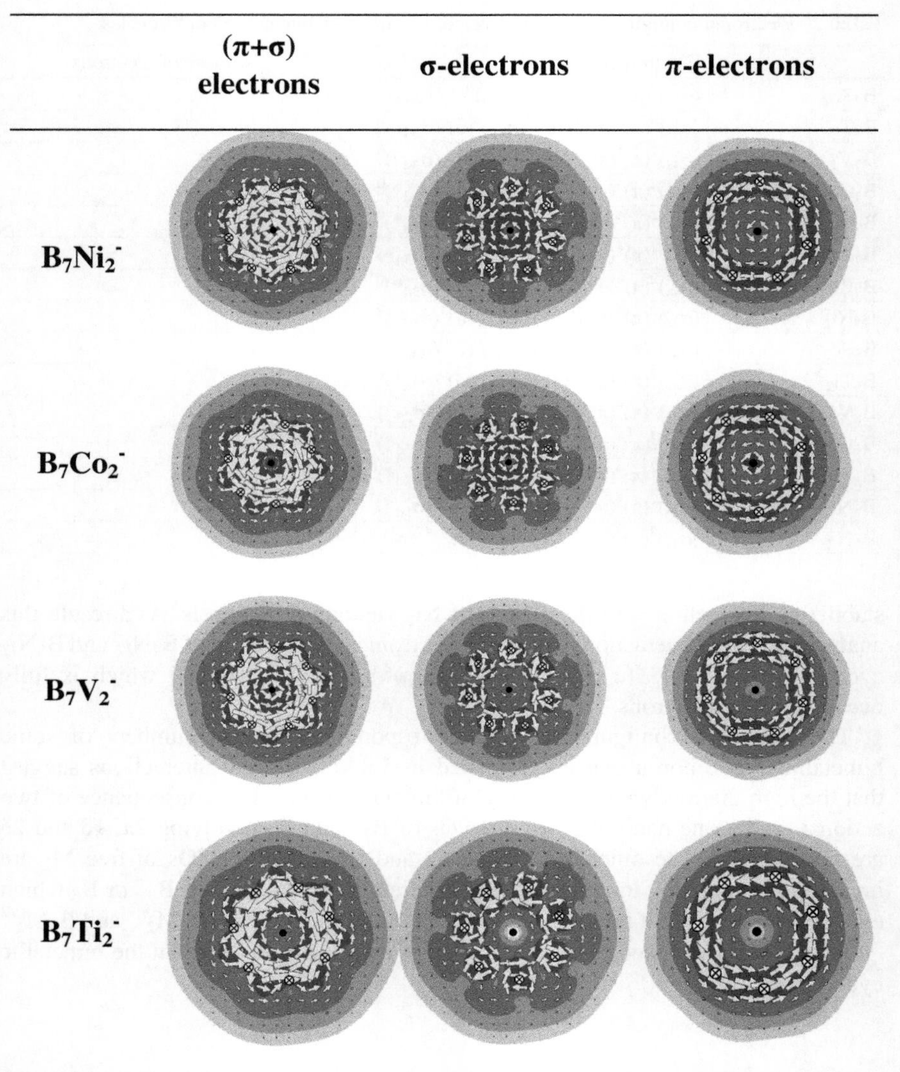


**Fig. 13** Total (π + σ), σ and π ring current maps of some $B_7M_2^-$ cyclic structures (B3LYP/6-311 + G(d)). *Black dot* is a metal atom

cycles. The ring current density of π and σ electrons are given in Figs. 15, 16 and 17, respectively.

For $B_7M_2^-$ boron cycles, except for Sc and Fe dopants, both π and σ electrons raise a diatropic current. As a consequence, the total current map, which is contributed by both π and σ electrons, has a strongly diatropic character. They can thus be classified as doubly aromatic species. The π electrons of Sc dopant induce a strong paratropic current indicating an anti-aromatic feature, whereas a diatropic

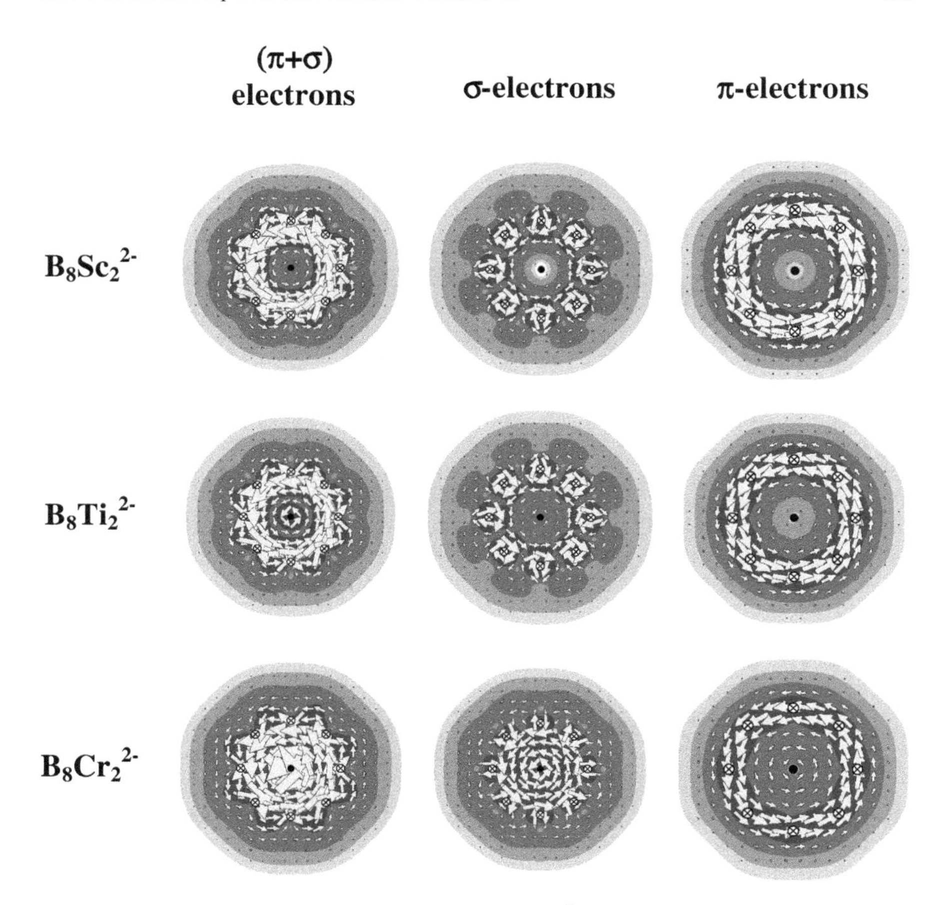


**Fig. 14** Total (π + σ), σ and π ring current maps of $B_8M_2^{2-}$ cyclic structures(B3LYP/6-311 + G (d)). Black dot is a metal atom

current is observed for σ electrons which give rise to aromaticity. $B_7Sc_2^-$ is thus considered as an σ-aromatic and π anti-aromatic compound. For their part, each of the $B_8M_2^{2-}$ cyclic structures with M = Sc, Ti, V and Cr exhibits a doubly aromatic character according to criteria of ring current.

To explore further the aromatic feature, the contribution of each orbital to the total ring current is performed and depicted in Figs. 15, 16 and 17. In most cases, the diatropic current of π electrons is mainly contributed by two doubly degenerate MOs, called π* and δ*-MOs. They result from an overlap between both π* and δ*-MOs of dimeric metal $M_2$ and different levels of $B_7$ and $B_8$ strings, and make significant contributions to the stabilization of bimetallic cycles, as indicated from MO analysis. The σ ring currents come from electrons which are populated in σ-MOs, and low-energy MOs that are contribute mainly by s-AOs of B. Electrons occupied in doubly degenerate δ-MOs formed by interaction of δ-MOs ($M_2$) and $B_n$

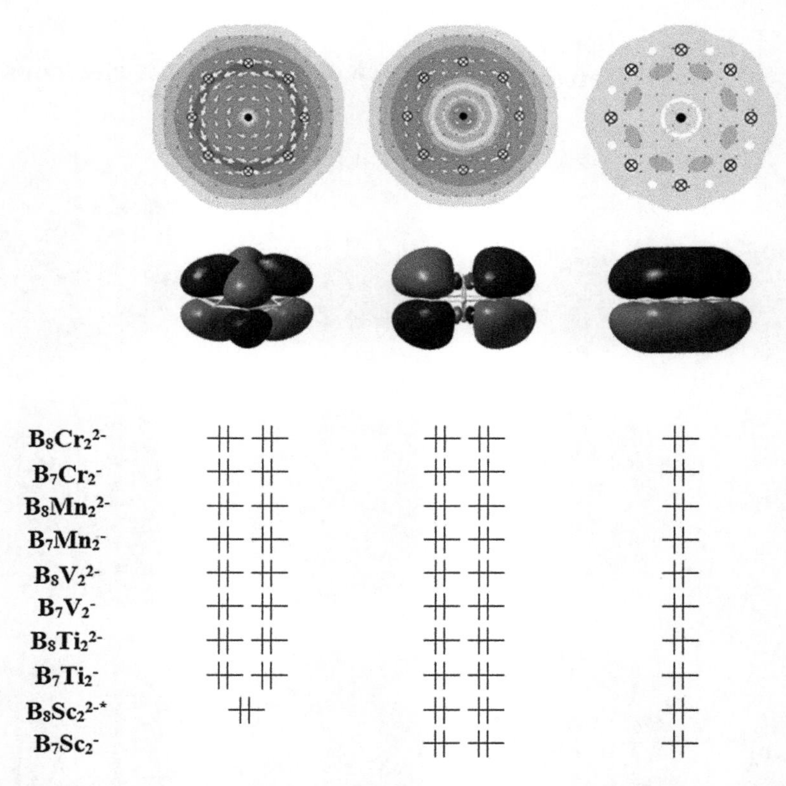


**Fig. 15** Current densities of π-MOs in bimetallic cycles. *: structure producing a paratropic ring current

string are mainly responsible for diatropic ring currents of $B_7M_2^-$ and $B_8M_2^{2-}$ cycles, whereas π-MOs are only slightly active in terms of magnetic response. The structures possess unfulfilled degenerate MOs and have paratropic ring current which thus can be assigned as anti-aromatic.

## 6 $B_{14}M_2$: When Bimetallic Tubular Structure Can Be Formed?

We now consider a novel class of boron clusters containing two transition metals in a tubular structure.

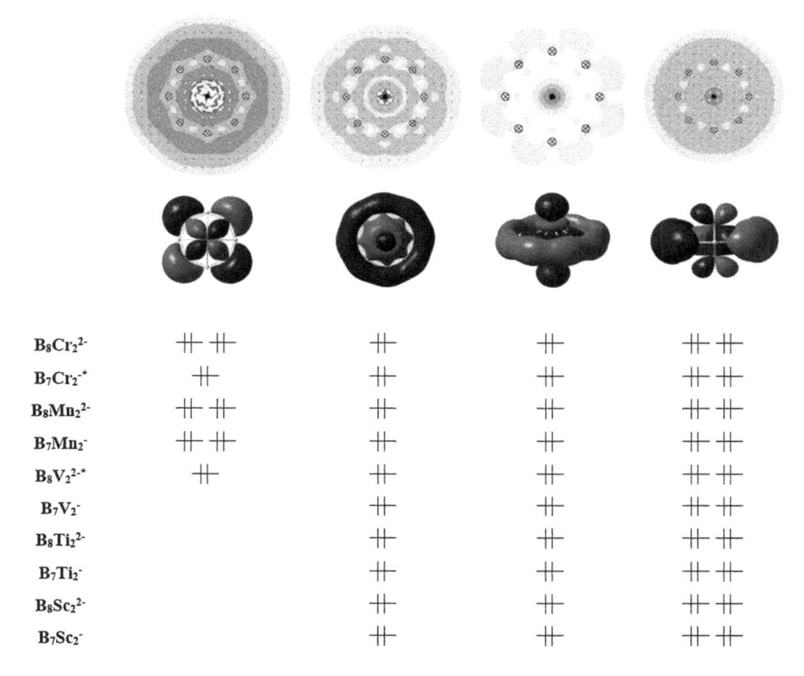


**Fig. 16** MO contributions to ring current of σ electrons of bimetallic cylces. *: structure raising a paratropic ring current

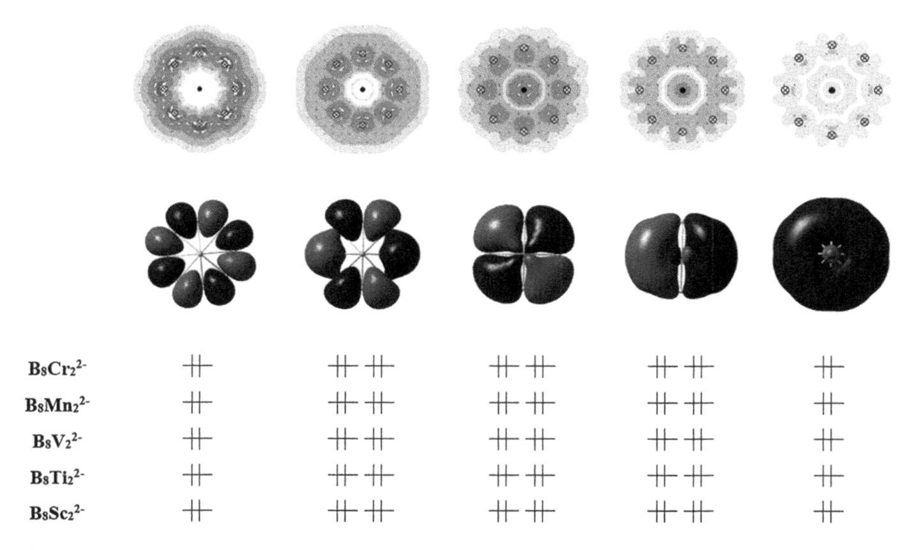


**Fig. 17** The ring current of s-MOs in different bimetallic cycles

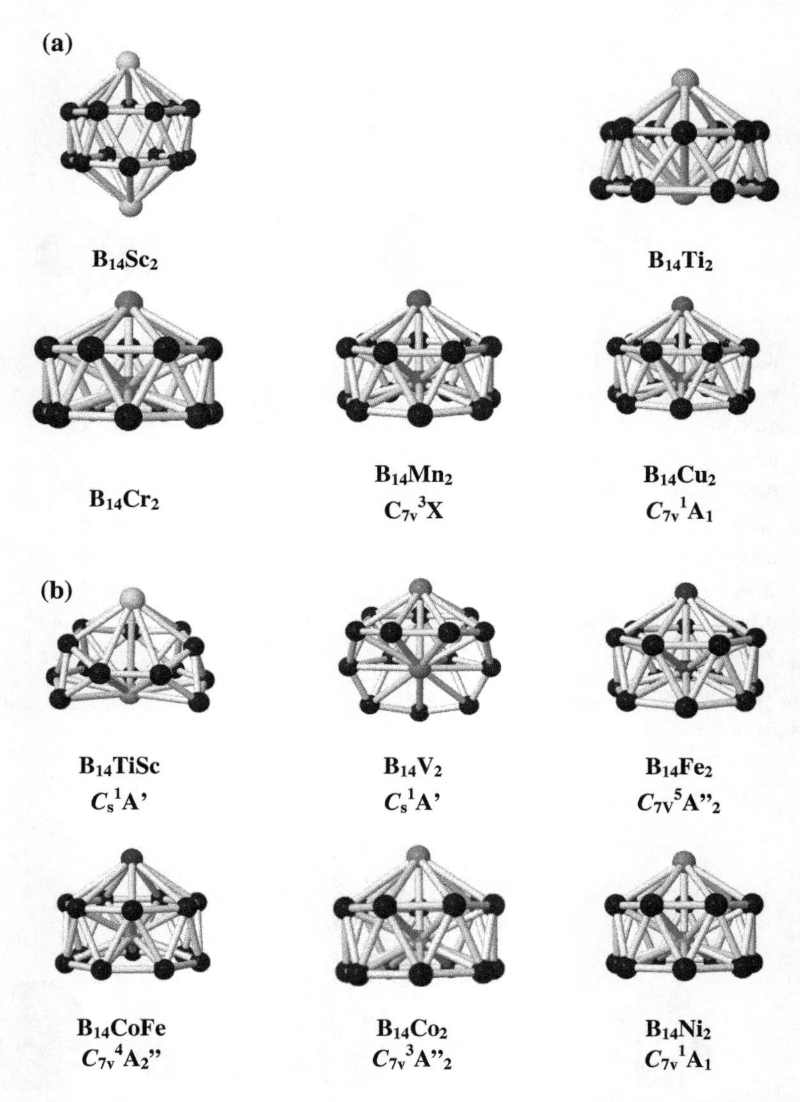


**Fig. 18** Shape of some bimetal doped $B_{14}$ clusters: **a** local minima **b** global minimum structure

## 6.1 Geometrical Identifications

Recent studies [37, 67] on multiply transition metal doped $B_{14}$ clusters showed the emergence of a new class of tube. The $B_{14}$ moiety in fact adopts a double ring tube superposing two $B_7$ strings in an anti-prism fashion. The $M_2$ dimer is vertically placed along the main axis of the tube. In other words, the two metallic atoms form the $C_7$ axis of the high symmetry $B_{14}M_2$ ($C_{7v}$) structure.

One metal atom is located in the central region of the tube, and the other is situated outside the tube.

Interestingly, the co-existence of Co and Fe also gives rise to a tube as the global minimum of $B_{14}CoFe$ in which Co is located inside and the iron atom at outside of the tube (Fig. 18). Such a tubular structure is equally observed in $B_{14}Ni_2$. Similar to the Fe and Co dopants, the presence of two Ni atoms produces the singlet and high symmetry ($C_{7v}$, $^1A_1$) $B_{14}Ni_2$ double ring which is again identified to be the lowest-energy on the potential energy surface (Fig. 18b). With Mn and Cu as dopants, the resulting $B_{14}Mn_2$ and $B_{14}Cu_2$ still possess a tubular form following geometry optimizations. However, it should be mentioned that for the identity of the global minima of $B_{14}Mn_2$, $B_{14}Cu_2$ could not be established yet.

Although the tubular structure is located following doping of two first-row transition metal atoms, from Mn to Cu, the formation of the double ring (DR) structural motif does not appear to be a general tendency of *3d* transition metals. As given in Fig. 18, the V dopant does not generate a DR form, but rather a 3D ground structure in which one V is located on a $B_{10}$ face and the other V is capped on the heptagonal face. While a typically hetero-bimetallic doping of Co and Fe successfully stabilizes the DR, a co-doping of Ti and Sc atoms on the $B_{14}$ cluster gives a distorted tube. Overall, for both homo- and hetero-bimetal doped $B_{14}M_2$ clusters, the DR form do not exist as a general structural motif; only a few 3*d* metals can stabilize a $B_{14}$ double ring. These results point out that a certain condition is required for the formation of a tube.

## 6.2 Orbital Interaction

A recent investigation on the tubular boron clusters $B_{2n}$ with n = 10–14 illustrated that the MOs of each DR tube can be classified into two distinct sets, namely the radial (r-MOs) and tangential (t-MOs) MO sets, on the basis of the orientation of the *p*-lobes [29]. Another MO set is identified as *s*-MOs that are constructed by a major contribution from *s*-AOs. The electron count rule is in this instance established as $(4N + 2M)$, with $M = 0$ and 1, depending on the number of non-degenerate MO for both r-MOs and t-MOs sets. In the case of the tubular $B_{14}$, with n = 7, their MOs are also classified into three sets involving radial (r-MOs), tangential (t-MOs) and s-MOs sets. Let us now analyze the orbital interactions of the t-MOs and r-MOs set of the $B_{14}$ DR with those of the metal-metal units.

**Orbital Interaction of t-MOs Set and $Ni_2$.** Interaction of the t-MOs set of the $B_{14}$ DR with the $[\sigma^2\ \pi^4\ \delta^4\ \delta^{*4}\ \sigma^{*2}]$ configuration of $Ni_2$ is depicted in Fig. 19. Combination of the vacant $(3 \pm 2\ 1)$ MOs of the DR and the anti-bonding $\delta^*$ orbital occupied by 4 electrons creates some stabilized MOs, being identified as the $1G_z2$ $(_x2{-}_y2)$ and $1G_{xyz}2$ electron shells (labelled in terms of the electron shell model). This interaction also introduces the doubly occupied MOs assigned by the $(3 \pm 2\ 1)$ level in terms of the hollow cylinder model [29]. The $\pi^*$ MOs of $Ni_2$ are stabilized upon overlapping with the doubly occupied $(3 \pm 1\ 1)$ MOs of the $B_{14}$

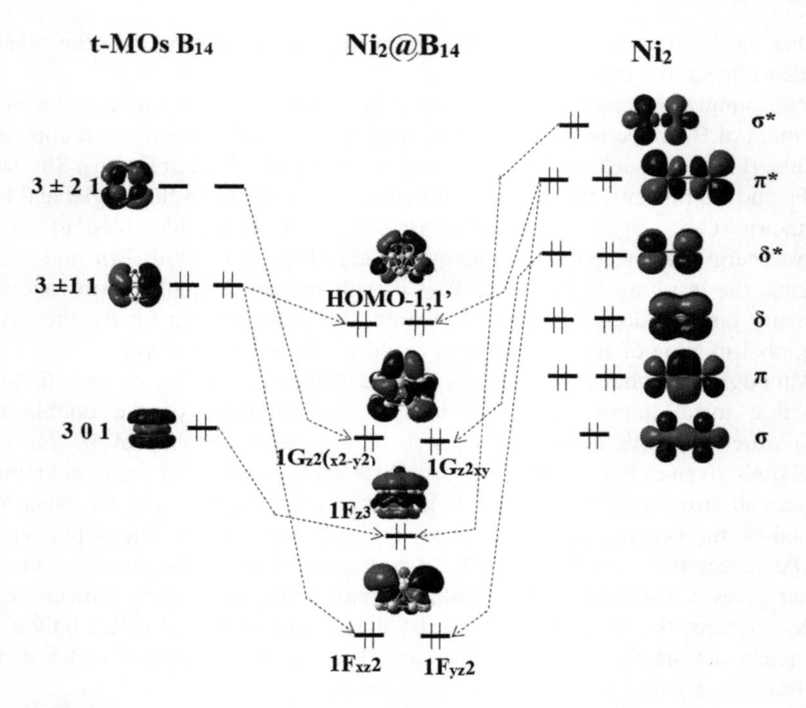


**Fig. 19** Orbital interaction of $Ni_2$ and the t-MO set of the $B_{14}$ double ring

DR, and form the $1F_{xz}2$ and $1F_{yz}2$ levels fully filled by 4 electrons. The doubly degenerate HOMO-1,1′ sub-shell is occupied by 4 electrons.

The two shells $1F_{xz}2$ and $1F_{yz}2$, as described in terms of the electron shell model, are also interacting with the doubly degenerate (3 ±1 1) MOs of the hollow cylinder model. The weak interaction of the non-degenerate (3 0 1) level and empty σ* MO results in the $2P_z$ level. Orbital overlaps of the metal dimer with the t-MO set of the $B_{14}$ tube end up in an enhanced stability.

The bimetallic tube $B_{14}Ni_2$ exhibits five fully filled MOs corresponding to the (3 0 1), (3 ±1 1) and (3 ±2 1) states of the hollow cylinder model, thus satisfying the electron count rule for the tangential MO set with 10 electrons. Due to the mixing of transition metal atoms, a modification of the potential occurs, and as a result, the energy ordering of the eigenvalues is actually changed.

**Orbital Interaction of the r-MOs Set of $Ni_2$**

Figure 20 displays the orbital interaction diagram between $Ni_2$ MOs and the radial MO set of the $B_{14}$ DR, in which the δ* MOs are not involved in any interaction. The lowest eigenvalue state (1 0 2) of $B_{14}$ DR can only interact with the bonding and anti-bonding σ-MOs of $Ni_2$ to yield the $1D_z2$ and 2S subshells (labelled in terms of electron shell model), respectively. The π* MOs of $Ni_2$ favorably overlap with the (1 ±2 2) eigenstates of $B_{14}$ and give rise to the $2D_{xz}$ and $2D_{yz}$ levels. Combination of π MOs and (1 ±2 2) levels introduces the $2P_x$ and $2P_y$

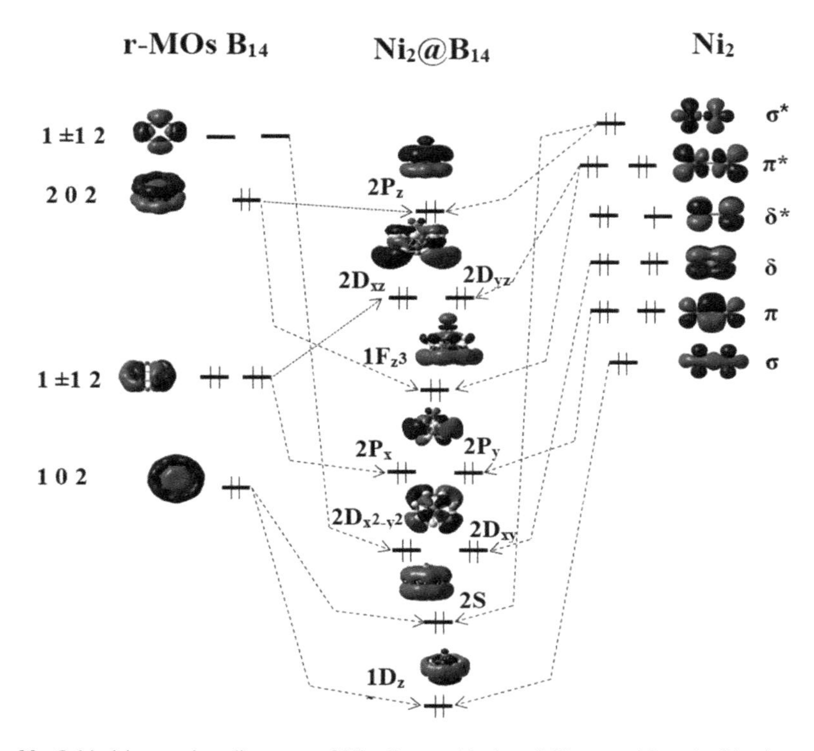


**Fig. 20** Orbital interaction diagrams of $Ni_2$ dimer with the r-MOs set of $B_{14}$ double ring

subshells. The σ* MO of $Co_2$ interacts with the doubly occupied state of (2 0 2) releases a fully filled MO assigned as the $2P_z$ subshell. Finally, the stability of δ bond is enhanced upon interaction with the empty (1 ±2 2) eigenstates of the DR, and consequently produces the $2D_x2_{-y}2$ and $2D_{xy}$ subshells.

The enhanced stability of bimetallic tubular clusters can thus in part be understood as the $M_2$ dimer introduces its electrons to fill up the radial MOs, the bare $B_{14}$ DR tube has only 8 radial electrons, populated in the (1 0 2), (1 ±1 2) and (2 0 2) eigenstates. The anti-bonding MOs of $M_2$ are stabilized by interacting with the r-MOs set of the $B_{14}$ tube. The δ bond is reinforced through combining with the empty (1 ±2 2) level of the $B_{14}$ tube to yield the 2D subshell. We again note that the presence of transition metal dimers modifies the potential of hollow cylinder model, and as a consequence, the energy ordering of eigenstates cannot be built as in the case of bare $B_{2n}$ DR tubes [77].

## 6.3 *Electronic Requirement*

Let us now analyze the electron count for $Ni_2@B_{14}$, as a typical tubular DR structure within the framework of the hollow cylinder model [29]. Electron count for the r-MOs set simply leads to (4*N*), due to the fact that all structures considered have two non-degenerate MOs, whereas the electrons in the t-MOs set amount to (4*N* + 2). On the one hand, because the MOs of the $M_2$ unit are formed from *d*-AOs or *sd*-AOs which has angular momentum $l = 2$, as a result they can only interact with the MOs of the cylinder model having quantum number $l$ not greater than 2. On the other hand, all possible states of the r-MO set, which can overlap with MO of $M_2$, include the levels of (1 0 2), (1 ±1 2), (1 ±2 2), (2 0 2), (2 ±1 2), (2 ±2 2) … In the current case, 12 electrons are found to be fully distributed in the (1 0 2), (1 ±1 2), (1 ±2 2) and (2 0 2) states. For the t-MOs set, only the (3 0 1), (3 ±1 1) and (3 ±2 1) levels can interact with the MO($M_2$) and they are fully occupied by 10 electrons perfectly corresponding to the (4*N* + 2) count rule for the t-MOs set having no degenerate MO.

Special geometrical feature in which one atom is encapsulated inside the $B_{14}$ double ring and the other is exposed outside appears to be made in order to gain advantages by stabilizing interactions of the anti-bonding of $M_2$. The σ* and π* MOs have the same irreducible representation as the (1 0 2), (1 ±1 2) and (2 0 2) MOs of the model, and consequently they induce additional interactions, producing the 2S, $1F_z3$, $2D_{xz}$, $2D_{yz}$ and 2Pz levels that are occupied by 10 electrons. Similarly, stable interaction is also observed for the t-MOs set. The doubly degenerate π* MOs interact with the (3 ±1 1) eigenstate of the $B_{14}$ DR and thereby give rise to the doubly degenerate HOMO-1,1 which are occupied by 4 electrons. Overall, a uniform electron count rule for all delocalized electrons cannot be established as in the bare $B_{2n}$ tubes, simply because of the involvement of extra MOs which also enhances the stability.

For each MO set, the total electron number involves not only electrons occupying the r-MOs and t-MOs sets, but also extra electrons. This raises a certain requirement for the electron number of the metal dimer to form a bimetallic tubular structure. In the case of $B_{14}Ni_2$, the dimeric $Ni_2$ contributes 4 electrons for each of the r-MOs and t-MOs sets of $B_{14}$ DR in order to have 12 and 10 electrons for the r-MOs and t-MOs sets of the bimetallic tubular structure, respectively. Additionally, the $Ni_2$ dimer gives 10 and 4 electrons in order to fill the extra interaction with the r-MOs and t-MOs sets, respectively. The $Ni_2$ dimer also provides valence electrons to fully occupy the levels of the $B_{14}$ tube and extra stabilized MOs. As a result, the resulting $B_{14}Ni_2$ tubular structure enjoys an enhanced stability.

Generally, to produce a bimetallic tubular structure with a $B_{14}$ DR moiety, the metallic dimer $M_2$ needs to have enough electrons to fill up the levels of both r-MOs and t-MOs sets of the hollow cylinder model, and also the extra-interacting MOs can sufficiently stabilize the resulting $B_{14}M_2$ tube.

The dopants $Co_2$, $Fe_2$ and CoFe can provide 18, 16 and 17 electrons, respectively, from the atomic configurations of Co: [Ar] $3d^7$ $4s^2$ and Fe: [Ar] $3d^6$ 4 $s^2$. These amounts of electrons significantly stabilize the $B_{14}$ moiety in a bimetallic tubular form, and subsequently the $B_{14}Co_2$, $B_{14}Fe_2$ and $B_{14}CoFe$ tube are located, as shown in Fig. 18, as the most stable isomer.

The $V_2$ and TiSc dopants have only 6 and 5 valence electrons, and as a result these dopants could not provide with enough electrons to significantly stabilize a $B_{14}$ DR form.

# 7 Chemical Bonding of Transition Metal Doped Boron Clusters

## 7.1 *Singly Doped $B_nM$*

The phenomenological shell model, or the Jellium model [68], has successfully rationalized the remarkable thermodynamical stability of several families of clusters. Within this rather simple model, the nuclei are ignored and replaced by a mean field, while electrons are considered to move freely within this mean field. The valence electrons are filled the S, P, D, F, … orbitals according to the angular momentum number L = 0, 1, 2, 3…. With a given quantum number L, the

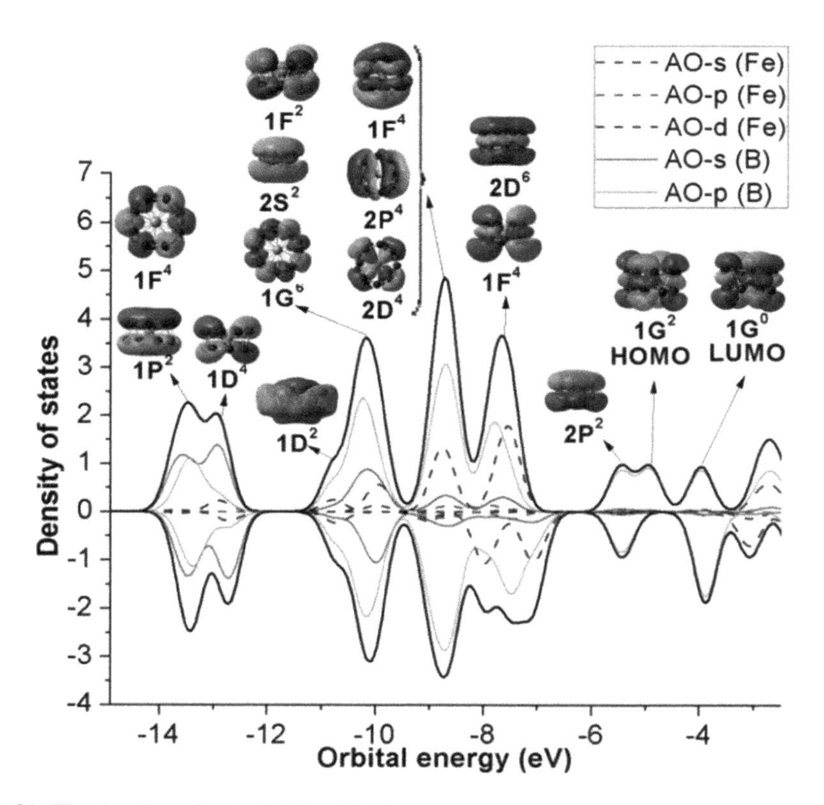


**Fig. 21** The densities of state (DOS) of $B_{16}Fe$ cluster

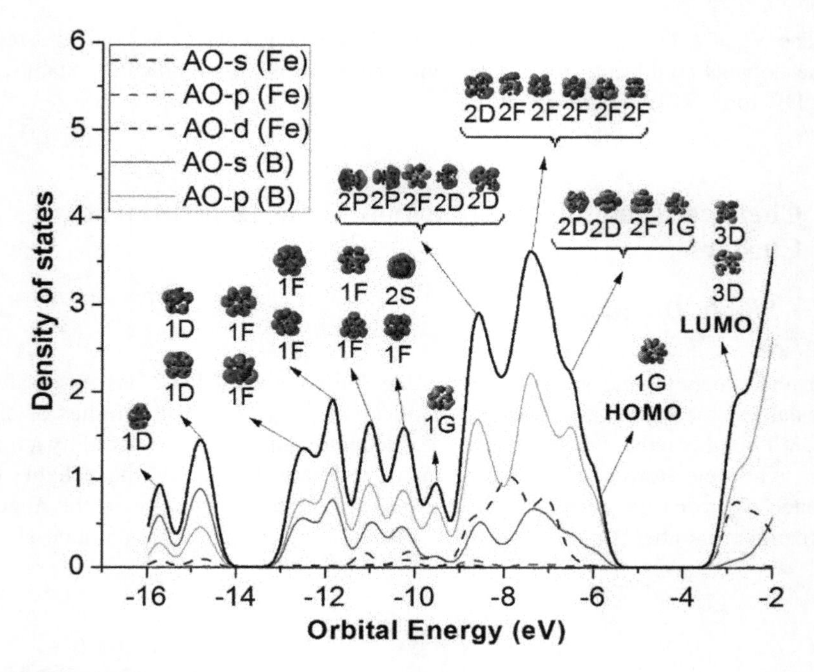


**Fig. 22** The densities of state (DOS) of $B_{20}Fe$ cluster

**Table 3** Electronic configuration of $B_{14}M_2$ bimetallic tubular structures

| $B_{14}M_2$ | Configuration | Number of electrons |
|---|---|---|
| $B_{14}Cr_2$ | $(\sigma_{4s})^2(\pi^*)^4(\delta)^4(\sigma_{3d})^2(\pi)^4(\delta^*)^4$ | 20 |
| $B_{14}Mn_2$ | $(\sigma_{4s})^2(\pi^*)^4(\delta)^4(\sigma_{3d})^2(\pi)^4\ (\delta^*)^4(\sigma_{3d}^*)^1$ | 21 |
| $B_{14}Fe_2$ | $(\sigma_{4s})^2(\pi^*)^4(\delta)^4(\sigma_{3d})^2(\pi)^4\ (\delta^*)^4(\sigma_{3d}^*)^1$ | 21 |
| $B_{14}Co_2$ | $(\sigma_{4s})^2(\pi^*)^4(\delta)^4(\sigma_{3d})^2(\pi)^4\ (\delta^*)^4(\sigma_{3d}^*)^1$ | 21 |
| $B_{14}Ni_2$ | $(\sigma_{4s})^2(\pi^*)^4(\delta)^4(\sigma_{3d})^2(\pi)^4\ (\delta^*)^4(\sigma_{3d}^*)^2$ | 22 |
| $B_{14}Cu_2$ | $(\sigma_{4s})^2(\pi^*)^4(\delta)^4(\sigma_{3d})^2(\pi)^4\ (\delta^*)^4(\sigma_{3d}^*)^2$ | 22 |

lowest-lying level has a principle number N = 1. In this electron shell model, the successive occupation of a level, giving a magic number, leads to a stabilized cluster. Within this family of clusters, the shell model rationalizes the stability of the singly Fe doped $B_nFe$, with n = 14, 16, 18 and 20 [33].

For an illustration, Figs. 21 and 22 display the densities of state (DOS) in combining with the MO shape of $B_{16}Fe$ and $B_{20}Fe$, respectively. The shell configuration of $1S^21P^61D^{10}1F^{14}2S^21G^62P^42D^{10}2F^{14}$ occupied by 68 electrons is assigned for $B_{20}Fe$ clusters. For the $B_{16}Fe$, the shell configuration is identified as $1S^21P^61D^{10}1F^{14}2S^21G^82P^42D^{10}$ with 52 electrons. We note that the theoretical model is based on spherical potential function, even though that the real shape of cluster is different from an ideal spherical form. As a result, the sublevels having the same value of N loose the degeneracy. Overall, according to the electron shell model, the Fe dopant provides valence electrons to fill up levels of shell configuration which is raised by overlap of the $B_n$ and dopant moieties.

## 7.2 *Bimetal Doped Boron Clusters: The Role of δ* MO*

This section is devoted to an analysis of the chemical bonding of bimetallic cyclic and tubular boron clusters on the basis of the M–M bond. Let us first summarize the essential bonding characteristics of metallic dimers. In the chemistry of bimetallic complexes, formation of the metal–metal bond is the result of a delicate balance of two factors, namely the valence orbitals involved must be sufficiently diffuse to afford a substantial diatomic overlap, and the competition in formation of bond with ligands should be avoided [69]. The intrinsic strength of M–M connection is contributed mainly by the number of available electrons and their radial and angular properties. The angular properties indicate the local symmetry of overlap between metals involving of σ, π and δ in which δ is unique for the systems having $l > 1$ (e. i. *d* or *f* orbitals). It is no doubt that the contribution of δ bonding to the M–M interaction is very small, and the σ and π symmetry components are expected to dominate the overall bond strength.

In a previous section, the construction of a molecular orbital configuration for bimetallic boron cycles is based on orbital interaction of dimeric metal and $B_n$ string. For a bimetallic boron tube, the interactions between metallic dimers and $B_{14}$ moiety are constructed and depicted in Figs. 19 and 20, and the MO configuration can be established.

On the one hand, the doubly degenerate δ and δ* MOs of bimetallic tube are formed by overlapping δ and δ* MOs of metallic dimer with the (1 ±2 2) and (3 ±2 1) levels of $B_{14}$ DR. The π MOs of bimetallic boron double ring, which is doubly degenerated, are a result of interaction between π-MOs of $M_2$ and (1 ±1 2) eigenstates of $B_{14}$ moiety. The combination of the (3 ±1 1) levels of $B_{14}$ DR with anti-bonding π* MOs of dimeric metal turns out to form the doubly degenerate π*-MOs for bimetallic boron tube. The $\sigma_{4s}$ and $\sigma_{3d}$ orbitals of bimetallic tubular structure are formed by interaction of $\sigma_{4s}$ and $\sigma_{3d}$ of $M_2$ with (1 0 2) level of $B_{14}$ DR, whereas the $\sigma_{3d}$* is constructed by overlapping between σ* of $M_2$ and (2 0 2) state of $B_{14}$ counterpart.

Figure 23 displays a comparison for the configuration of both $B_7Ni_2^-$ and $B_{14}Ni_2$ which are fully occupied by 22 electrons. The MO configurations of different tube are identified and listed in Table 3.

Geometrical identifications for $B_{14}M_2$ clusters with M = Sc–Cu point out that the bimetallic tubular motif are not observed in case of Sc, Ti and V, whereas clusters with other first-row transition metals as dopants exist in this shape. This result can be rationalized in terms of a full occupation of σ, π, π*, δ and δ* MOs

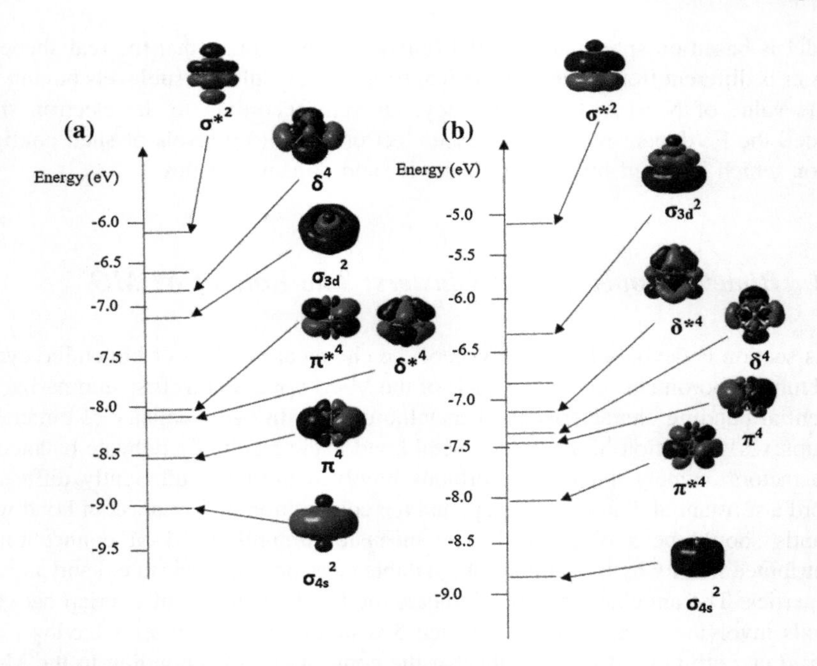


**Fig. 23** MO configuration of **a** $B_7Ni_2^-$ and **b** $B_{14}Ni_2$ double ring tubes

(the latter are doubly degenerate). As given in Table 3, the $B_{14}Cr_2$ cluster, having the smallest number electron with a MO configuration as $[(\sigma_{4s})^2\ (\pi^*)^4\ (\delta)^4\ (\sigma_{3d})^2\ (\pi)^4\ (\delta^*)^4]$, which has doubly degenerate π, π*, δ and δ* MOs, are fully occupied. In the other tubes, the $\sigma_{3d}$* MO is only singly or doubly filled. Accordingly, a full occupation of the σ, π, π*, δ and δ* MOs induces a stability for two metals doped $B_{14}M_2$ tube.

The Sc dopant possesses the smallest electron number and a MO configuration of $[(\sigma_{4s})^2\ (\pi^*)^4\ (\pi)^4\ (\delta^*)^2]$ in the case of $B_7Sc_2^-$, which is occupied by 12 electrons. The Ti dopant turns out to have a MO configuration as $[(\sigma_{4s})^2\ (\pi^*)^4\ (\pi)^4\ (\delta^*)^4]$ with 14 electrons. The other metals supply additional electrons to fill up the $\sigma_{3d}$, δ and $\sigma_{3d}$* MOs corresponding to 16, 18, 20 and 22 electrons.

Overall, formation of bimetallic boron tube requires an occupancy of at least 12 or 14 electrons for configuration $[(\sigma_{4s})^2\ (\pi^*)^4\ (\pi)^4\ (\delta^*)^2]$ or $[(\sigma_{4s})^2\ (\pi^*)^4\ (\pi)^4\ (\delta^*)^4]$ in cases of $B_7Sc_2^-$ and $B_8Sc_2^{2-}$, respectively, and at least 20 electrons that are populated in a orbital configuration of $[(\sigma_{4s})^2\ (\pi^*)^4\ (\delta)^4\ (\sigma_{3d})^2\ (\pi)^4\ (\delta^*)^4]$ of $B_{14}M_2$.. As a result, the M–M bond plays a vital role in the formation of bimetallic boron cycles and tubes. In particular, the π* and δ* MOs of dimeric metals enjoy stabilizing interactions with $B_7$, $B_8$ strings and $B_{14}$ double ring establishing bimetallic cyclic and tubular motif of structure. Most of the interactions of the bimetallic $M_2$ afford substantial stabilizing overlaps with the boron unit. According to such as bimetallic configuration model, the stability of cyclic and tubular structure arises from contributions of the occupation of metal-based MOs. The more electrons are populated in these MOs, the higher the overall thermodynamic stability.

# 8 Concluding Remarks

First-row transition metals introduce not only a new class of doped boron clusters but also particular growth patterns. The bimetallic cyclic motif, a novel type of structure in which two metals are vertically coordinated to the planar $B_n$ strings, provides us with a general form of bimetallic doped $B_7$ and $B_8$ clusters. The structural evolution was probed by systematic investigations of the $B_nTi_2$, $B_nSc_2$ and $B_nTiSc$ clusters in which the B atoms are added around a vertical M–M axis. When the $B_n$ string becomes large enough, the cyclic motif is operative. Superposition of two boron strings leads to a tube. The tubular motif is structurally characterized as having one metal atom located inside the $B_n$ double ring whereas the other metal occupies an exohedral position.

On the basis of M–M bond interaction, a bimetallic configuration model has been established to rationalize the electronic structure and stability for both bimetallic cyclic and tubular boron clusters. Within this simple model, formation of bimetallic tubular structure requires at least 20 electrons to populate the molecular orbital configuration of $[(\sigma_{4s})^2 (\pi)^4 (\pi^*)^4 (\delta^*)^4 (\sigma_{3d})^2 (\delta)^4]$. The anti-bonding $\pi^*$ and $\delta^*$-MOs of dimeric metals enjoy stabilizing interactions with the $B_7$, $B_8$ strings and the $B_{14}$ double ring inducing an enhanced stability for the whole doped cluster.

**Acknowledgments** We sincerely thank the Department of Science and Technology of Ho Chi Minh City, Vietnam, for supporting our work at ICST. We are indebted to KU Leuven Research Council (GOA program) and Flemish Fund for Scientific Research (FWO-Vlaanderen).

# References

1. Albert B, Hillebrecht H (2009) Angew Chem Int Ed 48:8640
2. Decker BF, Kasper JS (1959) Acta Crystallogr 12:503
3. Hoard JL, Sullenger DB, Kennard CHL, Hughes RE (1970) J Solid State Chem 1:268
4. Oganov AR, Chen J, Gatti C, Ma YZ, Ma YM, Glass CW, Liu Z, Yu T, Kurakevych OO, Solozhenko VL (2009) Nature 457:863
5. Liu F, Shen C, Su Z, Ding X, Deng S, Chen J, Xu N, Gao H (2010) J Mater Chem 20:2197
6. Ciuparu D, Klie RF, Zhu Y, Pfefferle L (2004) J Phys Chem B 108:3967
7. Bezugly V, Kunstmann J, Stock BG, Frauenheim T, Niehaus T, Cuniberti G (2011) ACS Nano 5:4997
8. Szwacki NG, Sadrzadeh A, Yakobson BI (2007) Phys Rev Lett 98:166804
9. Prasad DLVK, Jemmis ED (2008) Phys Rev Lett 100:165504
10. Sadrzadeh A, Pupysheva OV, Singh AK, Yakobson BI (2008) J Phys Chem A 112:13679
11. Wang X (2010) Phys Rev B 82:153409
12. Tang H, Ismail-Beigi S (2007) Phys Rev Lett 99:115501
13. Yang X, Ding Y (2008) Phys Rev B 77:041402

14. Tang H, Ismail-Beigi S (2009) Phys Rev B 80:134113
15. Tang H, Ismail-Beigi S (2010) Phys Rev B 82:115412
16. Penev ES, Bhowmick S, Sadrzadeh A, Yakobson BI (2012) Nano Lett 12:2441
17. Liu Y, Penev ES, Yakobson BI (2013) Angew Chem Int Ed 52:3156
18. Singh AK, Sadrzadeh A, Yakobson BI (2008) Nano Lett 8:1314
19. Schubert DM (2004) Boron hydrides, heteroboranes, and their metallic derivatives. Wiley-VCH Verlag GmbH & Co., KGaA
20. Fehlner TP, Halet JF, Saillard JY (2007) Molecular clusters: a bridge to solid-state chemistry. Cambridge University Press, Cambridge
21. Tai TB, Tam NM, Nguyen MT (2012) Chem Phys Letts 530:71
22. Tai TB, Nguyen MT (2015) Phys Chem Chem Phys 17:13672
23. Chen L (2012) J Chem Phys 136:104301
24. Duong LV, Pham HT, Tam NM, Nguyen MT (2014) Phys Chem Chem Phys 16:19470
25. Tai TB, Tam NM, Nguyen MT (2012) Theor Chem Acc 131:1241
26. Tai TB, Grant DJ, Nguyen MT, Dixon DA (2010) J Phys Chem A 114:994
27. Sergeeva AP, Popov IA, Piazza ZA, Li WL, Romanescu C, Wang LS, Boldyrev AI (2014) Acc Chem Res 47:1349
28. Tai TB, Ceulemans A, Nguyen MT (2012) Chem Eur J 18:4510
29. Pham HT, Long DV, Nguyen MT (2014) J Phys Chem C 118:24181
30. Pu Z, Ito K, Schleyer PR, Li QS (2009) Inorg Chem 48:10679
31. Ito K, Pu Z, Li QS, Schleyer PR, Li QS (2008) Inorg Chem 47:10906
32. Liao Y, Cruz CL, Schleyer PR, Chen Z (2012) Phys Chem Chem Phys 14:14898
33. Tam NM, Pham HT, Duong VL, Pham-Ho MP, Nguyen MT (2015) Phys Chem Chem Phys 17:3000
34. Guennic L, Jiao H, Kahlal S, Saillard JY, Halet JF, Ghosh S, Shang M, Beatty AM, Rheingold AL, Fehlner TP (2004) J Am Chem Soc 126:3203
35. Ghosh S, Beatty AM, Fehlner TP (2001) J Am Chem Soc 123:9188
36. Fokwa BPT, Hermus M (2012) Angew Chem Int Ed 51:1702
37. Pham HT, Nguyen MT (2015) Phys Chem Chem Phys 17:17335
38. Tai TB, Duong LV, Pham TH, Dang MTT, Nguyen MT (2014) Chem Comm 50:1558
39. Tai TB, Nguyen MT (2015) Chem Commun 51:7677
40. Pham HT, Duong LV, Tam NM, Pham-Ho MP, Nguyen MT (2014) Chem Phys Letts 608:295
41. Chen Q, Li WL, Zhao YF, Zhang SY, Hu HS, Bai H, Li HR, Tian WJ, Lu HG, Zhai HJ, Li SD, Li J, Wang LS (2015) ACS Nano 9:754
42. Tai TB, Nguyen MT (2015) Nanoscale 7:3316
43. Muya JT, Ramanantoanina H, Daul C, Nguyen MT, Gopakumar G, Ceulemans A (2013) Phys Chem Chem Phys 15:2829
44. Bean DE, Muya JT, Fowler PW, Nguyen MT, Ceulemans A (2011) Phys Chem Chem Phys 13:20855
45. Muya JT, Gopakumar G, Nguyen MT, Ceulemans A (2011) Phys Chem Chem Phys 13:7524
46. Muya JT, Sato T, Nguyen MT, Ceulemans A (2012) Chem Phys Letts 543:111
47. Gopakumar G, Nguyen MT, Ceulemans A (2008) Chem Phys Lett 450:175
48. Tai TB, Lee SU, Nguyen MT (2016) Phys Chem Chem Phys 18:11620
49. Tai TB, Nguyen MT (2016) Chem Commun 52:1653
50. Tai TB, Nguyen MT (2016) Chem Commun (To be published)
51. Jia J, Ma L, Wang JF, Wu HS (2013) J Mol Model 19:3255
52. Wang J, Jien J, Ma L, Wu H (2012) Acta Chim Sinica 70:1643
53. Jia J, Li X, Li Y, Ma L, Wu HS (2014) J Theor Comput Chem 1027:128
54. Wang ZYK, Jia WL, Wu JF, Shun H (2014) Acta Physica Sinica 63:233102
55. Pham HT, Nguyen MT. Non-published results on $B_nTiSc$ clusters
56. Yang Z, Xiong SJ (2008) J Chem Phys 128:184310
57. Liu X, Zhao GF, Guo LJ, Jing Q, Luo YH (2007) Phys Rev A 75:063201
58. Yuan Y, Cheng L (2012) J Chem Phys 137:044308

59. Cheng L (2012) J Chem Phys 136:104301
60. Yanagisawa S, Tsuneda T, Hirao K (2000) J Chem Phys 112:545
61. Barden CJ, Rienstra-Kiracofe JC, Schaefer HF (2000) J Chem Phys 113:690
62. Cotton FA, Nocera DG (2000) Acc Chem Res 33:483
63. Tai TB, Huong VTT, Nguyen MT (2014) Top Heterocyc Chem 38:161
64. Keith TA, Bader RFW (1993) Chem Phys Lett 210:223
65. Lazzeretti P, Malagoli M, Zanasi R (1995) J Chem Phys 102:9619
66. Steiner E, Fowler PW (2001) J Phys Chem A 105:9553
67. Pham HT, Tam NM, Nguyen MT. Non-published results on $B_nNi$ and $B_nNi_2$ clusters
68. Brack M (1993) Rev Mod Phys 65:67
69. McGrady JE (2015) Molecular metal-metal bonds: compounds, synthesis, properties. Wiley-VCH Verlag GmbH & Co., KGaA

# Silicate Nanoclusters: Understanding Their Cosmic Relevance from Bottom-Up Modelling

**Stefan T. Bromley**

**Abstract** In this chapter we provide a brief overview of bottom-up modelling of nanosilicate clusters, particularly with respect to their astronomical relevance. After providing some general background, we highlight the importance of computational modelling for obtaining unprecedentedly detailed insights into the low energy structures of nanosilicates of a range of compositions and sizes. Later we show how such insights can be useful in understanding the formation of nanosilicates through nucleation around stars, and their role as seeds for further nucleation of water ice.

**Keywords** Nanosilicates • $SiO_2$—silica • Silicates • Global optimisation • Cosmic dust • Nucleation • Astronomy

## 1 Introduction

### *1.1 The Origin of Silicates*

Silicon and oxygen can combine to make a wide range of $SiO_x$ binary compounds with chemical compositions spanning between $0 \leq x \leq 2$. Formally non-stoichiometric compounds, with $x < 2$, are known as silicon sub-oxides. The best characterised material of this class is SiO, or silicon monoxide, which is an amorphous material with semi-segregated regions with varying O-rich and O-poor nanoscale regions [1]. Fully oxidised bulk $SiO_2$, or silica, on the other hand, can form a range of both glassy and crystalline networks of O-linked tetrahedral $[SiO_4]^{4-}$ units, where every oxygen atom is shared between a linked pair of tetrahedra. When

S.T. Bromley (✉)
Departament de Ciencia de Materials i Química Física and Institut de Química Teòrica I Computacional (IQTCUB), Universitat de Barcelona, 08028 Barcelona, Spain
e-mail: s.bromley@ub.edu

S.T. Bromley
Institució Catalana de Recerca i Estudis Avançats (ICREA), 08010 Barcelona, Spain



M.T. Nguyen and B. Kiran (eds.), *Clusters*, Challenges and Advances in Computational Chemistry and Physics 23,
DOI 10.1007/978-3-319-48918-6_7

not all oxygen atoms are shared, these units can also form a variety of negatively charged rings, chains and sheets anions and even be found as isolated anion units within the more general family of silicate materials. This topological diversity is augmented by the vast potential choice of cations required to provide charge compensation in such materials, leading to thousands of known silicates [2]. Silicate minerals, such as olivine and pyroxene for example, are geologically very common, making up the approximately ninety percent of the Earth's crust. Olivine possesses isolated $[SiO_4]^{4-}$ species surrounded by a mixture of $Mg^{2+}$ and $Fe^{2+}$ cations with the general composition $Mg_{2-n}Fe_nSiO_4$ (where n = 0–2). Pyroxene is made up of $Mg^{2+}$/$Fe^{2+}$—compensated infinite linear chains of oxygen-sharing $[SiO_4]^{4-}$ tetrahedra (often described as $[SiO_3]_n^{2-}$) and has the chemical formula $Mg_{1-n}Fe_nSiO_3$ (where n = 0–1). Terrestrial olivine and pyroxene are typically very Mg-rich and thus often close to the n = 0 end members: forsterite and enstatite respectively (see Fig. 1). Such a situation is not specific to the Earth's intrinsic geology, with Mg-rich silicates being a common constituent of meteorites and further found in wide range of extra-terrestrial environments. Silicate dust has also been detected in a wide range of astronomical environments (e.g. comets, circumstellar disks, supernovae, quasars) [3]. The observed infrared spectra of cosmic silicate dust typically consists of two significantly broadened characteristic peaks with wavelengths at around 10 and 18 μm s (corresponding to Si–O and O–Si–O bending modes respectively) which are judged to correspond to non-crystalline silicate dust, usually with pyroxene or olivine chemical composition [3]. More detailed IR spectra characteristic of crystalline forsterite and enstatite have also been identified (e.g. in circumstellar shells), accounting for about 10 % of all observed cosmic dust. Typically, however, because of their broad spectral signatures, our knowledge of dust particles in space is more limited than that of specific molecules. Through a combination of lab characterization of pristine material [4–6] astronomical observations [7, 8] and analysis of spectral features [9], the general properties of dust particles in various environments have been discerned to a certain extent [10]. Fitting the observed spectra with a combination of lab spectra of materials of different crystallinity, shape, size and

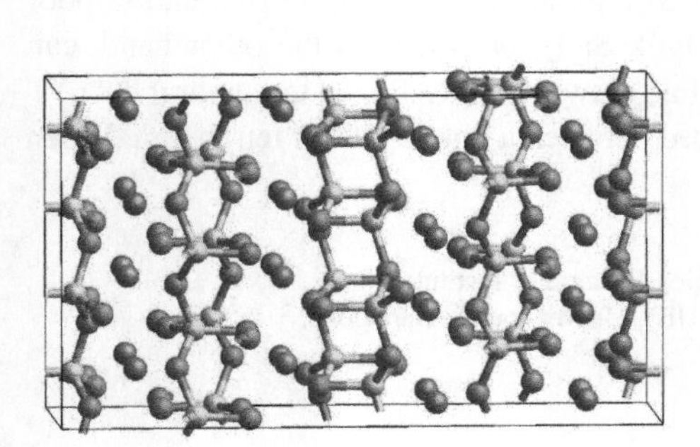

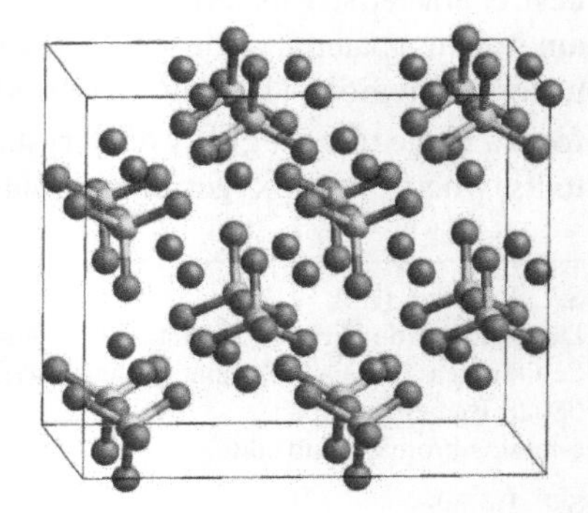

**Fig. 1** Crystal structures of $MgSiO_3$ enstatite (*left*) and $Mg_2SiO_4$ forsterite (*right*). Si atoms: *yellow*, Mg atoms: *blue*, O atoms: *red*

composition gives insight into the possible identity of dust particles in space [10, 11]. Such dust is found to be ubiquitous throughout the interstellar medium (ISM) and is the source of all cosmic and terrestrial silicate solids. In fact, silicates are thought to be the most abundant form of solid atomic matter (as opposed to the much higher fraction of gas phase atoms and molecules) in the known Universe. While dust particles take up only one mass percent of the total matter in the ISM, they play a crucial role in its chemical evolution [12, 13], catalyzing molecule formation and, through scattering the strong interstellar radiation, preventing molecular photodissociation in the denser regions.

As with all stable elements with atomic masses greater than boron, the constituent elements of silicate cosmic dust originate in nucleosynthesis processes in stars. These "heavy" elements form only about 2 % of all atomic matter in our galaxy with the rest being mainly in hydrogen (74 %) or helium (24 %). Of all the stellar-nucleosynthesised elements oxygen is the most abundant, constituting about 71 times less of the elemental mass fraction of our galaxy than hydrogen. Constituting approximately half the mass fraction of oxygen, carbon is the next most abundant element. After carbon, the six next most abundant elements in decreasing order are: Ne, Fe, N, Si, Mg, S each one constituting a mass fraction between 30 and 10 % of that of C. Thereafter, the elemental abundance starts to drop off more quickly with the heavier elements tending to be proportionally less plentiful. In all low-to intermediate-mass stars (0.6–10 solar masses), the thermal and gravitational conditions eventually permit the energetically favourable fusion of the $^{4}$He nuclei (or alpha particles) produced by the initial fusion of $^{1}$H. Stable $^{12}$C and $^{16}$O happen to be the most common products of nuclear reactions involving these alpha particles, via the triple-alpha and alpha processes respectively [14]. Some heavier elements with nuclei corresponding to integer multiples of alpha particles can also be formed by the alpha process (e.g. $^{20}$Ne, $^{24}$Mg and $^{28}$Si), but with significantly lower efficiencies. These reactions create a situation in which stellar nucleosynthesis produces large amounts of carbon and oxygen but only a small fraction of these elements is converted into neon and heavier elements. In the latter stages of an intermediate-mass star's life, when the supply of He is coming to an end, the star cools and burns any residual H and He in an erratic fashion driving convective mixing processes. At this "dredge-up" [15] stage some deep-lying carbon, oxygen and other heavier elements can reach the cooling expanding atmosphere. Such stars in this stage of their life (so-called Asymptotic Giant Branch or AGB stars) are typically found to have oxygen rich atmospheres. Stellar winds further expel elements in the atmosphere further from the star forming a circumstellar envelope within which chemical reactions between the elements can start. Close to the star's surface only very strongly bound molecules can form due to the high temperatures. One of the initial reactions to occur is simply that between C and O to form the extremely stable and rather chemically inert CO molecule. In the cooler regions further from the star, any excess oxygen which has not used up in CO formation can form other molecules. Of the remaining atomic species present, silicon forms the strongest bond with oxygen to form SiO molecules. Although it thus seems clear that this primary population of SiO monomers plays an essential role in the subsequent nucleation of silicates, there is much debate as to: (i) the transition in Si:O stoichiometry from 1:1 in SiO monomers to 1:2 in silicate dust, (ii) how Mg (and Fe) are incorporated into a SiO-based nucleation process to form silicate dust grains with pyroxene and olivine-type compositions, and (iii) what are the primary seeds which initiate silicate dust nucleation, (iv) the role of silicate grains are nucleation centres for water ice.

In this chapter we highlight the role that "bottom-up" (i.e. from the atomic scale upwards) computational modelling can play in providing important insights into the detailed structure, properties and formation of small silicate clusters. We focus on how such theoretical efforts are helping us to understand nanosilicates in astronomical environments and, in particular, with relation to the four points mentioned above.

## 1.2 Finding Low Energy Structures of Si–O-Based Clusters

Often simply by virtue of their size, nanosilicates are extremely difficult to analyse by experimental characterisation methods alone. In some recent studies, it has been demonstrated that detailed nanoscale information for well-defined laboratory-based nanosilicate systems can be accurately ascertained from a synergistic combination of computational modelling and experimental data [16]. For less-well-defined nanosilicate systems which are not prepared in the lab, such as atmospheric or cosmic silicates, computational modelling can play an essential role in providing realistic nanoscale structural models, which would be otherwise unobtainable. These candidate nano-models can subsequently be used to calculate properties (e.g. atomic and electronic structure, chemical reactivity, spectroscopy), which can be compared and used together with any relevant observational data to help explain the phenomenon under study. For example, quantum chemical calculations are particularly useful for providing (free) energies for nanocluster/molecular species and the barrier heights for their reactions for use in understanding nucleation. Even when no experimental/observational data are available, computational modelling can also provide detailed predictions and helpful insights which may help to guide future experimental research.

Although the various methods of computational modelling constitute a powerful and, often indispensable, toolkit, the theoretical treatment of nanoscale materials, or nanomaterials, is still not a routine task. The first obstacle is that of scale. Although, almost by definition, the nanoscale is small from a typical experimental materials perspective, even a nanocluster of only $\sim$10 nm diameter possesses thousands of atoms. Although, this is, in principle, of a size that is readily treated by computational modelling methods using rapidly evaluated classical interatomic potentials (IPs), the aim in many theoretical investigations of nanomaterials is to achieve an accurate quantum mechanical based description. Even for computataionally efficient *ab initio* electronic structure modelling approaches, nanomaterials consisting of a few thousand atoms are at a size and complexity that is still significant for routine calculations.

The second problem to be tackled is that of the high configurational complexity of many nanosystems. For small to moderate sized molecules and bulk crystals one often has sufficient structural information from experiment or chemical databases to start directly with *ab initio* quantum chemical calculations. Nanomaterials, on the other hand, tend to lie in a size regime between these two extremes for which, currently, there is a dearth of detailed structural information. Although knowledge of the chemical composition of a small molecule or a crystal with a small number of atoms in its unit cell can be sufficient for a rapid exhaustive search though all chemically sensible structures, for nanosystems this is typically not the case. For silicate nanoclusters, for example, the number of possible structural configurations for even modest sizes (e.g. 20–30 atoms) becomes very large and increases combinatorially with increasing cluster size. The multi-dimensional hyperspace of isomeric structures (depending on all the coordinates of all the atoms in a cluster) with respect to their energetic stability is often termed the potential energy surface (PES). Typically one wants to find the most stable nanocluster structures on the PES. The complexity of the PES, and thus in turn the ease of with which one can find low energy clusters, depends on many factors. A strong tendency to form atomically ordered closed packed clusters, for example, can simplify the PES. On the other hand, less dense systems with directional bonding and which have a high degree of structural freedom (e.g. bond angles, coordination number) and/or have multiple atom types which can permute positions, tend to have a PES which is much more difficult to search efficiently. In the latter case in particular, searches for low energy structures by hand, guided by chemical intuition, or simply randomly guessing, rapidly become untenable approaches and one must employ efficient global optimisation methods.

Global optimisation, when applied to cluster isomers, is a term which covers many different algorithmic approaches to finding the most energetically stable isomer structures on the PES. In particular, such searches aim to establish the single most stable global minimum energy cluster structure for a given composition. There are many global optimisation methods that have been applied to binary inorganic clusters and we refer the reader to a couple of more specialised reviews for a better overview of these methods and what they are capable of [17, 18]. In the specific case of Si–O-based clusters we note that many global optimisation investigations have been performed using the Monte Carlo Basin Hopping (MCBH) method, which was introduced by Li and Scheraga for protein folding [19] and subsequently adapted for clusters, by Wales and Doye [20].The MCBH algorithm uses a combination of Metropolis Monte Carlo sampling and local energy minimisations to sample the PES of cluster configurations. In a typical MCBH run two parameters need to be specified: the step size $\Delta$ (controlling the maximum change in structure per MCBH step) and the temperature $T$. These parameters need to be carefully tuned in order to maximise the sampling efficiency of the PES. If the temperature $T$ is too high the MCBH procedure is not sufficiently directed towards low energy structures, whereas if the temperature is too low there is a risk of remaining trapped in a local minimum. If the step size $\Delta$ is too small the sampling procedure is inefficient since many steps are needed to get an appreciable change in structure. If the step size is too large many of the changes will be rejected (according to a Metropolis [21] criterion).

An important issue for many global optimisation methods is the choice of the coordinates of the initial cluster, or set of clusters. In principle this choice should not matter, since if the run is sufficiently long all of the phase space of cluster configurations is sampled. In reality the length of each run is limited, and the form of the PES dictates the extent of the cluster isomer sampling achieved [22]. To overcome such problems a selection of different starting points (e.g. random, or deliberately inequivalent cluster structures) can be made to help ensure a more even and extensive phase space sampling [23]. Generally, the complexity of the PES (depending on cluster size, type of interatomic interactions, number of low energy minima, etc.) largely determines the ease with which any global optimisation method can find the global minimum. A more complex PES generally means a more difficult global search which in turn leads to more computation effort being needed in calculating energies and forces while searching the PES. Maximising computational efficiency while maintaining an adequate accuracy is essential for any global optimisation algorithm. Direct use of electronic structure methods (e.g. Density Functional Theory—DFT) for global optimisation provides high accuracy but at the expense of a slow exploration of the PES. At the time of writing, the state-of-the-art with respect to global optimisation of oxide binary clusters using direct DFT-evaluations of the PES can be found for $(MgO)_N$ clusters for $N \leq 16$ [24] and for $(TiO_2)_N$ for $N \leq 10$ [25] (i.e. in both cases up to approximately 30 atoms). For the complicated PES of the more covalent and directionally bonded silicates and silicon oxides, and for cluster sizes up to over 80 atoms in some cases, highly computationally efficient empirical IPs have often been used for global optimisation. As long as the IPs employed are adequately accurate, after an IP-based global optimisation search, the global minimum cluster can usually be found by checking and refining the resulting low energy candidate clusters using DFT.

## 2 Structure and Stability of Silicon Oxide Clusters

### 2.1 Silicon Sub-Oxide Clusters

Silicon sub-oxide clusters have been investigated using bottom-up computational modelling with direct relation to experimental observations in studies of silane oxidation [26], silicon nanowire formation [27, 28], and the possible nucleation of $(SiO)_N$ clusters in the ISM [29] and in circumstellar environments [30]. Purely theoretical, systematic computational modelling studies have investigated the structures and properties of a range of $Si_NO_M$ sub-oxide isomers structures of low energy isomers with $n \leq 7$ using both manual searches [31–33] and global optimisation [34]. (see Fig. 2). The neutral $(SiO)_N$ global minimum clusters in this size range display a structural transition from simple single rings of alternating Si and O

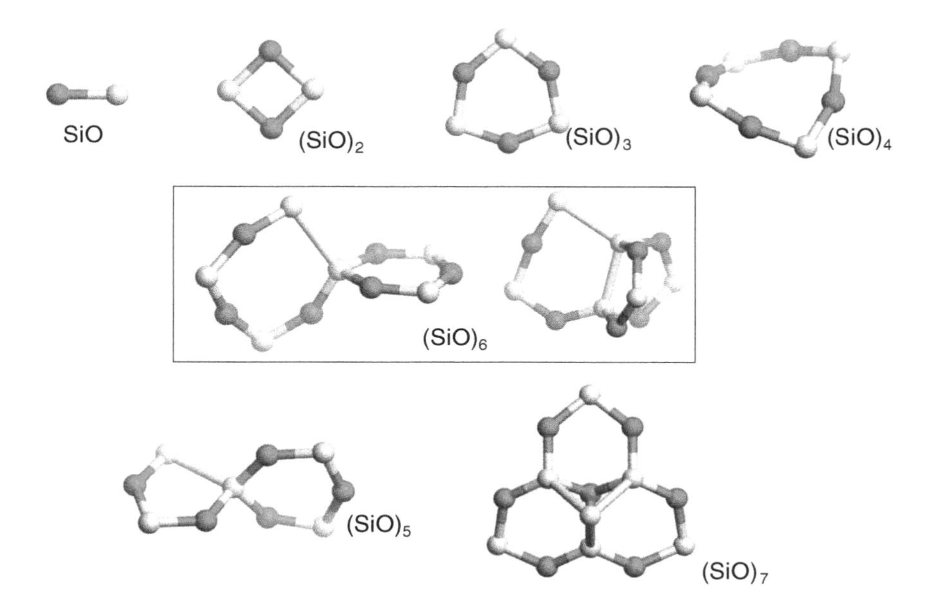


**Fig. 2** Theoretically proposed global minima $(SiO)_N$ cluster isomers for N = 1–7. The boxed isomers indicate that they are energetically nearly degenerate. Si atoms: *grey*, O atoms: *red*

atoms to more complicated topologies at N = 5. For $5 \leq N \leq 7$ the clusters structures display at least one Si–Si bond indicative of the initial stages of segregation into oxygen-rich and oxygen-poor regions as found in the bulk solid. We note that cationic $(Si_NO_M)^+$ clusters for N = 3–5 and M = N, N $\pm$ 1, have also been studied by cluster beam experiments together with global optimisation calculations, whereby the latter was used to assign isomer structures to observed IR absorption spectra [35, 36]. Here the tendency for Si–Si bond formation was found to start at $(SiO)_4^+$, which is also confirmed in another purely theoretical study [37]. Anionic $Si_NO_M$ clusters for $N \leq 5$ for a selection of M values have also been investigated by cluster beam experiments with the cluster structures assigned by calculations of measured electron affinities (EAs) and ionisation potentials (IPs) [38, 39]. The above studies generally confirm that from N = 1–7 for $Si_NO_M$ clusters Si–Si bonds tend to emerge with increasing N regardless of charge state and that Si–Si bonding is particularly favoured in more oxygen deficient (i.e. N < M) sub-oxide clusters. We note that in the $(SiO)_N$ case, due to the matched 1:1 stoichiometry, every Si and every O atom in all global minimum $(SiO)_N$ clusters have at least two bonds to other atoms. As we will see later, this is unlike the situation in silica with a numerically unbalanced oxygen rich 1:2 stoichiometry, where O-terminating defects are common.

Moderate sized $Si_MO_N$ oxygen-rich sub-oxide clusters (i.e. where M < N, with 20 < M + N < 30 atoms) were first proposed to energetically favour structures which exhibit linked $Si_2O_2$ "two-rings" and $Si_3O_3$ "three-rings" of alternating Si

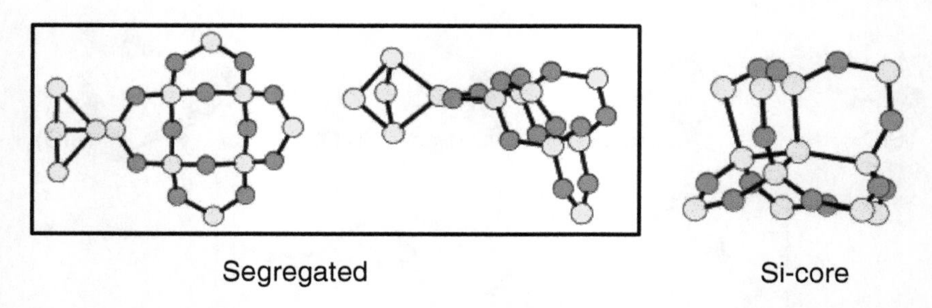


**Fig. 3** Comparison of a segregated $(SiO)_{12}$ isomer (*left*) with a higher energy Si-core type $(SiO)_{12}$ isomer (adapted from ref. [44]). Si atoms: *yellow*, O atoms: *red*

and O atoms with no Si–Si bonding [40]. For N = M, however, similar sized $(SiO)_N$ clusters were first proposed to be based on Si–Si bonded "cores" surrounded by silicon sub-oxide sheaths to be lowest in energy for the size ranges N = 6–21 [41], N = 14–26 [42], and N = 5–12 [43]. In 2008 an alternative and more energetically stable "segregated" motif for silicon monoxide clusters was put forward for $(SiO)_N$ N = 12–18 whereby the cluster structures are formed by a small silicon sub-cluster attached to a silicon oxide sub-cluster in a side-by-side manner [44]. In Fig. 3 we show a "segregated" versus a "Si-cored" isomer structure for $(SiO)_{12}$ where the former is calculated to be 0.86 eV lower in energy. Clearly, the computational modelling of small silicon sub-oxide clusters shows that they have a strong propensity to form anisotropic structures with O-rich and Si-rich regions, as is known experimentally to be a feature of bulk SiO [1]. In ref [45] updated lower energy global minimum $(SiO)_N$ candidates have been reported for the size range N = 8–20 based upon searches using the segregated model and initially "over" oxidised structures from global optimisation of $(SiO_2)_M$.

The nucleation of SiO monomers with respect to its possible relevance to the formation of circumstellar silicate dust has been the focus of a number of "top down" theoretical studies using classical nucleation theory (CNT) [46–48]. Here the atomistic structures of the growing cluster isomers are not taken into account and their energetic stabilities are replaced by scaled bulk-derived average values. Such approaches thus overlook the inherent tendency for SiO to segregate, even in very small clusters, which may have some astronomical relevance. In circumstellar environments the nucleation of SiO molecules could potentially form segregated nanoparticles containing 100 s of atoms. Subsequent fragmentation of these species into nearly (i) pure Si nanoparticles and, (ii) $SiO_x$ nanoparticles with x close to 2 (i.e. silica-like) one could provide an attractive account of the initial stages of silicate dust growth (i.e. via oxygen enrichment of SiO) and the source of a suggested carrier for a spectroscopic feature known as the extended red emission (ERE) [49, 50]. A couple of studies using bottom-up computational modelling together with cluster beam experiments have suggested that growth and fragmentation of small silicon suboxide clusters ($Si_NO_M$ with N = 1–17, M = N ± 1) could be relevant for silicate dust formation and possibly also contribute to the ERE [43, 51]. Work taking into account the calculated free energies of segregated low energy $(SiO)_N$ clusters, with their nucleation taken into account by a bottom-up kinetic approach, has strongly suggested that pure SiO nucleation is only relevant to astronomical silicate dust formation in relatively low temperature environments (<650 K) [45]. This result confirms that pure SiO nucleation would be unfeasible in most circumstellar dust forming environments (having typical temperatures >1000 K). This general view also been confirmed in other CNT-based studies using recent experimental vapour pressure data, where SiO nucleation is predicted to occur only for temperatures <800 K around specific Mira-type variable stars [52].

## 2.2 $(SiO_2)_N$ Clusters

Following the proposed scenario in the previous sub-section, one product of the nucleation of SiO could be silica which has the astronomically observed 1:2 Si–O stoichiometry and brings us closer to Mg-containing silicate dust. In fact, although relatively less common, pure silica dust has been observed around some stars [53, 54]. The vast majority of the many crystalline and amorphous bulk phases of pure $SiO_2$ are constructed from frameworks of 4-coordinated silicon atoms joined together by bridging oxygen atoms. Although the oxygen atoms around each silicon centre have a strong tendency to maintain O–Si–O angles close to 109.47° (*i.e.* to be near-ideally tetrahedral), the Si–O–Si linkages between silicon centres that are approximately two orders of magnitude more flexible. This structural property, allows for highly complex three-dimensional networks spanning a wide range of densities and topologies. A number of mathematical-based searches have thus far, accumulated approximately 100,000 periodic tetrahedral nets [55], thousands of which have been shown by calculations of their energies to be stable as silica materials [56]. The complexity of silica has also been compared with that of water, another tetrahedrally ordered material. Calculations have shown that water clusters exhibit a particularly complex PES characteristic of a so-called ‘strong’ liquid [57]. From comparative investigations of the structures and stabilities of low energy silica and water clusters, topological analogies between water and silica are found to persist down to the nanocluster size range [58] indicating a similar level of complexity. One important difference between water and silica, however, is that the former is a discrete molecular-based system, whereas silica is a continuously bonded material. Although water nanoclusters have an inherently closed-shell electronic structure by virtue of being constituted by discrete water molecules, those of silica must invoke structural and electronic reconstructions to avoid energetically costly terminating open-shell ‘dangling’ bonds. The structure and properties of nanoclusters in general are largely governed by their inherently high proportion of “surface” atoms. Knowledge of how silica deals with surface terminations, in particular with respect to its “excess” oxygen atoms, is thus key to understanding silica at the nanoscale.

Stoichiometric silica nanoclusters are found to maintain closed-shell electronic structures at their surfaces via the formation of: (**a**) three coordinated silicon centres terminated by singly coordinated silanone oxygen centres, (**b**) closed $Si_2O_2$ rings, or "two-rings", and (**c**) pairs of terminating singly-coordinated oxygen centres and three-coordinated oxygen centres. We note that although silanones are often written as Si = O implying a double bond, there is strong experimental and theoretical evidence that a more ionic $Si^+$–$O^-$ description is more appropriate [59–61]. Silanones, along with the topological two-ring defect, tend to be highly reactive with water to spontaneously give silanol Si–OH groups. From, theoretical studies, it is also known that if a silica network is not unduly strained, the presence of two silanones is thermodynamically unstable with respect to two-ring formation [27, 62]. Type (**c**) defects involve a single electron charge transfer from a three-coordinated oxygen centre to a singly coordinated oxygen centre (*i.e.* $\equiv O^+ \cdots O^-$) This type of defect was first proposed to be present in bulk amorphous silica [63, 64] in the 1970's based on Mott's Valence Alternation Pair (VAP) model [65]. In low energy silica nano-clusters, both the singly terminated $O^-$ species and the donating $\equiv O^+$ centre are present exclusively as a surface species [66, 67]. Such surface-terminating VAP defects (or Compensated Non-bridging Oxygen (CNBO) defects [66]) have also been observed in modelling studies of metastable surfaces of alpha-quartz [68, 69]. Furthermore, it is notable that when the $\equiv O^+$ centre is in a sub-surface site, the remaining terminating $O^-$ species is unreactive with water [66] and, much like the related but more reactive terminating silanolate group Si–$O^-$, may even have the propensity to order water molecules [70] or act as a catalyst [71]. In Fig. 4 we show the structures of the three closed-shell surface defect types (a-c).

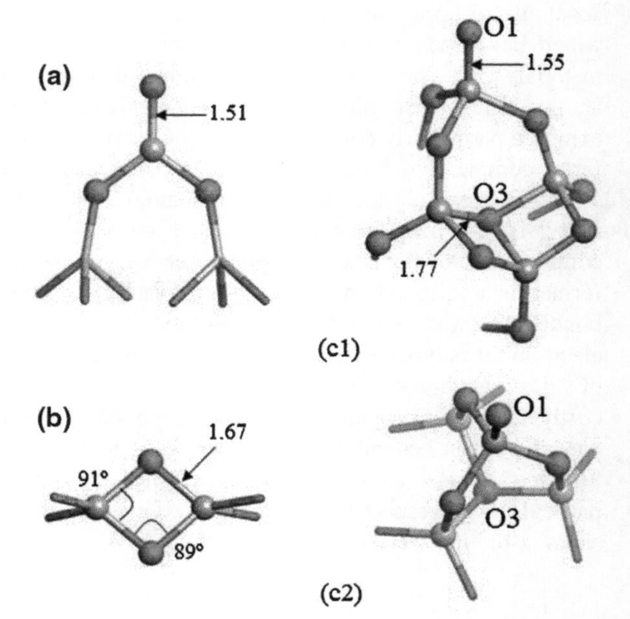


**Fig. 4** Surface defects exhibited by low energy silica clusters: **a** silanone, **b** $Si_2O_2$ two-ring, **c1** and **c2** two examples of a surface-VAP defect. Distances are in Angstroms. O1 and O3 correspond to singly and triply coordinated oxygen atoms. Si atoms: *grey*, O atoms: *red*

The candidate global minimum $(SiO_2)_{-N}$ clusters in the range $N$ = 2–27 [67, 72–74] shown below were all found by extensive global optimisation searches using MCBH. Although the MCBH algorithm is thought to be one of the least hindered global optimisation methods with respect to the specific topology of the PES [23] its success with respect to clusters of a real material, as with all global optimisation methods, relies further on a sufficiently accurate yet efficient representation the PES of that material at the nanoscale. For the smaller $(SiO_2)_N$ clusters we report (approx. $N < 16$) an interatomic potential set was used that was specifically parameterised to accurately predict the energies and structures of silica nanoclusters [72]. For the larger clusters (approx. $N \geq 16$) the TTAM silica potential [75] was also employed, which although parameterized for bulk silica, has also been shown also to be of some use in global optimization studies of silica clusters [73, 76]. From the resulting large number of isomers from these global optimizations (typically a few hundred for each $(SiO_2)_N$ cluster size) the 20–30 lowest energy structures and selected higher energy isomers with high symmetry were taken for energetic and structural refinement using DFT calculations employing no symmetry constraints. The post-optimisations employed the B3LYP functional [77] and a 6–31G(d) basis set which has been shown in numerous studies to be a suitable level of theory for calculating reliable structures and relative energetics of silica nanoclusters [78]. The set of candidate global minimum $(SiO_2)_N$ isomers resulting from this procedure are shown in Figs. 5 and 6. In following we briefly describe the evolution of the

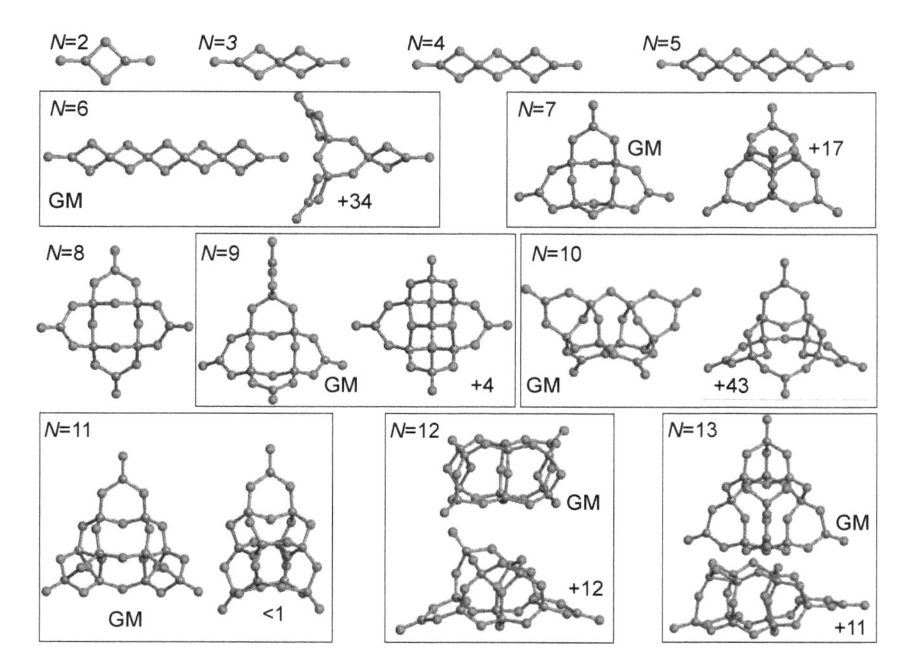


**Fig. 5** Low energy $(SiO_2)_N$ clusters for $N$ = 2–13. Candidate global minima are labelled GM and the relative total energy of the energetically next nearest isomer is shown in kJ mol$^{-1}$ when <50 kJ mol$^{-1}$ higher in energy. Si atoms: *grey*, O atoms: *red*

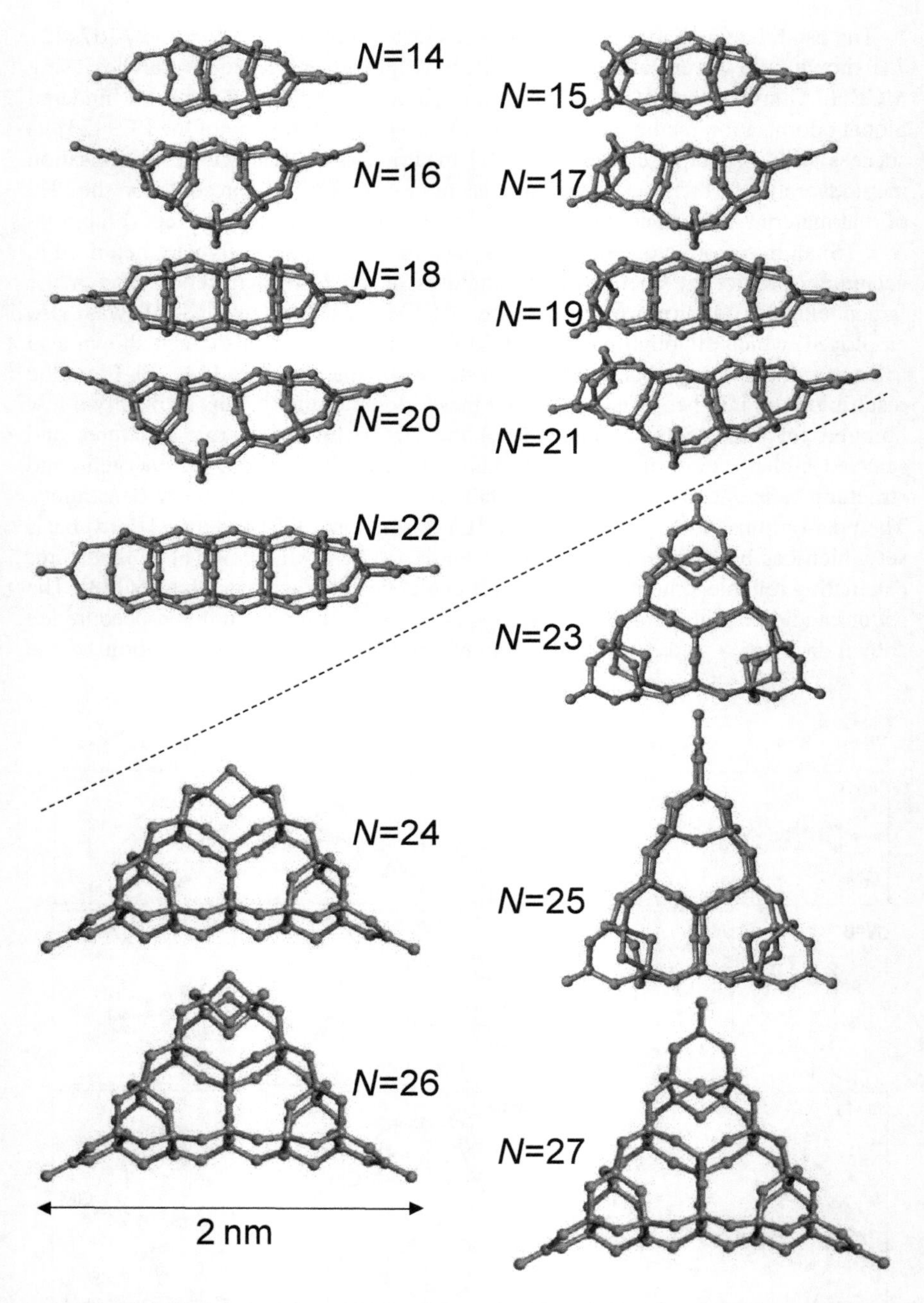


**Fig. 6** Candidate global minima $(SiO_2)_N$ clusters for N = 14–27. The *dashed line* indicated a transition between columnar and disk-like growth. Si atoms: *grey*, O atoms: *red*

structures of the $(SiO_2)_N$ nanosilica global minima with increasing size, noting the occurrence of the three defect types.

For $(SiO_2)_N$ clusters for small values of $N$ the low energy isomer spectrum contains relatively few isomers and can be explored to a relatively high degree of thoroughness by manual construction of isomers meaning that global optimization is largely unnecessary. Early studies following this intuitive approach were the first to suggest that the global minima for $(SiO_2)_N$ clusters N = 2–5 are silanone-terminated two-ring chains [79, 80] (see Fig. 5, note the relaxed linear $SiO_2$ monomer is not shown). This theoretical prediction was later confirmed by global optimisation studies [67] and also supported by cluster beam experiments on anionic $(SiO_2)_N$ species for N = 2–4 [39]. For $(SiO_2)_6$, the global minimum is still theoretically predicted to be the two-ring chain at the B3LYP/6–31G(d) level of theory but other non-linear isomers incorporating $(Si_3O_3)$ three-ring become energetically competing structures. The PES of the $(SiO_{2-})_6$ system has also been explored by *ab initio* molecular dynamics (AIMD) revealing many low energy isomers [81]. Optimisation of this set of isomers at the B3LYP/6–311 + G(d) level of theory indicates that the linear two-ring chain may be slightly higher in energy than the low energy 3-ring-containing $(SiO_2)_6$ isomer shown in Fig. 5. This cluster size marks a crossover point in the evolution of silica cluster structure with increasing size where we pass from a preference for one-dimensional growth to more complex growth trends for $(SiO_2)_N$ $N > 6$. It is noted that this increase in structural complexity is mirrored in the increased difficulty in searching the low energy $(SiO_2)_6$ PES.

The $(SiO_2)_7$ global minimum has a $C_{3v}$ symmetric trigonal structure and is energetically almost degenerate with a $C_s$ symmetric isomer. The former cluster isomer is the first global minimum candidate to contain a surface VAP defect in addition to its three silanone terminations. The lowest energy $(SiO_2)_8$ cluster isomer is a $D_{2d}$ symmetric cross-like structure containing four silanone terminations and no other defect centres. The cluster's highly symmetric form, seems to be strongly energetically favoured with respect to other $(SiO_2)_8$ cluster isomers, with the next lowest energy $(SiO_2)_8$ isomer being 99 kJ mol$^{-1}$ higher in energy. For $(SiO_2)_9$, the two lowest energy structures found take advantage of the low energy symmetric core of the $(SiO_2)_8$ global minimum cluster. The lowest energy cluster adds one $SiO_2$ unit to one of the terminating silanone centres of the $(SiO_2)_8$ global minimum forming a pendant two-ring and lowering the symmetry to $C_s$. The next lowest energy $(SiO_2)_9$ cluster results from adding a $SiO_2$ unit onto the centre of the $(SiO_2)_8$ global minimum, creating two compensating-pair VAP defects and a structure having $C_{2v}$ symmetry. Both these $(SiO_2)_9$ structures are exceedingly close in energy.

For the two lowest energy $(SiO_2)_{10}$ clusters there appears to be a tendency to move away from cluster structures built upon smaller low-energy forms. Both $(SiO_2)_{10}$ clusters shown have complex three-dimensional structures predominately formed from interlocking $(Si_3O_3)$ three-rings. Both these clusters display four silanone centres, and have $C_2$ and $C_s$ symmetries respectively. The two lowest energy $(SiO_2)_{11}$ clusters have a similar three-dimensional triangular structural form with two VAP defects and one silanone, and are essentially degenerate in energy, differing by only 4 kJ mol$^{-1}$. Both $(SiO_2)_{12}$ lowest energy isomers shown in Fig. 5 have a compact elongated form containing numerous four-rings. The global minimum isomer contains a compensating-pair VAP termination at each end, joined together at a single four-ring. The second lowest energy isomer, in constrast, has three oxygen terminations: two silanones and one compensating-pair VAP defect.

With regard to $(SiO_2)_N$ cluster structure the size N = 13, as for N = 6, marks a transitional size for cluster structural preference. For $(SiO_2)_N$ N = 6–13 the clusters seem to have no well defined structural type but can be roughly characterized as possessing a large proportion of three-rings, typically three or four defective oxygen terminations, and often having pyramidal/trigonal-like structures. Only for $(SiO_2)_{12}$ does this trend appear to be opposed with the lowest energy cluster isomers having more compact elongated forms with two or three terminations and more four-rings. For $(SiO_2)_{13}$ we see structures of both types in the two lowest energy isomers.

The structural growth trend for $(SiO_2)_N$ N = 14–22 clusters is found to be remarkably simple and proceeds via the addition of $Si_4O_4$ four-rings to make progressively longer compact elongated clusters (hereafter referred to as columnar clusters [73]) of a similar form to that of seen for the low energy isomers for $(SiO_2)_{12}$ and $(SiO_2)_{13}$. This four-ring addition growth pattern can be seen through N = 14, 18, 22 and N = 15, 19 in Fig. 6. For nanoclusters between these values of N the number of $SiO_2$ units is not sufficient to make a column of complete four-rings and instead a $Si_2O_2$ two-ring is inserted into the side of the cluster (e.g. N = 16, 17, 20, 21). For all columnar clusters the dominant defect termination is via a silanone, either at both ends of the column for even N, or only at one end for odd N. For clusters N = 15–21, the odd number of $SiO_2$ units does not allow for a two-fold symmetric termination with only silanone groups, and the odd-N columnar clusters are instead terminated by one silanone defect and a less energetically favorable compensating-pair defect. In all these cases, however, the connected four-ring columnar skeleton is maintained, which still appears to be the energetically favoured structural route to obtain the lowest energy for this size range.

Although persistent, the energetic preference for one-dimensional columnar growth is overcome after a $(SiO_2)_N$ cluster size of N = 22 by the emergence of a more compact two-dimensional disk-like structure based on the centrally symmetric sharing of three double $(Si_5O_5)$ five-ring cages. For the N-odd $(SiO_2)_N$ clusters for N $\geq$ 23, it appears that the having one silanone termination and one less energetically favored compensating-pair termination in a columnar cluster is out-weighed by having three silanone terminations and a more compact structure. For even-N for N > 23 the energetic benefit of having a disk-like form as opposed to a columnar form appears to solely be structural with both forms having the same double silanone defect termination.

Considering the range of candidate global minimum $(SiO_2)_N$ clusters between $n = 2 - 27$, some trends with increasing size can be identified. A common way to analyse models of annealed amorphous bulk silica glasses is in terms of the average $(SiO)_R$ ring size which tends to have a distribution centred around $<R> \approx 6$. Rings with $R < 6$ are intrinsically more strained and thus their presence is energetically

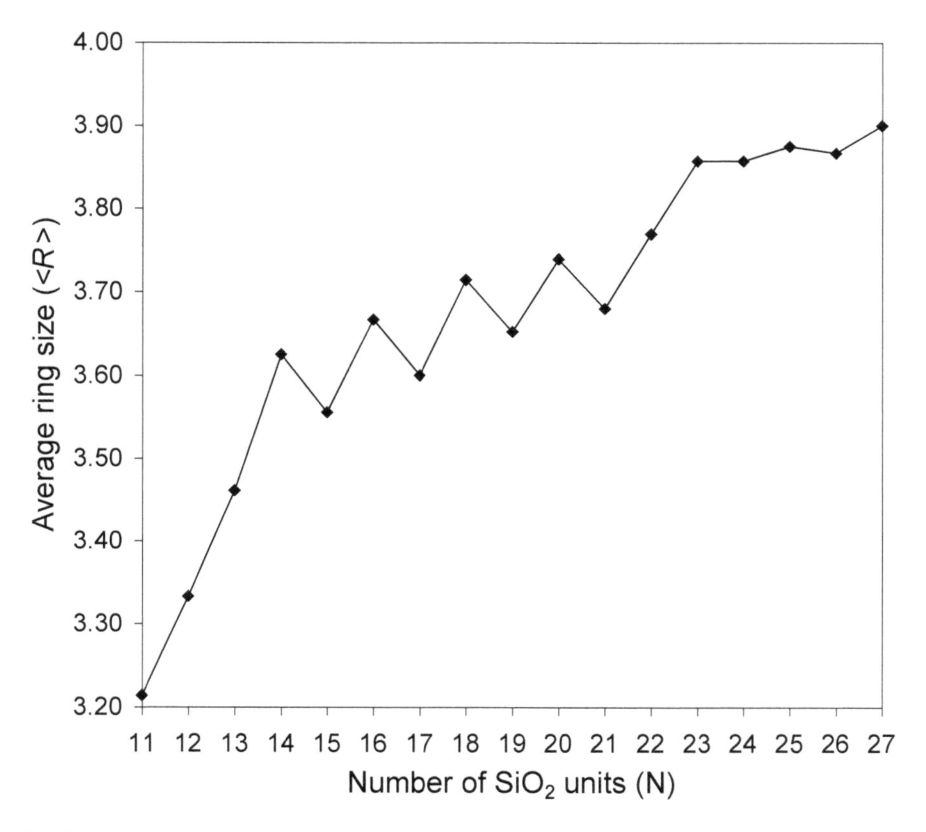


**Fig. 7** Variation in average $(SiO)_N$ ring size in global minima $(SiO_2)_N$ cluster isomers from N = 13–27

destabilising. As seen in Fig. 7, the dominant ring size in silica cluster global minima gradually increases from $R = 2$ for the smallest two-ring chain-based clusters to $R = 5$ for the largest. Evidently, the clusters in this size range are still some way from being simple cuts from annealed silica glass.

Cluster evolution can also be analysed by the change in relative total energy per $SiO_2$ unit ($E_{rel}$) with increasing size. In Fig. 8 we show the gradual decrease in $E_{rel}$ (with respect to the energy per unit of alpha-quartz which is set to zero) of the $(SiO_2)_N$ cluster global minima for $N = 1$–27. For dense spherical clusters, we expect this energy difference to decrease proportionally to $N^{-1/3}$ [82]. Although the silica clusters reported herein can hardly be considered as corresponding to this idealised model, it is interesting that for $N > 13$ the cluster binding energies fit well to such a trend, possibly reflecting the near regular columnar and disk-like growth pattern in this size range (see Fig. 6). We note that by extrapolating this trend we expect to reach typical bulk like energies per $SiO_2$ unit for silica glass at a cluster size of approximately $N = 7000$.

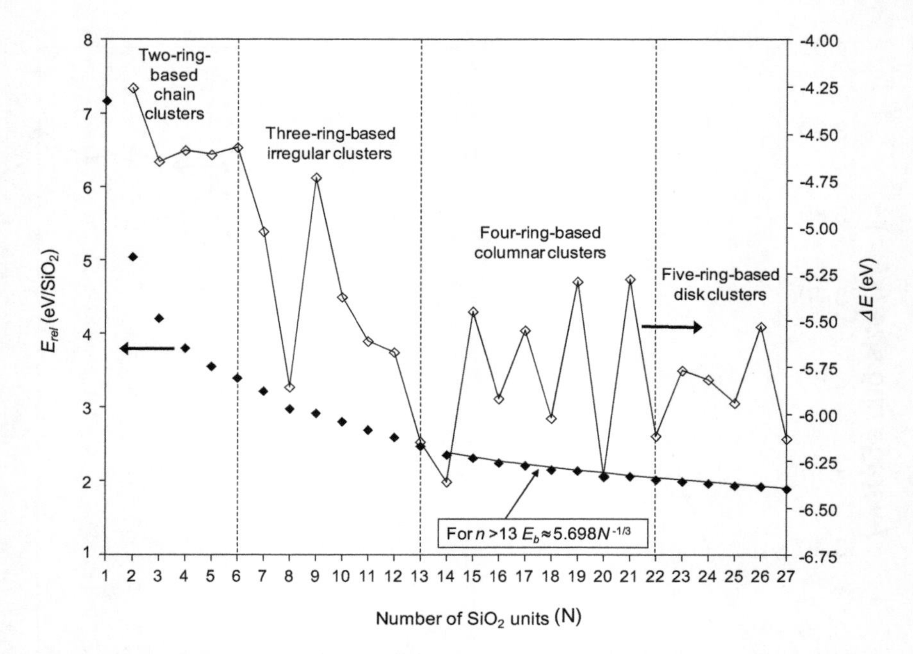


**Fig. 8** *Left-hand axis* decrease in total energy difference of $(SiO_2)_N$ cluster global minima with respect alpha-quartz with increasing size (filled diamonds). *Right-hand axis* first-order energy difference i.e. $\Delta E = E[(SiO_2)_N]–E[(SiO_2)_{N-1}]–E(SiO_2)$ (open diamonds)

Another measure of cluster stability, also shown in Fig. 8, is the first-order energy difference, ΔE, which measures of the energetics of monomeric $SiO_2$ addition: $\Delta E = [E_N–(E_{N-1} + E_{SiO2})]$. The initial $(SiO_2)_N$ global minima for $N = 2–6$ show evident regular linear chain growth based on two rings (see Fig. 5) which is reflected in the high but stable values of ΔE. The global minimum clusters $(SiO_2)_N$ $N = 7–9$ are based around a common structural motif build upon four connected three rings, which is most clearly and symmetrically exhibited for the N = 8 cluster. The pronounced dip in ΔE for growth to, and then away from the $N = 8$ cluster is indicative of its particular energetic stableness or "magicness". Due to this property it has been speculated that this cluster could act a building block for new bulk polymorphs [83] and nanowires [84]. For $N > 9$ until $N = 13$ the clusters continue to be based mainly on three-rings with a monotonic decrease in ΔE. For N > 13 until $N = 22$ the clusters form columnar structures based upon four rings (see Fig. 6) with odd-even fluctuations in ΔE caused by the types of defect terminations allowed for even-*n* (two silanones) versus odd-*n* (silanone and VAP defect). In the last region considered the clusters for $22 > N > 28$ are based upon five ring disk-line clusters (see Fig. 6) for which the ΔE values are relatively smaller and stable indicating a gradual trend towards bulk-like behaviour.

For $(SiO_2)_N$ clusters with N > 27 it is increasingly more difficult for global optimization methods to find low energy global minima due, mainly, to the exponentially increasing complexity of the PES with increasing cluster size. In an attempt to predict the likely structure of larger silica clusters, on can instead expand upon an idea planted in a study of the size-dependent transition from two silanone defects to a two-ring in a model nanochain-to-nanoring system [62]. In this study, fully coordinated (FC) rings of two rings were found to be energetically favoured over defective chains after a certain size was reached. Other DFT calculations of larger non-globally optimised compressed nanoslabs of 36–48 $SiO_2$ units have also observed a competition between silanones and two-rings as surface-energy-reducing mechanisms [85]. As the reaction between two separated silanones to make a FC two-ring is an energetically favourable and barrier-less process [27], depending on the constraints of the bonding topology of the cluster it is expected that all silanone terminations will eventually lose out to FC two-rings. Structures of cage-like FC silica clusters consisting of 12–60 $SiO_2$ units have been theoretically proposed in the literature and found to be structurally stable and moderately energetically stable. [86, 87] Using a algorithm specifically designed to explore the PES of low energy fully-reconstructed directionally bonded nanosystems (e.g. FC nanoclusters), however, it has been shown that as silica nanoclusters grow in size complex non-cage-like FC structures become increasingly energetically stabilised [88]. Using this approach it has been predicted that FC clusters to become the most stable form of nanosilica beyond a system size of approximately 100 atoms (i.e. $(SiO_2)_N$ of approx. N > 27) and before the eventual emergence of bulk crystalline structures. An example of a selection of FC $(SiO_2)_{24}$ nanoclusters, with respect to the silanone-terminated global minimum is shown in Fig. 9. With increasing cluster size (>100 atoms) global optimisation methods increasingly become intractable due to the sheer complexity of the PES. Based upon the above noted structural and energetic trends, however, for such systems it is expected that the percentage of terminating defects in the lowest energy clusters will significantly reduce (i.e. larger clusters will tend towards having FC topologies). Theoretical design of FC sililca nanoclusters of up to 360 atoms in size have indicated that fullerene cage-like clusters, although not the most stable cluster type, may be viable synthetic targets [89, 90]. As the property of being fully-coordinated mirrors in a finite way the bulk crystalline property of an extended unbroken bonded network (also leading to an associated large bulk-like band gap) such fully-coordinated nanoclusters can also be useful theoretical models for bulk or nano-particulate surfaces. [91, 92]

In summary, bottom-up computational modelling studies find that low energy $(SiO_2)_N$ isomers containing up to $\sim 100$ atoms are structurally rich and all unlike the bulk. Within this size range, although at small sizes low energy silica clusters are very defective for larger sizes FC clusters may be energetically favoured. The low energy nanosilica clusters found in these studies form a stability baseline which is fundamental for understanding the chemical, physical and structural properties of nanosilicates in general.

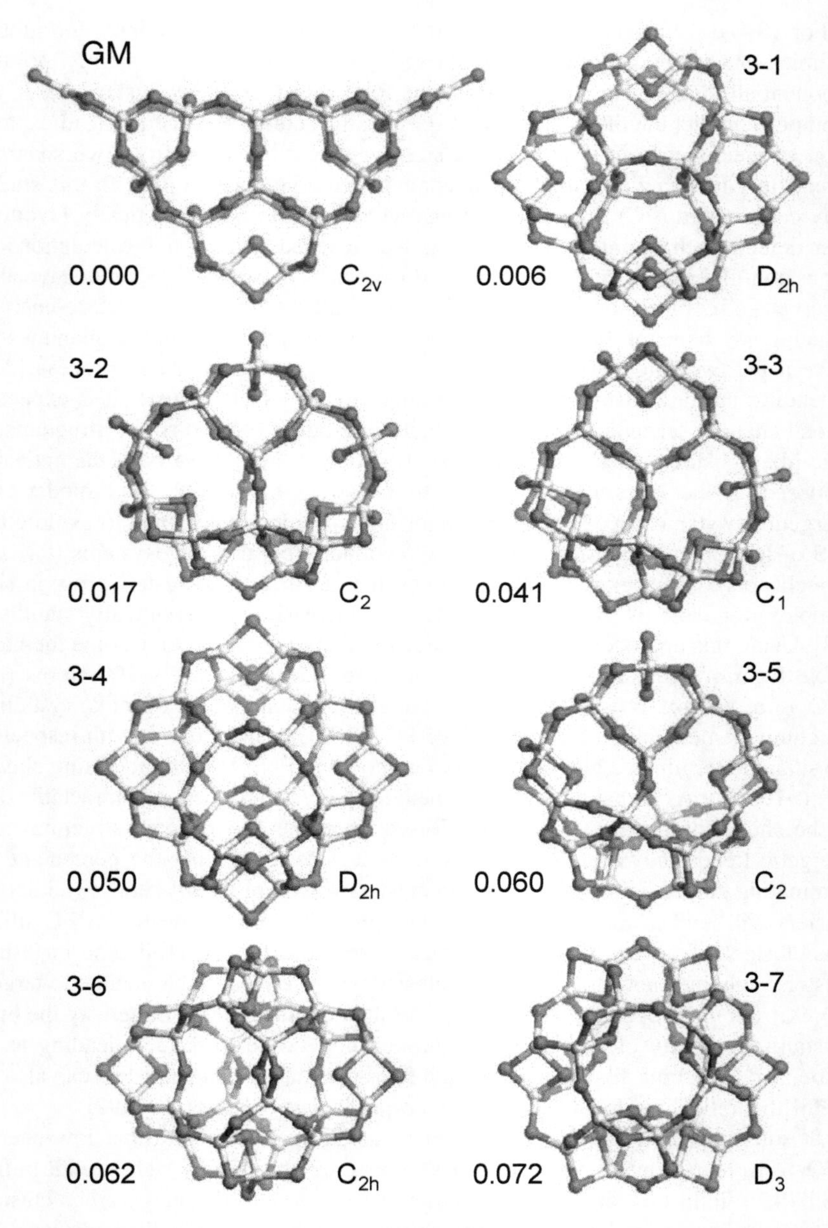


**Fig. 9** Structures, symmetries and total energies ($eV/SiO_2$) of lowest energy $(SiO_2)_{24}$ FC nanoclusters (3–1 to 3–7) relative to the global minimum (GM). Si atoms: *grey*, O atoms: *red*

# 3 Nanosilicates Clusters Around Stars

## *3.1 Nucleation of Silicate Dust*

Although SiO monomers are astronomically abundant, it is clear that the aggregation of SiO cannot be the only process contributing to the observed high temperature production of Mg-rich silicates around stars. Nevertheless, as mentioned above, the nucleation of SiO monomers with relation to circumstellar silicate dust formation has been investigated theoretically in a number of studies using classical nucleation theory (CNT) [46–48]. However, the puzzle of how the O:Si ratio increases and how Mg is incorporated, must also be an integral and essential part of any full explanation of silicate dust nucleation. Through extensive explorations of many possible intermediate nanoclusters with variable $Mg_xSi_yO_z$ compositions and their calculated stabilities, bottom-up computational modelling has begun to shed some light on this complex problem [93]. Gail and Sedlmayr [94, 95] argue that in the silicate dust condensation zone (1200–1000 K, 0.1–0.001 Pa), O is locked up in $H_2O$ and SiO, while Mg is atomic. From this premise, one can begin to theoretically search for clusters composed by adding these components (assuming $H_2$ release with oxidative $H_2O$ additions), and which are stable under typical circumstellar conditions. We note that such a search quickly becomes impracticable with the increase in both cluster size and chemical complexity and global optimisation techniques (such as MCBH as discussed above) are required. For condensation processes the dominating contribution to the entropy of reaction is generally the loss of translation entropy, which is only partially compensated for by an increase in rotational and vibrational entropy. At the high temperatures under consideration this huge entropy loss weighs heavily on the Gibbs free energy of reaction: $\Delta G_{rxn} = \Delta H_{rxn} - T\Delta S_{rxn}$, with $T\Delta S_{rxn}$ of the order of 200–300 kJ $mol^{-1}$ for any bimolecular addition reaction at 1000 K. Consequently, only very exothermic reactions ($\Delta H_{rxn} \ll 0$) can occur under these conditions. While the homomolecular nucleation of SiO ends effectively at the dimer stage, subsequent oxidation and Mg incorporation steps are found to be sufficiently exothermic to compensate for the enormous entropic costs involved in the condensation processes at these high temperatures [93]. This bottom/up approach strongly suggests that homogeneous homomolecular condensation of SiO is unfavourable in the dust condensation zone of stellar winds [45], while homogeneous heteromolecular nucleation of magnesium silicates could be feasible [93]. One of many possible routes from the set of monomeric species (SiO, Mg, $H_2O$) to a small pyroxene cluster is shown is Fig. 10 (see ref. 93 for more details). The validity of this general scheme of silicate nucleation from three basic monomeric units has been confirmed through its incorporation into detailed models of dust formation in supernovae [96] and AGB stars [97].

The second question arises due to the fact that the production of silicate dust around states is observed to begin at temperatures as high as 1200 K whereas calculated CNT-based predictions of the onset temperature for SiO condensation,

$$1.\quad 2\,SiO \rightarrow Si_2O_2$$

$$2.\quad Si_2O_2 + H_2O \rightarrow Si_2O_3 + H_2$$

$$3.\quad Si_2O_3 + \left\{ \frac{Mg \rightarrow MgSi_2O_3}{H_2O \rightarrow Si_2O_4H_2} \right\}$$

$$4.\quad \left\{ \frac{MgSi_2O_3 + H_2O}{Si_2O_4H_2 + Mg} \right\} \rightarrow MgSi_2O_4 + H_2$$

$$5.\quad MgSi_2O_4 + \left\{ \frac{Mg \rightarrow Mg_2Si_2O_4}{H_2O \rightarrow MgSi_2O_5 + H_2} \right\}$$

$$6.\quad \left\{ \frac{Mg_2Si_2O_4 + H_2O \rightarrow Mg_2Si_2O_5 + H_2}{MgSi_2O_5 + H_2O \rightarrow MgSi_2O_6H_2} \right\}$$

$$7.\quad \left\{ \frac{Mg_2Si_2O_5 + H_2O}{MgSi_2O_6H_2 + Mg} \right\} \rightarrow Mg_2Si_2O_6 + H_2$$

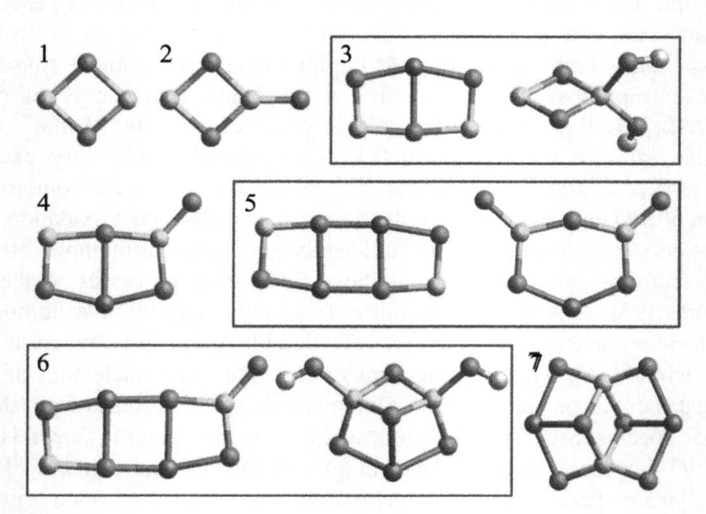


**Fig. 10** A seven step 2 SiO → $Mg_2Si_2O_6$ (pyroxene dimer) nucleation pathway (*above*) with corresponding global minimum geometries the cluster species involved (*below*). Si atoms: *yellow*, *Mg* atoms: *blue*, O atoms: *red*, and H atoms: *white*

using data extrapolated from laboratory measurements of the vapour pressure of SiO, are at least 200 K lower [46–48]. More recent work using bottom-up approach [45] and more advanded CNT-based models [52] have further confirmed that astronomical nucleation of pure SiO is only feasible at temperatures below 650–800 K. Even if we reject the idea that SiO molecules homogeneously nucleate and adopt a richer heteronuclear approach to silicate nucleation, as described above, the very first species to condense at ~1200 K is likely to be more stable than any silicate. The presence of such extremely stable refractory species, could also act as nucleation seeds which could strongly favour the heteronuclear nucleation of magnesium silicates.

Clearly, an understanding of nanosized silicate clusters, or nanosilicates, is essential for understanding the initial stages of silicate dust nucleation around stars. We also note that, although, AGB stars are commonly thought to be by far the main producers of silicate dust, brown dwarf stars, and particularly supernovae also can contribute to the total silicate dust quotient. Moving away from a dust-producing AGB star the average size of typical silicate grains quickly becomes significantly larger than a few nanometers and soon the dust grain population becomes dominated by larger grains. These dust grains are increasingly processed by shocks and sputtering in the ISM, where they are amorphized and become mostly olivinic. Throughout these stages of dust condensation, destruction and coagulation, nano-sized silicates could play an important role. Indeed, a substantial (~10 per cent) mass fraction of the silicate grain population in the diffuse interstellar medium (ISM) could be in the form of very small nanosilicates with <1.5 nm diameters. [98].

While dust particles make up only one mass per cent of the total matter in the ISM, they play a crucial role in its chemical evolution [99] where they catalyse molecule formation. The formation of molecular $H_2$, for example, is thought to be primarily produced through the reactions on dust grain surfaces. As diffuse clouds of gas and dust contract to form denser clouds, silicate dust also scatters the strong interstellar radiation, preventing molecular photodissociation. In such regions silicate dust grains gradually acquire icy mantles, largely formed from the aggregation of $H_2O$ molecules. [100] The role of nanosilicates as initiators of heterogeneous nucleation of water-ice is also relevant for terrestrial atmospheric processes. Computational modelling allows us to follow in detail the initial steps of $H_2O$ condensation on nanosilicates as shown in the next sub section.

Due to the difficulties in reconciling the estimated upper temperature limit for pure SiO condensation with the highest temperatures observed for silicate dust formation around stars other candidate constituents for condensation nuclei have been sought. Although Ti is approximately 300 times less abundant than Si in stellar outflows, solid $Ti_xO_y$ oxides are typically very stable and are potentially good candidates for forming the primary CNs. Following this line of thinking, the nucleation of pure $(TiO_2)_N$ has been theoretically studied [95, 101]. Considering, the high abundance of SiO molecules we may also consider reactions with $TiO_2$ and SiO. The smallest possible species of this type is the $SiTiO_3$—molecule which has been predicted to be a possible kick-starter for subsequent Mg-rich silicate dust condensation [102]. We note that another possible titanium-based compound that could possibly be present in a stellar outflow and act as silicate nucleation seed is $CaTiO_3$ [103]. This proposal is supported by calculations of condensation sequences of bulk solids from high temperatures, where $CaTiO_3$ (the protypical perovskite) usually appears first [95].

## 3.2 *Nanosilicates as Nuclei for $H_2O$ Ice Condensation*

From their essential role in dissolution and nucleation in terrestrial geologic and atmospheric processes, to the formation and growth of icy dust grains in astronomical environments, hydroxylated silicate nanoclusters, are of ubiquitous fundamental importance. From a technological perspective such species are also inherently involved in the synthesis of widely used nanoporous silicate materials such as zeolites. Despite their widespread significance, the experimental determination of the structures and properties of silicate nanoclusters is hindered by their structural and dynamic complexity. Typically, however, modelling studies have focused upon the energetics and structures of very small hydroxylated pure silica species, $(SiO_2)_M(H_2O)_N$, such as dimers (M = 2), up to tetramers (M = 4) [104–106] and their oligomerisation reactions. [105, 107, 108] In this section we see how bottom-up computational modelling can be employed to follow the step-wise hydroxylation of both a set of pure silica nanoclusters of different sizes, and of a small magnesium silicate cluster of pyroxene-type composition.

Silica is increasingly being employed in a wide array of functionalised nanostructures [109] and bio-inspired (nano)materials [110] where hydroxylation reactions at its surfaces are key. Computational modelling methods have contributed greatly to the understanding of the complex silica-water system with uniquely detailed microscopic insights into the mechanisms of silica hydroxylation reactions [111]. From such studies water molecules are generally predicted to preferentially attack the surfaces of bulk silica at terminating point defects and strained Si–O–Si sites (see Sect. 2.3) with modest reaction barriers [112–114]. On the contrary, regular non-defective crystalline silica surfaces are known from experiment to be often very resistant to reactions with water (e.g. thin ordered films [115], zeolite interiors [116]). As shown above, even the most stable silica nanoclusters up to at least $\sim 100$ atoms inherently exhibit terminating surface defects and strained $(SiO)_n$ rings and thus they are expected to be easily and energetically favourably hydroxylated. Below we concisely summarise some of the main results from a number of computational modelling studies and explore and compare the structures and stabilities of silicate clusters: $(SiO_2)_M(H_2O)_N$, M = 4, 8, 16, 24 through a systematic step-wise molecular hydroxylation from their anhydrous state until a N: M ratio ($R_{N/M}$) of $\geq$ 0.5.

The candidate global minima for each of the silica clusters shown were derived from a two step approach. Firstly low energy clusters of composition $(SiO_2)_M(H_2O)_N$, M = 4, 8, 16, 24, (in each case from N = 0 until $R_{N/M} \geq 0.5$) were comprehensively searched for using the MCBH global optimisation algorithm (see above) and specifically parameterized interatomic potentials [72, 117]. Secondly, the energies and geometries of the 10–15 lowest energy clusters were re-optimised using DFT, employing the B3LYP hybrid functional [77] and a 6–31G(d,p) basis set with no symmetry constraints. This approach has been shown to provide a combination of accuracy and computational efficiency for hydroxylated nanosilicate systems [118–121], even when compared to very high level methods [107]. Throughout each series the energy of a suitable number of water molecules is added as required so that all quoted energies for a fixed M correspond to the same chemical composition. When comparing energies of clusters in different series, total energies are normalised by dividing by the number of $SiO_2$ units in the respective cluster. The average deviation of the O–Si–O angle with respect to the optimal unstrained value of 109.47 is employed as a measure of tetrahedral distortion. This measure is only calculated for clusters for which all Si atoms have four oxygen neighbours and is thus not given for some clusters with low $R_{N/M}$ values which have three-coordinated Si centres.

Firstly, the energetic stability and structure is compared for each series for three representative degrees of hydroxylation corresponding to $R_{N/M} = 0.0, 0.25, 0.5$. In Fig. 11 the normalised energetic stability of the lowest energy cluster isomer for each series and the three values of $R_{N/M}$ are plotted. For $R_{N/M} = 0$ the increase in cluster size leads to an expected drop in energy per unit following an inverse power law dependence on the cluster size [82]. With increasing $R_{N/M}$, however, this decreasing energetic trend becomes less pronounced until for $R_{N/M} = 0.5$ the normalised energies for cluster sizes M = 8, 16, 24 are almost identical, with only the M = 4 cluster being higher in energy. For $R_{N/M} = 0.5$ the clusters have two Si atoms per hydroxylating water molecule and thus, potentially, one OH group per Si atom. This latter situation naturally provides a means for the formation of cages with N Si–OH vertices perhaps suggesting an underlying structural rational for the increased similarity in normalised energetic stability for $R_{N/M} = 0.5$. From extensive global searches, however, it is found that at $R_{N/M} = 0.5$ such cages are only energetically favoured for clusters of size M = 4, 8, where a tetrahedron and a cube are obtained respectively. For M = 16, 24 and for $R_{N/M} = 0.5$, instead of cages, it is found that free space clusters energetically prefer to form a dense amorphous cluster core with a discrete water molecule hydrogen-bonded to the surface Si–O–H groups of the cluster. For M = 16 this structural phenomenon first occurs for $R_{N/M} = 0.5$, whereas for M = 24 it first happens at a slightly lower hydroxylation level of $R_{N/M} = 0.458$. In Fig. 11 the gradual change in structure of the lowest energy cluster isomers found for a selection of $R_{N/M}$ values for $R_{N/M} \leq 0.5$ is shown. Clearly the water-induced structural evolution of the smaller clusters with N = 4, 8 is quite distinct to that of the larger clusters with N = 16, 24. Although in the former, each addition of a water molecule leads to an energetically favourable hydroxylation and appears to be proceeding to the full dissolution natural limit of N $Si(OH)_4$ monomeric units, the latter clusters are inherently resistant to reaching this limit.

In Fig. 12 a more general overview of the energetics of hydroxylation is shown for each $(SiO_2)_M(H_2O)_N$, M = 4, 8, 16, 24 cluster series through a plot of

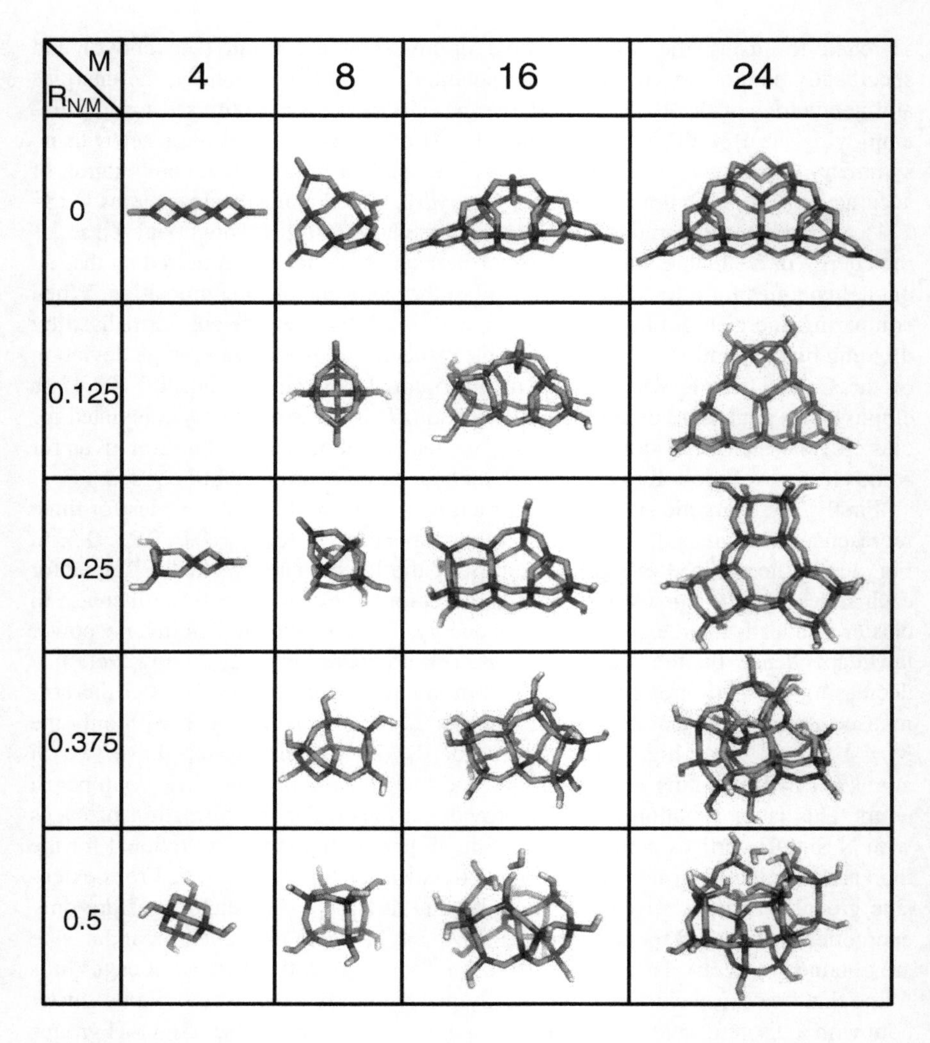


**Fig. 11** Candidate global minimum cluster isomer structures of $(SiO_2)_M(H_2O)_N$, M = 4, 8, 16, 24 clusters for $R_{N/M}$ = 0–0.5

normalised cluster energy versus $R_{N/M}$ for all clusters found in our investigation. From the anhydrous N = 0 clusters in each series we see an initial steep drop in energy with increasing step-wise hydroxylation which then becomes less pronounced and eventually reaches a point where each successive energy drop is very similar and at its lowest value. The transition from large energy decreases at lower hydroxylations to a linear minimal energy decreasing regime with increasing hydroxylation is suggestive of a threshold being reached at which subsequent hydroxylation is no longer so effective. The first few hydroxylation steps effectively

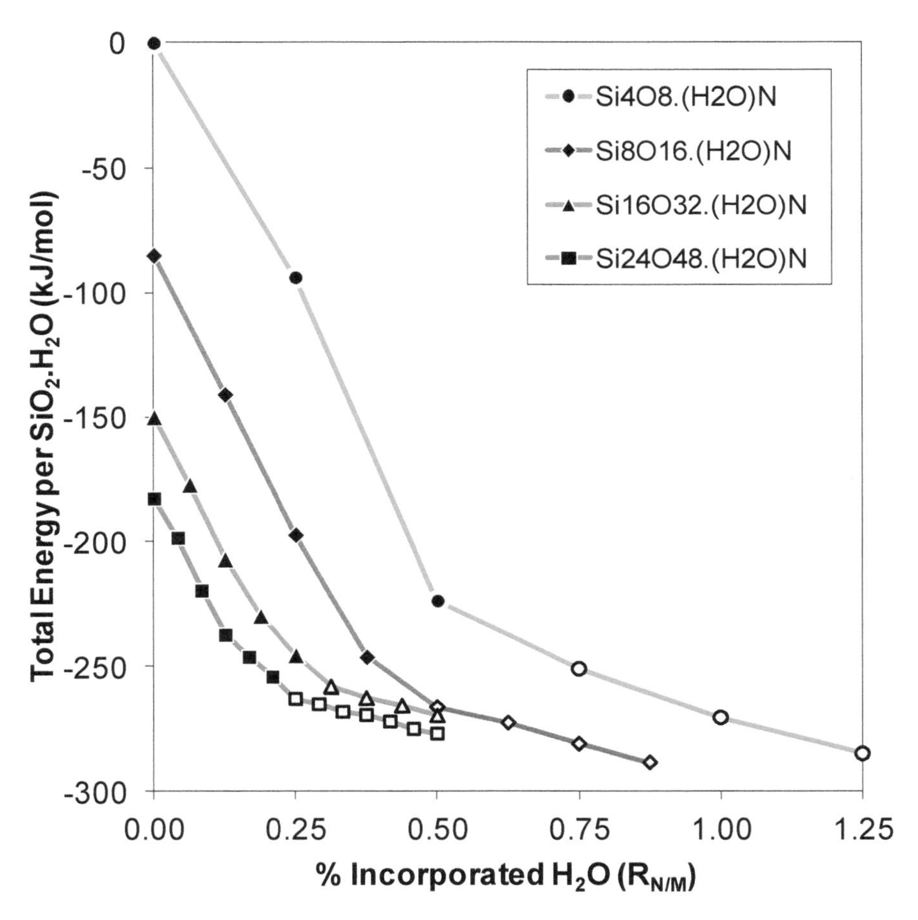


**Fig. 12** Relative total energies of $(SiO_2)_M(H_2O)_N$, M = 4, 8, 16, 24 clusters with respect to $R_{N/M}$. Open symbols indicate the linearly decreasing energy regime

convert surface defects (e.g. terminal silanone groups, strained small $(SiO)_2$ two-rings), which are common on anhydrous clusters, into more relaxed four-coordinated centres. The effects of such structural reconstructions are not localised to the surface-most atoms of a structure but also permit greater structural relaxation throughout a cluster. Once a cluster has been sufficiently hydroxylated to optimise the structural relaxation, any further hydroxylation is ineffective at reducing strain, and, for smaller clusters, may even cause the structure to start to be broken open. Thus, in the latter stages of hydroxylation each energy drop is by a fairly fixed amount mainly due to adsorption of $H_2O$ without any extra energetic stabilisation due to structural relaxation. As the size of the cluster increases, proportionally fewer atoms are involved in strained ("surface") defects and so the linear hydroxylation regime is encountered at progressively smaller $R_{N/M}$ values. This effect is analysed in detail in Jelfs et al. [121].

In order to examine whether the interaction water molecules with magnesium silicate nanoclusters follows that predicted for pure silica nanoclusters, the step-wise addition of $H_2O$ molecules to a $Mg_4Si_4O_{12}$ nanocluster of pyroxene-type composition has also been followed using the same computational methodology [122]. In order to compare the results with those reported above for pure silica nanoclusters the number of oxygen atoms in the magnesium silicate cluster is used to determine an effective number of $SiO_2$ units. In the case of the $Mg_4Si_4O_{12}$ cluster we take this to be 6 (i.e. 12/4) effective Si cations. In this way, after normalisation with respect to the number of effective and/or real Si centres, we can compare the impact of having two $Mg^{2+}$ cations for each "replaced" $Si^{4+}$ cation on the energetics of hydroxylation.

In Fig. 13 the normalised stabilisation energies of the $Mg_4Si_4O_{12}$ cluster due to interaction with $(H_2O)_N$, for N = 1–5 is compared with that calculated for the pure silica clusters: $Si_8O_{16}$ and $Si_{16}O_{32}$. The most stable product of the interaction of a single water molecule with the $Mg_4Si_4O_{12}$ nanocluster is also a hydroxylated species with a stabilisation energy per effective $SiO_2$ unit of 52 kJ $mol^{-1}$. This is very similar to the values of 56 kJ $mol^{-1}$ and 54 kJ $mol^{-1}$ calculated for $Si_8O_{16}$ and $Si_{16}O_{32}$ respectively, indicating that a similar strain release/defect healing may also be occurring. Looking at the structure of the $Mg_4Si_4O_{12}$ nanosilicate cluster before and after this initial water addition step (see Fig. 13) we, indeed, see that two of the Si centres become hydroxylated and a strained $(SiO)_3$ ring in the anhydrous cluster is no longer present after hydroxylation. As the hydroxylated nanosilicate structure was found by global optimisation, however, it does not provide any information with respect to possible reaction pathways from the original anhydrous state. From the previous section we know that adding discrete water molecules to a small

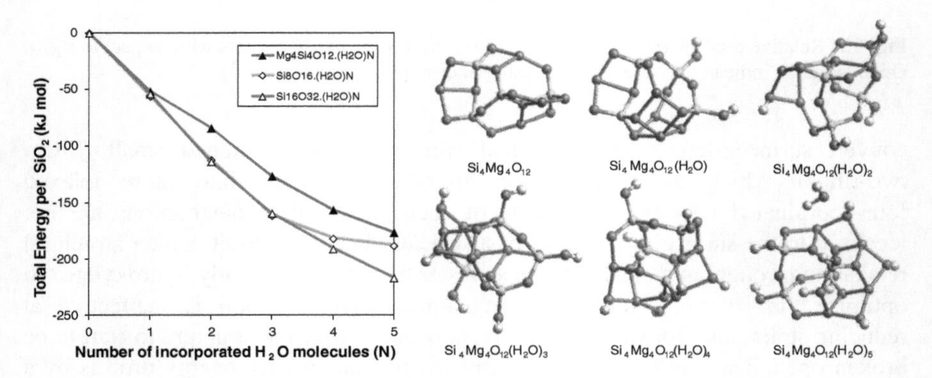


**Fig. 13** *Left* Total relative energies of $Si_8O_{16}(H_2O)_N$, $Si_{16}O_{32}(H_2O)_N$ and $Si_4Mg_4(H_2O)_N$ for N = 0–5. All energies are w.r.t. the corresponding anhydrous cluster and 5 discrete water molecules which is set to zero in all three cases. All energies are also normalised w.r.t. the effective number of $SiO_2$ units (where each MgO counts as 0.5). *Right* The most energetically stable structural configurations for the $Si_4Mg_4(H_2O)_N$ nanoclusters for N = 0–5

magnesium silicate species does not (initially) spontaneously lead to hydroxylation and thus a barrier from the physisorbed state to the chemisorbed state exists. Whether this barrier is "submerged" (i.e. lower that the energy gained from physisorption), which would facilitate hydroxylation, is currently unknown.

The addition of a second water molecule to the $Mg_4Si_4O_{12}$ nanocluster further hydroxylates the cluster but stabilises it by 20 kJ $mol^{-1}$ per $SiO_2$ unit less than the first hydroxylation step. Conversely, the respective stabilisation energies for the second $H_2O$ addition for $Si_8O_{16}$ and $Si_{16}O_{32}$ stay constant. Unlike pure silica, only half the cations in a pyroxyene-type nanocluster are associated with network-forming $SiO_2$ centres, with Mg cations helping to bind these centres together and saturate potential Si–O terminating defects. Considering this, it may be expected that the observed strong initial stabilisation by water addition to be limited to relatively fewer water molecule additions than for pure silica. The lowest energy structure found for two water molecules hydroxylating the $Mg_4Si_4O_{12}$ nanocluster has two hydroxyls on two Si centres and two OH groups each bound to two Mg centres (see Fig. 13). This isomer is found to be 18 kJ $mol^{-1}$ more stable than the most stable isomer with four Si–OH groups. Although, in general, Mg-hydroxylation is likely to be less stabilising than Si-hydroxylation due to the smaller ionic charge of the Mg centres, one must also consider the destabilising structural distortion induced by increasing Si–OH formation (see above) which may help explain the pattern of hydroxylation at this stage. The third and forth water molecules further hydroxylate the $Mg_4Si_4O_{12}$ nanocluster with a gradually reduction in the stabilisation energy for each successive $H_2O$ molecule addition. For the subsequent addition of a fifth water molecule the lowest energy nanosilicate cluster isomer is found to exhibit a hydrogen bonded $H_2O$ molecule. This isomer is found to be only ~6 kJ $mol^{-1}$ more stable than the lowest energy fully hydroxylated isomer found. Although the stabilisation energy associated with the addition of the hydrogen bonded fifth water molecule is only 20 kJ $mol^{-1}$ per $SiO_2$, this follows the gradual decrease in stabilisation with an increasing number of added water molecules and does not appear to mark a change to a linear stabilisation regime as seen for pure silica nanoclusters (see above). For the $Si_8O_{16}$ cluster, for example, a levelling off of the stabilisation curve starting at four added water molecules (see Figs. 12 and 13) is observed. The smoothly decaying stabilisation trend of the $Mg_4Si_4O_{12}$ nanocluster for 2–5 added water molecules is more akin to that displayed by larger clusters such as $Si_{16}O_{32}$. Further, the tendency of the $Mg_4Si_4O_{12}$ nanocluster to energetically prefer to be only hydroxylated up to a certain number of added water molecules (and beyond which hydrogen bond discrete water molecules) is also a characteristic only of larger silica clusters.

The fact that relatively small magnesium silicate nanoclusters appear to act more like relatively larger silica clusters with respect to their interactions with water may make them: (i) relatively stable to hydrolysis into hydroxylated monomers, (ii) able to absorb relatively high numbers of water molecules while maintaining reasonable absorption energies. These properties could make small magnesium silicate species significant for water ice nucleation in the upper atmosphere and, in an astronomical context, in dense molecular clouds. Ice formation in interstellar clouds only becomes possible once they are opaque enough to prevent water photodesorption from the intense UV field in the interstellar medium. Prior to ice formation, highly hydroxylated species such as $Mg_4Si_4O_{12}(H_2O)_4$ could be interesting due to their relatively high oxygen to metal ratio (in this case O/M equals 2). This ratio is interesting in view of the recent observations that oxygen depletes more strongly than other elements towards high total depletion regions. [123, 124] In the denser regions more oxygen ($\sim$2.4 $\times$ $10^{-4}$ $n_H$) seems to be tied up in solids than can be incorporated in silicates (Mg + Si abundance $\sim$8 $\times$ $10^{-5}$ $n_H$) with a bulk O/M ratio of 1.33 (olivines) to 1.5 (pyroxenes). If indeed, as is estimated, 10 % of the interstellar silicate mass consists of nano-sized particles [98] and these could incorporate extra O atoms up to a O/M ratio of 2, substantially more O could be tied up in these superoxygenated nano-silicates. Nano-silicates, thus, by virtue of their higher surface to volume ratio, can adsorb relatively more water molecules than bulk silicates [125]. Although it is likely that there are other unidentified solid particulates or molecules that also contribute to the strong depletion of oxygen, nanosilicates could be a significantly contributing unidentified depleted oxygen carrier [122, 124].

## 4 Summary and Outlook

In this chapter we have attempted to provide a brief overview of bottom-up modelling of nanosilicates, particularly with respect to their astronomical relevance. We have described how such modelling approaches are beginning to provide detailed insights into the formation and structure of nanosilicates in circumstellar environments and their subsequent physico-chemical properties. Our chapter primarily deals with studies of rather small nanosilicate species largely due to the dearth of theoretical and experimental data regarding the structures and properties of such species in the size regime between clusters possessing $\sim$50 atoms and significantly larger bulk-like particles. Nevertheless, even within this limited size range, we have shown the importance of computational modelling for obtaining unprecedentedly detailed insights into the low energy structures of nanosilicates, their formation through nucleation, and their role as seeds for further nucleation of water ice. Further work in the burgeoning field of cosmic nanosilicates from laboratory experiment, observation and computational modelling will be key to further understanding such species. Specifically, from a computational viewpoint, we see many opportunities for further work and a particular need for modelling studies on: (i) spectroscopic properties (e.g. IR and UV adsorption), (ii) bridging the gap between the properties and structure of small clusters and bulk-like nanosilicates, and (iii) investigations into more complex chemical processes on nanosilicates. We hope that this chapter provides some inspiration for further studies in these areas.

## References


1. Schulmeister K, Mader W (2003) J Non-Cryst Solids 320:143
2. Liebau F (1985) Structural chemistry of silicates, structure, bonding, and classification. Springer, Berlin Heidelberg
3. Henning T (2010) Annu Rev Astron Astrophys 48:21
4. Messenger S, Keller LP, Lauretta DS (2005) Sci 309:737
5. Nittler LR (2010) Sci 328:698
6. Burkea DJ, Brown WA (2010) Phys Chem Phys 12:5947
7. Abraham P, Juhasz A, Dullemond CP, Kospal A, Van Boekel R, Bouwman J, Henning T, Moor A, Mosoni L, Sicilia-Aguilar A, Sipos N (2009) Nat 459:224
8. Sloan GC, Matsuura M, Zijlstra AA, Lagadec E, Groenewegen MAT, Wood PR, Szyszka C, Bernard-Salas J, van Loon JT (2009) Sci 323:353
9. Bradley JP, Keller LP, Snow TP, Hanner MS, Flynn GJ, Gezo JC, Clemett SJ, Brownlee DE, Bowey JE (1999) Sci 285:1716
10. Henning T (2010) Astromineralogy lecture notes Physics 815 Springer-Verlag, Berlin Heidelberg
11. Min M, Waters LBFM, de Koter A, Hovenier JW, Keller LP, Markwick-Kemper F (2007) Astron Astrophys 462:667
12. Williams DA, Herbst E (2002) Surf Sci 500:823
13. Bromley ST, Goumans TPM, Herbst E, Jones AP, Slater B (2014) Phys Chem Chem Phys 16:18623–18643
14. Ostlie DA, Carroll BW (2007) An introduction to modern stellar astrophysics. Addison Wesley, San Francisco
15. Kwok S (2000) The origin and evolution of planetary nebulae. Cambridge University Press, Cambridge
16. Weissenrieder J, Kaya S, Lu JL, Gao HJ, Shaikhutdinov S, Freund HJ, Sierka M, Todorova TK, Sauer J (2005) Phys Rev Lett 95:076103
17. Hartke B (2002) Issue. Angew Chem Int Ed 41:1468
18. Catlow CRA, Bromley ST, Hamad S, Mora-Fonz M, Sokol AA, Woodley SM (2010) Phys Chem Chem Phys 12:786
19. Li Z, Scheraga HA (1987) Proc Natl Acad Sci USA 84:6611
20. Wales DJ, Doye JPK (1997) J Phys Chem A 101:5111
21. Metropolis N, Rosenbluth AW, Rosenbluth MN, Teller AH, Teller E (1953) Chem Phys 21:1087
22. Wales DJ, Miller MA, Walsh T (1998) Nat 394:758
23. Doye JPK, Wales DJ (1998) Phys Rev Lett 80:1357
24. Haertelt M, Fielicke A, Meijer G, Kwapien K, Sierka M, Sauer J (2012) Phys Chem Chem Phys 14:2849
25. Marom N, Kim M, Chelikowsky JR (2012) Phys Rev Lett 108:106801
26. Zachariah MR, Tsang W (1993) Aerosol Sci Tech 19:499
27. Avramov PV, Adamovic I, Ho KM, Wang CZ, Lu WC, Gordon MS (2005) J Phys Chem A 109:6294
28. Zhang R-Q, Lifshitz Y, Lee S-T (2003) Adv Mater 15:635
29. Krasnokutski SA, Rouille G, Jager C, Huisken F, Zhukovska S, Henning Th (2014) Astrophys J 782:15
30. Pimentel AS, das Lima FCA, da Silva A (2006) J Phys Chem A 110:13221
31. Lu WC, Wang CZ, Nguyen V, Schmidt MW, Gordon MS, Ho KM (2003) J Phys Chem A 107:6936
32. Zang QJ, Su ZM, Lu WC, Wang CZ, Ho KM (2006) J Phys Chem A 110:8151
33. Zang QJ, Su ZM, Lu WC, Wang CZ, Ho KM (2006) Chem Phys Lett 430:1

34. Caputo MC, Oña O, Ferraro MB (2009) J Chem Phys 130:134115
35. Garand E, Goebbert D, Santambrogio G, Janssens E, Lievens P, Meijer G, Neumark DM, Asmis KR (2008) Phys Chem Phys 10:1502
36. Savoca M, Langer J, Harding DJ, Palagin D, Reuter K, Dopfer O, Fielicke A (2014) J Chem Phys 141:104313
37. Du H-B, Huang S-P, Sarkar AD, Fan W-J, Jia Y, Zhang R-Q (2014) J Phys Chem A 118:8893
38. Wang L-S, Nicholas JB, Dupuis M, Wu H, Colson Steven D (1997) Phys Rev Lett 78:4450
39. Wang L-S, Desai SR, Wu H, Nichloas JB (1997) Z Phys D 40:36
40. Lu WC, Wang CZ, Ho KM (2003) Chem Phys Lett 378:225
41. Zhang RQ, Zhao MW, Lee ST (2004) Phys Rev Lett 93:095503
42. Hu S-X, Yu J-G, Zeng EY (2010) J Phys Chem A 114:10769
43. Reber AC, Clayborne PA, Reveles JU, Khanna SN, Castleman AW, Ali A (2006) Nano Lett 6:1190
44. Wang H, Sun J, Lu WC, Li ZS, Sun CC, Wang CZ, Ho KM (2008) J Phys Chem C 112:7097
45. Bromley ST, Gómez Martín JC, Plane JC (2016) Phys Chem Chem Phys 18:26913
46. Gail HP, Wetzel S, Pucci A, Tamanai A (2013) Astron Astrophys 555:A119
47. Paquette JA, Ferguson FT, Nuth JA III (2011) Astrophys J 732:62
48. Nuth JA III, Ferguson FT (2006) Astrophys J 649:1178
49. Ledoux G, Ehbrecht M, Guillois O, Huisken F, Kohn B, Laguna BMA, Nenner I, Paillard V, Papoular R, Porterat D, Reynaud C (1998) Astron Astrophys 333:L39
50. Witt AN, Gordon KD, Furton DG (1998) Astrophys J Lett 501:L111
51. Reber AC, Paranthaman S, Clayborne PA, Khanna SN, Castleman AW (2008) ACS Nano 2:1729
52. Gail H-P, Scholz M, Pucci A (2016) Astron Astrophys 591:A17
53. Sargent B, Forrest WJ, D'Alessio P, Li A, Najita J, Watson DM, Calvet N, Furlan E, Green JD, Kim KH, Sloan GC, Chen CH, Hartmann L, Houck JR (2006) Astrophys J 645:395
54. Lisse CM, Chen CH, Wyatt MC, Morlok A, Song I, Bryden G, Sheehan P (2009) Astrophys J 701:2019
55. Treacy MMJ, Rivin I, Balkovsky E, Randall KH, Foster MD (2004) Micro Meso Mat 74:121
56. Foster MD, Treacy.MMJ http://www.hypotheticalzeolites.net
57. Wales DJ, Miller MA, Walsh TR (1998) Nat 394:758
58. Bromley ST, T S, Bandow B, Hartke B (2008) J Phys Chem C 112:18417
59. Zwijnenburg MA, Sokol AA, Sousa C, Bromley ST (2009) J. Chem. Phys 131:034705
60. Avakyan VG, Sidorkin VF, Belogolova EF, Guselnikov SL, Guselinikov LE (2006) Organometallics 25:6007
61. Epping JD, Yao S, Karni M, Apeloig Y, Driess M (2010) J Am Chem Soc 132:5443
62. Bromley ST, Zwijnenburg MA, Maschmeyer Th (2003) Phys Rev Lett 90:035502
63. Greaves GN (1978) Phil Mag B 37:447
64. Lucovsky G (1979) Phil Mag B 39:513
65. Street RA, A R, Mott NF (1975) Phys Rev Lett 35:1293
66. Hamad S, Bromley ST (2008) Chem Comm 44:4156
67. Flikkema E, Bromley ST (2004) J Phys Chem B 108:9638
68. Rignanese GM, De Vita A, Charlier JC, Gonze X, Car R (2000) Phys Rev B 61:13250
69. Murashov VV (2005) J Phys Chem B 109:4144
70. Du Q, Freysz E, Shen YR (1994) Phys Rev Lett 72:238
71. Goumans TPM, Catlow CRA, Brown WA (2008) J Phys Chem C 112:15419
72. Flikkema E, Bromley ST (2003) Chem Phys Lett 378:622
73. Bromley ST, Flikkema E (2005) Phys Rev Lett 95(185505)

74. Bromley ST, Illas F (2007) Phys Chem Phys 9:1078–1086
75. Tsuneyuki S, Tsukada M, Aoki H, Matsui Y (1988) Phys Rev Lett 61:869–872
76. Bromley ST (2006) Phys Stat Sol A 203:1319–1323
77. Stephens PJ, Devlin FJ, Chabalowski CF, Frisch MJ (1994) J Phys Chem 98:11623–11627
78. Chu TS, Zhang RQ, Cheung HF (2001) J Phys Chem B 105:1705–1709
79. Harkless JAW, Stillinger DK, Stillinger FH (1996) J Phys Chem 100:1098–1103
80. Nayak SK, Rao BK, Khanna SN, Jena P (1998) J Chem Phys 109:1245–1250
81. Zhong A, Rong C, Liu S (2007) J Phys Chem A 111:3132
82. Johnston RL (2002) Atomic and molecular clusters. Taylor and Francis, London
83. Wojdel JC, Zwijnenburg MA, Bromley ST (2006) Chem Mater 18:1464
84. Zhang D, Zhang RQ (2006) J Phys Chem B 110(3):1338–1343
85. Kuo C-L, Lee S, Hwang GS (2008) Phys Rev Lett 100(076104)
86. Bromley ST (2004) Nano Lett 4:1427
87. Zhang D, Wu J, Zhang RQ, Liu C (2006) J Phys Chem B 110:17757–17762
88. Flikkema E, Bromley ST (2009) Phys Rev B 80(035402)
89. Zwijnenburg MA, Bromley ST, Flikkema E, Maschmeyer Th (2004) Chem Phys Lett 385:389
90. Zhang D, Zhang RQ, Han Z, Liu C (2006) J Phys Chem B 110:8992–8997
91. Bromley ST, Zwijnenburg MA, Maschmeyer Th (2003) Surf Sci Lett 539:L554
92. de Leeuw N, Du Z, Li J, Yip S, Zhu T (2003) Nano Lett 3:1347
93. Goumans TPM, Bromley ST (2012) Mon Not Roy Astron Soc 420:3344–3349
94. Gail HP, Sedlmayr E (1986) Astron Astrophys 166:225–236
95. Gail HP, Sedlmayr E (1998) Faraday Discuss 109:303–319
96. Sarangi A, Cherchneff I (2013) Astrophys J 776:107–126
97. Gobrecht D, Cherchneff I, Sarangi A, Plane JMC, Bromley ST (2016) Astronom Astrophys 585:A6
98. Li A, Draine BT (2001) Astrophys J 550:L213–L217
99. Williams DA, Herbst E (2002) Surf Sci 500:823–837
100. Ferriere KM (2001) Rev Mod Phys 73:1031–1066
101. Jeong KS, Winters JM, Le Bertre T, Sedlmayr E (2003) Astronom Astrophys 407:191–206
102. Goumans TPM, Bromley ST (2013) Phil Trans Roy Soc A 371(20110580)
103. Plane JMC (2013) Phil Trans R Soc A 371(20120335)
104. Mora-Fonz MJ, Catlow CRA, Lewis DW (2007) J Phys Chem C 111:18155–18158
105. Trinh TT, Jansen APJ, van Santen RA, Meijer EJ (2009) J Phys Chem C 113:2647–2652
106. White CE, Provis JL, Kearley GJ, Riley DP, van Deventer JSJ (2011) Dalt Trans 40: 1348–1355
107. Schaffer CL, Thomson KT (2008) J Phys Chem C 112:12653–12662
108. Trinh TT, Rozanska X, Delbecq F, Sautet P (2012) Phys Chem Chem Phys 14:3369–3380
109. Mochizuki D, Shimojima A, Imagawa T, Kuroda K (2005) J Amer Chem Soc 127: 7183–7191
110. Sanchez C, Arribart H, Guille MMG (2005) Nature Mat 4:277–288
111. Yang J, Wang EG (2006) Curr Op Sol State Mat Sci 10:33–39
112. Walsh TR, Wilson M, Sutton AP (2000) J Chem Phys 113:9191–9201
113. Masini P, Bernasconi M (2002) J Phys Cond Matt 14:4133–4144
114. Du MH, Kolchin A, Cheng HP (2003) J Chem Phys 119:6418–6422
115. Wendt S, Frerichs M, Wei T, Chen MS, Kempter V, Goodman DW (2004) Surf Sci 565:107–120
116. Demontis P, Stara G, Suffritti GB (2003) J Phys Chem B 107:4426–4436
117. Hassanali AA, Singer SJ (2007) J Phys Chem B 111:11181–11193
118. Bromley ST, Flikkema E (2005) J Chem Phys 122(114303)
119. Jelfs KE, Flikkema E, Bromley ST (2012) Chem Comm 48:46–48
120. Flikkema E, Jelfs KE, Bromley ST (2012) Chem Phys Lett 554:117–122

121. Jelfs KE, Flikkema E, Bromley ST (2013) Phys Chem Chem Phys 15:20438–20443
122. Goumans TPM, Bromley ST (2011) Mon Not Roy Astron Soc 414:1285–1291
123. Jenkins EB (2009) Astrophys J 700:1299–1348
124. Whittet DCB (2010) Astrophys J 710:1009–1016
125. Carrez P, Demyk K, Cordier P, Gengembre L, Grimblot J, D'Hendecourt L, Jones AP, Leroux H (2002) Meteorit Planet Sci 37:1599–1614

# Magnetic Anisotropy Energy of Transition Metal Alloy Clusters

**Nabil M.R. Hoque, Tunna Baruah, J. Ulises Reveles and Rajendra R. Zope**

**Abstract** The binary clusters of transition metal atoms form an interesting platform for studying the effects of shape, size, chemical compositions, and ordering on its magnetic properties. Notably, mixed clusters often show higher magnetic moments compared to pure elemental clusters. Due to the reduced dimension of the clusters, they tend to behave as single domain particles. One important parameter of their magnetic behavior is the magnetic anisotropy energy. In this chapter we review previous works on the magnetic anisotropy energy of binary alloy clusters, along with a density functional theory based method to compute the anisotropy energy with applications to binary metal clusters. The clusters of transition metal atoms often show high spin moments but generally are also reactive with the environment. Passivation of the surface atoms can lead to more stable clusters. We have explored one such avenue for passivation in this work. We consider the $As@Ni_{12}@As_{20}$ cluster which in the neutral state has a magnetic moment of 3 $\mu_B$. We dope this cluster by substituting various numbers of Ni atoms by Mn atoms. The substitutional doping leads to spin moments located mostly on the Mn atoms. The doping also leads to symmetry breaking and as a consequence the number of structural isomers and spin ordered states for each isomer becomes very large. We have investigated all possible ferromagnetic isomers for a given number of dopants and subsequently all the possible anti-ferromagnetic states for the lowest energy structure were examined. The results show that the encapsulation within the $As_{20}$ cage stabilizes the clusters and the atomization energy of the clusters increases as the number of dopant increases. These clusters have small energy barrier for reversal of magnetization and also have rich variation in configuration and spin states with many low-lying spin states.

N.M.R. Hoque · T. Baruah · J. Ulises Reveles · R.R. Zope (✉)
Department of Physics, University of Texas El Paso, El Paso, TX 79968, USA
e-mail: rzope@utep.edu

J. Ulises Reveles
Department of Physics, Virginia Commonwealth University, Richmond, VA 23284, USA



M.T. Nguyen and B. Kiran (eds.), *Clusters*, Challenges and Advances in Computational Chemistry and Physics 23,
DOI 10.1007/978-3-319-48918-6_8

# 1 Introduction

The atomic clusters of transition metals (TM) have been studied extensively over the last three decades. The main interest in the transition metal clusters arises from their magnetic and catalytic properties. The super-paramagnetic behavior of magnetic nanoparticles has been known since mid-twentieth century [1]. Magnetic nanoparticles have been used for recording purposes and are also being experimentally used for medical purposes where their magnetic properties are exploited for drug delivery or for bio-imaging. More novel applications of the magnetic particles may emerge in future.

The nanoparticles used in data storage or for biomedical engineering are typically large extending from tens to hundreds of nanometers [2, 3]. The atomic clusters containing tens of atoms on the other hand fall in the small size regime of the nanoparticles [4]. The atomic clusters in this range often show very different properties from the bulk materials [5]. The reduction in size results in larger variation in coordination numbers (compared to bulk) that usually leads to structures significantly different from bulk fragments. As the cluster size grows the structure starts to resemble bulk fragments. However, even for larger clusters, significant fraction of the atoms are still on the surface and are under coordinated which leads to interesting chemical properties [5]. For example, the TM clusters are also used to attach and transport other molecules to targeted areas in biomedicine [2].

The transition metal clusters typically have non-zero spin moment [5–12]. Clusters smaller in size than the typical bulk magnetic domains behave as single magnets. The spin-orbit interaction leads to the magneto crystalline anisotropy that results in an energy barrier in flipping the direction of the spin. As the cluster size increases the magnetic anisotropy energy (MAE) decreases and can become comparable to thermal energy for large particles. Such large particles ($\sim$10 nm) can undergo super-paramagnetic relaxation. The super-paramagnetic behavior of the clusters is exploited in applications as contrasting agent whereas for the data storage purposes an energy barrier between the magnetic states higher than the ambient temperature is highly desirable. For a cluster to possess higher energy barrier it needs to have a large spin moment. It is to be noted that a large value of spin moment does not necessarily result in higher anisotropy barrier. It is found that the alloy clusters often have larger magnetic moment compared to clusters of one element. For example, in mixed Co–Rh clusters the Rh atoms are found to have induced spin moments [13]. The alloying offers the possibility of achieving a high variability in composition and structure, which in turn govern the electronic properties.

In this chapter, we focus on the discussion of anisotropy energy barrier in the transition metal alloy clusters. The metal clusters are typically more reactive due to under coordinated surface atoms. Passivation of the clusters can saturate the bonds and make the cluster chemically stable. Another way is to encapsulate the metal clusters in an inorganic cage to stabilize these clusters. This chapter provides a brief review of earlier work done on magnetic anisotropy in bare TM alloy clusters with a discussion of recent work on encapsulated TM clusters.

## 2 Magnetic Anisotropy in Transition-Metal Alloy Clusters

The transition metal alloy clusters have been studied both theoretically and experimentally for their magnetic and chemical properties [14–29]. Enhancing a particular property through dopants is an effective way in materials science to investigate new species with potential applications [25, 30–33]. The vast variability of size, structure and stoichiometry in clusters requires detailed systematic study. In this work we consider only bimetallic alloy clusters. The alloy clusters can form mixtures or segregated clusters mostly due to their different sizes [34]. The mixing pattern in such clusters was classified by Johnston et al. as core-shell, mixed, subcluster segregated, and multi-shell [4]. The magnetic properties of a few types of mixed clusters have been studied experimentally. A large number of early theoretical calculations mainly focused on the spin alignment in the clusters. Early on it was noted from theoretical calculations that the mixed 3d metal clusters have higher spin moments than the elemental clusters [35]. It was found that embedding clusters of 3d transition metal (Fe) in a non-magnetic solid (Ag) can lead to giant magneto resistance [36]. This interesting experimental result led to many works on alloy clusters of 3d transition metal atoms with 4d or 5d metal atoms. The 3d materials have high spin moment and the 4d materials show strong spin-orbit coupling. When combined these two effects can lead to systems with high magnetic anisotropy energy. Moreover, enhancement of the spin moments on the 4d/5d atoms also leads to higher magnetic moment of the cluster. One such system is Pd coated Ni nanoparticles which were found to possess a ferromagnetic Ni core surrounded by ferromagnetic and paramagnetic Pd layer. However, the magnetization showed steep saturation in hydrogen atmosphere [37]. On the other hand, intermixed Ni–Pd alloy nanoparticles were found to be superparamagnetic with a blocking temperature of 290 K [38]. Co–Rh alloy particles have shown strong enhancement of magnetization compared to bulk alloys. This effect is attributed to a combination of size reduction and coupling with a magnetic 3d element (Co), leading to an enhanced induced electronic spin polarization of the 4d (Rh) atoms, while retaining the magnetism due to the Co atoms [39–41]. In these clusters there is an induced magnetic moment on the Rh atoms with increasing Co concentration, which results in an increase in the average magnetic moment per atom. Rohart et al. studied the magnetic anisotropy energy of Co–Pt mixed clusters embedded in various matrix [19, 42]. By comparing different chemical compositions they found that a small amount of Pt induces increase in volume anisotropy. Tournus et al. have noted the dispersion of the anisotropy constant in size selected CoPt alloy clusters which shows the chemical environment of the Co atoms is important as it directly determines the anisotropy. Luis et al. [42] have reported significant increase in blocking temperature and coercive field for Co clusters capped by a thin Au layer [20]. Detailed theoretical studies have been done on the magnetic anisotropy energy (MAE) of mixed clusters of Co–Rh [13, 43, 44] CoPd [41, 45], Pt covered Co, Fe and Ni clusters [46], Fe–Co mixed clusters [28], Co–Mo mixed clusters [47]. Theoretical works point out the contribution from the orbital magnetic moment, and also the importance of chemical composition or environment of the 3d metal atoms. The mixed clusters with 4d or 4d non-magnetic atoms showed enhanced spin moment on the 4d or 5d atoms. The Pt covered Co clusters retained the magnetization properties of the Co cluster in core-shell structure but increased the volume anisotropy in case of mixed clusters. On the other hand, Au capped Co clusters showed enhanced magnetization. From these studies it is evident that alloying helps in enhancement of magnetization but also depends on the structure. Materials assembled from such clusters have vast potential for novel applications but more studies are needed. In the following we describe a density functional based method for calculation of the magnetic anisotropy energy in molecules and clusters followed by some recent applications to mixed TM clusters.

## 3 Magnetic Anisotropy Energy

Magnetic materials are said to have magneto crystalline anisotropy if it takes a stronger field to magnetize in a specific direction compared to the others. Depending on the orientation of the field with respect to the crystal lattice one would need a lower or higher magnetic field to reach magnetic saturation, a characteristic that leads to the definition of two types of axes, the easy and the hard one, that arise from the interaction of the spin magnetic moment with the crystal lattice. The easy axis is the direction inside a crystal, along which a small applied magnetic field, is sufficient to reach the magnetic saturation. Finally, the magneto-crystalline anisotropy energy is the energy needed to deflect the magnetic moment in a single crystal from the easy to the hard direction.

The magnetic anisotropy energy barrier is related to the zero-field splitting of the spin states due to spin-orbit coupling. A few algorithms have been formulated to compute magnetic anisotropy within density functional theory [48, 49]. The discussion below follows an approach due to Pederson and Khanna [48], Pederson and Baruah [50]. In the classical explanation of spin-orbit coupling an electron moving with velocity **v** and accounting for the fact that electron is not spinless, the interaction energy is given by,

$$U(\boldsymbol{r},\boldsymbol{p},\boldsymbol{S}) = -\frac{1}{2c^2}\boldsymbol{S}.\boldsymbol{p} \times \nabla\varphi(\boldsymbol{r}), \quad (1)$$

where $\varphi(\boldsymbol{r})$ is the Coulomb potential, $\boldsymbol{p}$ is the momentum operator. The determination of spin-orbit coupling matrix element is a necessary ingredient to the numerical solution of the Schrodinger equation. To determine the generalized spin-orbit interaction from Eq. (1) it is necessary to calculate matrix element of the form,

$$\begin{aligned} U_{j,\sigma,k,\sigma'} &= \langle f_j\chi_\sigma|U(\boldsymbol{r},\boldsymbol{p},\boldsymbol{S})|f_k\chi_{\sigma'}\rangle \\ &= \sum\nolimits_x \frac{-1}{i2c^2}\langle f_j|[\nabla\times\nabla\varphi(\boldsymbol{r})]_x|f_k\rangle\langle\chi_\sigma|S_x|\chi_{\sigma'}\rangle \\ &= \sum\nolimits_x \frac{1}{i}\langle f_j|V_x|f_k\rangle\langle\chi_\sigma|S_x|\chi_{\sigma'}\rangle, \end{aligned}$$

with the operator $V_x$ defined according to

$$\langle f_j|V_x|f_k\rangle = \frac{-1}{2c^2}\left\langle f_j\left|\frac{d}{dy}\frac{d\varphi}{dz} - \frac{d}{dz}\frac{d\varphi}{dy}\right|f_k\right\rangle. \tag{2}$$

Using the identity,

$$\left\langle f_i\frac{d\varphi}{dy}\frac{d}{dz}f_j\right\rangle = \int d^3r\frac{d}{dy}\left[f_i\varphi\frac{df_j}{dz}\right] - \left\langle\frac{df_i}{dy}|\varphi|\frac{df_j}{dz}\right\rangle - \left\langle f_i|\varphi|\frac{d^2f_j}{dzdy}\right\rangle,$$

we get

$$\langle f_j|V_x|f_k\rangle = \frac{1}{2c^2}\left(\left\langle\frac{df_j}{dz}|\varphi|\frac{df_k}{dy}\right\rangle - \left\langle\frac{df_j}{dy}|\varphi|\frac{df_k}{dz}\right\rangle\right). \tag{3}$$

The matrix elements for $V_y$ and $V_z$ are determined by the cyclic permutation.

Let us assume that, in the absence of a magnetic field and spin-orbit coupling, we have determined the wave functions $\psi_{i\sigma}$ within a self-consistent field (SCF) approximation. The SCF wave functions satisfy

$$H|\psi_{i\sigma}\rangle = \epsilon_{i\sigma}|\psi_{i\sigma}\rangle,$$

where $|\psi_{i\sigma}\rangle$ is a product of a spatial function and spinor according to $|\psi_{i\sigma}\rangle = \varphi_{i\sigma}(\boldsymbol{r})\chi_\sigma$.

With the inclusion of spin-orbit coupling and the introduction of a magnetic of field the perturbed wave functions satisfy,

$$\left[H + \left(\frac{V}{i} + \frac{1}{c}\boldsymbol{B}\right)\cdot\boldsymbol{S}\right]\psi'_{i\sigma}\rangle = \epsilon'_{i\sigma}|\psi'_{i\sigma}\rangle.$$

Here, the operator $\boldsymbol{V}$ is defined according to Eq. (3) and the magnetic field (**B**) is assumed to be uniform.

Now let $\boldsymbol{W} = \left(\frac{V}{i} + \frac{1}{c}\boldsymbol{B}\right)$. According to the second order perturbation theory, the Hamiltonian matrix is perturbed by the following equation,

$$\Delta = \Delta_1 + \Delta_2.$$

In the absence of applied magnetic field the first order energy shift is,

$$\Delta_1 = \sum_{i\sigma} S_i^{\sigma\sigma} \sum_k \langle \varphi_{k\sigma} | W_i | \varphi_{k\sigma} \rangle.$$

The first order energy shift vanishes due to the operator $-i\boldsymbol{V} \cdot \boldsymbol{S}$ and the first order correction to the orbital is purely imaginary. The second order energy shift is,

$$\Delta_2 = \sum_{i\sigma'} \sum_{ij} W_{ij}^{\sigma\sigma'} S_i^{\sigma\sigma'} S_j^{\sigma'\sigma}, \tag{4}$$

where, $S_i^{\sigma\sigma\prime} = \langle \chi_\sigma | S_i | \chi'_{\sigma'} \rangle$,
and, $W_{ij}^{\sigma\sigma'} = W_{ji}^{\sigma\sigma'*} = \sum_{ij} \frac{\langle \varphi_{k\sigma} | W_i | \varphi_{l\sigma'} \rangle \langle \varphi_{k\sigma'} | W_j | \varphi_{l\sigma} \rangle}{\epsilon_{k\sigma} - \epsilon_{l\sigma'}}$.

In Eq. 4 here the 1st sum is running over spin-up and spin-down states. In the 2nd sum we are running over all the co-ordinate levels (x, y, z). The *W* matrices are simplified to,

$$W_{ij}^{\sigma\sigma'} = -\sum_{kl} \frac{\langle \sigma_{k\sigma} | V_i | \sigma_{l\sigma'} \rangle \langle \sigma_{l\sigma'} | V_j | \sigma_{k\sigma} \rangle}{\epsilon_{k\sigma} - \epsilon_{l\sigma'}} + \frac{B_i B_j}{c^2} \sum_{kl} \frac{< \sigma_{k\sigma} | V_i | \sigma_{l\sigma'} \rangle \langle \sigma_{l\sigma'} | V_j | \sigma_{k\sigma} \rangle}{\epsilon_{k\sigma} - \epsilon_{l\sigma'}}$$

In the limit of B approaching zero, the second term vanishes and the W matrix becomes as follows

$$W_{ij}^{\sigma\sigma'} \rightarrow M_{ij}^{\sigma\sigma'} = -\sum_{kl} \frac{< \sigma_{k\sigma} | V_i | \sigma_{l\sigma'} > < \sigma_{l\sigma'} | V_j | \sigma_{k\sigma} >}{\epsilon_{k\sigma} - \epsilon_{l\sigma'}}. \tag{5}$$

In the above equations $\chi_\sigma$ and $\chi'_\sigma$ are any set of spinors.$\varphi_{k\sigma}$ and $\varphi_{l\sigma'}$ are the occupied and unoccupied states respectively. $\epsilon' s$ are the corresponding energies.

The second-order shift in the energy of the system in the absence of a magnetic field can be rewritten as, $\Delta_2 = \sum_{ij} \gamma_{ij} \langle S_i \rangle \langle S_j \rangle$. By diagonalizing the anisotropy tensor ($\gamma$), the anisotropy energy can be determined. The value of $\gamma$ can be calculated within the density functional framework using second-order perturbation theory and in terms of Kohn-Sham orbitals, it is given by,

$$\gamma = \left( \frac{2}{\Delta N^2} \right) \left( M_{zz}^{11} + M_{zz}^{22} + M_{xx}^{12} + M_{xx}^{21} - M_{xx}^{11} - M_{xx}^{22} - M_{zz}^{12} - M_{zz}^{21} \right),$$

*w*here ΔN is the number of unpaired electrons. Once the anisotropy tensor is diagonalized the second-order energy shift can be rewritten as,

$$\Delta_2 = \frac{1}{3}(\gamma_{xx}+\gamma_{yy}+\gamma_{zz})S(S+1) + \frac{1}{3}\left[\gamma_{zz} - \frac{1}{2}(\gamma_{xx}+\gamma_{yy})\right] \times \left[3S_z^2 - S(S+1)\right] + \frac{1}{2}(\gamma_{xx}-\gamma_{yy})(S_x^2 - S_y^2). \quad (6)$$

The anisotropy Hamiltonian splits the 2S + 1 spin states and when the isotropi term is ignored, it can be expressed as

$$H = DS_z^2 + E\left(S_x^2 - S_y^2\right), \quad (7)$$

The value of D and E parameters in the above equation can be directly obtained from the $\gamma_{xx}$, $\gamma_{yy}$, $\gamma_{zz}$ and therefore the magnetic anisotropy energy can be obtained.

The method described above is implemented in the NRLMOL suite of software [51–56]. Below we present results of previous calculations on several single molecule magnets and compare them with experimental values for benchmark purpose.

The values of the axial anisotropy parameter D in Eq. 7 are available from a number of experiments for different single molecule magnets (SMM). Several first-principle calculations have been carried out with the use of the NRLMOL code on SMMs such as $Mn_{12}$ [48, 57, 58], $Mn_{10}$ [59], Ferric star, $Fe_4$, $Fe_8$ [60–62], $Mn_4$ [63], $Co_4$ [64], $Mn_9$ [65] etc. These results are summarized in Table 1. In general a good agreement of the spin ordering is found between theoretical calculations and experiments. The broken symmetry approach is needed for systems with antiferromagnetic spin ordering. Moreover, the calculated D parameters for $Mn_{12}$, $Mn_{10}$, $Mn_9$, the ferric star $Fe_4$ and Cr-amide molecular magnets are in excellent agreement with experimental values. The only discrepancy is found for $Fe_8$, a system that seems to pose complications for the DFT treatment. Apparently DFT may be unable to predict the ground state density accurately enough due to important electronic correlations beyond the mean-field treatment or missing Madelung stabilization

**Table 1** Comparison of experimental and NRLMOL calculated magnetic anisotropy parameters (D)

| Molecule | S | MAE (K) | |
|---|---|---|---|
| | | Experiment | Theory |
| $Mn_{12}O_{12}(O_2CH)_{16}(H_2O)_4$ | 10 | −0.56 | −0.56 [48] |
| $[Fe_8O_2(OH)_{12}(C_6H_{15}N_3)_6Br_6]^{2+}$ | 10 | −0.30 | −0.53 [60] |
| $[Mn_{10}O_4(2,2'\text{-biphenoxide})_4Br_{12}]^{4-}$ | 13 | −0.05 | −0.06 [59] |
| $Co_4(CH_2C_5H_4N)_4(CH_3OH)_4ACl_4$ | 6 | −0.7 to −0.09 | −0.64 [66] |
| $Fe_4(OCH_2)_6(C_4H_9ON)_6$ | 5 | −0.57 | −0.56 [67] |
| $Cr[N(Si(CH_3)_3)_2]_3$ | 3/2 | 2.66 | −1.15 [67] |
| $Mn_9O_{34}C_{32}N_3H_{35}$ | 17/2 | −0.32 [65] | −0.33 [65] |
| $Mn_4O_3Cl_4(O_2CCH_2CH_3)_3(NC_5H_5)_3$ | 9/2 | −0.72 | −0.58 [63] |

(absent in the isolated system). The SMMs listed in Table 1 are in general characterized by a high spin ground-state. However, a high spin state does not necessarily correlate with a high anisotropy barrier. The parameter D is also very important. In order to increase the barrier one has to understand and control D, which will be the main goal of future research in this area. In all cases where the E parameter is not zero by symmetry it has been predicted with similar accuracy as D. These results provide confidence in the predictive ability of the formalism.

## 4 Applications to Transition Metal Mixed Clusters

(a) **$Fe_nCo_m$Clusters**:

Kortus et al. used NRLMOL to study $Fe_nCo_m$ ($n + m = 5$ and 13) binary clusters with bipiramidal and icosahedral symmetries shown in Fig. 1, and investigated the effects of alloying on its magnetic moment and MAEs [28]. The dopant atoms were placed along the molecular axis. Their density-functional study showed that many alloy clusters have moments comparable to or higher than those present in pure clusters of Fe or Co, ranging from 13 to 41 $\mu_B$. They found these systems have very low anisotropies, 1.8 to −63 K (Cf. Table 2) making them ideal candidates for soft magnetic materials. The bulk Co has higher anisotropy than pure Fe or mixed Fe–Co alloys however the Co clusters have the lowest anisotropy. This work showed that a high magnetic anisotropy requires a strong coupling between occupied and unoccupied states close to the Fermi energy, and that one possible way to accomplish this may be by generating unreactive, compositionally ordered uniaxial clusters with small gaps. It also highlights the importance of shape, composition and compositional ordering in mixed atom clusters.

(b) **$Co_nMo_m$NANOCLUSTERS**

Cobalt dimer has a large magnetic anisotropy [68–71]. Garcia-Fuente et al. studied $Mo_{4-x}$ $Fe_x$ and concluded that these clusters are good candidates for molecular electronic devices [72]. It has also been shown that $Mo_2X_2$ (X = Fe, Co, Ni) clusters are able to act as spin filters [73]. These observation led Liebing et al. to

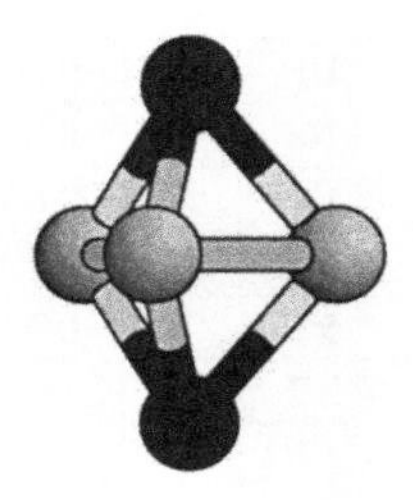

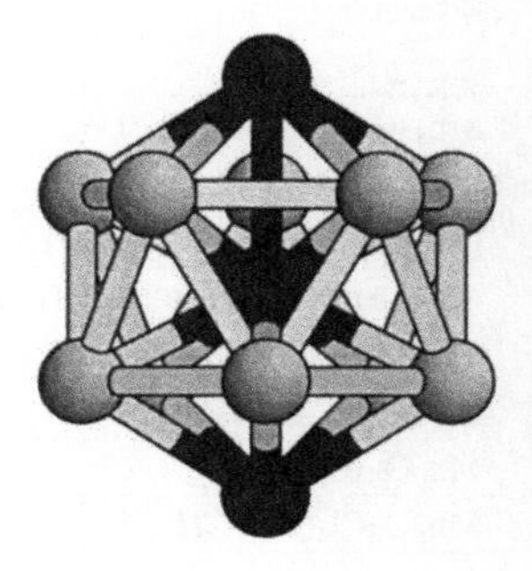

**Fig. 1** The uniaxial geometrical configurations for the 5 and 13 atoms cluster, triangular bipyramid geometries and distorted icosahedrons. The dopant atoms were placed along the uniaxial axis. Reprinted with permission from [28]. Copyright 2002, AIP Publishing LLC

**Table 2** Magnetic moment and MAE for the relaxed 5-atom and 13-atom clusters

| Cluster | Moment ($\mu_B$) | MAE (K) |
|---|---|---|
| $Co_5^a$ | 13 | 6 |
| $Co_3Fe_2$ | 13 | 27/15 |
| $Co_2Fe_3$ | 16 | 21 |
| $Fe_5$ | 16 | 14 |
| $Co_{13}$ | 21 | 0 |
| $Co_{10}Fe_3$ | 30 | 63 |
| $Fe_{10}Co_3$ | 41 | 9 |
| $Fe_{13}$ | 44 | 41 |



**Table 3** Magnetic ground state S and magnetic anisotropy energy MAE for the $Co_nMo_m$ clusters

| Cluster | S | MAE (K) | Cluster | S | MAE (K) |
|---|---|---|---|---|---|
| $Co_2$ | 2 | −5.6 | $Co_4Mo$ | 4 | −1.5 |
| CoMo | 3/2 | 13.4 | $Co_3Mo_2$ | 5/2 | −1.6 |
| $Co_3$ | 5/2 | −6.5 | $Co_2Mo_3$ | 2 | 6.1 |
| $Co_2Mo$ | 2 | 12.5 | $Co_6$ | 7 | −0.02 |
| $Co_4$ | 5 | 1.0 | $Co_5Mo$ | 9/2 | −0.6 |
| $Co_3Mo$ | 5/2 | 10.7 | $Co_4Mo_2$ | 3 | −2.7 |
| $Co_2Mo_2$ | 2 | −5.9 | $Co_3Mo_3$ | 5/2 | 3.6 |
| $CoMo_3$ | 3/2 | 8.5 | $Co_2Mo_4$ | 2 | −10.8 |
| $Co_5$ | 11/2 | −1.2 | – | – | – |



investigate $Co_nMo_m$ nanoclusters [47] with $n + m = x$ and $2 <= \times <= 6$. They studied pure cobalt and molybdenum clusters and compared their properties with those of its respective mixed species for each cluster size $x$. They found that the magnetic moment of a given cluster is mainly determined by the Co content and increases with increasing $n$. The magnetic anisotropy on the other hand becomes smaller for larger magnetic moments as shown in Table 3. They reported the variation of the electronic properties of the clusters as a function of size. An increase in binding energy, electron affinity, and average bond length, and a decrease in ionization potential, chemical potential, molecular hardness, and the HOMO–LUMO gap was found for increasing cluster size.

(c) Encapsulated Clusters

Cluster based materials offer the potential for new novel properties, which can be tuned by changing the cluster structure, composition, size, coordinating ligands etc. One of the reasons that gas phase metal cluster properties cannot be interpreted as indicator of the cluster based bulk materials is that the clusters often have unsaturated bonds and therefore are chemically active. Such gas-phase clusters may change their size and properties when in a bulk environment. However, these clusters could be protected using organic/inorganic ligands. One such example is the $[As@Ni_{12}@As_{20}]^{3-}$ ion that was isolated in 2004 by Eichhorn and coworkers [74]. This ion has perfect icosahedral symmetry and has a structure that contains an icosahedral $Ni_{12}$ cluster embedded in the dodecahedral $As_{20}$ cage. The inner icosahedral $Ni_{12}$ cluster has an arsenic atom at its center. The two interlocked cages result in a polyhedron containing 60 triangular faces with icosahedral symmetry. Subsequent density functional calculations showed that the bonding between the $As_{20}$ and $Ni_{12}@As$ units is about 26 eV and that the cluster was vibrationally stable in the gas phase [75, 76]. Recently, we carried out DFT calculations on the magnetic anisotropy energies of $As@Ni_{12}@As_{20}$ cluster doped with Mn atoms. The doping is substitutional in that the dopant Mn atoms replace Ni atoms in the cluster. The substitutional doping thus resulted in the $Ni_{12-x}Mn_x$ alloy clusters encapsulated within the $As_{20}$ cage. We doped the parent cluster by up to six Mn atom. Depending on the number of dopant Mn atoms, the substitutional doping can result in various isomers as there are multiple possible configurations of doping sites. We have considered all possible sites for substitutional doping. For every possible configuration of Mn dopants in the parent clusters, the lowest energy ferromagnetic isomer was obtained by geometry optimization. Subsequently, search for the lowest energy spin configuration was carried out by considering all possible ferrimagnetic isomers. We begin with the optimized geometry of the neutral $As@Ni_{12}@As_{20}$ cluster which we will henceforth refer to as the parent cluster. The geometrical parameters of the parent cluster optimized using the PBE functional were found to be in good agreement with experiment [76]. This cluster is highly symmetric with a magnetic moment of 3 μB. The high symmetry of the molecule rules out any anisotropy of magnetization. The $As@Ni_{12}@As_{20}$ cage is symmetric and as a result does not have any magnetic anisotropy. In this cage each Ni atom is coordinated with five As atoms. The Ni atoms form an icosahedral inner cage and thus each Ni atom is also coordinated to five other Ni atoms. The doping is substitutional in nature, that is, Ni atoms are replaced by the Mn atoms. As a result, the Mn atoms in these clusters also have similar coordination with As and Ni/other Mn atoms. The substitutional doping of the Ni atoms by multiple Mn atoms can result in various different isomers. We considered all possible substitutional ordering/permutations, that is, for a given x in $As@Ni_{12-x}Mn_x@As_{20}$ all possible isomers arising due to different configurations of substitutional doping were considered. The resultant geometry of every such an isomer was optimized simultaneously with spin optimization to find most stable isomers with ferromagnetic spin ordering. These results indicated that the spin charges are mainly located on the Mn atoms. Due to this reason, the search of possible ferrimagentic spin ordering involved spin orientations of only Mn atoms.

The calculations were carried out at the all-electron level using generalized gradient approximation of Perdew, Burke and Ernzerhof (PBE96) to describe the exchange-correlation effects [8]. The electronic orbitals and eigenstates are determined using a linear combination of Gaussian atomic type orbital molecular orbital (LCGTO) approach as implemented in the NRLMOL code developed by Pederson and co-workers [9–11]. NRLMOL uses an optimized large basis set that include supplemental diffuse *d*-type polarization function [12]. The integrals are accurately and efficiently calculated using a variational mesh and an analytic solution of Poissons equation is implemented to accurately determine the self-consistent potentials, secular matrix, total energies, and Hellmann-Feynman-Pulay forces.

In the following we discuss the structural and electronic properties of each of the encapsulated alloy clusters in detail.

### $As@Ni_{11}Mn@As_{20}$

In the parent $As@Ni_{12}@As_{20}$ system, all Ni atoms that form inner icosahedron are equivalent by symmetry. Likewise, all the As atoms that form outer dodecahedral cage are also symmetrically equivalent. Since all Ni atoms are equivalent, one only needs to replace one of the 12 Ni atoms by Mn atom. Thus, there is only one resultant structure. The optimized structure of $As@Ni_{12}@As_{20}$ is shown in Fig. 2 below.

As evident from the figure, the $As_{20}$ cage in this case is distorted due to the expansion of the As–As bond lengths near Mn. The doping however leads to further stabilization of the cluster with a larger atomization energy of 4.93 eV compared to the 4.80 eV of the parent cluster. The gap between the highest occupied molecular orbital (HOMO) and the lowest unoccupied molecular orbital (LUMO) is 0.16 eV at the PBE level. The net spin moment in this molecule is $6\mu_B$. Our calculation on the charge density inside a sphere placed on the Mn shows that the spin charge on the Mn ion is 3.2 $\mu_B$. The spin moment on the Ni atoms are small within a range of 0.2 $\mu_B$. The presence of the Mn atom as well as the distortion of the cage breaks the symmetry of the system, resulting in second order MAE of 10.4 K. It forms a tri-axial system with a D parameter of 1.15 K and has E parameter of 0.06 K.

### $As@Ni_{10}Mn_2@As_{20}$

In the parent $As@Ni_{12}@As_{20}$ system, two Mn atoms were introduced by replacing two of the 12 Ni atoms. There are three unique resultant possible structures of the $As@Ni_{10}Mn_2@As_{20}$ which were first optimized in the ferromagnetic state. The

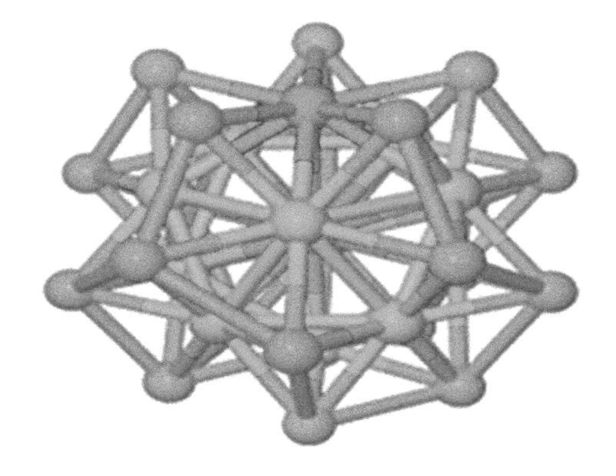

**Fig. 2** Structure of $As@Ni_{11}Mn@As_{20}$ cluster

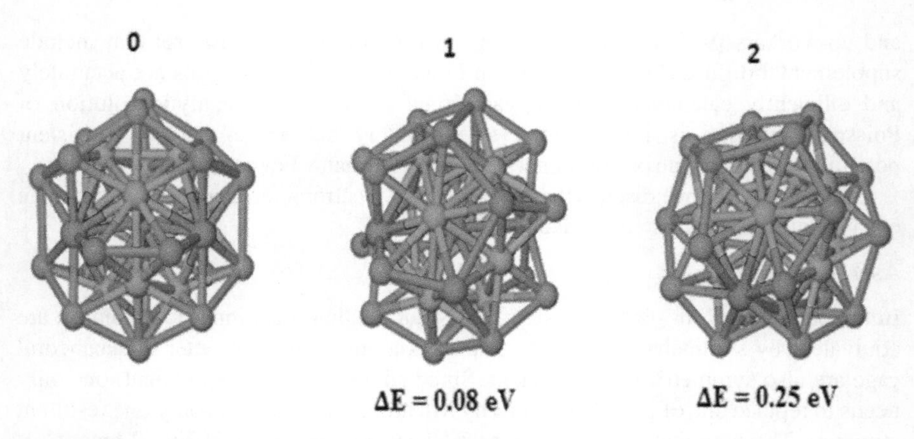


**Fig. 3** Three distinct isomers of $As@Ni_{10}Mn_2@As_{20}$

structures of the three ferromagnetic isomers are shown in Fig. 3. The lowest energy isomer has a net spin of 9 $\mu_B$ whereas the low-lying isomers both have spin moment of 7 $\mu_B$. The net spin on each of the Mn ions is 3.15 $\mu_B$ and the average spin on the Ni ions is 0.25 $\mu_B$, which in the ferromagnetic state leads to a net moment of 9 $\mu_B$ for the lowest energy structure. The lowest energy cluster has MAE on the order of 20 K. The cluster also has a gap of 0.25 eV, which indicates that the cluster is chemically more stable compared to the single Mn doped system.

The lowest ferromagnetic isomer was further optimized in an anti-ferromagnetic (AFM) state. The AFM spin configuration was achieved by applying a potential to force the spin orientation in the first iteration of the SCF cycle and later releasing it. Since there are only two Mn atoms in this cluster, only one antiferromagnetic ordering is possible. The antiferromagnetic spin ordering results in a structure that is higher in energy by 0.12 eV with a total magnetic moment of 1 $\mu_B$. We find that the Mn atoms induce similar spin orientation in the nearby Ni atoms.

**$As@Ni_9Mn_3@As_{20}$**

The optimization of the three Mn doped $As_{21}Ni_{12}$ cage resulted in nine different isomers as shown in Fig. 3 below. All the isomers show distortion of the parent cluster from the highly symmetric structure due to alloying. There are a few low-lying isomers within 0.1 eV of the lowest energy structure. The magnetic moment in all clusters except isomer 7 is 12 $\mu_B$. The magnetic moment is 10 $\mu_B$ for this isomer in which all the three Mn ions are adjacent. Due to the breaking of symmetry of the parent cage, magnetic anisotropy energy becomes significant. The largest MAE is 37 K for the lowest energy ferromagnetic structure in which the distances between the Mn ions are 0.264, 0.536, 0.472 nm. The spin moments on the Mn atoms in this state are 3.1 $\mu_B$.

We have investigated various ferrimagnetic states of the lowest energy alloy cluster. There are three possible ferromagnetically ordered spin states and the cluster was optimized in all these states. We find one AFM state that is about

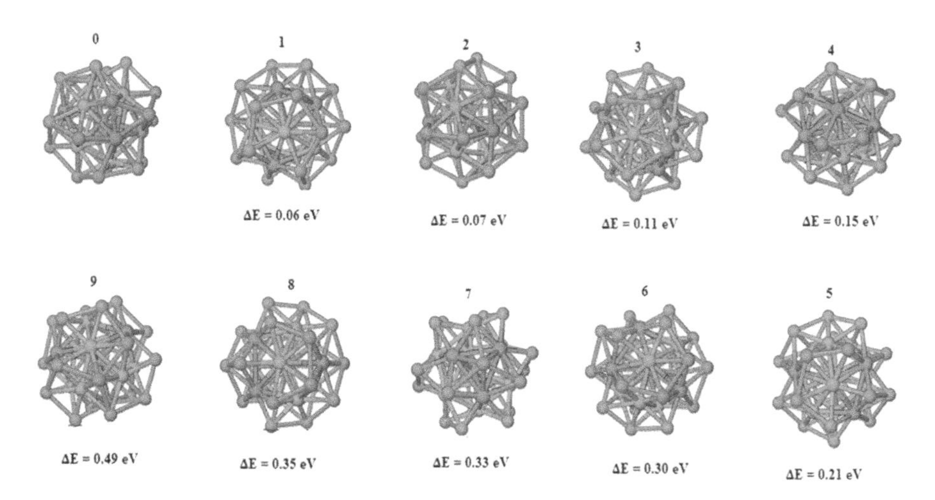


**Fig. 4** Isomers of $As@Ni_9Mn_3@As_{20}$

0.15 eV below the ferromagnetic state and forms the ground state of this isomer. This spin of this state is S = 2. The depiction of the anti-ferromagnetically ordered spin system is shown in Fig. 4. There is also another spin ordered state 0.05 eV above the lowest AFM state. The lowest energy AFM state shows an energy barrier of 18.6 K for reversal of magnetization. The three Mn doped system forms an easy axis with a D parameter of 4.5 K and an E parameter of 0.68 K. The corresponding parameters for the ferromagnetic state are 1.0 K and 0.1 K. The cluster undergoes structural changes in the AFM state in which the bonds with the nearest As atoms are shortened in two cases by as much as 0.11 Angstrom. The shortening of the bonds increases the crystal field as a result of which the D and E parameters are higher in the AFM state. However, due to the lower net spin of the cluster the overall MAE is lower for the AFM state (Fig. 5).

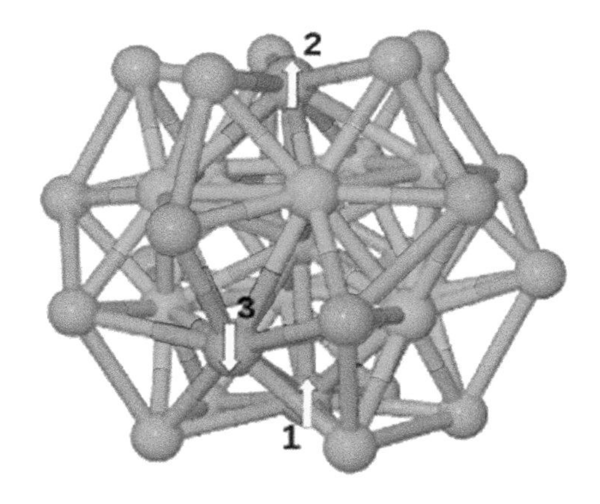


**Fig. 5** Spin ordering of Mn atom in $As@Ni_9Mn_3@As_{20}$

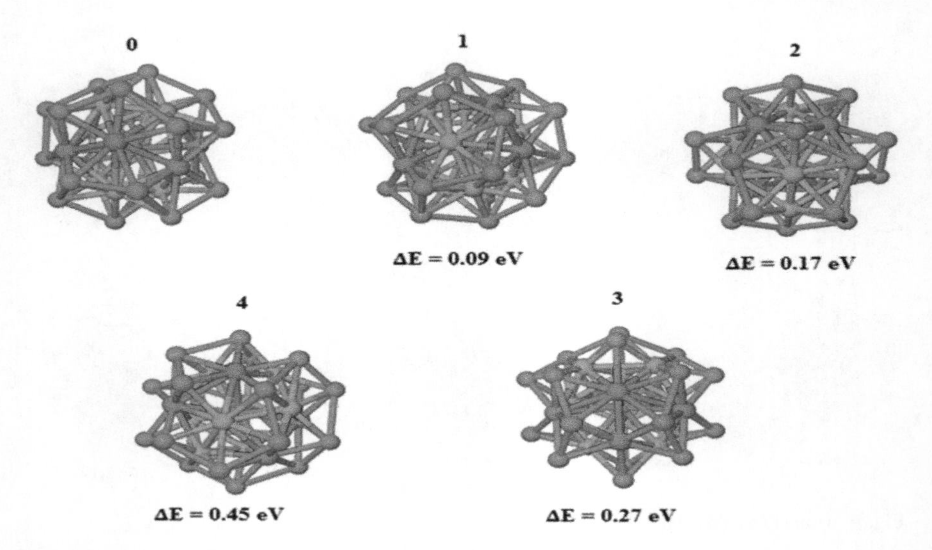


**Fig. 6** All the possible combination of $As@Ni_8Mn_4@As_{20}$ system

### $As@Ni_8Mn_4@As_{20}$

There are five possible unique structures which can be generated from $As@Ni_8Mn_4@As_{20}$ system by replacing four Ni atoms by Mn atoms. The optimized structures of $As@Ni_8Mn_4@As_{20}$ in the ferromagnetic state are shown in Fig. 6. The isomers are within a range of 0.09–0.45 eV above the lowest energy isomer. The spin magnetic moments in these isomers range from 13 to 15 $\mu_B$. The spin ordering is shown in Fig. 7. The local moments on the Mn atoms in the lowest FM isomer range from 2.4 to 3.1 $\mu_B$. The magnetic anisotropy energy in this cluster is 33 K. However, a study of the optimized spin-ordered clusters showed a ferromagnetic cluster with magnetic moment of 1 $\mu_B$ is found to be lower than the ferromagnetic cluster by 0.2 eV. The physical picture of the spin ordering is shown in Fig. 6.

### $As@Ni_7Mn_5@As_{20}$

We find seven stable isomers of the $As@Ni_7Mn_5@As_{20}$ clusters which were optimized first in the ferromagnetic state (Fig. 8). The lowest energy ferromagnetic structure has the five Mn atoms occupying adjacent positions in a pentagonal pyramid. The energies of the other isomers are within a range of 0.04–0.87 eV. We optimized all the possible spin ordered states of the lowest energy ferromagnetic (FM) isomer. In total, there are 12 such distinct ferrimagnetic states. We found one state which is lower in energy by only 0.03 eV compared to the FM state. There are two other AFM states within 0.03 eV from the lowest energy AFM state. Thus this cluster has several low-lying states, which differ in their structural as well as spin configuration. The HOMO–LUMO gap of the lowest energy AFM state is 0.32 eV,

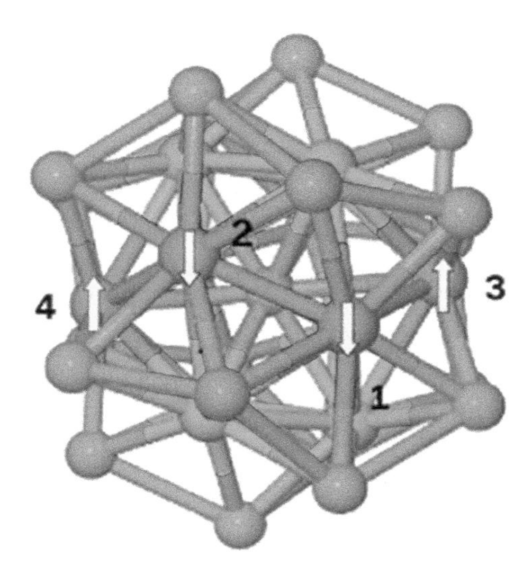


**Fig. 7** Spin ordering in the lowest energy structure of $As@Ni_8Mn_4@As_{20}$

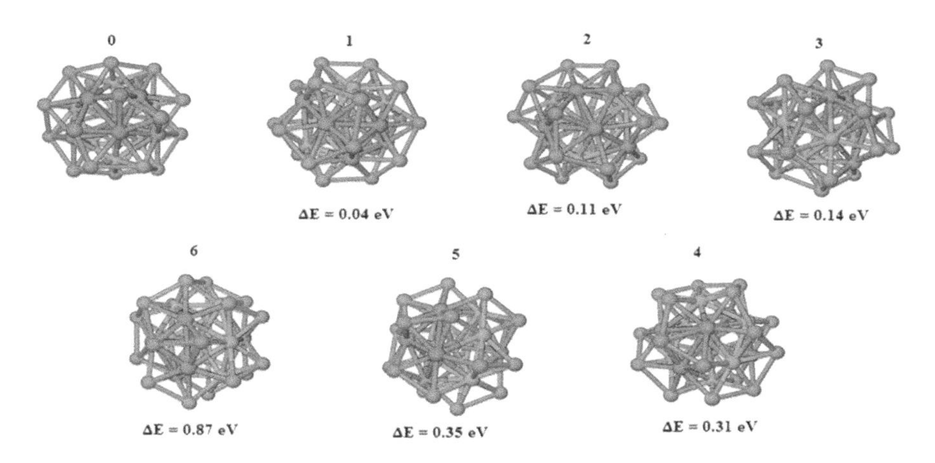


**Fig. 8** All the possible combination of $As@Ni_7Mn_5@As_{20}$ cluster according to their energy

which has spin moment 2 μB. The D and E parameters in the lowest energy AFM state are 6.4 and 4.7 K respectively. On the other hand, these parameters are much smaller in the FM state with values of 0.27 and 0.1 K for the D and E respectively (Fig. 9).

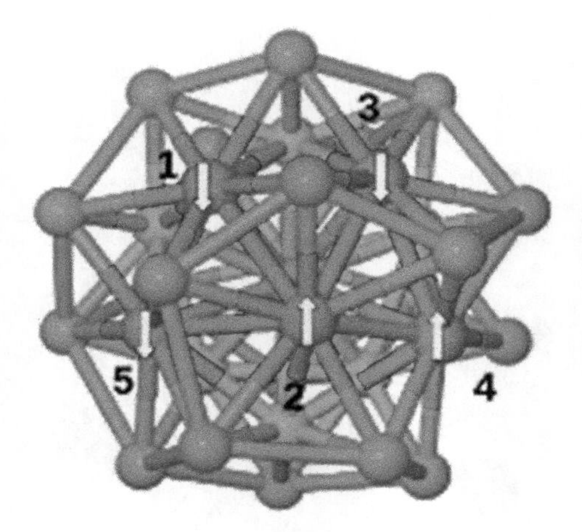


**Fig. 9** Spin ordering of Mn atoms in the lowest energy structure of As@$Ni_7Mn_5$@$As_{20}$ system

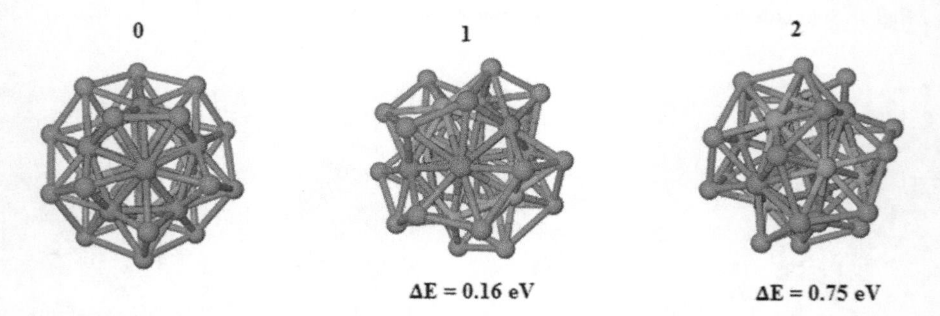


**Fig. 10** All the possible combination of As@$Ni_6Mn_6$@$As_{20}$ cluster according to their energy

### As@$Ni_6Mn_6$@$As_{20}$

In the parent As@$Ni_{12}$@$As_{20}$ system six Mn atoms are doped by replacing any six of the 12 Ni atoms. The optimized structures of ferromagnetic As@$Ni_6Mn_6$@$As_{20}$ isomers are shown in the Fig. 10.

The spin magnetic moment of these isomers varies from 15 to 21 $\mu_B$ in these clusters. The lowest structure has six Mn atoms forming a pentagonal pyramid. The MAE of the lowest energy isomer is 7 K. The binding energy of the six Mn doped cluster increases to 5.48 eV from 4.81 eV in the parent cluster. The gap of 0.25 eV suggests the cluster to be chemically stable. The presence of six Mn atoms in the cluster leads to 32 spin states in total. The most favorable ferromagnetic cluster is chosen for studying different spin ordering of the Mn atoms. To reduce the computational costs, the calculations on the anti-ferromagnetic states are first done as single point calculations. The lowest 10 spin ordered states of the cluster are selected and optimized in order to find the lowest energy spin-ordered clusters.

We find six AFM isomers that have lower energy compared to the ferromagnetic case. The energetically most favorable spin ordered cluster is shown in Fig. 11. This isomer has magnetic anisotropy energy barrier of 12.6 K. Similar to the other clusters studied here, the $As@Ni_6Mn_6@As_{20}$ cluster also has several closely spaced spin states with nearly same magnetic moments. It can be mentioned here that the bare $Mn_6$ cluster also has a ferrimagnetic spin arrangement but the preferred structure is octahedral with a magnetic moment of 9 $\mu_B$. Experimentally determined value of magnetic moment of the bare $Mn_6$ cluster ranges from 2 to 4 $\mu_B$ at temperature 50 K. The MAE of the bare $Mn_6$ cluster is 0 K compared to the 12 K of the $As@Mn_6Ni_6@As_{20}$ cluster. The spin moment on the individual Mn ions in $As@Ni_6Mn_6@As_{20}$ ranges from 2–3 $\mu_B$ compared to the 3.7 $\mu_B$ in the bare $Mn_6$ cluster. These comparisons show that the bonding of the Mn ions with the As and Ni atoms reduces the individual spin moments. Moreover, the chemical environment around the Mn atoms is important in the determination of the anisotropy energy.

The calculated results on the lowest energy FM and AFM states of the various doped $As@Ni_{12-x}Mn_x@As_{20}$ clusters are summarized in Table 4. We find that the doping by Mn distorts the geometry of the parent cluster but also increases the atomization energy of the whole cluster. The atomization energy of the parent $As@Ni_{12}@As_{20}$ cage is 4.8 eV which increases to 5.48 eV for the $As@Ni_6Mn_6@As_{20}$ cluster. The As–Mn bonds are shorter than the As–Ni bonds and the overall cluster size is reduced with the dopants. The lowest energy isomers have HOMO–LUMO gaps higher than 0.25 eV at the PBE level. The gaps also indicate higher chemical stability of the clusters. In general our study which included a large number of isomers indicated that the $Ni_{12\text{-}x}Mn_x$ alloy clusters within the $As_{20}$ cage exhibit a rich phase space with a large number of low-lying spin states. This is particularly true for clusters with larger number of Mn atoms. The clusters with ferrimagentic spin ordering are found to be more stable compared to those with ferromagnetic ordering. It can be mentioned here that in the pure Mn

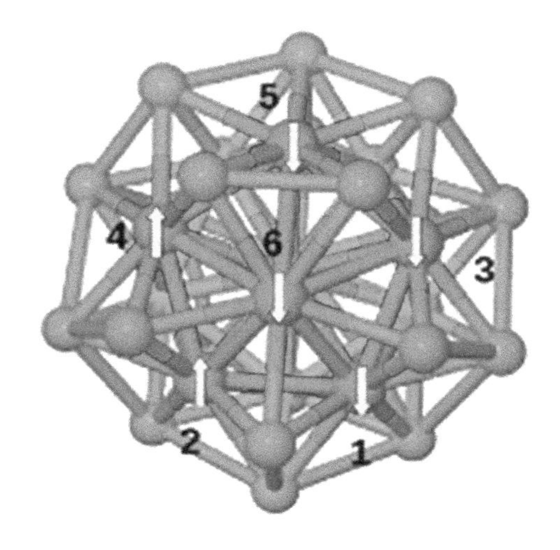


**Fig. 11** Spin ordering of the Mn atoms in the lowest energy spin state of $As@Ni_6Mn_6@As_{20}$

**Table 4** The lowest ferro- and anti-ferromagnetic states of the Mn doped $As@Ni_{12}@As_{20}$ cluster

| Systems | State | MAE (K) | M.M ($\mu_B$) | H–L gap (eV) | ΔE (eV) |
|---|---|---|---|---|---|
| $As_{21}Ni_{11}Mn$ | FM | 10.4 | 6 | 0.07 | 0.0 |
| $As_{21}Ni_{10}Mn_2$ | FM | 20.4 | 9 | 0.25 | 0.0 |
| | AFM | 0.0 | 1 | 0.23 | 0.12 |
| $As_{21}Ni_9Mn_3$ | FM | 37.4 | 12 | 0.33 | 0.15 |
| | AFM | 18.6 | 4 | 0.47 | 0.0 |
| $As_{21}Ni_8Mn_4$ | FM | 32.8 | 15 | 0.16 | 0.18 |
| | AFM | 0.0 | 1 | 0.45 | 0.0 |
| $As_{21}Ni_7Mn_5$ | FM | 17.2 | 14 | 0.22 | 0.03 |
| | AFM | 11.0 | 2 | 0.32 | 0.0 |
| $As_{21}Ni_6Mn_6$ | FM | 7.1 | 15 | 0.25 | 0.29 |
| | AFM | 12.6 | 5 | 0.32 | 0.0 |

clusters also anti-ferromagnetic or ferrimagentic spin ordering is preferred. The spin charges on the Mn atoms range from 2–3 $\mu_B$ in the alloy clusters, which are lower than the spin charge in pure Mn clusters. The doping introduces changes in the environment around the Mn atoms resulting in anisotropy energy.

# 5 Conclusions

The transition metal alloy clusters offer the possibility of tuning their magnetic properties through size, shape, and chemical composition. The magnetic properties show strong variation with cluster size. Reduction in size leads to enhancement of the magnetic moments. Alloying of 3d transition metal atoms with 4d or 4d non-magnetic metals results in larger magnetic moment for the whole cluster. The structure of the cluster is also important in this regard. For example mixed clusters show enhanced magnetization but similar enhancement is not seen in core-shell structures. This field is growing and needs very systematic analysis. The large variability of cluster shape, size, and composition presents vast potential for development novel materials. At the same time, a large amount of effort is required to examine all possible structures. The energetic preference of spin ordering of the systems poses a significant challenge.

For the development of cluster-based materials, precise control over particle growth and composition is required. The salt of $As@Ni_{12}@As_{20}$ is one example of cluster based material with precise structure. Theoretical studies on such clusters with various compositions can help in pinpointing the materials with novel properties, which can be targeted for synthesis. Another advantage of such materials is the clear passivation of the surface metal atoms, which results in higher chemical stability compared to bare gas-phase clusters. Our calculations presented here show that doping with magnetic atoms could lead to enhanced magnetic anisotropy of these clusters. More studies need to be done by varying the composition.

**Acknowledgments** This work was partially supported by DOE Basic Energy Sciences. This work was partially supported by the DOE Basic Energy Science under award numbers DE-SC0006818, and DE-SC0002168 and the NSF PREM program between UCSB and UTEP (DMR-1205302). The authors thank the Texas Advanced Computing Center (TACC) from the National Science Foundation (NSF) (Grant no. TG-DMR090071) and NERSC for the computational time.

## References

1. Néel L (1949) Ann Géophys 5:99
2. Jun YW, Seo JW, Cheon A (2008) Acc Chem Res 41:179
3. Lim EK, Kim T, Paik S, Haam S, Huh YM, Lee K (2015) Chem Rev 115:327
4. Ferrando R, Jellinek J, Johnston RL (2008) Chem Rev 108:845
5. Reddy BV, Khanna SN, Dunlap BI (1993) Phys Rev Lett 70:3323
6. Dunlap BI (1990) Phys Rev A 41:5691
7. Dunlap BI (1991) Z Phys D Atom Mol Cl 19:255
8. Chretien S, Salahub DR (2002) Phys Rev B 66
9. Vega A, Balbas LC, Dorantesdavila J, Pastor GM (1994) Phys Rev B 50:3899
10. Pastor GM, Dorantesdavila J (1995) Phys Rev B 52:13799
11. Alvarado P, Dorantes-Davila J, Pastor GM (1998) Phys Rev B 58:12216
12. Pastor GM (1998) Curr Probl Conden Matter 161
13. Munoz-Navia M, Dorantes-Davila J, Zitoun D, Amiens C, Jaouen N, Rogalev A, Respaud M, Pastor GM (2009) Appl Phys Lett 95
14. Borras-Almenar JJ, Coronado E, Clemente-Juan JM, Palii AV, Tsukerblat BS (2003) Polyhedron 22:2521
15. Jones NO, Beltran MR, Khanna SN, Baruah T, Pederson MR (2004) Phys Rev B 70
16. Jones NO, Khanna SN, Baruah T, Pederson MR (2004) Phys Rev B 70
17. Datta S, Saha-Dasgupta T (2013) J Phys-Condens Mat 25
18. Sahoo S, Islam MF, Khanna SN (2015) New J Phys 17
19. Rohart S, Raufast C, Favre L, Bernstein E, Bonet E, Dupuis V (2006) Phys Rev B 74
20. Luis F, Bartolome J, Bartolome F, Martinez MJ, Garcia LM, Petroff F, Deranlot C, Wilhelm F, Rogalev A (2006) J Appl Phys 99
21. Xie YN, Blackman JA (2006) Phys Rev B 74
22. Bornemann S, Minar J, Staunton JB, Honolka J, Enders A, Kern K, Ebert H (2007) Eur Phys J D 45:529
23. Oda T, Yokoo Y, Sakashita H, Tsujikawa M (2009) J Comput Theor Nanosci 6:2603
24. Boufala K, Fernandez-Seivane L, Ferrer J, Samah M (2010) J Magn Magn Mater 322:3428
25. Islam MF, Khanna SN (2014) J Phys-Condens Mat 26
26. Giguere A, Foldeaki M, Dunlap RA, Chahine R (1999) Phys Rev B 59:431
27. Jamet M, Negrier M, Dupuis V, Tuaillon-Combes J, Melinon P, Perez A, Wernsdorfer W, Barbara B, Baguenard B (2001) J Magn Magn Mater 237:293
28. Kortus J, Baruah T, Pederson MR, Ashman C, Khanna SN (2002) Appl Phys Lett 80:4193
29. Aguilera-Granja F, Vega A (2009) Phys Rev B 79
30. Chen HX, Shi DN, Qi JS, Wang BL (2011) J Magn Magn Mater 323:781
31. Fink K (2006) Chem Phys 326:297
32. Sattler K (1986) Z Phys D Atom Mol Cl 3:223
33. Li SF, Xue XL, Jia Y, Zhao GF, Zhang MF, Gong XG (2006) Phys Rev B 73
34. Mukherjee S, Moranlopez JL (1987) Surf Sci 189:1135
35. Vega A, Dorantesdavila J, Pastor GM, Balbas LC (1991) Z Phys D Atom Mol Cl 19:263
36. Sumiyama K, Suzuki K, Makhlouf SA, Wakoh K, Kamiyama T, Yamamuro S, Konno TJ, Xu YF, Sakurai M, Hihara T (1995) J Non-Cryst Solids 193:539

37. Manago T, Otani Y, Miyajima H, Akiba E (1996) J Appl Phys 79:5126
38. Brayner R, Coradin T, Fievet-Vincent F, Livage J, Fievet F (2005) New J Chem 29:681
39. Zitoun D, Respaud M, Fromen M-C, Casanove MJ, Lecante P, Amiens C, Chaudret B (2002) Phys Rev Lett 89:037203
40. Dennler S, Morillo J, Pastor GM (2004) J Phys-Condens Mat 16:S2263
41. Munoz-Navia M, Dorantes-Davila J, Pastor GM (2004) J Phys-Condens Mat 16:S2251
42. Tournus F, Blanc N, Tamion A, Hillenkamp M, Dupuis V (2010) Phys Rev B 81
43. Muñoz-Navia M, Dorantes-Dávila J, Respaud M, Pastor GM (2009) Eur Phys J D 52:171
44. Blanc N, Diaz-Sanchez LE, Ramos AY, Tournus F, Tolentino HCN, De Santis M, Proux O, Tamion A, Tuaillon-Combes J, Bardotti L, Boisron O, Pastor GM, Dupuis V (2013) Phys Rev B 87
45. Guirado-Lopez R, Villasenor-Gonzalez P, Dorantes-Davila J, Pastor GM (2003) Eur Phys J D 24:73
46. Sahoo S, Hucht A, Gruner ME, Rollmann G, Entel P, Postnikov A, Ferrer J, Fernandez-Seivane L, Richter M, Fritsch D, Sil S (2010) Phys Rev B 82
47. Liebing S, Martin C, Trepte K, Kortus J (2015) Phys Rev B 91
48. Pederson MR, Khanna SN (1999) Phys Rev B 60:9566
49. Blonski P, Hafner JJ (2011) Phys-Condens Mat 23
50. Pederson MR, Baruah T (2007) Molecular magnets: phenomenology and theory, chapter 9 in handbook of magnetism and advanced magnetic materials. Wiley, Hoboken
51. Pederson MR, Jackson KA (1990) Phys Rev B 41:7453
52. Jackson K, Pederson MR, Erwin SC (1990) Forces and geometry optimization in 1st-principles atomic cluster calculations, vol 193
53. Jackson K, Pederson MR (1990) Phys Rev B 42:3276
54. Pederson MR, Klein BM, Broughton JQ (1988) Phys Rev B 38:3825
55. Pederson MR, Porezag DV, Kortus J, Patton DC (2000) Phys Status Solidi B-Basic Res 217:197
56. Pederson MR, Baruah T, Allen PB, Schmidt C (2005) J Chem Theory Comput 1:590
57. Pederson MR, Khanna SN (1999) Chem Phys Lett 307:253
58. Pederson MR, Bernstein N, Kortus J (2002) Phys Rev Lett 89
59. Kortus J, Baruah T, Bernstein N, Pederson MR (2002) Phys Rev B 66
60. Kortus J, Pederson MR, Hellberg CS, Khanna SN (2001) Eur Phys J D 16:177
61. Baruah T, Kortus J, Pederson MR, Wesolowski R, Haraldsen JT, Musfeldt JL, North JM, Zipse D, Dalal NS (2004) Phys Rev B 70
62. Pederson MR, Kortus J, Khanna SN (2002) J Appl Phys 91:7149
63. Park K, Pederson MR, Richardson SL, Aliaga-Alcalde N, Christou G (2003) Phys Rev B 68
64. Baruah T, Pederson MR (2002) Chem Phys Lett 360:144
65. Piligkos S, Rajaraman G, Soler M, Kirchner N, van Slageren J, Bircher R, Parsons S, Gudel HU, Kortus J, Wernsdorfer W, Christou G, Brechin EK (2005) J Am Chem Soc 127:5572
66. Baruah T, Pederson MR (2002) Chem Phys Lett 360:144
67. Kortus J, Pederson MR, Baruah T, Bernstein N, Hellberg CS (1871) Polyhedron 2003:22
68. Gambardella P, Rusponi S, Veronese M, Dhesi SS, Grazioli C, Dallmeyer AIC, Zeller R, Dederichs PH, Kern K, Carbone C, Brune H (2003) Science 300:1130
69. Xiao RJ, Fritsch D, Kuz'min M D, Koepernik K, Eschrig H, Richter M, Vietze K, Seifert G (2009) Phys Rev Lett 103
70. Fritsch D, Koepernik K, Richter M, ESchrig H (2008) J Comput Chemis 29:2210
71. Strandberg TO, Canali CM, MacDonald AH (2008) Phys Rev B 77
72. Garcia-FuenteA, Vega A, Aguilera-Granja F, Gallego L (2009) J Phys Rev B 79
73. Garcia-Fuente A, Garcia-Suarez VM, Ferrer J, Vega A (2012) Phys Rev B 85
74. Moses MJ, Eichhorn B, Fettinger J (2003) Abstracts Papers Am Chem Soc 225:U85
75. Baruah T, Zope RR, Richardson SL, Pederson MR (2004) J Chem Phys 121:11007
76. Baruah T, Zope RR, Richardson SL, Pederson MR (2003) Phys Rev B 68

# Growth Pattern and Size-Dependent Properties of Lead Chalcogenide Nanoclusters

**Ann F. Gill, William H. Sawyer, Kamron Salavitabar, Boggavarapu Kiran and Anil K. Kandalam**

**Abstract** In this chapter we review the structural evolution of lead-chalcogenide $(PbX)_n$ ($X$ = S, Se, and Te; $n$ = 1–32) nanoclusters and how various properties of these clusters change with increasing cluster size. We first give an overview of different experimental techniques that are used to synthesize or generate lead sulfide clusters. The growth mechanism and size-dependent electronic structure and stabilities of these clusters obtained from density functional theory (DFT) based computational studies are also discussed in detail. The importance of the synergy between computational study and experiments is demonstrated by focusing on lead sulfide clusters.

## 1 Introduction

Semiconductor nanocrystals have attracted considerable interest in the past three decades due to their unusual size-tunable electronic and optical properties. As the physical size of the material shrinks (to nanometer range) below its exciton Bohr radius ($a_B$), a strong quantum confinement occurs with the electron-hole pair in effectively a one-dimensional potential well. Typically straddling a range from 1 to 20 nm, these particles are larger than a single molecule yet small enough for their size to affect their optical properties and large enough that their size can be adjusted to effectively tune their optical behavior. As a result, these Quantum Dots (QD) as they have come to be known [1] have been extensively studied since the 1980s and early 90s for their ability to exhibit a "tunable" florescence capability.

A.F. Gill · W.H. Sawyer · K. Salavitabar · A.K. Kandalam (✉)
Department of Physics, West Chester University, West Chester, PA 19383, USA
e-mail: akandalam@wcupa.edu

B. Kiran
Department of Chemistry, McNeese State University, Lake Charles, LA 70609, USA
e-mail: kiran@mcneese.edu





M.T. Nguyen and B. Kiran (eds.), *Clusters*, Challenges and Advances in Computational Chemistry and Physics 23,
DOI 10.1007/978-3-319-48918-6_9

In the early 1983, Bruss and co-workers reported [2] the quantum size effects (QSE) in the excited electronic properties of CdS nanocrystals in aqueous solutions. Shortly after, in 1985 Ekimov et al. reported [3] size-dependent absorption spectra of Group I–VII and Group II–VI microcrystals grown in a glassy dielectric matrix. In the years following the work reported by Bruss in 1983, multiple techniques have been developed to synthesize and characterize semiconductor QDs of various sizes. Although most work centered primarily on CdS [4] and CdSe [5–9] nanocrystals, numerous methods have been devised to synthesize lead-chalcogenide (mostly PbS and PbSe) QDs in various sizes, shapes, and in diverse environments such as solution [10–20], embedded in zeolites [21, 22], glasses [23, 24], in polymers [25–27], and on carbon nanotubes [28]. The motivation to study lead-chalcogenides came from the fact that, unlike Group II–VI and III–V systems, PbS and PbSe QDs have a large exciton Bohr radii with equal electron-hole contributions [29]. The larger exciton Bohr radii of PbS and PbSe QDs (18 nm for PbS, 46 nm for PbSe versus 5.3 nm for CdSe) make them ideal materials for stronger quantum confinement effects at relatively larger sizes than CdSe. For example, decreasing the size of PbS from the bulk to nanometer dimensions results in a blue-shift in the absorption spectrum, all the way from the near-IR to the near-UV range (0.41–4 eV) [30]. Thus, PbS systems exhibit a wide range of size-tunable electronic and optical properties. Moreover, Kang et al. [31] have pointed out that PbS nanocrystal- doped glasses have band structures that make them very attractive saturable absorbers in the near infrared. Wundke et al. used these features to demonstrate their ability to be used as absorbers for picosecond pulse generation in lasers [32]. Thus, the lead-chalcogenide QDs have a wide range of possible applications such as photovoltaics, near IR sensors, electroluminescent devices, and biomedical industries.

Since the publication of Bruss's Effective Mass Model (EMM) [33], there have been significant advances in both the analytical tools available for developing theoretical models and the ability to compare their results with experimentation. However, the majority of these studies focused on lead chalcogenide nanocrystals, quantum dots, and nanorods, relatively very few studies of lead chalcogenide *clusters* exist. Clusters form an important transition stage between molecules and their bulk materials. Clusters, at the sub-nano scale regime, often exhibit structural and electronic properties that differ from those of their bulk counterparts. Most importantly, the QSE in these clusters could be different from those of their QDs. Thus, a systematic study of lead chalcogenide clusters is of a great interest. While spectroscopic investigations of PbS monomers and dimers in matrices and gas-phase [34–36] were carried out more than 40 years ago, experimental studies focusing on gas-phase lead sulfide $(PbS)_n$ clusters beyond $n = 2$ were not reported until the early 1980s. Two mass spectrometric studies of $(PbS)_{1-4}$ cluster cations [37, 38] were reported in early 1980s, while a photoelectron spectroscopic study of the $(PbS)^-$ molecular anion [39] was reported in 2002. In 1996, Silbey and co-workers carried out a semi-empirical tight-binding based calculations to study the electronic structure of size-selected $(PbS)_n$ ($n$ = 8, 32, 56, 88, 160, 208, 280, 552, and 912) clusters [30]. It was found from this study that the densities of states in these nanoclusters did approach the bulk density of states with increasing cluster size, while the band gaps predicted by tight-binding method was in reasonable agreement with the experimental data. Despite this seminal work, there were no published experimental or theoretical studies until 2005 on the structural growth pattern of lead sulfide *clusters*.

In 2005, Schelly and co-workers reported synthesis of un-capped PbS QDs in the molecular size regime via electroporation of synthetic vesicles [13]. These QDs were smaller than 1 nm in size and corresponded to $(PbS)_n$ ($n$ = 1–9) clusters. Most importantly, the time-dependent UV spectra of these small clusters exhibited an oscillating red and blue shifts with increasing cluster size from $n$ = 1–9. Beyond $n$ = 9, a monotonic red shift was reported. These authors have also noted that this oscillating behavior is unique to PbS and it is in contrast to the trend observed in other semi-conducting materials, such as, AgBr and CdS clusters. This oscillating behavior was attributed to significant covalent bonding between the Pb and S atoms of the cluster. Marynick and co-workers published the very first density functional theory (DFT) based study of small $(PbS)_n$ ($n$ = 1–9) clusters [40]. This study revealed that PbS clusters start to exhibit cubical structures (similar to its bulk counterpart) at small cluster sizes, starting at $n$ = 4. Most importantly, this DFT study showed that binding energies and HOMO-LUMO gaps of these stoichiometric clusters oscillated with increasing cluster size. Another DFT study focusing on the size-dependent stability and optical properties of $(PbS)_n$ ($n$ = 1–16) clusters was reported by He et al. [41]. Our group recently reported [42] a combined experimental and computational study of $(PbS)_n^-$ ($n$ = 1–10) clusters and neutral $(PbS)_n$ ($n$ = 1–15) clusters that added significantly to the fundamental understanding of their structures and stability. Using the photoelectron spectra to determine the vertical detachment energies (VDEs) for the cluster anions and the electron affinities (EA) for their neutral cluster counterparts we uncovered a pattern in the EA and VDE values. Clusters with even values of $n$ up to 8 exhibited lower EA and VDE values than those with odd values of $n$. In addition, our computations found that neutral clusters with even $n$ were thermodynamically more stable than their immediate (odd $n$) neighbors and uncovered a pattern in their HOMO–LUMO gaps that is consistent with the previous studies [13, 40] on lead-sulfide clusters. The structural evolution of lead sulfide clusters revealed a two-dimensional stacking of cubic moieties. A more detailed discussion of the structural evolution and energetics of neutral lead-sulfide clusters is given in the later part of this chapter. As a follow up to this work, our group reported another study on $(PbS)_{4n}$ ($n$ = 1–8) clusters and beyond [43]. Our computations revealed that $(PbS)_{32}$ is the cubic cluster containing bulk-like coordination for the inner Pb and S atoms and can be replicated to form bulk lead-sulfide. In the accompanying experiments, $(PbS)_{32}$ clusters were soft-landed on a graphite surface and imaged using a Scanning Tunneling Microscopy (STM). The STM images revealed formation of nano-block aggregates with $(PbS)_{32}$ as a building block. Our predicted dimensions of these nano-blocks were confirmed by these experiments. Again, the details of these results are discussed in the later part of the chapter. Recently Gupta et al. [44] reported a theoretical study focused on the electron transport properties of bare $(PbS)_{32}$ QD deposited on Au (001) and (110) surfaces. Using density functional theory (DFT) they predicted a relatively high tunneling current despite a noticeable band gap (~2 eV) affirming the bright images of $(PbS)_{32}$ reported in the earlier work [43]. The *I-V* characteristics of $(PbS)_{32}$ were found to be substrate dependent, where $(PbS)_{32}$ on Au (001) exhibited a molecular diode-like behavior with an unusual negative differential resistance.

Interestingly, PbSe and PbTe clusters have *not* been studied as extensively as PbS clusters. To our knowledge, there exist no experimental studies on neutral or charged PbSe and PbTe *clusters*. On the computational front, however, a few studies do exist. Argeri et al. reported a study [45] on the effect of surface passivation on the shape and composition of PbSe nanoclusters by using pure periodic surfaces as well as finite clusters as models. In this DFT based study, it was found that neutral ligands had the greatest affinity for the (110) surface, followed by the (100) and the (111) surfaces, while the anionic ligand (propanoate) preferred the (111) surface. Recently, Zeng et al. [46] reported the structural evolution and optical properties of $(PbSe)_n$ ($n$ = 1–10) clusters using first-principles molecular dynamics approach. In the case of lead telluride nanoclusters, the structural growth pattern of PbTe clusters has been subject of two theoretical studies recently [47, 48]. It should be noted here that even though there exist other computational studies on lead-chalcogenide clusters, such as supported lead sulfide clusters [44] and surface-passivated lead selenide nanoclusters [45], in this chapter we focus only on the structural growth pattern, electronic structure, and stabilities of *bare* lead-chalcogenide clusters obtained from computational and experimental studies.

The remainder of this chapter is organized as follows: An overview of various experimental techniques that have been used to synthesize or generate bare and capped lead-sulfide clusters and the characterization tools are discussed in the next section (Sect. 2). The structural growth mechanism, and various size-dependent properties of bare lead-chalcogenide clusters obtained by our group and other groups are presented and discussed in Sect. 3. We conclude with a summary and an outlook for future work in Sect. 4.

## 2 Experimental Analysis of Pb*X* (*X* = S, Se, and Te) Clusters

### 2.1 Early Work

As with much of the early experimental work investigating nanoparticles, initial Pb*X* (*X* = S, Se, and Te) research centered on liquid fabrication methods. Principal among these were colloidal solutions. Shortly thereafter additional techniques incorporating sol gels and block copolymers were applied in order to contain the Pb*X* clusters. By the mid-1990s PbS Quantum Dots (QDs) were being produced in glass substrates with a very narrow size distribution [29]. Particle size was usually determined using TEM and analysis of the broadening of the X-ray diffraction peak using the Scherrer-Jones equation for clusters above 25 Å. IR absorption spectroscopy in conjunction with lattice spacing measurements were used to verify the composition of the clusters and compare them with bulk Pb*X* (*X* = S, Se, and Te). By the turn of the century, investigators had begun to employ molecular beam and vapor deposition techniques to gain greater insight into the structure of clusters and their ions. As instrumentation was developed and refined, it became possible to compare experimental measurements from Raman spectroscopy and photo-electron spectroscopy with theoretical predictions based on quantum mechanical modeling methods, of which DFT has been an important tool.

In 1985 Rossetti and his colleagues [49] at Bell laboratories, reported the optical properties of "tiny single PbS crystals" in colloidal solutions. By maintaining the temperature of the colloid just above its freezing point they were able to slow the aggregation process sufficiently to preferentially produce crystallites on the order of 25 Å. From analysis using both TEM and X-Ray diffraction they found the crystallite lattice spacing to be comparable to that of bulk PbS. When they investigated the absorption spectra of these clusters as a function of temperature down to 130 K they found the spectra were consistent with each other but they did not have the characteristic near IR absorption of bulk PbS.

Two years later Wang and his group reported [21] the synthesis of PbS clusters encapsulated in a zeolite matrix that were smaller than 13 Å. Their selection of a zeolite matrix was due to the fact that zeolites form an open porous framework with internal dimensions from 3 to 13 Å. As a result they were able to produce smaller, more uniform clusters than was typically the case with colloids. In their words [21] "Zeolites can be thought of as providing a solid solvent for the semiconductor clusters." Equally important, by encapsulating the clusters in zeolite they were able to significantly reduce and control aggregation of the clusters. In water, however, the zeolite released the PbS clusters, which in turn led to aggregation of these clusters into larger particles. In their conclusion, despite some of the issues they confronted, the authors proposed that many additional types of zeolite "containers" should be possible. A prescient observation which has led to the development of a broad range of investigative techniques and applications.

In a follow-on paper later that year Wang et al. [50] focused on the impact of the size on the band gap to examine the transition of PbS from bulk form to the molecular form using polymer films. The clusters were produced by exchanging $Pb^{++}$ into the polymer film and then reacting it with $H_2S$. They controlled the size of the particles by regulating the initial $Pb^{++}$ concentration. Heating was used to mobilize the clusters enabling them to migrate and form larger particles. The average particle size was determined by applying the Scherrer equation to the FWHM of the X-Ray diffraction peak and then confirmed with TEM. With this method their analysis found the particle size ranged from 25 Å to approximately 100 Å; X-Ray diffraction could not be used to determine particle sizes less than 25 Å. The X-Ray diffraction for 58 Å clusters produced by their method was essentially the same as that produced by bulk PbS suggesting a transition to bulk structure for clusters formation on this order and larger.

The near IR absorption spectra of these PbS particles showed that the absorption edge shifts to higher energy and became significantly sharper as the particle size decreased from 125 Å to less than 13 Å (Fig. 1). For very small particle sizes, estimated to be less than 13 Å, the authors observed discrete absorption bands. Their results however did not provide good agreement with the classical "electron-hole in a box, effective mass model". The lack of agreement was particularly significant for smaller particles. Based on these findings the authors introduced a modified effective mass model based on two assumptions. First, the lowest excitation of the PbS lattice resulting from and excitation energy equal to the band gap could be approximated by a simple electron capture from an $S^-$ by a $Pb^+$ and second, only two bands are important for the calculation. By incorporating these assumptions they were able to obtain good agreement for all but the smallest particles in their study. With a further modification model they were able to explain the results for particle diameters as small as 25 Å.

Silbey and co-workers [51] took advantage of the micro-domain formation exhibited by selected block co-polymers to create small PbS clusters. Using a mild ROMP (ring opening metathesis polymerization) catalyst they were able to create small PbS clusters by treating copolymer films containing regularly distributed microdomains of a lead polymer (poly[$(C_7H_9CH_2C_5H_4)_2Pb$]) in a polynorbornene matrix with $H_2S$. The resulting crystallites had an average domain size between 20 and 40 Å as determined using TEM and STEM. The composition of the clusters was confirmed using X-Ray fluorescence in a STEM and separately using powder diffraction.

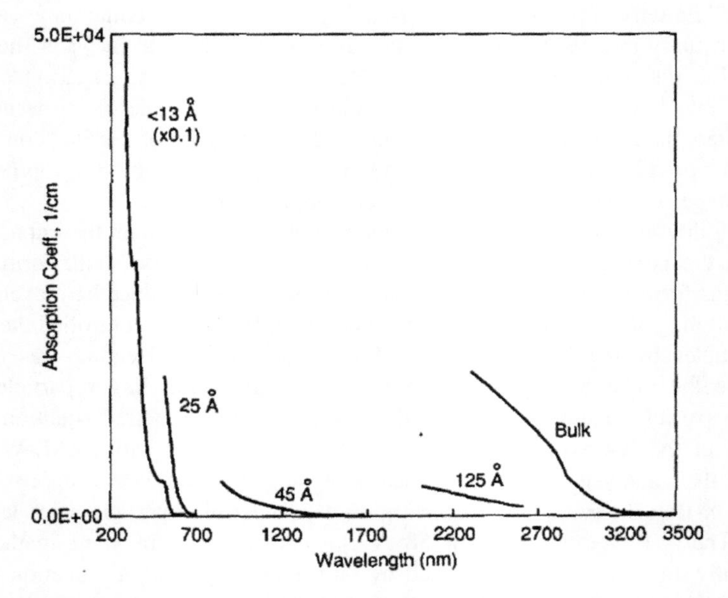


**Fig. 1** The absorption spectra of PbS particles in E-MAA polymer film as function of particle size (Reproduced from [50])

In a follow-up study, Silbey, together with Kane et al. [52] reported the use of a block copolymer nanoreactor technique developed by Cohen and Yue [53] to produce PbS clusters. Their technique selectively placed the metals into preformed microdomains of the block copolymer film. In the case of PbS, the clusters were created "within the microphase separated films of diblock copolymers containing carboxylic acid units in one of the blocks." Tetraethyllead was used to selectively place lead into the domains containing the acid. The thin films were subsequently treated with $H_2S$ to produce PbS clusters. For a given domain size, the cluster size could be controlled by $H_2S$ exposure, and temperature. Since in this method the reactive carboxylic acid sites are regenerated, larger clusters can be created by repeated applications of $H_2S$. Clusters produced using this procedure tended to be similar in size. Specimens were characterized using HRTEM, and X-Ray fluorescence using STEM. Wide angle X-Ray diffraction was also used to determine the composition of the clusters and, in conjunction with the Scherrer equation, to confirm the average particle size.

## 2.2 *Experimental Analysis of Small Clusters*

In order to study the properties and structure of quantum dots as a function of size below 5 nm, methods need to be employed that will reliably produce clusters of a uniform size. Molecular beam mass spectroscopy is one way to obtain reliable size isolation. While their yields are relatively small, molecular beam mass spectroscopy offers mass control enabling the ability to make detailed comparisons between predicted structure and experiment. Since the early days of experimental work in nanomaterials and specifically quantum dots much effort has been directed toward this goal.

In 2004, Schelly and co-workers reported [13] a unique technique for controlling the growth of sub-nanometer un-capped PbS clusters using synthetic unilamellar vesicles. In this study, they reported the largest cluster size to be $(PbS)_9$. Their method began by creating multilamellar vesicles (MLVs) containing $Pb^{2+}$ ions in an aqueous solution using a synthetic phospholipid. After drying the lipid in a mild vacuum to create a thin film, MLVs containing $Pb^{2+}$ ions were then created by rehydrating the lipid film in a dilute molar solution of $Pb(NO_3)_2$. The MLVs were subsequently extruded at high pressure multiple times through stacked poly carbonate filters to create unilamellar vesicles containing $Pb^{2+}$ ions. The $Pb^{2+}$ ions on the outside of the vesicles were then removed by introducing an aqueous solution of $Na_2S$ into the solution containing the vesicles. The precipitate and most of the loaded vesicles was removed by centrifugation. The remaining clear, dilute (<0.4 mg/ml) solution of loaded vesicles was used for the experiment. In order to create very small PbS clusters (<10 Å) and control their size, the authors employed a process called electroporation. In this process, the vesicles were exposed to a strong electric field causing them to elongate in the direction of the field. This full reversible elongation opened micro-pores at ends of the elongated vesicles. When

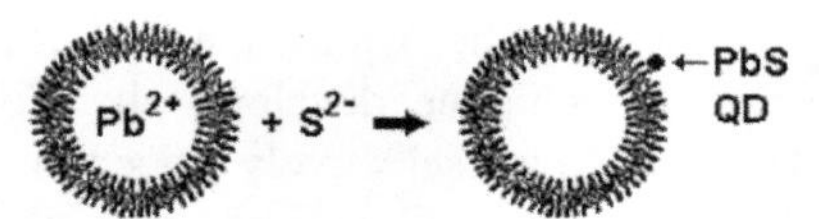


**Fig. 2** Formation of PbS quantum dots (QD) via electroporation (Reproduced from Wu et al. (2004) Langmuir 21:686–691)

the pores were large enough $Pb^{2+}$ ions were forced out of the vesicles by an increase in the interior pressure (resulting from the deformation) and combine with the $S^{2-}$ ions in the external solution, as shown in Fig. 2. The PbS molecules adsorbed on to the outside of the vesicles where very slow aggregation took place. As a result, aggregation of the PbS monomers occurred over a period ranging from hours to days. The authors were further able to control the growth by varying the concentration and temperature of the solution. Together with the ability to control the release of the $Pb^{2+}$ ion, the binding to the surface of the vesicles greatly slowing the aggregation process was critical to the success of the experiment.

The authors also measured the absorption spectra of the solution containing PbS-coated vesicles with a high resolution (<0.5 nm) UV-Vis spectrophotometer at regular intervals over a period of several days. As a result, they were able to observe the evolution of the characteristic absorption band over time. They reported an oscillating trend in the absorption band of PbS clusters with the increase in the cluster size (see Fig. 3).

After the first day, following electroporation the initial absorption band was found to reach a maximum at 237.5 nm. It subsequently began to diminish and a clear but broad peak at 282 nm began to appear reaching its peak 3 days after electroporation. By day 4 the 282 peak had receded to be replaced by narrower peak at 232 nm. After 5 days the 232 nm peak was gone and a broader peak at 281 nm. After 7 days the peak had again shifted to the blue with its peak at 234.5 nm. By the 9th day the absorption was centered on 278 nm and continued to shift slightly to the red so that after 20 days it was centered on 280 nm. These authors noted that, semi-conducting clusters such as AgBr and CdS clusters in the same size range as PbS clusters, namely in the order of 1 nm and larger, have showed a monotonic red shift with increased cluster size.

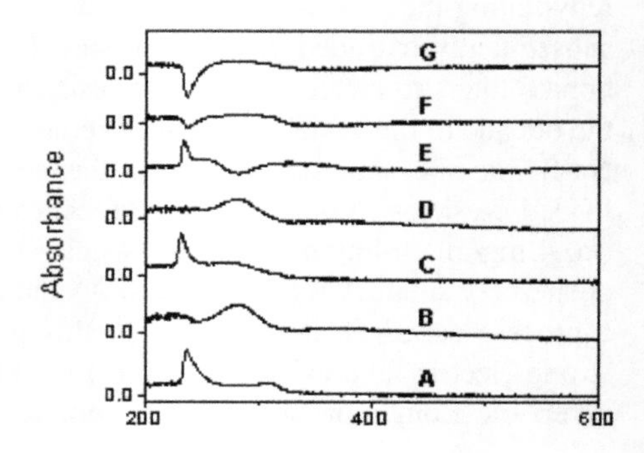


**Fig. 3** Oscillating shift of the absorption band of growing $(PbS)_n$ clusters. Time elapsed after electroporation: 1 day (*A*), 3 days (*B*), 4 days (*C*), 5 days (*D*), 7 days (*E*), 9 days (*F*), and >20 days (*G*) (Reproduced from Wu et al. (2004) Langmuir 21, 686–691)

## 2.3 Cluster Beam Analysis

In order to make detailed comparisons between theoretical models of small clusters and experimental results it is useful to be able to work with isolated clusters of well controlled size and molecular weight. Although much can and has been learned from examination of nanoparticles in solutions, gels, glasses, and other matrices, by the nature of the setting they are influenced by their molecular environment. Gas-phase environments allow for examination of particles/clusters without significant nearest neighbor influences. Much work has been done in this area using sputter sources, plasmas and hot vapor deposition to name only a few [54]. An array of techniques employing various instruments and methodologies has been employed to hone in on one or more specific variables to allow analysis and comparison with theory. These include multiple types of beam producing ion sources designed to facilitate particular cluster materials ranging from metals and semiconductors to insulators. Structure analysis, while utilizing a variety of techniques frequently employs Photo electron or Raman Spectroscopy.

Mass selected beams of charged clusters have been in use for some time. In 1988 McHugh et al. [55] reported the development of a "Smoke Ion Source" with the ability to produce a continuous, intense beam of either positive or negative cluster ions. In their design an electron beam formed by a biased hot wire filament was mounted immediately outside the orifice of the condensation cell and just before the extraction lens such that the electron beam would interact with the densest portion of neutral cluster beam and the largest portion of the resulting ions would be collected.

In 2011 Koirala et al. [42] used this source to perform a combined experimental and theoretical study of the electronic structure of both neutral and negatively charged lead sulfide clusters $(PbS)_n$ clusters with $n$ ranging from 2 to 10. In this experiment, a negatively charged mass selected beam of PbS anions was crossed with a fixed frequency photon beam. The resulting photoelectron spectrum was used to determine the electron binding energy (EBE) by subtracting the photoelectron kinetic energy from the incident photon energy. The EBE was then compared with the theoretical vertical detachment energy (VDE).

The following year Kiran et al. [43] published the results of a combined theoretical and experimental work which determined the smallest cubic cluster for which the inner $(PbS)_4$ core has bulk-like coordination. To test the validity of these computational predictions, mass-selected lead-sulfide clusters were deposited on highly ordered pyrolytic graphic (HOPG) and imaged by STM for their dimensions. The magnetron ion source consisted of a lead target mounted in front of a permanent magnet which created a cylindrical field with its axis perpendicular to the target. The target was biased at −300 V and a mixture of argon gas with 3 % $H_2S$

was bled into the source until the pressure reached approximately 2 mbars. The argon ions from resulting discharge sputtered lead clusters from the target which in turn reacted with the $H_2S$ to form lead sulfide clusters. The resulting cluster beam was accelerated using 565 V and passed through a sector magnet. The mass analyzed clusters were then decelerated to a kinetic energy of less than 0.1 eV per atom and soft landed at room temperature on freshly cleaved highly ordered pyrolytic graphite (HOPG). A schematic diagram of their apparatus is shown below in Fig. 4.

The dimensions of these deposited clusters were then compared with the theoretical predictions. Results of these two experiments [42, 43] are discussed in the next section where the structural growth pattern of lead sulfide clusters is discussed in detail.

In order for applications to be developed using PbS clusters they will have to be suspended within some form of bulk structure. Nonetheless, in order to gain a firm understanding of a cluster's structure particularly for small sizes techniques that allow them to be examined in isolation without the influence of ligands or surrounding matrices and the results compared with those predicted by theory are essential. Once this information is firmly in hand it is possible to examine in detail the impact of the surrounding matrix. Gas phase analysis while expensive to implement and maintain provides this capability in a unique way.

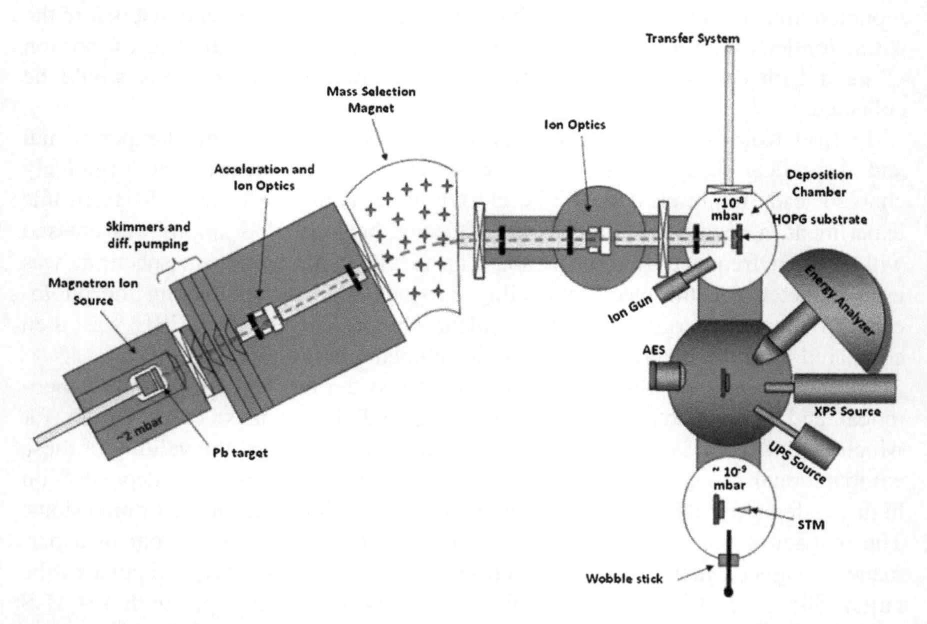


**Fig. 4** Schematic of the apparatus used determined the smallest cubic cluster for which the inner $(PbS)_4$ core has bulk-like coordination (Reproduced from [43])

# 3 Structural Growth Pattern and Energetics of $(PbX)_n$ ($X$ = S, Se, and Te) Clusters

In this section, we start our discussion on the structural growth pattern of lead chalcogenide $(PbX)_n$ ($X$ = S, Se, and Te) clusters, followed by a detailed discussion on the stability, energetics, and other properties of these clusters.

## *3.1 Growth Mechanism*

### A. $(PbS)_n$ Clusters

The structural growth pattern of lead sulfide $(PbS)_n$ clusters is one of the most thoroughly studied among the lead chalcogenide clusters. Computational studies play an important role in identifying the growth mechanisms of small gas-phase nanoclusters and in understanding the experimental observations. For example, Zeng et al. reported the geometries of neutral lead sulfide $(PbS)_n$ ($n$ = 1–9) clusters using density functional theory based calculations [40]. They showed that small PbS clusters prefer to adopt cubic structures, similar to that of the rock salt structure of bulk PbS. Most importantly, it was shown that the HOMO-LUMO gap of these clusters oscillate with increasing cluster size, which was in agreement with their previously observed [13] behavior of the UV absorption spectra of ultra-small PbS clusters (Fig. 5).

In a subsequent DFT based computational study, He et al. reported the size-dependent stability and optical properties of neutral $(PbS)_n$ ($n$ = 1–16) clusters [41]. Using B3LYP density functional and SBKJC/6-31 + G* basis sets for Pb and S atoms, it was shown that the even number clusters show higher stability than their odd number neighboring clusters, with magic numbers appearing at $n$ = 4, 8, 10,

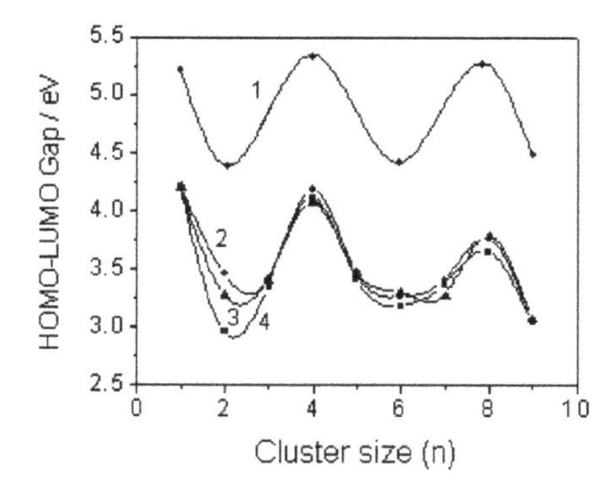


**Fig. 5** Comparison of the observed experimental transition energy (1) of $(PbS)_n$ clusters with the HOMO-LUMO gaps calculated by the use of different basis sets on lead (2–4). Basis sets used: 6-31 + G* on sulfur, and SBKJC + d (2), LANL2DZ (3), and SBKJC (4) on lead (Reproduced from [13])

and 14. Recently, a joint anion photoelectron spectroscopy (PES) and computational study was reported [42] by our group, in which a systematic investigation of growth pattern, stabilities, and HOMO-LUMO gaps of $(PbS)_n$ ($n$ = 1–15) nanoclusters were carried out. An accurate prediction of the lowest energy isomers is vital in understanding the growth pattern and stabilities of these clusters. Our combined experimental and computational study provided such an opportunity. The PES of the size-selected anionic $(PbS)_n^-$ ($n$ = 1–10) clusters (Fig. 6) revealed a resemblance to the PES of sodium fluoride clusters with an odd-even alteration of the VDE/ADE values, where the odd $n$ clusters (i.e., $n$ = 1, 3, 5, and 7) exhibited higher VDE/ADE values than the even $n$ clusters (i.e., $n$ = 2, 4, 6, and 8). In addition, in cluster sizes $n$ = 4–10, the lowest VDE and ADE values were observed for the $(PbS)_4^-$ cluster, indicating the stability of its neutral counterpart, i.e., $(PbS)_4$ cluster.

Comparison of the computed electron detachment energies with the anion PES showed that the PW91PW91 functional along with the SBKJC basis set for Pb atoms and 6–311 + G* basis set for S atoms provides a better agreement compared to the previously used B3LYP functional and SDD basis set. A graph comparing the experimental and calculated VDE values are shown in Fig. 7.

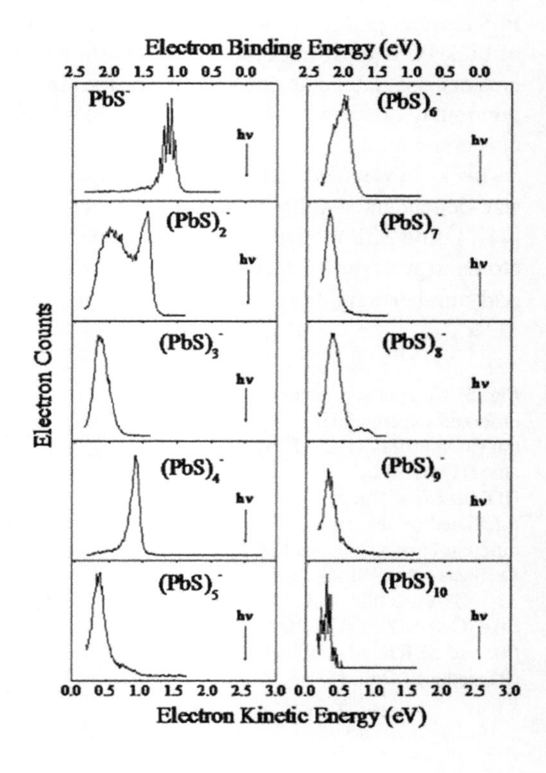


**Fig. 6** The 488 nm photoelectron spectra of lead sulfide $(PbS)_n^-$ ($n$ = 1–10) clusters (Reproduced from [42])

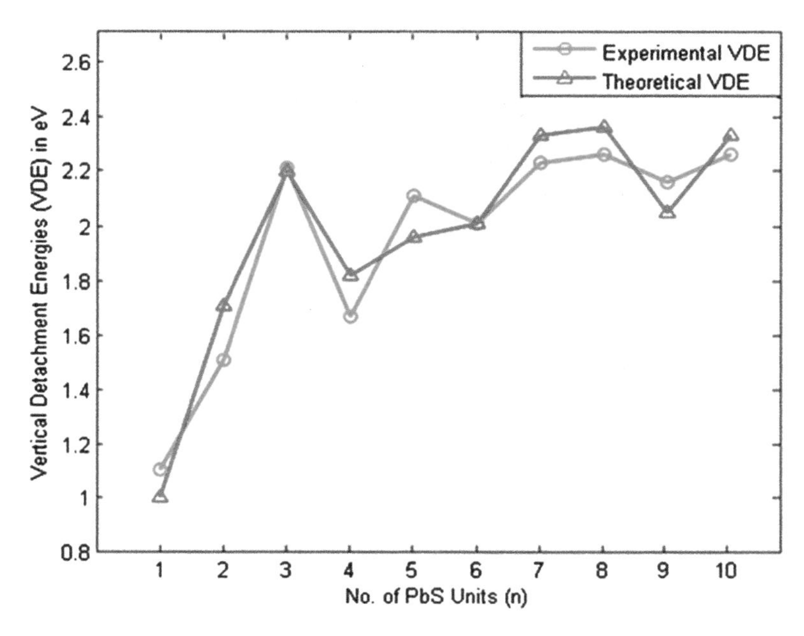


**Fig. 7** A comparison of experimental and calculated VDE values of negatively charged $(PbS)_n$ ($n$ = 1–10) clusters

The calculated geometrical structures of neutral and negatively charged $(PbS)_n$ ($n$ = 1–10) clusters are displayed in Figs. 8 and 9, respectively. The calculated results for $n$ = 1–5 showed some interesting trends. For example, in contrast to the earlier reported studies [40, 41], in $(PbS)_3$ cluster a trigonal bipyramidal structure with a bridged S atom (see Fig. 3) was preferred over an edge-removed cubic/folded rectangular structure (ΔE = 0.53 eV). A comparison of the calculated VDE and ADE values of these two structures with the measured value clearly indicated that the $(PbS)_3$ cluster forms an edge-capped trigonal bipyramidal structure. The $(PbS)_4$ cluster was found to be the first cubic structure in this growth pattern. In the case of a $(PbS)_5$ cluster, a bi-capped octagonal structure was preferred over the previously reported structures.

Beyond the cluster size, $n$ = 5, the $(PbS)_4$ cube formed the basis (seed) for the structural growth pattern of $(PbS)_n$ clusters. For example, $(PbS)_6$ cluster corresponds to a structure containing slightly-distorted face-sharing cubes with an average Pb–S bond length of 2.69 Å. In the case of the corresponding anion, the electron infused distortion was more pronounced with five different Pb–S bond lengths, ranging from 2.60 to 2.89 Å. Starting from the cluster size $n$ = 7 and beyond, calculated results revealed the possibility of multiple-isomers for lead sulfide clusters. These isomers mostly varied on how cubic units are stacked (one dimensional or two-dimensional stacking) and also if they are edge-sharing or face-sharing cubes. In the case of the $(PbS)_7$ cluster, there were two iso-energetic

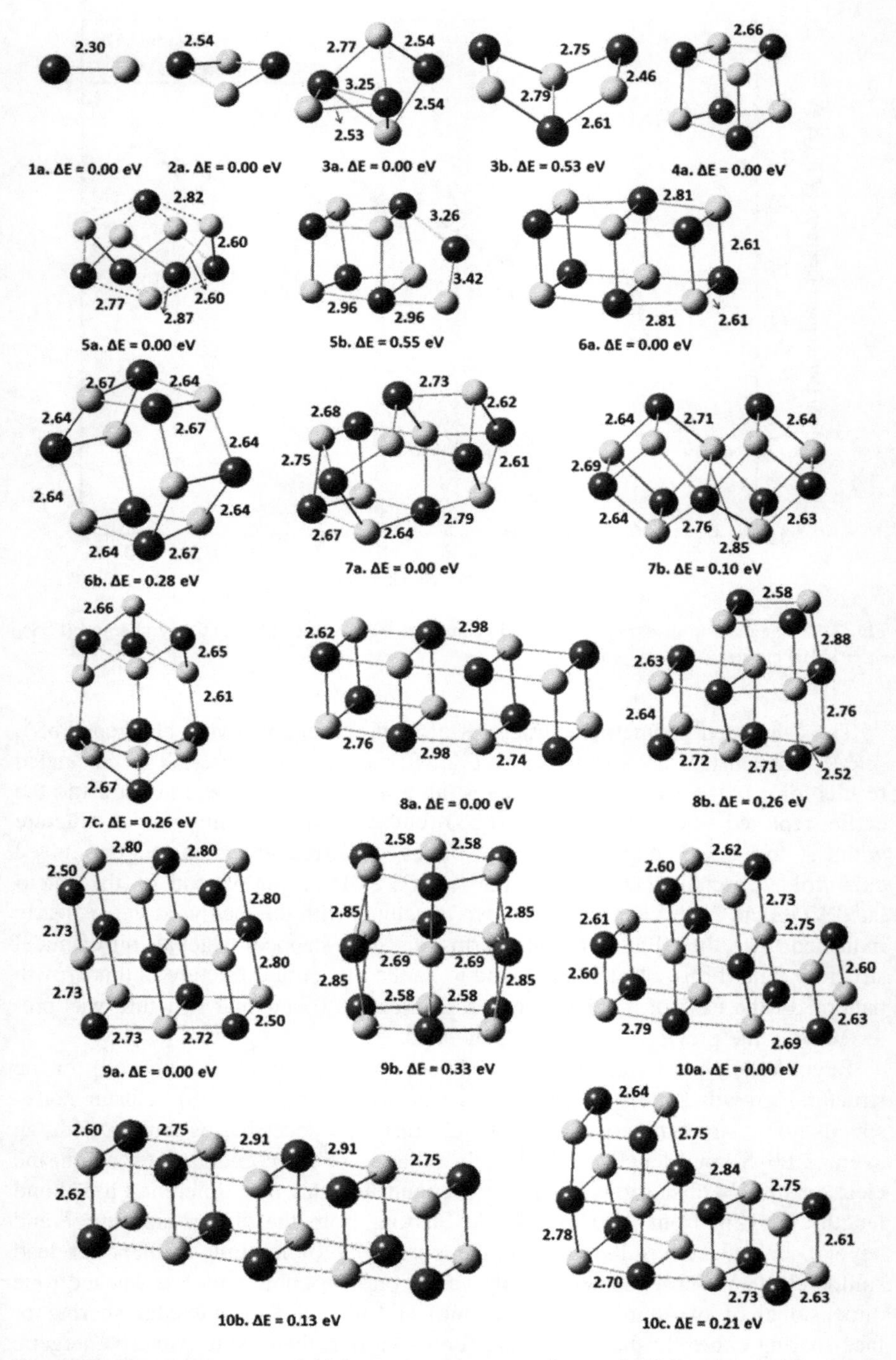


**Fig. 8** The lowest energy and higher energy isomers of neutral $(PbS)_n$ ($n$ = 1–10) clusters (Reproduced from [42])

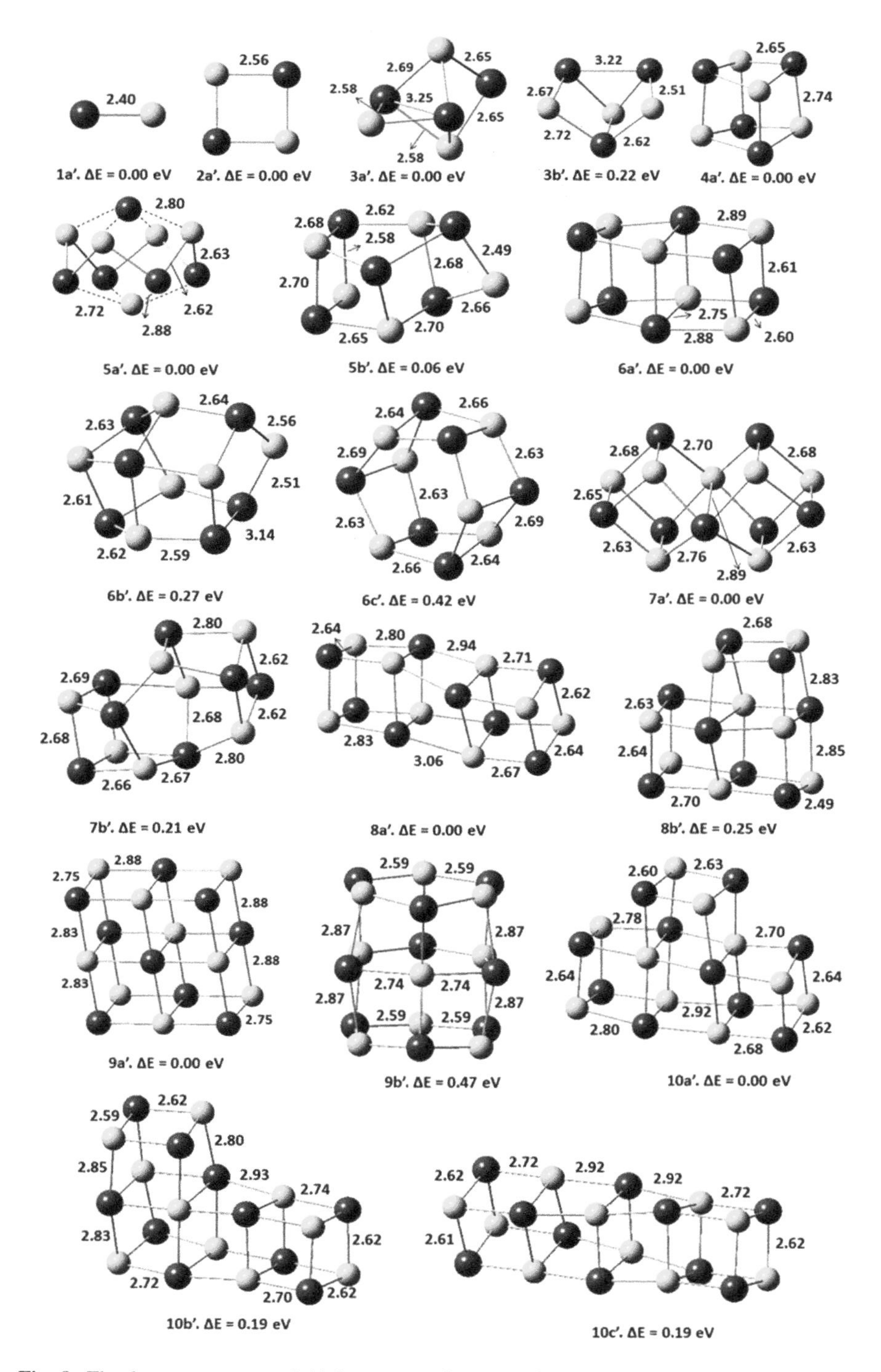


**Fig. 9** The lowest energy and higher energy isomers of anionic $(PbS)_n$ ($n$ = 1–10) clusters (Reproduced from [42])

structures: one is a highly distorted open-cage structure, which can also be described as two distorted cubes fused on the sides; the other is a more symmetric structure in which two cubes are sharing an edge. However, comparison of their calculated VDE and ADE values with that of the corresponding experimental values showed that only the structure containing edge-shared cubes was observed in the experimental cluster beam. The lowest energy isomer of the neutral $(PbS)_8$ cluster is a result of a one-dimensional stacking of two $(PbS)_4$ cubic units, while another isomer resulting from a two-dimensional stacking of face-sharing cubes was found to be 0.26 eV higher in energy. In the lowest energy structure of the $(PbS)_8$ cluster, there were three distinct Pb–S bonds, with the Pb–S bond length between the two cubes being significantly longer at 2.98 Å. Note that this bond length is closer to the Pb–S bond length in the lead sulfide crystal (2.957 Å) [55]. As observed in the $(PbS)_6$ cluster, a significant structural distortion was observed in the structures of negatively charged $(PbS)_8$ cluster. The presence of the extra-electron weakened the interaction between the two $(PbS)_4$ cubic units, which resulted in an elongation of the Pb–S bond length between the two cubes to 3.06 from 2.98 Å. Just as in the case of neutral cluster, the isomer with two-dimensional stacking was found to be 0.25 eV higher in energy than the lowest energy isomer of the anion. Again, a comparison of the calculated electron detachment energies with the measured photoelectron spectra revealed the possibility of the presence of both one-dimensional stacked (linear) structure and two-dimensional stacked structure in the anionic cluster beam of the experiment. Similarly, the lowest energy structure of $(PbS)_9$ cluster can be built by extending the two-dimensional $(PbS)_8$ cluster or from a two-dimensional stacking of four face-shared cubic units. The calculated results for $(PbS)_{10}$, the largest lead sulfide cluster for which the anion PES was reported, showed three isomers within an energy difference of 0.21 eV. All these structures were a result of either one or two dimensional stacking of cubic units. The lowest energy isomer of the $(PbS)_{10}$ cluster is a symmetric structure obtained by a two-dimensional stacking of face-sharing lead sulfide cubes. On the other hand, a structure containing a one-dimensional stacking of face-sharing cubic units (quadrangular prism structure), followed by another isomer with two-dimensional stacking of cubes (L-shaped structure) were found to be 0.13 and 0.21 eV higher in energy, respectively. It was inferred that the photoelectron spectrum of anionic $(PbS)_{10}$ cluster is a representative of all three isomers. Thus, the structural growth pattern of these clusters for sizes $n = 5$–10 is dominated by a two-dimensional (albeit, one–dimensional in some cases) stacking of face-sharing lead sulfide cubical units, with the possibility of existence of multiple isomers for certain cluster sizes.

A further increase in the size of $(PbS)_n$ clusters beyond $n = 10$ revealed that the lowest energy structures of these larger clusters can be directly traced back to the smaller size clusters. For example, the lowest energy isomer of $(PbS)_{12}$ cluster can be built by adding a $(PbS)_2$ unit to the lowest energy structure of $(PbS)_{10}$ cluster; while, the next higher energy isomer can be built by the addition of PbS molecule to the lowest energy isomer of $(PbS)_{11}$ cluster. The lowest energy and the next higher energy isomers of $(PbS)_n$ ($n = 11$–15) clusters are displayed in Fig. 10.

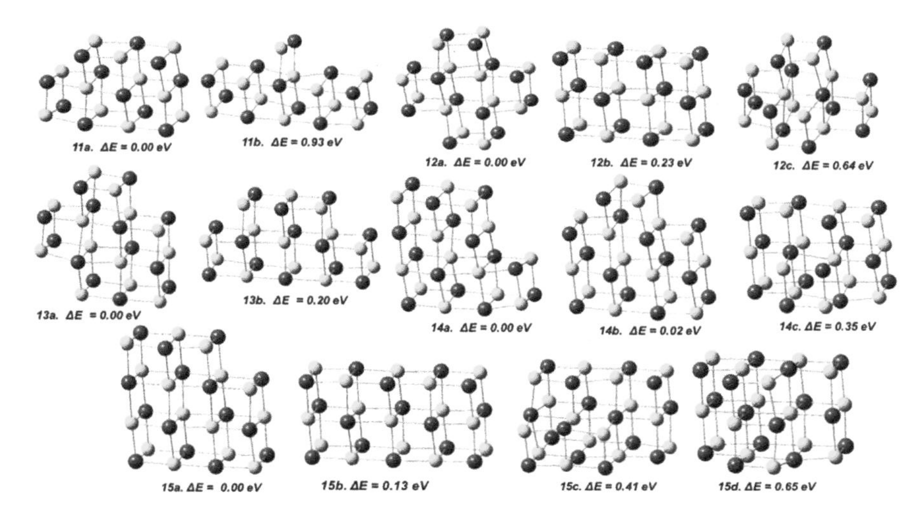


**Fig. 10** The lowest energy and higher energy structures of neutral $(PbS)_n$ ($n$ = 11–15) clusters

One of the interesting features of the lowest energy structures of these larger clusters was the coordination number for lead and sulfur atoms. Unlike in the bulk lead sulfide, none of these clusters ($n$ = 11–15) preferred to form structures with a six-fold coordination for either lead or sulfur atoms. In these clusters, the largest coordination number for these constituent atoms was found to be five. On the other hand, structures containing hexa-coordinated Pb or S atoms, were found to be very high in energy. For example, in the case of the $(PbS)_{14}$ cluster, hexa-coordinated structure was 0.35 eV higher in energy than the lowest energy isomer; while the energy difference was 0.41 eV in the case of $(PbS)_{15}$ cluster. While some of our results agree with the earlier theoretical studies [40, 41], there are significant differences in the structure and stability of various isomers at cluster sizes: $n$ = 3, 5, 7, 10, 12, 13, and 14. For instance, the $(PbS)_{12}$ and $(PbS)_{14}$ cluster isomers are found to be 0.5 eV higher in energy than isomers in our study. In addition, in the earlier theoretical study [41], the growth pattern of the PbS clusters was not consistent, varying abruptly from a one-dimensional stacking to two-dimensional stacking of cuboids.

To summarize, the structural studies of these clusters revealed that the potential energy surface of small $(PbS)_n$ clusters ($n$ = 1–5) are not only dominated by cuboidal structures, but also non-cuboidal structures. On the other hand, for cluster sizes $n$ = 6–15, a two-dimensional arrangement of face-sharing cuboidal units is preferred over a three-dimensional arrangement. Due to this preference, in all these clusters the maximum coordination attained by lead and sulfur atoms was five, rather than the six-fold coordination, as found in the bulk. Since at these sizes, lead sulfide clusters are too small to adopt bulk-like six-fold coordination, what is the smallest $(PbS)_n$ cluster size that can do so and that can also be replicated to form the bulk material?

Another combined theoretical and experimental study was reported by our group [43] to answer the above question. In this study, two computational methods were employed. The PW91PW91 functional with SBKJC basis set of lead and 6-311G* basis for sulfur were employed for small $(PbS)_n$ ($n$ = 4–32) clusters. For larger clusters, namely, composite clusters containing multiple units of $(PbS)_{32}$ we have used the DMol3 program suite [56, 57]. For these larger clusters, PW91 functional [58] along with Double Numerical plus d-functional (DND) and density functional semi-core pseudopotentials (DSPP) basis sets [59] were used. Since $(PbS)_4$ is the primitive cell of crystalline lead sulfide and is also a *magic* cluster, it was hypothesized that the successive dimerization of $(PbS)_{4n}$ clusters, starting from $(PbS)_4$ (i.e., $4n = 4 \rightarrow 8 \rightarrow 16 \rightarrow 32$) will eventually lead us to the bulk crystal. Our calculated results have shown that while, $(PbS)_4$ and $(PbS)_8$ clusters form a cube and a di-cube, respectively, the lowest energy isomer of $(PbS)_{16}$ is the dimer of a $(PbS)_8$ units which are arranged to form a two dimensional square structure. The dimerization of $(PbS)_{16}$ led to $(PbS)_{32}$, which is a large three-dimensional cube ($T_d$ symmetry) with an edge length of 8.32 Å; it is the result of the vertical stacking of $(PbS)_{16}$ units (see Fig. 11). In $(PbS)_{32}$, the geometrical environment of the inner cube is identical to that of the bulk crystal.

On the other hand, computations for $(PbS)_{4n}$ ($n$ = 5, 6, and 7) clusters have revealed that the on-set of six-fold coordinated structures occurs at size $n$ = 6, i.e., $(PbS)_{24}$ is the first cluster in which a preference for six-fold coordinated structure was observed. However, $(PbS)_{32}$ is the first in this series of clusters to not only attain a six-fold coordination for the inner $(PbS)_4$ cube, but also lead to bulk material upon replication. Therefore, $(PbS)_{32}$ was termed a *baby crystal* of PbS bulk material—the smallest possible stable cluster that attains bulk-like coordination (inner cube) which under aggregation/replication leads to the bulk structure.

On the experimental side, PbS clusters were generated in gas-phase and mass-selected $(PbS)_{32}$ clusters were soft-landed onto Highly Ordered Pyrolytic Graphite (HOPG) surface. The experimental setup was discussed in the earlier sections. A typical STM image (100 nm × 100 nm) of low coverage $(PbS)_{32}$ clusters on HOPG is shown in Fig. 12.

Even though, most of these clusters exhibit square/rectangular-like shapes consistent with the cuboid structures, none of the clusters had dimensions of the deposited $(PbS)_{32}$ clusters (0.88 nm × 0.88 nm × 0.88 nm). For example, in

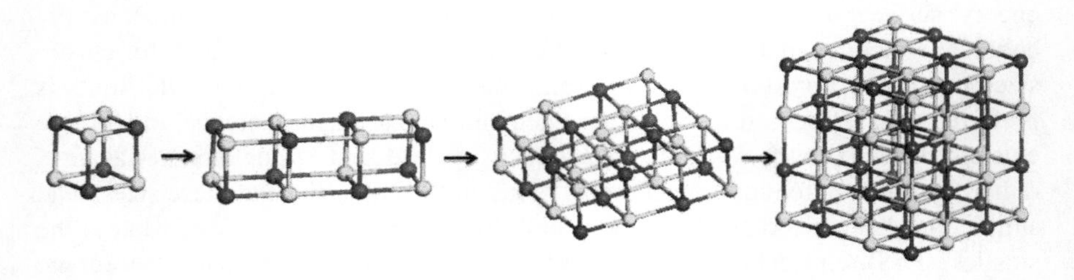

**Fig. 11** The dimerization of $(PbS)_{4n}$ units, starting from $(PbS)_4$ leading to $(PbS)_{32}$, baby crystal

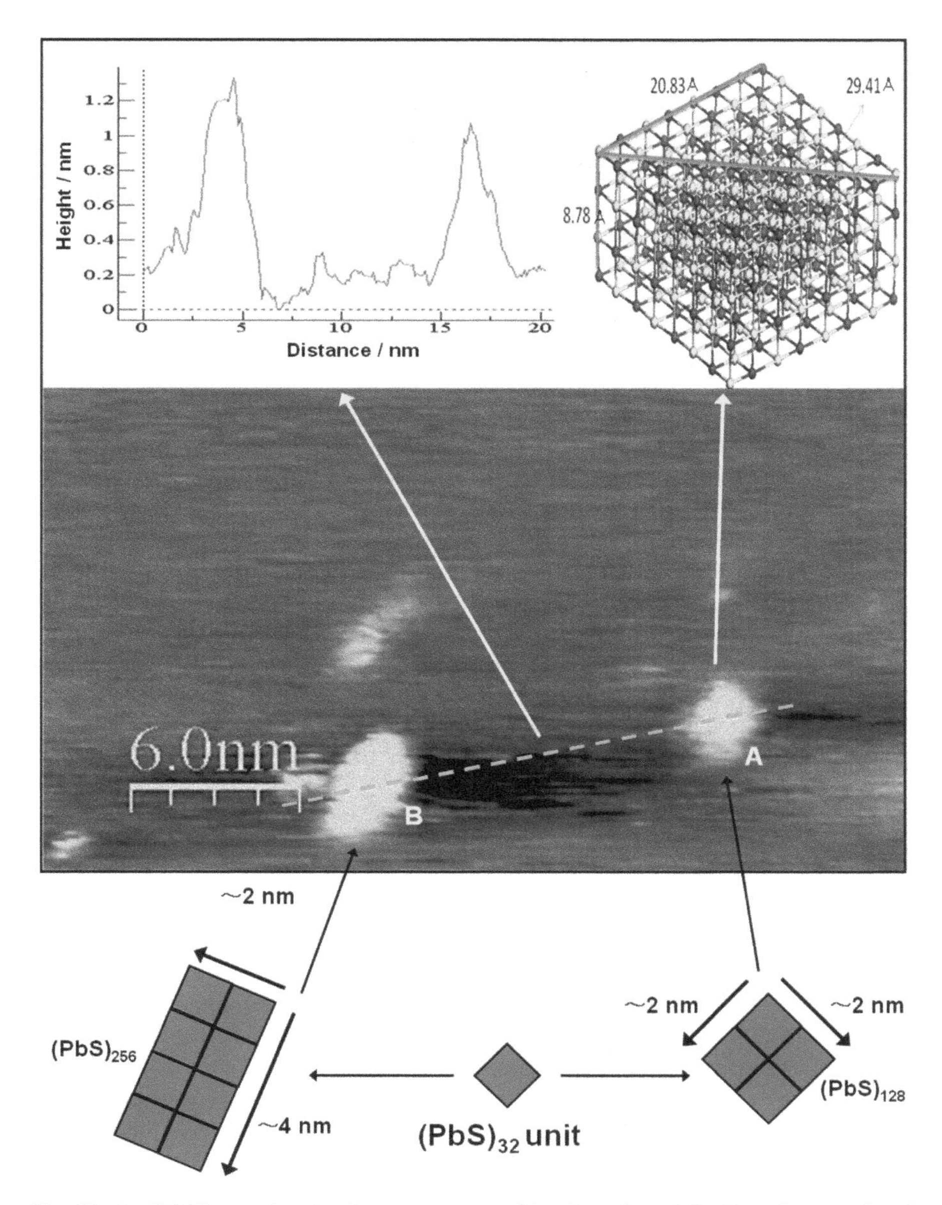


**Fig. 12** An STM image showing the aggregates resulting from the soft-landing of mass-selected $(PbS)_{32}$ onto a HOPG surface (Reproduced from [43])

Fig. 12, two aggregations with dimensions of $\sim$2 nm × 2 nm × 1 nm and a larger aggregation of 4 nm × 2 nm × 1 nm size are highlighted as position A and B (see highlighted clusters in Fig. 12). Since the deposited clusters can move on the substrate without a large diffusion energy barrier on HOPG, the highlighted images correspond to a random aggregation of $(PbS)_{32}$ clusters.

Further calculations, using the DMol3 program suite, showed that the dimer of $(PbS)_{32}$, i.e., the $(PbS)_{64}$ cluster has dimensions of 2.07 nm × 0.87 nm × 0.87 nm. The dimerization of $(PbS)_{64}$ along the width resulted in the required dimensions of 2 nm × 2 nm × 1 nm as observed in the STM image (position A in Fig. 12). Indeed, the calculated structure of $(PbS)_{128}$ has dimensions of 2.08 nm × 2.08 nm × 0.88 nm. This indicated that the aggregated clusters in the STM were all multiples of $(PbS)_{32}$ units. Thus, based on the dimensions of the aggregation, the nature of aggregation as well as the number of $(PbS)_{32}$ units involved in the aggregation were predicted. For example, the aggregation at position B in Fig. 12 corresponds to a *dimer* of $(PbS)_{128}$.

The calculated dimensions of the $(PbS)_{32}$ cluster, thus provided a rubric to understand the formation of 'nano-blocks' when these clusters were deposited on HOPG. The realization that clusters can exhibit bulk-like features was proposed before [60, 61], and $(PbS)_n$ clusters are not the only clusters thought to exhibit bulk-like structures at similar sizes. Calculations on $(MgO)_n$ and $(AgCl)_n$ clusters have also shown them to have bulk-like geometries with sixfold coordination for $n \geq 32$, [62, 63]. However, to our knowledge, only the growth pattern of cubic $(PbS)_n$ clusters predicted by computations have a direct experimental verification. This work underpins the importance of the synergy between well-defined computational studies and the experiments in the deposition of size-selected clusters, with precise composition and structure, on a suitable surface. Compared to the traditional synthetic methodologies, this procedure provides more control over the nature of the deposited cluster.

### B. $(PbSe)_n$ and $(PbTe)_n$ Clusters

There exist several theoretical studies focusing on the structural, electronic, and optical properties of bulk [64–67], nanocrystals [68], quantum dots [69–71] nanorods, and nanowires of lead-selenide (PbSe) [72]. Compared to the PbS clusters however, the structural evolution of PbSe *nanoclusters* has not been subject of many studies. Zeng et al. [46] reported the structural evolution and optical properties of $(PbSe)_n$ ($n$ = 1–10) clusters using first-principles molecular dynamics approach. The effects of spin-orbit coupling (SOC) corrections on the geometries, binding energies, and optical properties were also studied in this work. It was reported that the SOC correction exerts a significant influence on the binding energies, HOMO-LUMO gaps, and most importantly optical properties of $(PbSe)_n$ clusters, while it has no affect on the structures and relative stabilities. Recently, our group has carried out a systematic study of the structural evolution and stabilities of $(PbSe)_n$ ($n$ = 1–15, 16, 20, 24, 28, and 32) clusters using DMol3 program. In this study, PW91 functional along with the DNP/DSPP basis set were employed for all the calculations. Note that these computational parameters were also used in our earlier study on lead sulfide clusters [43]. The lowest energy isomers and higher energy isomers of $(PbSe)_n$ ($n$ = 1–10) clusters are shown in Fig. 13.

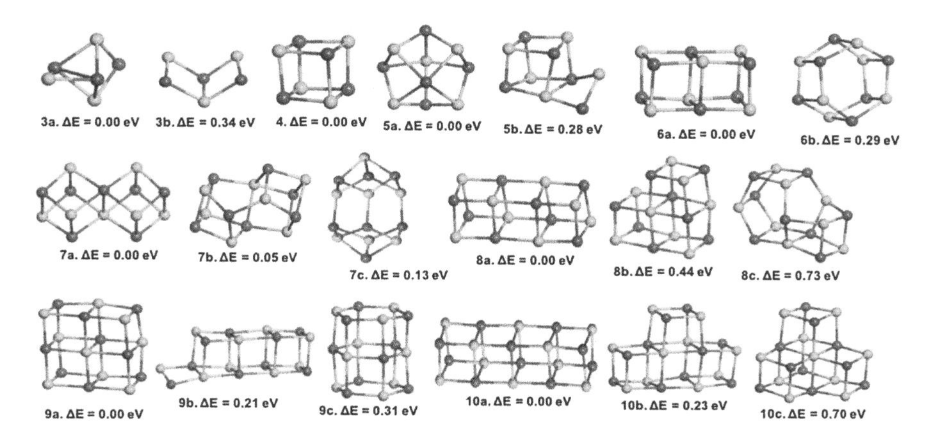


**Fig. 13** The lowest energy and higher energy isomers of $(PbSe)_n$ ($n$ = 1–10) clusters

The lowest energy structures of $(PbSe)_n$ ($n$ = 1–9) are similar to the structures preferred by the corresponding $(PbS)_n$ clusters with the cubic unit acting as a seed for clusters with size, $n > 5$. These structures are also in agreement with the previously reported structures of lead selenide [46], with an exception at $n = 7$. In our case, $(PbSe)_7$ has two iso-energetic structures: a symmetric structure with two edge-shared cubes and a distorted open cage like structure. However, Zeng et al. reported [46] only the open cage structure. In contrast to the $(PbS)_{10}$ cluster, which prefers to form a symmetric structure resulting from a *two-dimensional* stacking of cubic units, the $(PbSe)_{10}$ cluster prefers a *one-dimensional* stacking of cubic units, thereby forming a quadrangular prism structure. The structure resulting from two-dimensional stacking is 0.23 eV higher in energy. It is noteworthy here that, similar to the PbS cluster, the structure resulting from a three-dimensional stacking of cubes was not a preferred structure (ΔE = 0.70 eV). As expected, the PbSe clusters were more distorted compared to the PbS clusters due to a stronger surface relaxation in PbSe. The average Pb–Se bond length in these clusters was found to be 2.96 Å.

The lowest energy and higher energy isomers of $(PbSe)_n$ ($n$ = 11–15) clusters are displayed in Figs. 14 and 15. Interestingly, as the PbSe cluster size increased beyond $n = 10$, a competition between one-dimensional and two-dimensional stacking of face-sharing cubic units was observed at $n$ = 12 and 14. This trend is in contrast to the $(PbS)_n$ clusters, where the two-dimensional arrangement of face-sharing cubes was clearly the preferred growth pattern in the size range of $n$ = 10–15. On the other hand, following the trend that was observed in lead sulfide clusters, structures resulting from a three-dimensional stacking of cubic units were not preferred in $(PbSe)_n$ ($n$ = 11–15) clusters. The Pb–Se bond lengths in $(PbSe)_n$ ($n$ = 11–15) clusters varied from 2.70 to 3.35 Å, with an average bond length of 3.00 Å in most of these clusters. Note that the Pb–Se bond length in bulk PbSe was reported [73] to be 3.06 Å.

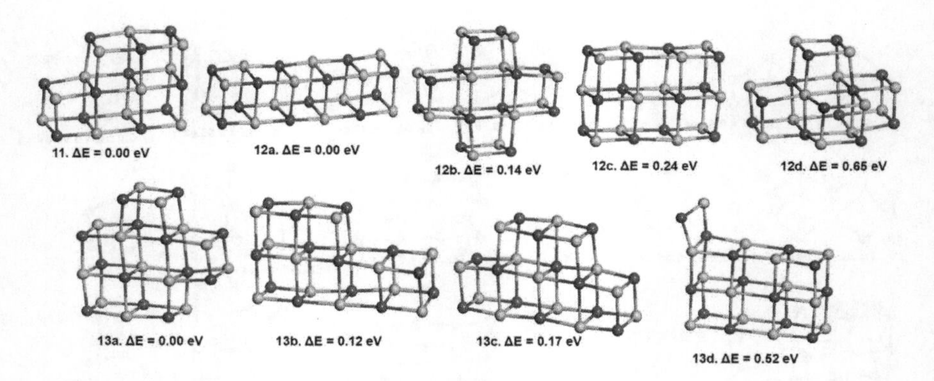


**Fig. 14** The lowest energy isomer and higher energy isomers of $(PbSe)_n$ ($n$ = 11–13) clusters

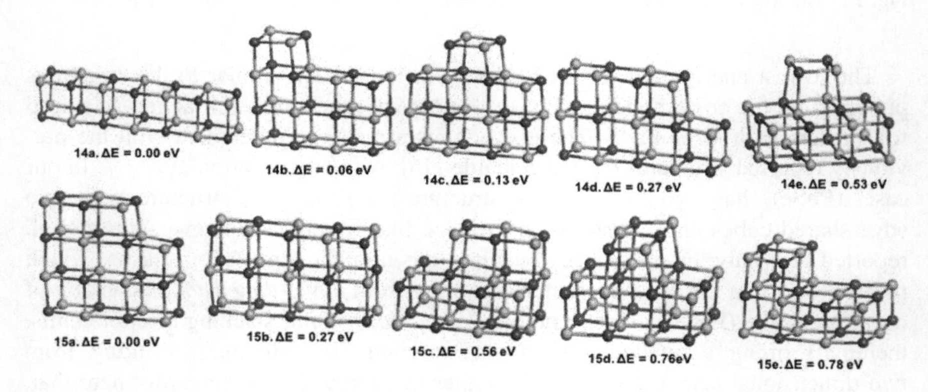


**Fig. 15** The lowest energy and higher energy isomers of $(PbSe)_n$ ($n$ = 14, 15) clusters

In order to understand the growth pattern on $(PbSe)_n$ clusters beyond $n$ = 15, we have adopted a strategy similar to as the one employed in $(PbS)_n$ clusters; namely, considering only $(PbSe)_{4n}$ clusters starting from $n$ = 4. The lowest energy and higher energy isomers of $(PbSe)_{4n}$ clusters, where $n$ = 4–6 and $n$ = 7–8 clusters are shown in Figs. 16 and 17, respectively.

Interestingly, the $(PbSe)_{24}$ cluster preferred to form a structure with a two-dimensional arrangement of face-sharing cubic units, in which the maximum coordination for Pb and Se atoms is five. This is in complete contrast to the trend observed in $(PbS)_{4n}$ clusters, where $(PbS)_{24}$ was the first cluster in which a six-fold coordination for Pb and S atoms was observed. This preference for two-dimensional stacking continued up to even $(PbSe)_{28}$ cluster, in which the structure containing the three-dimensional arrangement of cubes was found to be 0.36 eV higher in energy. Thus, in the case of lead-selenide clusters, the first instance of a six-fold coordination can be observed in $(PbSe)_{32}$ ($T_d$ symmetry), which is its *baby crystal*.

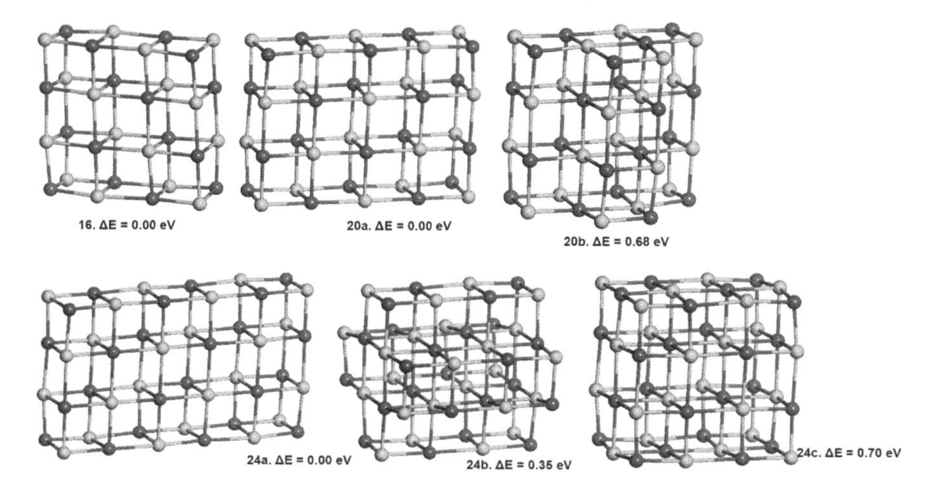


**Fig. 16** The lowest energy and higher energy isomers of $(PbSe)_{4n}$ ($n$ = 4–6) clusters

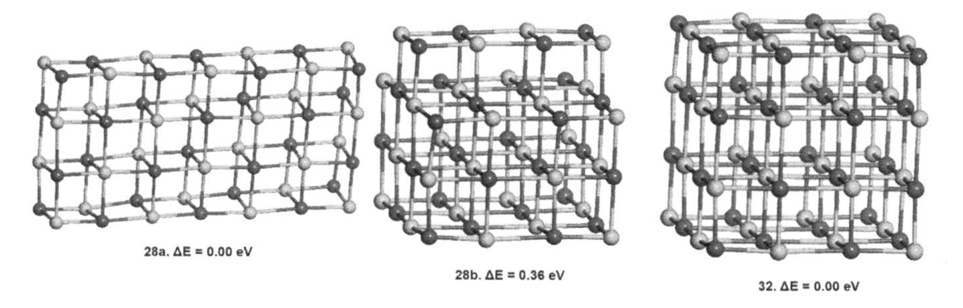


**Fig. 17** $(PbSe)_{28}$ and $(PbSe)_{32}$ clusters

Recently, the structural growth pattern of lead telluride (PbTe) clusters has been subject of two theoretical studies [47, 48]. In these studies, the structural properties of neutral $(PbTe)_n$ ($n$ = 1–45) clusters were studied using the local density approximation (LDA) for exchange and correlation functionals and ultra-soft pseudopotentials for electron-ion interactions. It was reported that the $(PbTe)_{24}$ cluster is the smallest cluster in which the central atoms have a six-fold coordination, while $(PbTe)_{32}$ is a perfect cubic cluster similar to that of the primitive cell of lead telluride crystals. It was also reported that $(PbTe)_{24}$, $(PbTe)_{32}$, and $(PbTe)_{40}$ clusters were more stable than their neighboring clusters, thus “magic clusters”. These studies on $(PbTe)_n$ were an imitation of our previous works [42, 43] on $(PbS)_n$ ($n$ = 1–32, 64, and 128) clusters, but using LDA instead of GGA density functionals. It is to be noted here that the reliability of PW91 functional and DNP/DSPP numerical basis sets for lead-chalcogenide clusters was well established in our studies on lead sulfide clusters, while the reliability of LDA functionals in estimating their stabilities is not clear. Moreover, it was shown in our lead sulfide work that the choice of the basis set is important in predicting the VDE and ADE values of these clusters. Therefore, as an extension up to our work on lead sulfide and lead-selenide clusters and to compare the similarities and differences in the growth pattern and stabilities of all of the lead-chalcogenide clusters at the same theoretical level, we have carried out a systematic study of the structural evolution and stabilities of neutral, stoichiometric $(PbTe)_n$ ($n$ = 1–16, 20, 24, 28, and 32) clusters using the PW91 functional with DNP/DSPP basis sets. The lowest energy structures of $(PbTe)_n$ ($n$ = 4–15) clusters obtained from our calculations are shown in Fig. 18. In contrast to the structural growth mechanism of $(PbS)_n$ clusters [42], where there is a clear preference for a two-dimensional stacking of face-shared cubes beyond $n$ = 10, in the case of $(PbTe)_n$ clusters for even values of $n$ ($n$ = 6, 8, 10, 12, and 14) structures containing one-dimensional stacking of face-shared cubic units is preferred, while for odd values of $n$ ($n$ = 9, 11, 13, and 15), two-dimensional arrangement of face-sharing cubes is preferred. Thus, the structure of a $(PbTe)_n$ ($n$ = 6–15) cluster can be built by adding $(PbTe)_2$ unit to $(PbTe)_{n-2}$ cluster. For instance, $(PbTe)_8$ cluster can be built from $(PbTe)_6$, while $(PbTe)_{11}$ cluster can be built from $(PbTe)_9$ cluster. Note that a similar trend was observed in $(PbSe)_n$ ($n$ = 4–15) as well. Compared to PbS and PbSe clusters, the PbTe clusters were more distorted with Pb–Te bond lengths ranging from 2.90 to 3.40 Å, depending on the size of the cluster. For example, a $(PbTe)_8$ cluster has three different Pb–Te bond lengths, with the largest being 3.44 Å between the $(PbTe)_4$ cubic units.

Following the same trend as $(PbSe)_{4n}$ clusters, $(PbTe)_{4n}$ ($n$ = 4–6) clusters preferred to form a structure with a two-dimensional arrangement of face-sharing cubic units, in which the maximum coordination for Pb and Te atoms is five (see Fig. 19). This preference for two-dimensional stacking continues to even $(PbTe)_{28}$ cluster, in

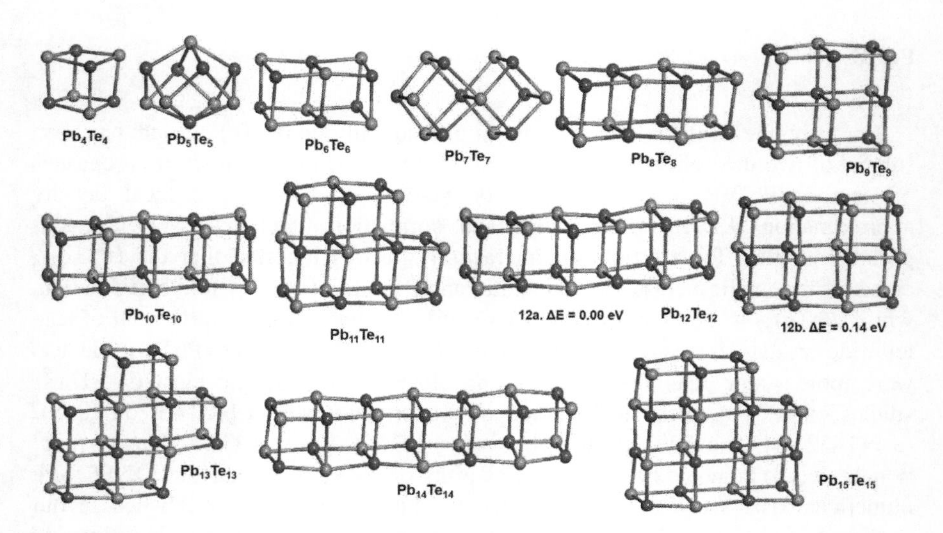


**Fig. 18** The lowest energy isomers of $(PbTe)_n$ ($n$ = 4–15) clusters

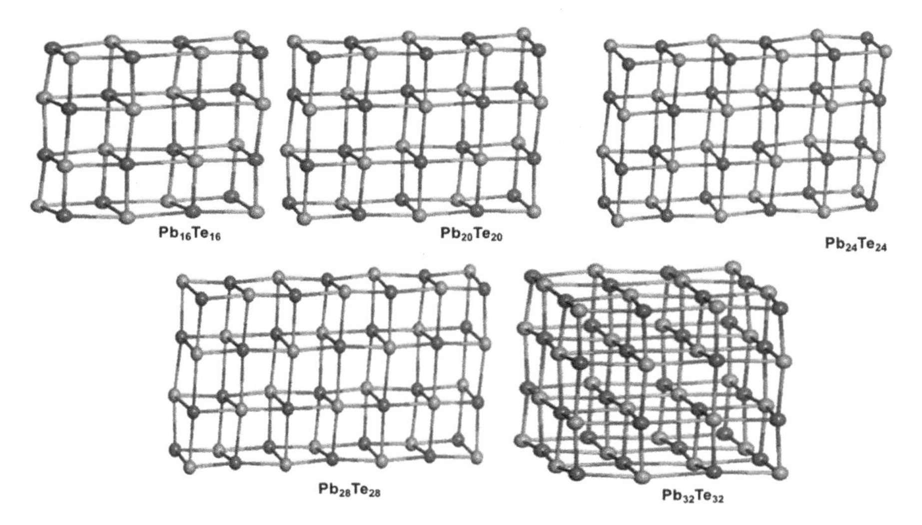


**Fig. 19** The lowest energy isomers of $(PbTe)_{4n}$ ($n$ = 4–8) clusters

which the structure containing the three-dimensional arrangement of cubes was found to be 0.26 eV higher in energy. Thus, in the case of lead-telluride clusters, the first instance of a six-fold coordination can be observed in $(PbTe)_{32}$ ($T_d$ symmetry).

## 3.2 *Stability and Electronic Structure*

The thermodynamic stabilities, variation in the binding energies, and HOMO-LUMO gaps of lead chalcogenide clusters as a function of the size of the cluster are reported in this section.

The binding energy per unit Pb$X$ ($X$ = S, Se, Te) of neutral $(PbX)_n$ ($n$ = 1–15) clusters is calculated from the following equation:

$$E_b = \frac{-\left[E(PbX)_n - nE(PbX)\right]}{n} \tag{1}$$

A plot of these $E_b$ values as a function of $n$ is shown in Fig. 20. The $E_b$ values of $(PbX)_n$ clusters have shown a dramatic increasing trend between $n$ = 2–4 values, with the $(PbS)_4$ cluster having the largest $E_b$ value (2.24 eV) among the $(PbX)_n$ ($n$ = 2–4; $X$ = S, Se, and Te) clusters. Beyond $n$ = 4 however, as the cluster size increased, the *change* in the binding energy values, $\Delta E_b$, was minimal. For example, as the cluster size increased from $n$ = 4 to $n$ = 15, the $E_b$ values of $(PbS)_n$ clusters changed from 2.24 eV in $(PbS)_4$ to 2.50 eV in $(PbS)_{15}$. Similarly, the maximum $\Delta E_b$ value for $(PbSe)_n$ and $(PbTe)_n$ clusters as the cluster size increased from $n$ = 4 to $n$ = 15 was less than 0.20 eV. The nearly saturated binding energies,

$E_b$, beyond $(PbX)_4$ cube can be traced back to the fact that structures of $(PbX)_n$ clusters for $n = (4–15)$ can be obtained either by one-dimensional or two-dimensional arrangements of $(PbX)_4$ cuboidal units. Among the lead-chalcogenide clusters considered in this study, $(PbS)_n$ clusters consistently have the higher $E_b$ values compared to the corresponding $(PbSe)_n$ and $(PbTe)_n$ clusters. This trend can possibly be due to weaker bonding between the $(PbX)_4$ cuboidal units in PbSe and PbTe clusters compared to that in PbS clusters. It is to be noted here that in an earlier work by Zeng et al., it was reported [46] that the binding energies (*BE*s) of $(PbSe)_n$ ($n$ = 1–10), calculated using various DFT functionals, showed a significant decrease (by 1.0–1.3 eV) when SO corrections were included in their calculations (see Fig. 21). The formula used to calculate the *BE*s by these authors is: $BE = E(\mathrm{Pb}) + E(\mathrm{Se}) – E(\mathrm{PbSe})_n/n$.

The thermodynamic stabilities of $(PbX)_n$ ($X$ = S, Se, Te) clusters against fragmentation into a monomer, dimer, and a tetramer were calculated using the following equations, respectively:

$$\text{Fragmentation losing Pb}X := -\left[E(\mathrm{Pb}X)_n - E(\mathrm{Pb}X)_{n-1} - E(\mathrm{Pb}X)\right], \quad (2)$$

$$\text{Fragmentation losing }(\mathrm{Pb}X)_2 := -\left[E(\mathrm{Pb}X)_n - E(\mathrm{Pb}X)_{n-2} - E(\mathrm{Pb}X)_2\right], \quad (3)$$

$$\text{Fragmentation losing }(\mathrm{Pb}X)_4 := -\left[E(\mathrm{Pb}X)_n - E(\mathrm{Pb}X)_{n-4} - E(\mathrm{Pb}X)_4\right]. \quad (4)$$

The ground state energies of both the parent and product clusters were considered in these calculations.

Fragmentation energies as a function of $n$ for all three of the fragmentation pathways is plotted in Fig. 22. Among the three pathways considered,

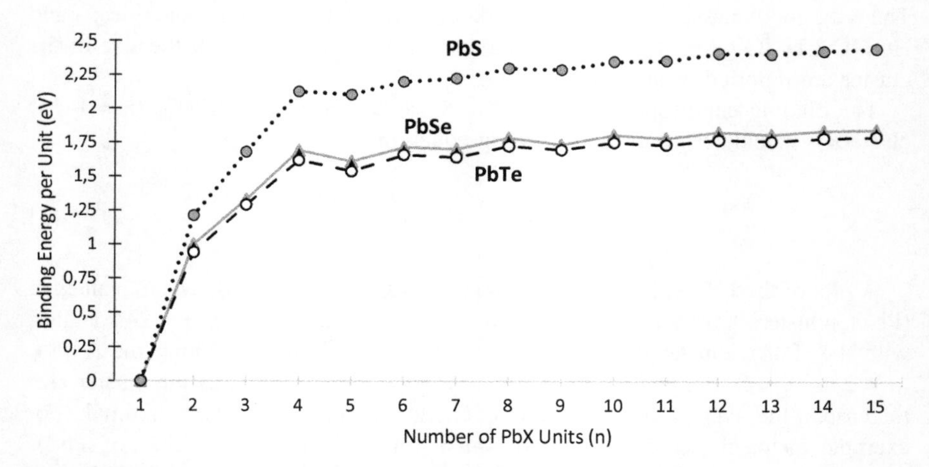


**Fig. 20** The binding energy per Pb$X$ ($X$ = S, Se, Te) unit of neutral $(PbX)_n$ clusters ($n$ = 1–15) as a function of $n$, the number of Pb$X$ units

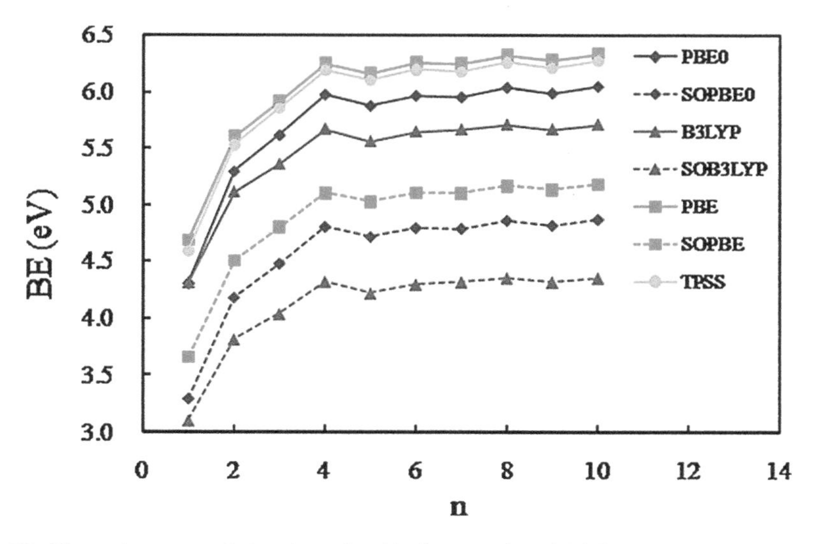


**Fig. 21** The exchange-correlation dependent binding energies of $(PbSe)_n$ ($n$ = 1–10) clusters as a function of $n$ (Reproduced from [46])

fragmentation into $(PbS)_4$ and $(PbS)_{n-4}$ product clusters is the most preferred pathway for all $(PbS)_n$ cluster. The fragmentation pathway #2, leading to PbS and $(PbS)_{n-1}$ products, exhibits an even-odd alteration, with even $n$ clusters being more stable than their odd $n$ neighbors. Among all the $(PbS)_n$ ($n$ = 2–15) clusters, the $(PbS)_8$ cluster was found to be the least stable one, with the energy required to fragment into two $(PbS)_4$ clusters being 1.39 eV. This is consistent with the fact that the $(PbS)_8$ cluster is made of two weakly bound $(PbS)_4$ clusters.

Fragmentation energies of PbSe and PbTe clusters for all three of the above equations are given in Figs. 23 and 24, respectively. Similar to the PbS fragmentation pathways, for PbSe and PbTe clusters, the fragmentation into $(PbX)_4$ and $(PbX)_{n-4}$ product clusters was found to be the least energy pathway, with $(PbX)_8$ ($X$ = Se and Te) requiring the least amount of energy to dissociate into two $(PbX)_4$ cubiod units.

Since in the case of smaller $(PbX)_n$ ($n \leq 15$) clusters the fragmentation into $(PbX)_4$ and $(PbX)_{n-4}$ products was the most preferred pathway, for larger clusters ($n$ = 16, 20, 24, 28, and 32), we calculated the fragmentation energy for only one pathway using the following expression:

$$\text{Energy to remove}(\text{Pb}X)_4 := -\left[E(\text{PbX})_{4n} - E(\text{PbX})_{4n-4} - E(\text{PbX})_4\right]. \quad (5)$$

Figure 25 shows these fragmentation energies as a function of the number of Pb$X$ units in a given cluster.

All $(PbX)_{4n}$ clusters exhibit a smooth increase in their stability against fragmentation into $(PbX)_4$ and $(PbX)_{4n-4}$ clusters up to $4n$ = 20. However, distinct

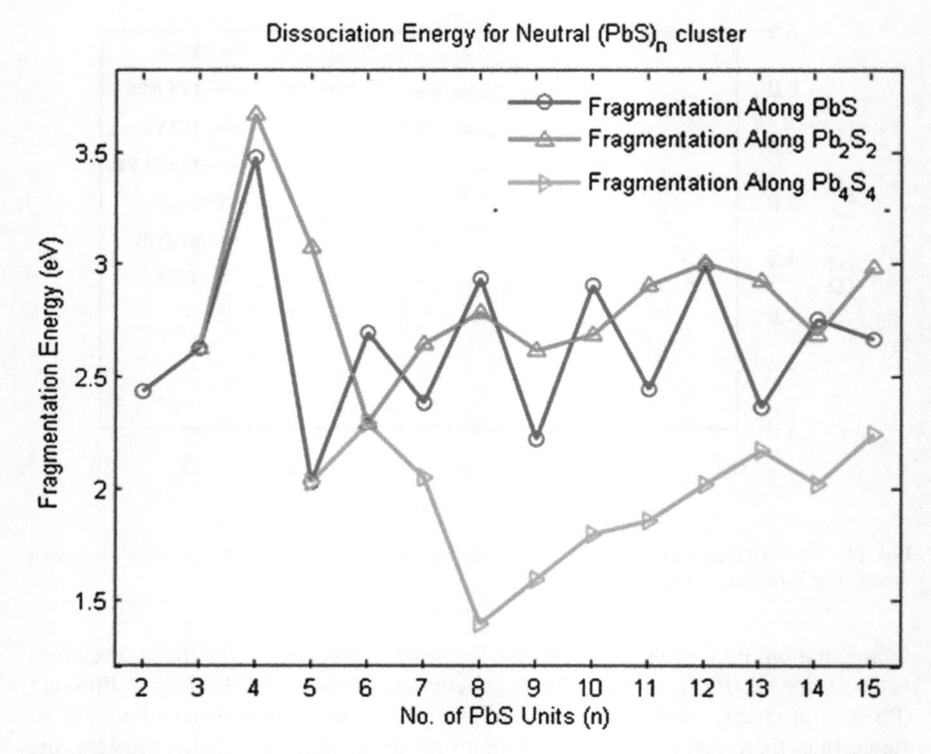


**Fig. 22** The fragmentation energies of $(PbS)_n$ ($n$ = 1–15) clusters as a function of the number of PbS units, $n$ (Reproduced from [42])

differences between PbS and the remaining chalcogenide clusters were observed beyond $4n = 20$. In the case of PbS clusters at $4n = 24$ there was a sharp increase in the fragmentation energy. This sudden increase in the stability of $(PbS)_{24}$ can be traced back to the first structural transformation from two-dimensional stacking to a three-dimensional stacking of the $(PbS)_4$ cuboids at this size. On the other hand, in PbSe and PbTe clusters, no such sudden increase was observed at $4n = 24$. This is again consistent with the fact that in $(PbX)_{4n}$ ($X$ = Se and Te) clusters the first instance of structural transformation from two-dimensional to three-dimensional arrangement of $(PbX)_4$ cuboids happened only at $4n = 32$. A second maximum (a sharp increase) for PbS clusters and the first sharp increase for PbSe and PbTe clusters was observed at the cluster size $4n = 32$, indicating the fact that the cubic $(PbX)_{32}$ structure is highly stable against fragmentation.

The electronic structure of lead-chalcogenide clusters was studied by calculating their HOMO-LUMO (*H-L*) gaps. Our calculated *H-L* gaps of $(PbX)_n$ ($X$ = S, Se, and Te; $n$ = 1–15) clusters are shown in Fig. 26. The *H-L* gaps of $(PbS)_n$ and $(PbSe)_n$ clusters exhibit an oscillating behavior with cluster size up to $n = 13$, with the even $n$ sizes in general having a larger *H-L* gap than their immediate odd $n$ size

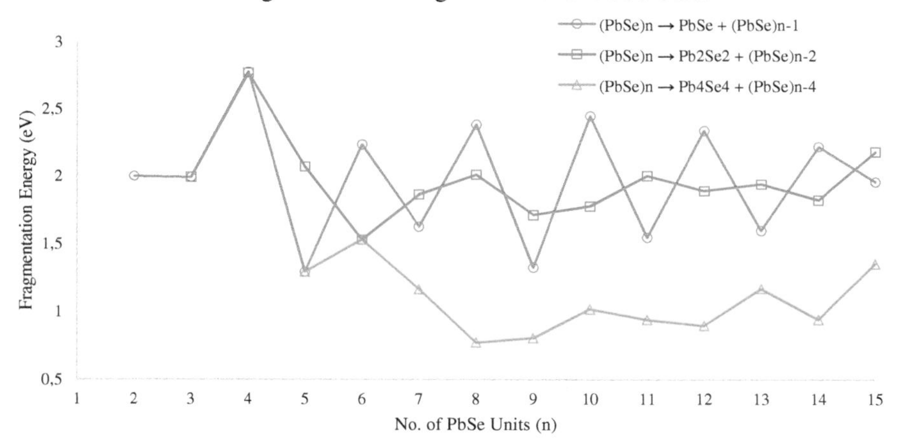


**Fig. 23** The fragmentation energies of $(PbSe)_n$ ($n$ = 1–15) clusters as a function of the number of PbSe units, $n$

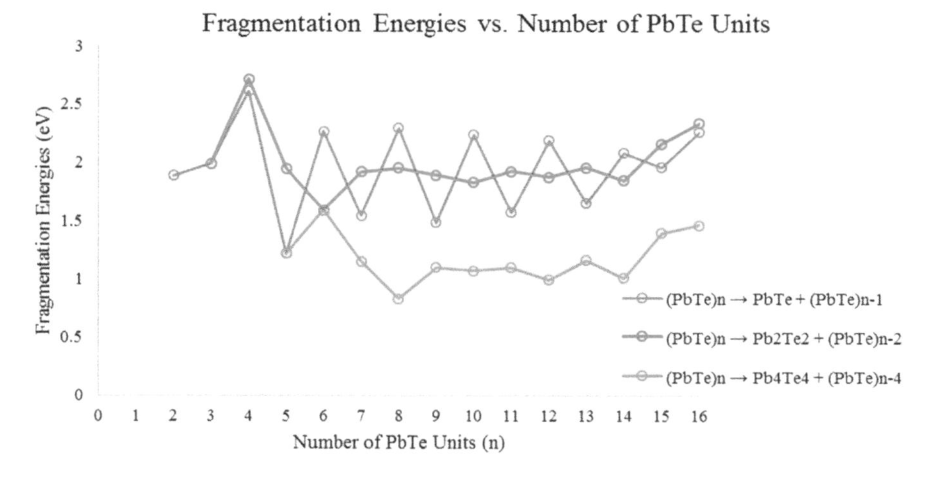


**Fig. 24** The fragmentation energies of $(PbTe)_n$ ($n$ = 1–15) clusters as a function of the number of PbTe units, $n$

neighbors. Interestingly, these variations decreased beyond $n = 9$. On the other hand, the *H-L* gaps of $(PbTe)_n$ clusters did not change significantly beyond $n = 5$, with a maximum change of 0.20 eV at $n = 12$. In addition, only $(PbTe)_{4,8,12}$ have higher HOMO-LUMO gaps than their odd numbered unit neighbors. In each of the $(PbX)_n$ ($n$ = 2–15) clusters, $(PbX)_4$ cuboid has the largest *H-L* gap with $(PbS)_4$: 2.96 eV, $(PbSe)_4$: 2.64 eV, and $(PbTe)_4$: 2.30 eV. It is to be noted here that Schelly and co-workers [13] (see Sect. 2.2 of this chapter) have reported a time-dependent,

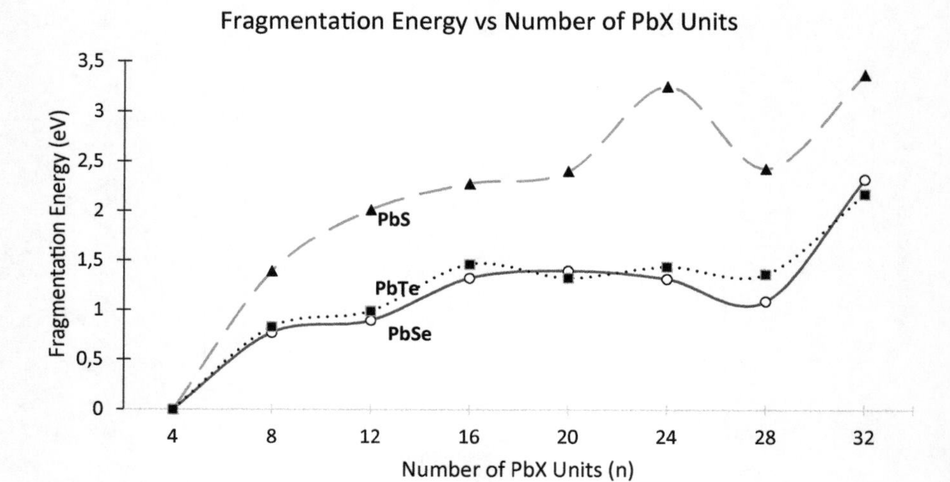


**Fig. 25** The fragmentation energies of $(PbX)_{4n}$ ($X$ = S, Se, Te), ($n$ = 1–8) clusters as a function of the number of Pb$X$ units, $n$

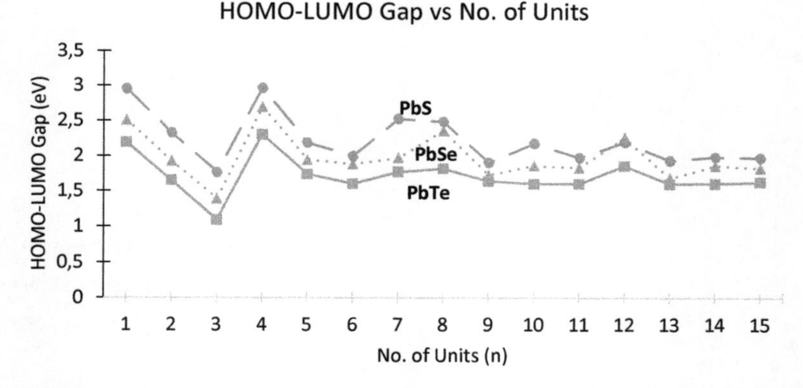


**Fig. 26** HOMO-LUMO gaps of $(PbX)_n$ ($n$ = 1–15; $X$ = S, Se, Te) clusters as a function of $n$, the number of Pb$X$ units

oscillating red and blue-shifts in the UV absorption bands of $(PbS)_n$ ($n$ = 1–9) clusters. Thus, the variation in the calculated *H-L* gaps of PbS clusters with increasing size is in qualitative agreement with the measured spectra.

As discussed above, Zeng et al. studied [46] the effects of including SO corrections on the electronic structure of $(PbSe)_n$ ($n$ = 1–10) clusters. The authors reported a significant change in the *H-L* gaps of PbSe clusters when the SO corrections were included in their calculations. Their calculated *H-L* gaps with different DFT functionals are shown in Fig. 27.

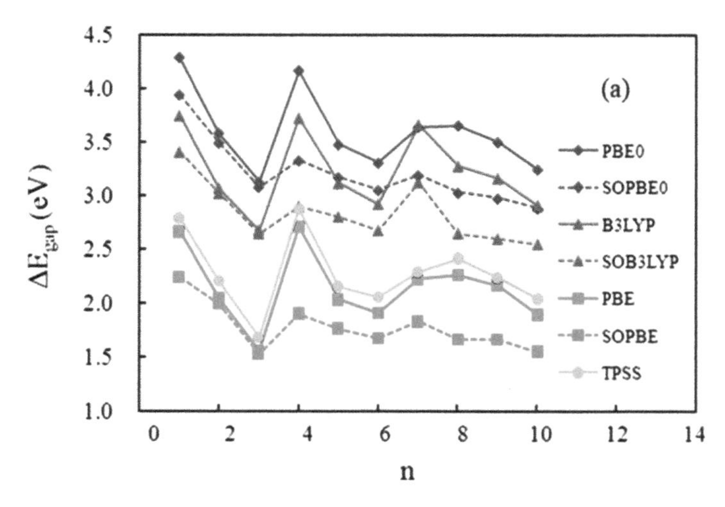


**Fig. 27** HOMO-LUMO gaps of $(PbSe)_n$ ($n$ = 1–10) clusters calculated using various exchange-correlation functionals (Reproduced from [46])

As expected, the calculated H-L gaps of a given cluster varied with the choice of the DFT functional, with PBE0 and B3LYP predicting a larger gap compared to TPSS and PBE (without any SO corrections). It is evident from the above figure that inclusion of SO corrections leads to significant decrease in the *H-L* gaps of PbSe clusters. For example, in the case of $(PbSe)_4$ cluster, the *H-L* gap decreased by 1.0 eV, irrespective of the DFT functional employed. These changes in the H-L gaps were reportedly due to the rearrangement of the energy levels due to SO corrections. The authors reported that the SO corrections led to the lowering of LUMO energy levels while having a minimal impact on the HOMO energy levels, thereby reducing the H-L gaps of all these clusters. Interestingly, the *H-L* gaps of PbSe clusters obtained from our calculations (computed at PW91/DNP, DSPP level) are similar to the *H-L* gaps obtained by Zeng et al. from the SO corrected PBE functional (SOPBE in Fig. 27). In addition, the authors have also concluded that the SO correction terms significantly impact the UV-vis spectra of these clusters.

## 4 Summary and Outlook

In this chapter, we have presented a review of past and recent experimental and computational studies on technologically important lead-chalcogenide nanoclusters. Various experimental works on the synthesis of these clusters, both in solution and gas-phase, were discussed. In addition, recent anion photoelectron spectroscopy studies of gas-phase lead-sulfide and size-selected lead sulfide nanoclusters deposited on the HOPG surface, carried out by our experimental collaborators were also discussed in detail. On the computational front, we focused on our recent work on the structural growth pattern of neutral $(PbX)_n$ ($X$ = S, Se, and Te; $n$ = 1–32) clusters. We have shown that the structural evolution of $(PbX)_n$ ($X$ = S, Se, and Te; $n$ = 1–32) clusters was mostly dominated by a two-dimensional arrangement of $(PbX)_4$ cuboids, up to $n$ = 24 for PbS, and up to $n$ = 32 for PbSe and PbTe clusters. Using computations, we have demonstrated that $(PbX)_{32}$ cluster is the smallest *cubic* cluster for which its inner $(PbX)_4$ unit has the bulk-like (six-fold) coordination, thus termed as "*baby crystal*". In addition, in the case of lead sulfide nanoclusters, we have demonstrated the significance of the collaboration between experimental and computational studies in understanding the mechanisms involved in the formation of lead sulfide nanoblocks and most likely bulk materials. The size dependent stabilities and HOMO-LUMO gaps of these clusters were also discussed in detail.

There are, however, several important issues that were not discussed in the current chapter. A recent computational study [46] has shown the importance of the spin-orbit coupling (SOC) effect in understanding the electronic structures of PbSe nanoclusters. This work leads us to a few unanswered questions, such as the SOC effects in PbS and PbTe nanoclusters. Another topic that needs further inquiry is the effect of dopant atoms and surface-passivation on the structural and electronic properties of lead-chalcogenide clusters in the size-ranges of $n$ = 4–32. In a recent computational study, Argeri et al., studied surface passivated PbSe nanoclusters by using large clusters ($Pb_{50}Se_{50}$) to model (100), (110), and (111) surfaces. However, as to our knowledge, there are no computational studies focusing on the ligated lead chalcogenide clusters in size-ranges of $n$ = 4–32. On the other hand, by doping bare lead chalcogenide nanoclusters with appropriate transition metal (*TM*) atoms, one can make these clusters magnetic. In addition, at appropriately large cluster sizes, one can dope these clusters with multiple *TM* atoms and study how the size of the cluster affects the magnetic coupling between the dopants. A computational study in this direction is currently underway in our research group.

**Acknowledgments** The authors acknowledge all students and collaborators who contributed to our original papers on the lead-sulfide nanoclusters: Pratik Koirala, Rameshu Rallabandi, Xiang Li, Xin Tang, Yi Wang, Gerd Gantefoer, and Kit H. Bowen. The authors (AFG, WHS, and AKK) are thankful to West Chester University, College of Arts and Sciences for supporting part of this work.

## References

1. Reed MA, Randall JN, Aggarwal RJ, Matyi RJ, Moore TM, Wetsel AE (1988) Observation of discrete electronic states in a zero-dimensional semiconductor nanostructure. Phys Rev Lett 60:535–537
2. Ekimov AI, Efros L, Onushchenko AA (1985) Quantum size effect in semiconductor microcrystals. Solid State Commun 56:921–924
3. Rossetti R, Nakahara S, Brus LE (1983) Quantum size effects in redox potentials, resonance Raman spectra, and electronic spectra of CdS crystallites in aqueous solutions. J Chem Phys 79:1086

4. Vossmeyer T, Katsikas L, Giersig M, Popovic IG, Diesner K, Chemseddine A, Eychmuller A, Weller H (1994) CdS nanoclusters: synthesis, characterization, size dependent oscillator strength, temperature shift of the exitonic transition energy, and reversible absorbance shift. J Phys Chem 98:7665–7673
5. Wang C, Wehrenberg BL, Woo CY, Guyot-Sionnest P (2004) Light emission and amplification in charged CdSe quantum dots. J Phys Chem B 108(26):9027–9031
6. Neeleshwar S, Chen CL, Tsai CB, Chen YY, Chen CC, Shyu SG, Seehra MS (2005) Size-dependent properties of CdSe quantum dots. Phys Rev B 71:201307(R)
7. Al-Salim N, Young AG, Tilley RD, McQuillan AJ, Xia J (2007) Synthesis of CdSe nanocrystals in coordinating and noncoordinating solvents: Solvent's role in evolution of the optical and structural properties. Chem Mater 19(21):5185–5193
8. Kilina SV, Kilin DS, Prezhdo OV (2009) Breaking the phonon bottleneck in PbSe and CdSe Quantum dots: Time-domain density functional theory of charge carried relaxation. ACS Nano 3(1):93–99
9. Kilina SV, Velizhabin KA, Ivanov S, Prezhdo OV, Tretiak S (2012) Surface Ligands increase photoexcitation relaxation rates in CdSe quantum dots. ACS Nano 6(7):6515–6524
10. Rosetti R, Hull R, Gilson JM, Brus LE (1985) Hybrid electronic properties between the molecular and solid state limits: lead sulfide and silver halide crystallites. J Chem Phys 83:1406
11. Nozik AJ, Williams F, Nenadovic MT, Rajh T, Micic OI (1985) Size quantization in small semiconductor particles. J Phys Chem 89:397–399
12. Nenadovic MT, Comor MI, Vasic V, Micic OI (1990) Transient bleaching of small PbS colloids. influence of surface properties. J Phys Chem 94:6390–6396
13. Wu S, Zeng H, Schelly ZA (2005) Preparation of ultrasmall, uncapped PbS quantum dots via electroporation of vesicles. Langmuir 21:686–691
14. Cao H, Wang G, Zhang S, Zhang X (2006) Growth and photoluminescence properties of PbS nanocubes. Nanotechnology 17:3280–3287
15. Peterson JJ, Krauss TD (2006) Fluorescence spectroscopy of single lead sulfide quantum dots. Nano Lett 6:510–514
16. Peterson JJ, Huang L, Delerue C, Allan G, Krauss TD (2007) Uncovering forbidden optical transitions in PbSe nanocrystals. Nano Lett 7(12):3827–3831
17. Moreels I, Fritzinger B, Martins JC, Hens Z (2008) Surface chemistry of Colloidal PbSe nanocrystals. J Am Chem Soc 130(45):15081–15086
18. Koupanou E, Ahualli S, Glatter O, Delgado A, Krumeich F, Leontidis E (2010) Stabilization of lead sulfide nanoparticles by polyamines in aqueous solutions. A structural study of dispersions. Langmuir 26(22):16909–16920
19. Cademartiri L, Montanari E, Calestani G, Migliori A, Guagliardi A, Ozin GA (2006) Size-dependent extinction coefficients of PbS quantum dots. J Am Chem Soc 128:10337
20. Sykora M, Koposov AY, McGuire JA, Schulze RK, Tretiak O, Pietryga JM, Klimov VI (2010) Effect of air exposure on surface properties, electronic structure, and carrier relaxation in PbSe nanocrystals. ACS Nano 4(4):2021–2034
21. Wang Y, Herron N (1987) Optical properties of cadmium sulfide and lead(II) sulfide clusters encapsulated in zeolites. J Phys Chem 91:257–260
22. Moller K, Bein T, Herron N, Mahler W, Wang Y (1989) Encapsulation of lead sulfide molecular clusters into solid matrixes. Structural analysis with x-ray absorption spectroscopy. Inorg Chem 28:2914–2919
23. Krauss TD, Wise FW (1997) Coherent acoustic phonons in a semiconductor quantum dot. Phys Rev Lett 79:5102
24. Okuno T, Lipovskii AA, Ogawa T, Amagai I, Masumoto Y (2000) Strong confinement of PbSe and PbS quantum dots. J Lumin 491:87–89
25. Bakueva L, Musikhin S, Hines MA, Chang TWF, Tzolov M, Scholes GD, Sargent EH (2003) Size-tunable infrared (1000–1600 nm) electroluminescence from PbS quantum-dot nanocrystals in a semiconducting polymer. Appl Phys Lett 82:2895–2897

26. McDonald SA, Konstantatos G, Zhang S, Cyr PW, Klem EJD, Levina L, Sargent EH (2005) Solution-processed PbS quantum dot infrared photodetectors and photovoltaics. Nat Mater 4:138–142
27. Deen GR, Hara M (2005) Preparation and Characterization of PbS nanoclusters made by using a powder method on ionomers. Polymer 46:10883
28. Na YJ, Kim HS, Park J (2008) Morphology-controlled lead selenide nanocrystals and their *insitu* growth on carbon nanotubes. J Phys Chem C 112(30):11218–11226
29. Wise FW (2000) Lead salt quantum dots: the limit of strong quantum confinement. Acc Chem Res 33:773–780
30. Kane RS, Cohen RE, Silbey R (1996) Theoretical study of the electronic structure of PbS nanoclusters. J Phys Chem 100:7928–7932
31. Kang K, Daneshvar K, Tsu R (2004) Size dependence saturation and absorption of PbS quantum dots. Microelectron J 35:629–633
32. Wundke K, Potting S, Auxier J, Schulzgen A, Peyghambarian N, Borrelli NF (2000) PbS quantum-dot-doped glasses for ultrashort-pulse generation. Appl Phys Lett 76:10–12
33. Brus LE (1986) Electronic wave functions in semiconductor clusters: experiment and theory. J Phys Chem 90:2555–2560
34. Marino CP, Guerin JD, Nixon ER (1974) Infrared spectra of some matrix-isolated germanium, tin, and lead chalcogenides. J Mol Spectrosc 51:160–165
35. Teichman RA, Nixon ER (1975) Vibronic spectra of matrix-isolated lead sulfide. J Mol Spectrosc 54:78–86
36. Colin R, Drowart J (1962) Thermodynamic study of tin sulfide and lead sulfide using a mass spectrometer. J Chem Phys 37:1120–1125
37. Saito Y, Mihama K, Noda T (1983) Binary compound clusters formed by free jet expansion. Jpn J Appl Phys 22:L179
38. Saito Y, Suzuki M, Noda T, Mihama K (1985) Study on fragmentation of ionized atom-clusters by time-of-flight mass spectrometry. Jpn J Appl Phys 25:L627
39. Fancher CA, de Clercq HL, Bowen KH (2002) Photoelectron spectroscopy of PbS. Chem Phys Lett 366:197–199
40. Zeng H, Schelly ZA, Ueno-Noto K, Marynick DS (2005) Density functional study of the structures of lead sulfide clusters $(PbS)_n$ $(n = 1 - 9)$. J Phys Chem A 109:1616–1620
41. He J, Liu C, Li F, Sa R, Wu K (2008) Size-dependence of stability and optical properties of lead sulfide clusters. Chem Phys Lett 457:163–168
42. Koirala P, Kiran B, Kandalam AK, Fancher CA, de Clercq HL, Bowen KH (2011) Structural evolution and stabilities of neutral and anionic clusters of lead sulfide: joint anion photoelectron and computational studies. J Chem Phys 135:134311
43. Kiran B, Kandalam AK, Rallabandi R, Koirala P, Li X, Tang X, Wang Y, Fairbrother H, Gantefoer G, Bowen K (2012) $(PbS)_{32}$: a baby crystal. J Chem Phys 136:024317
44. Gupta SK, He H, Banyai D, Kandalam AK, Pandey R (2013) Electron tunneling characteristics of a cubic quantum dot, $(PbS)_{32}$. J Chem Phys 139:244307
45. Argeri M, Fraccarollo A, Grassi F, Marchese L, Cossi M (2011) Density Functional Theory modeling of PbSe nanoclusters: effect of surface passivation on shape and composition. J Phys Chem C 115:11382–11389
46. Zeng Q, Shi J, Yang M, Wang F, Chen J (2013) Structures and optical absorptions of PbSe clusters from ab initio calculations. J Chem Phys 139:094305
47. Mulugeta Y, Woldeghebriel H (2014) Structural evolution and stabilities of $(PbTe)_n$ $(n = 1$–$20)$ clusters. Comput Theor Chem 1039:40
48. Mulugeta Y, Woldeghebriel H (2014) Size effects on the structural and electronic properties of lead telluride clusters. J Quant Chem, Int. doi:10.1002/qua.24817
49. Rosetti R, Hull R, Gibson JM, Brus LR (1985) Hybrid Electronic properties between the molecular and solid state limits: lead sulfide and silver halide crystallites. J Chem Phys 83:1406–1410
50. Wang Y, Suna A, Mahler W, Kasowski R (1987) PbS in polymers. From molecules to bulk solids. J Chem Phys 87:7315–7322

51. Sankaran Y, Cummins CC, Schrock RR, Cohen RE, Silbey RJ (1990) Small PbS clusters prepared via ROMP block copolymer technology. J Am Chem Soc 112:6858–6859
52. Kane RS, Cohen RE, Silbey R (1996) Synthesis of PbS nanoclusters within block copolymer nanoreactors. Chem Mater 8:1919–1924
53. Yue J, Cohen RE (1994) Nanoreactors for inorganic cluster synthesis. Supramol Sci 1:117–120
54. Claridge SA, Castleman AW, Khanna SN, Murray CB, Sen A, Weiss PS (2009) Cluster-assembled materials. ACS Nano 3:244–255
55. Noda Y, Masumoto K, Ohba S, Saito Y, Toriumi K, Iwata Y, Shibuya I (1987) Temperature dependence of atomic thermal parameters of lead chalcogenides, PbS, PbSe, and PbTe. Acta Cryst C 43:1443
56. Delley B (1990) An all-electron numerical method for solving the local density functional for polyatomic molecules. J Chem Phys 92:508
57. Delley B (2000) From molecules to solids with $DMol^3$ approach. J Chem Phys 113:7756 ($DMol^3$ is available as part of Materials Studio)
58. Perdew JP (1991) Generalized gradient approximations for exchange and correlation: A look backward and forward. Physica B 172:1–6
59. Delley B (2002) Hardness conserving semilocal pseudopotentials. Phys Rev B 66:155125
60. Sun Q, Rao BK, Jena P, Stolcic Kim Y, Gantefor G, Castleman A (2004) Appearance of bulk properties in small tungsten oxide clusters. J. Chem. Phys. 121:9417–9422
61. Catlow CRA, Bromley ST, Hamad S, Mora-Fonz M, Sokol AA, Woodley SM (2010) Phys Chem Chem Phys 12:786
62. Veliah S, Pandey R, Li YS, Newsam JM, Vessal B (1995) Density functional study of structural and electronic properties of cube-like MgO clusters. Chem Phys Lett 235(1–2):53–57
63. Glaus S, Calzaferri G (1999) Silver chloride clusters and surface states. J Phys Chem B 103:5622
64. Wei SH, Zunger A (1997) Electronic and structural anomalies in lead chalcogenides. Phys Rev B 55:13605
65. Albanesi EA, Okoye CMI, Rodrigues CO, Peltzer y Blanca EL, Petukhov AG (2000) Electronic structure, structural properties, and dielectric functions of IV-VI semiconductors: PbSe and PbTe. Phys Rev B 66:16589
66. Hummer K, Gruneis A, Kresse G (2007) Structural and electronic properties of lead chalcogenides from first principles. Phys Rev B 75:195211
67. Ekuma CE, Singh DJ, Jarrell M (2012) Optical properties of PbTe and PbSe. Phys Rev B 85:085205
68. Franceschett A (2008) Structural and electronic properties of PbSe nanocrystals from first principles. Phys Rev B 78:075418
69. An JM, Franceschetti A, Dudiy SV, Zunger A (2006) The peculiar electronic structure of PbSe quantum dots. Nano Lett 6:2728–2735
70. Kamisaka H, Kilina SV, Yamashita K, Prezhdo OV (2008) Ab initio study of temperature and pressure dependence of energy and phonon-induced dephasing of electronic excitations in CdSe and PbSe quantum dots. J Phys Chem C 112:7800–7808
71. Gai Y, Peng H, Li J (2009) Electronic properties of nonstoichiometric PbSe quantum dots from first principles. J Phys Chem C 113:21506
72. Bartnik AC, Efros Al L, Koh W-K, Murray CB, Wise FW (2010) Electronic states and optical properties of PbSe nanorods and nanowires. Phys Rev B 82:195313
73. Davlen (1969) A review of the semiconductor properties of PbTe, PbSe, PbS, and PbO. Infrared Phys 9:141

# Chemical Reactivity and Catalytic Properties of Binary Gold Clusters: Atom by Atom Tuning in a Gas Phase Approach

**Sandra M. Lang and Thorsten M. Bernhardt**

**Abstract** Industrial heterogeneous catalysts are complex multi-component systems which typically contain different transition metal particles supported on porous materials. For the future design of new tailor-made catalytic materials, a molecular level insight into the reaction mechanisms, energetics, and kinetics of the catalytic processes are mandatory. Furthermore, the detailed investigation of the nature of the interaction between different elements in alloy materials and their influence on the catalytic properties is essential. Free clusters in the gas phase represent simplified but suitable model systems which allow to obtain insight into catalytic processes on a rigorously molecular level. In this chapter we summarize experimental and theoretical studies on the reactivity and catalytic activity of free gold clusters and the change of their chemical properties caused by doping these clusters with transition metal atoms. In particular, we focus on three selected catalytic reactions, the oxidation of carbon monoxide, the conversion of methane, and the coupling of methane and ammonia, which have all been shown to be catalyzed by small binary gold clusters.

**Keywords** Binary gold cluster · Gas phase · CO oxidation · Activation of oxygen · Activation of methane · C–N coupling

## 1 Introduction

Industrial heterogeneous catalysts are complex multi-component systems which usually comprise different transition metal particles supported on porous materials of metal-, alkaline earth metal-, aluminum- or silicon-oxides. For example, the three-way catalyst for car exhaust conversion consists of an alumina washcoat

S.M. Lang (✉) · T.M. Bernhardt
Institute of Surface Chemistry and Catalysis, Ulm University, Ulm, Germany
e-mail: sandra.lang@uni-ulm.de



M.T. Nguyen and B. Kiran (eds.), *Clusters*, Challenges and Advances in Computational Chemistry and Physics 23,
DOI 10.1007/978-3-319-48918-6_10

typically decorated with palladium, platinum, and rhodium nanoparticles [1]. In contrast, industrial catalysts for the production of acrylonitrile via oxidation of propylene contain several transition metals such as nickel, cobalt, iron, and molybdenum supported on silica [1]. These examples demonstrate that the elemental composition is essential for the activity and selectivity of catalytic materials.

The simplest models of such multi-component systems are bimetallic (or binary) model catalysts. The interaction of the different metals in bimetallic systems are usually discussed in terms of four different concepts [2]: (1) the "ligand effect", i.e. a change in the electronic structure due to the electronic interaction between the different materials; (2) the "ensemble effect" which describes the blocking of active sites which may increase the selectivity for certain reaction channels; (3) the "geometric" or "strain effect" caused by a the change of the interatomic distance in bimetallic systems; (4) the "bifunctional mechanism" which describes the ability of different neighboring metal atoms to activate different molecules and thus supply the precursors for a catalytic reaction.

In order to design new tailor-made binary or even multi-component catalytic materials or to optimize the properties of existing catalytic materials, a molecular level insight into the reaction mechanism, energetics, and kinetics of the catalytic processes are mandatory. Furthermore, detailed knowledge of the nature of the interaction between different elements in alloy materials and their influence on the catalytic properties is essential. However, the detailed investigation of the catalytic processes on an atomic and molecular level is often impeded by the high complexity of the catalytic processes in conjunction with the high complexity of the heterogeneous catalysts. Thus, to nevertheless gain an essential understanding, model systems must be developed which are on the one hand simple enough to be understood in detail and on the other hand are able to mimic the essential features of the real catalysts.

The most simple but highly suitable model systems to gain insight on a rigorous molecular level are small free clusters in the gas phase. Of course due to the small size and the charge (in most cases singly charged anionic or cationic clusters are investigated) as well as the isolated nature (i.e. the absence of counterions, solvation, or supports) of such free clusters, these model systems cannot account for a number of complex processes occurring in catalytic processes on solid surfaces or in solution [3]. Nevertheless, gas phase studies on isolated cluster ions provide ideal means to model the catalytically active centers, i.e. the location where the actual chemical transformations involved in the catalytic reactions take place. Such active sites are usually characterized by a spatial extension of a few atoms and may involve defect sites, under-coordinated atoms, or unsaturated bonds, which may be appropriately modeled with the use of small clusters [4].

Hence, gas phase studies of free clusters can provide an attractive complementary approach to condensed phase model systems and deliver fundamental conceptual insight into, e.g., reactivity patterns, the nature of reaction intermediates, aspects of the electronic structure, charge transfer processes, but also basic parameters which determine the catalytic properties such as particle size, charge state, and in particular the particle elemental composition. In this respect the combination of gas phase experiments with theoretical simulations have been proven to be an especially powerful approach to identify factors that are of fundamental importance to distinct catalytic systems [4] and which might guide the development of new advanced low temperature catalyst materials.

Although the chemical and catalytic properties of small gas phase clusters represent an active field of research since the advent of intense cluster sources, both experimental and theoretical studies on binary metal clusters are still rather scarce. After the presentation of experimental techniques typically employed in gas phase reactivity studies (Sect. 2) the following chapter focuses on three catalytic reactions: the oxidation of CO (Sect. 3.1), the conversion of $CH_4$ (Sect. 3.2), and the coupling of $CH_4$ and $NH_3$ (Sect. 3.3). All three catalytic reactions have been shown to be catalyzed by small binary gold clusters in the gas phase at thermal conditions. Each section starts with a short introduction into the relevance and particular problems of the catalytic reaction at hand before the reactivity and catalytic properties of pure gold clusters in the gas phase are presented. Finally, the change of these properties by doping the gold clusters with other transition metals is discussed. For the sake of completeness, a short overview of other studied reactions is given in the last section (Sect. 3.4).

# 2 Experimental Techniques of Gas Phase Reactivity Studies

To gain insight into the reactive and catalytic properties of free clusters a variety of experimental techniques have been adapted or newly developed. Figure 1 displays a scheme of a simple generalized setup for reactivity and catalysis studies in the gas phase which typically consists of three regions for (1) production, (2) reaction, and (3) analysis. First the (metal) clusters are produced in a cluster source which might be followed by a mass filter in order to select a specific cluster size for further studies. Reactivity and catalysis studies are then performed in a reaction cell which acts as a test tube for gas phase reactions. The formed reaction intermediates and final products can then be analyzed in a mass filter in conjunction with an ion detector.

For analysis usually standard mass spectrometric methods such as time-of-flight, quadrupole, or ion cyclotron resonance mass spectrometers are used in combination with electron multipliers. These methods have been described in detail previously (see for example references [5, 6]) and will not be further addressed here. Methods for cluster production, in particular of binary clusters, as well as for reaction and catalysis studies will be described in the following subsections.

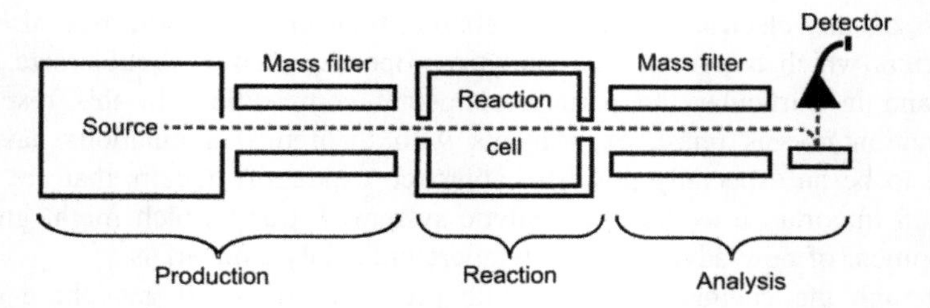


**Fig. 1** Scheme of a typical setup for reactivity and catalysis studies in the gas phase

## 2.1 *Production of Metal Clusters*

Small metal clusters comprising only a few (up to a few tens of) atoms are usually produced by one of two fundamentally different methods: (1) sputtering of surfaces or (2) gas condensation (ablation-thermalization principle) [6, 7].

In a sputter source, a primary ion beam (usually of rare gases) with an energy of up to some keV is focused onto a metal target which leads to the ejection of neutral and charged clusters. Figure 2a displays a schematic drawing of a typical sputter source [7, 8]. Here the primary ion beams (indicated by the red bold line) are generated by ionization with electrons which are ejected from resistively heated filaments. The ions are then extracted, accelerated, and focused onto four metal targets by a series of three cylindrical lenses. This leads to the production and ejection of neutral and charged clusters. However, due to the direct ejection of the clusters without subsequent aggregation this cluster source is limited to the production of very small clusters comprising only a few atoms as illustrated in Fig. 2b by the example of cationic gold clusters.

In contrast, in all cluster sources working according to the ablation-thermalization principle (gas condensation), the metal of interest is evaporated into a buffer gas, which cools the initially hot atoms until they condense and form clusters. The mixture of buffer gas and clusters of different sizes then expands into the vacuum. The different cluster sources working according to this principle mainly differ by the method how the metal vapor is generated and by the pressure of the buffer gas which determines, if the expansion is supersonic or not. Typically, the metal vapor is generated by means of intense laser pulses [9], electrical discharges [10], or magnetron discharges [11]. Due to the cluster formation via gas condensation, the cluster size is usually adjustable and considerably larger clusters can be produced than in a pure sputter source.

All these methods have been developed, optimized, and used for the production of monometallic clusters for more than two decades. However, the combination of two metals in order to produce binary clusters is more challenging and requires in some cases even a modification of the common cluster sources. The simplest way to

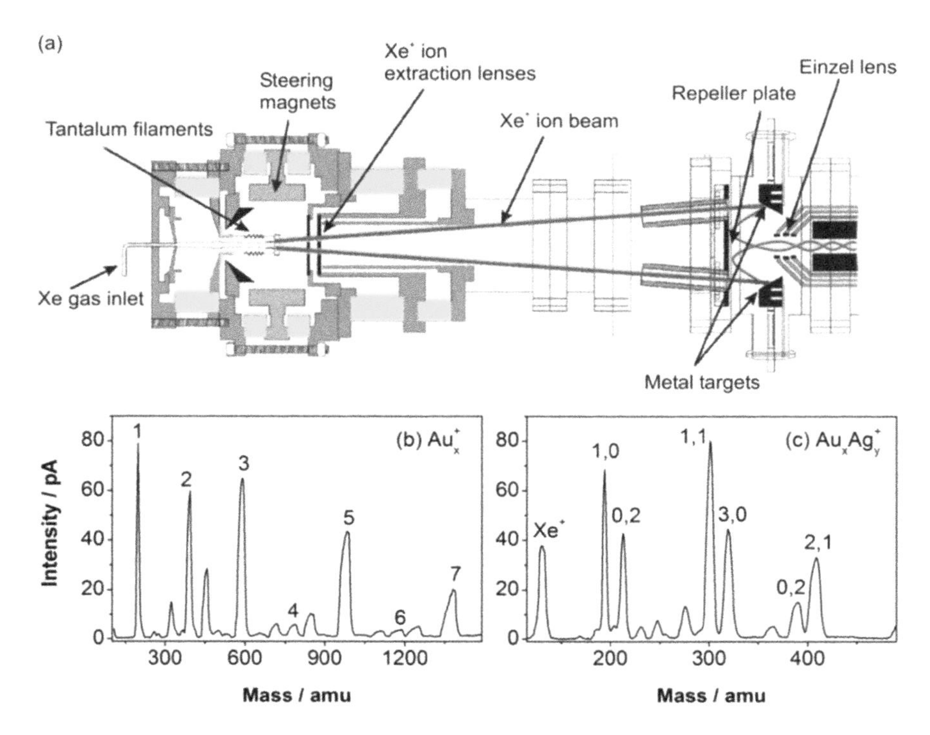


**Fig. 2** **a** Schematic drawing of a cold reflex discharge ion source (CORDIS) with adjacent sputter region [7, 8] and typical cluster distributions obtained by sputtering **b** pure gold and **c** binary gold-silver targets in this sputter source. Reprinted from Int. J. Mass Spectrom. 387, 56 (2015), Copyright (2015), with permission from Elsevier

produce binary clusters is the use of binary metal targets. Figure 2c displays as an example a mass distribution of binary gold–silver clusters generated in a sputter source by using a mixed gold–silver target. However, by using mixed targets the possible compositions of the produced clusters are usually determined by the composition of the targets. Thus, to cover the whole composition range of interest different targets must be used which can make this method extremely expensive.

A very innovative cluster source offering the highest possible flexibility for cluster size and composition has been developed by Lievens et al. [12]. This cluster source is based on the principle of a laser vaporization source [9], i.e. the metal vapor is generated by an intense laser pulse and clusters are formed in an helium atmosphere and cooled in a cluster formation region with subsequent supersonic expansion. For controlled production of binary clusters this source has been converted into a dual-target dual-laser source as schematically shown in Fig. 3a. Two rectangular targets of different metals are placed next to each other and are irradiated by the beams of two different lasers. The produced metal vapors are cooled by pulses of He gas and the binary clusters are formed in the cluster formation chamber prior to supersonic expansion into the vacuum. Thereby the production of

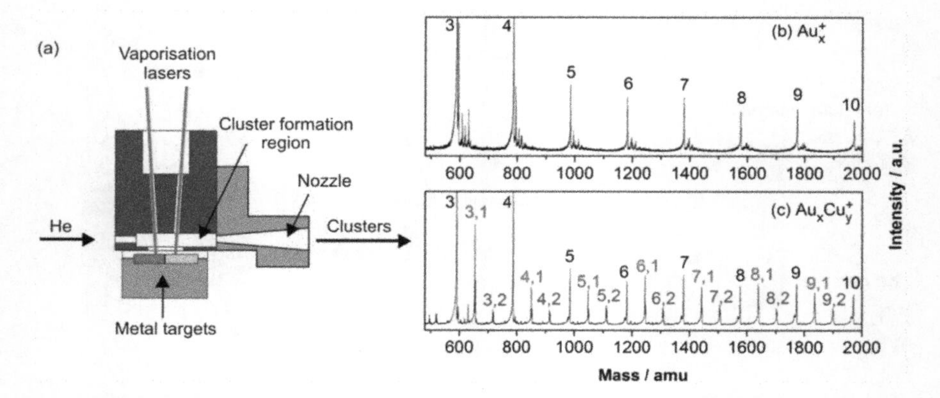


**Fig. 3** **a** Schematic drawing of a dual-target dual laser evaporation source [12] and typical cluster distributions obtained by laser ablating **b** pure gold and **c** a combination of gold/copper targets in this cluster source

the bimetallic clusters is achieved due to the spatial and temporal overlap of the two plumes of the vaporized materials. The use of two different ablation lasers allows for the precise adjustment of the delay time between the material plume formation which has been determined to represent the critical parameter for bimetallic cluster formation. With this dual-target dual-laser vaporization source not only the size range of the produced cluster can be adjusted but also the controlled atom by atom doping of the clusters is possible. As an example Fig. 3b and c display a section of a typical gold cluster distribution (Fig. 3b) in comparison with a cluster distribution obtained by using a combination of a gold and a cooper target (Fig. 3c).

An alternative approach to produce binary or even multi-component clusters according to the ablation-thermalization principle has been developed by Yasumatsu et al. [13]. In this source, an array of magnetron sputtering sources is mounted inside a gas-aggregation cell. Each magnetron sputtering source evaporates a different material and the ejected atoms and ions aggregate into cluster ions via collisions with buffer gas leading to the production of multi-component clusters.

## 2.2 *Techniques for Reactivity Studies*

Reaction cells represent the test tubes for gas phase investigations of the reactivity and catalytic properties of metal clusters. During the last decades several different types of reaction cells have been developed that permit the investigation of (catalytic) reactions at pressure conditions and reaction times covering a wide range of several orders of magnitude.

Figure 4a displays an illustration of typical reaction time and pressure ranges applied in the most commonly used reaction cells: flow tube reactors [14, 15], collision cells [16, 17], drift cells [18], ion guides [19], linear ion traps [20], and

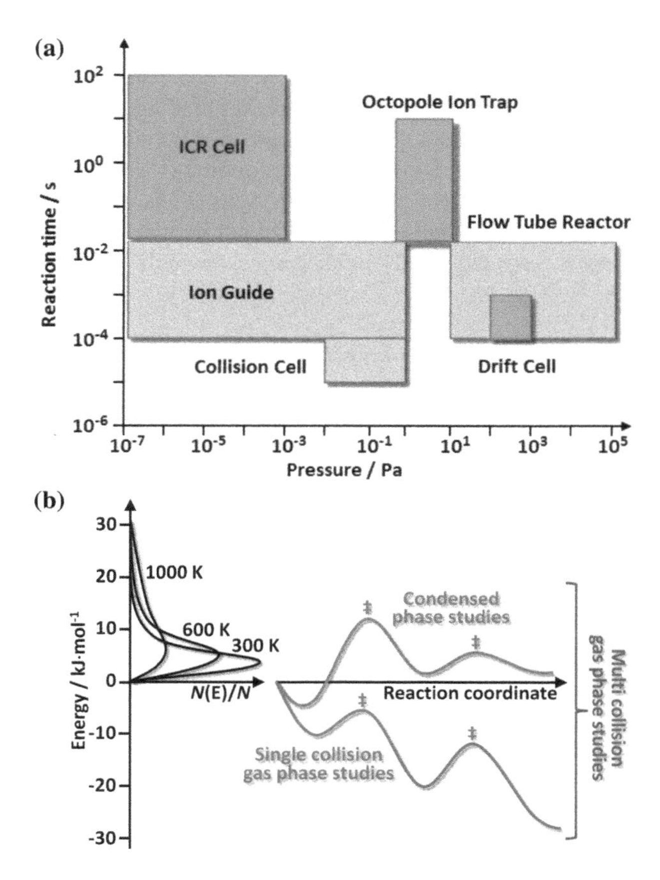


**Fig. 4** **a** Compilation of typical reaction time and pressure values applied in different reaction cells. Methods operated at a fixed reaction time (determined by the geometry of the setup) are indicated by gray boxes while methods which allow for the variation of the reaction time are indicated by blue boxes. **b** *Left side* Relative number of particles N(E)/N as a function of the energy following a Boltzmann distribution. *Right side* Possible reaction pathways for gas phase and condensed phase studies [4]. Reproduced from Ref. [4] with permission from the PCCP Owner Societies

ICR (ion cyclotron resonance) cells [21, 22]. The fact, if the different techniques allow for the variation of the reaction time is indicated by the different colors. The variation of the pressure over several orders of magnitude for different methods arises from the fact that some reaction cells are operated in the presence of reactive gases only while others require the addition of inert buffer gas (typically helium) which results in considerable differences in the reaction conditions varying from single and few collision conditions to multi-collision conditions.

Closely related to the pressure and collision conditions is the possible energy transfer between the clusters and the neutral molecules present in the reaction cell [4].

The reaction of a cluster with a neutral molecule first leads to an encounter complex with a certain internal energy (internal energy of the reactants before reaction, potential energy gained in forming the complex, i.e. binding energy and the energy contained in the translational and rotational degrees of freedom which is converted into internal energy upon reaction). This internal energy defines the total energy available for all subsequent chemical transformations. Thus, at single collision conditions (on average one collision is possible on the time scale of the experiment) all possible reactions must energetically be directed "downhill" as shown by the blue potential profile in Fig. 4b. Such experimental conditions have to be termed micro-canonical [23].

In contrast, under multi-collision conditions (typically $>10^4$ collisions per second) energy can be dissipated or provided by collisions with buffer gas atoms or also other reactants. Thus, the available energy is no longer determined by the energy gained upon reaction but is determined by the Boltzmann distribution of particle energies (displayed on the left side of Fig. 4b for different temperatures). Such energy transfer is similar to the canonical conditions under which also condensed phase reactions occur (cf. red potential profile in Fig. 4b). As a consequence, collisions with He atoms of the high energy tail of the Boltzmann distribution [23] can provide enough energy to overcome reaction barriers above the initial reactant energy. Thus, also activated reactions become accessible to gas phase model investigations if they are performed at multi-collision conditions, which is extremely important for the investigation of catalytic reactions. In this case temperature dependent studies can provide insight into complex thermal catalytic cycles and the corresponding energetics [4].

# 3 Tuning the Reactive and Catalytic Properties of Binary Gold Clusters

## *3.1 Catalytic CO Oxidation*

The catalytic oxidation of carbon monoxide CO

$$CO + \frac{1}{2}O_2 \rightarrow CO_2 \tag{1}$$

is of utmost scientific and industrial interest and thus represents one of the most widely studied reactions in catalysis research. The industrial interest originates from the serious environmental problem of toxic CO, which is produced in large amounts during (petrochemical) industrial processing as well as in automobiles. In addition to the industrial relevance, the scientific interest also arises from the apparent simplicity of this reaction which makes it an ideal model reaction to gain a basic understanding of heterogeneous catalysis.

Although CO oxidation catalysts have already been established for several decades the optimization of these catalysts and the design of new materials are still subject to intensive research activities. The aim of most of these activities is the development of low temperature catalytic materials in order to significantly reduce the typically high operation temperature of the commercially employed catalysts. Such materials can on the one hand make the catalytic conversion more energy efficient and can on the other hand also avoid the inactivity of the catalysts in the so called "start-up" period [24] which is particularly important for applications in car exhaust systems.

Typically, the overall reaction (1) is described by a Langmuir-Hinshelwood type reaction mechanism involving the following elementary reaction steps

$$O_{2,gas} \rightarrow 2\,O_{ad} \tag{2a}$$

$$CO_{gas} \rightarrow CO_{ad} \tag{2b}$$

$$O_{ad} + CO_{ad} \rightarrow CO_{2,ad} \tag{2c}$$

$$CO_{2,ad} \rightarrow CO_{2,gas} \tag{2d}$$

This mechanism involves the dissociative adsorption of gaseous $O_{2,gas}$ (reaction 2a) and the adsorption of gaseous $CO_{gas}$ on the catalyst (reaction 2b) as well as the subsequent formation (reaction 2c) and desorption (reaction 2d) of $CO_{2,ad}$ into the gas phase. Previous studies on palladium model systems identified two of these reaction steps to determine the efficiency of the reaction [25, 26]: (1) the dissociative chemisorption of oxygen on the catalyst surface which requires (without a catalyst) an energy of 498 kJ/mol [27] to break the strong O=O double bond (reaction 2a) and (2) the adsorption strength of carbon monoxide (reaction 2b) [26]. In particular, the usually strong binding of CO was found to determine the high reaction temperature since it causes a fast covering of the catalyst surface with CO which inhibits $O_2$ coadsorption at low temperatures and thus leads to catalyst poisoning [26, 28]. Consequently, the rate limiting reaction step is believed to be the desorption of CO molecules to liberate $O_2$ adsorption sites which in turn determines the rate of oxygen coadsorption [29].

For a long time gold has been considered as an uninteresting material for catalysis due to the weak binding of small molecules and the typically high activation energy for their dissociation on gold surfaces [30]. Only in the late 1980s the interest in gold as catalytic material started when Haruta et al. reported the high catalytic activity of oxide supported gold nanoparticles in the oxidation of CO at temperatures as low as −70 °C [31]. Since then gold and binary gold nanoparticle based catalysis, with a particular focus on oxidation catalysis, has evolved into a very active field of research (see for example [32]). Apart from condensed phase model system studies also numerous investigations with gold clusters in the gas phase have been performed to gain insight into the elementary reaction steps and their energetics.

### 3.1.1 CO Oxidation Mediated by Gold Clusters

As outlined above, the reaction steps which typically determine the activity and operation temperature of a CO oxidation catalyst are the (dissociative) adsorption of $O_2$ (reaction 2a) and the desorption of CO (reverse of reaction 2b). Thus, in the following these two reaction steps will be discussed separately before summarizing the results for CO oxidation.

**Adsorption and activation of $O_2$** The reaction between neutral and charged (cationic and anionic) gold clusters and molecular oxygen at thermal conditions has been experimentally studied by several groups [33–42].

Cationic gold clusters $Au_x^+$ were found to be completely non-reactive toward $O_2$ at thermal condition [33, 34, 40, 41]. Similarly, anionic gold clusters $Au_x^-$ containing an odd number of atoms appear to be non-reactive while, in marked contrast, even sized negatively charged gold clusters do adsorb $O_2$ [33, 34, 37–39]. This odd–even alternation in the reactivity is displayed in Fig. 5 (black squares) together with the vertical electron detachment energies (VDE) of the gold clusters (red circles) [43] which exhibit opposite odd–even variations, i.e. clusters with low VDE are those with high reactivity toward $O_2$. This anti-correlation can be understood qualitatively by considering the frontier orbital interactions: since gold exhibits an [Xe]$5d^{10}6s^1$ electron configuration, singly negatively charge clusters $Au_x^-$ consisting of an even number x of atoms have one unpaired valence electron in their highest occupied molecular orbital (HOMO) and thus exhibit a low VDE. This electron can be easily donated to the oxygen molecule upon formation of the complex $Au_xO_2^-$ (x = even) yielding a high reactivity of even x clusters toward $O_2$. In contrast, $Au_x^-$ consisting of an odd number of atoms comprise an even number,

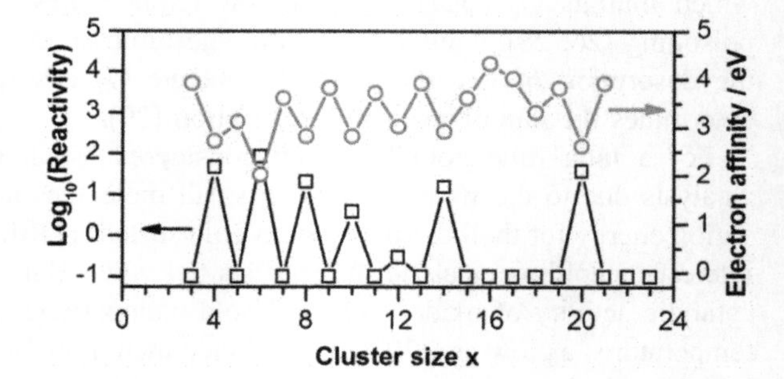


**Fig. 5** Cluster size dependent reactivity of anionic gold clusters toward $O_2$ (*black squares*) [33] and corresponding vertical electron detachment energies of the clusters (*red circles*) [43]. Reprinted from Int. J. Mass Spectrom. 243, 1 (2005), Copyright (2005), with permission from Elsevier

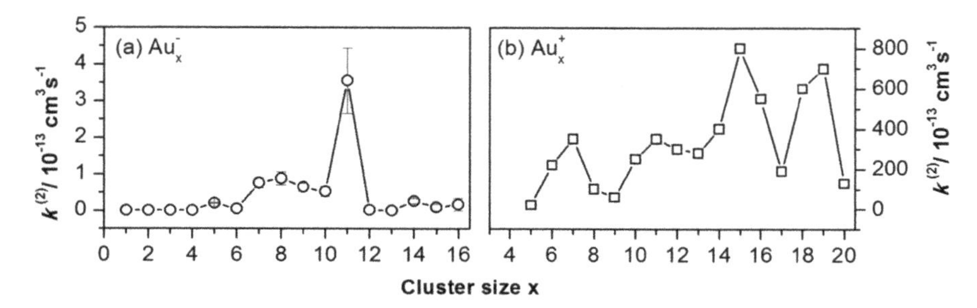


**Fig. 6** Cluster size dependent bimolecular rate constants $k^{(2)}$ for the adsorption of a first CO molecule on **a** anionic [21] and **b** cationic [22] gold clusters

spin paired valence electron structure with high VDE. Consequently, the reaction with $O_2$ leading to $Au_xO_2^-$ (x = odd) is hampered or even inhibited. Most importantly, photoelectron [35] and infrared spectroscopic [42] investigations revealed that $O_2$ does not dissociate on the clusters but binds molecularly while, however, the oxygen molecule is activated and represents a superoxo ($O_2^-$) moiety [42]. Furthermore, both end-on binding and bridging coordination of the $O_2$ molecule have been observed in these studies.

**Adsorption of CO** The adsorption of CO on small gold clusters in the gas phase has been investigated in several experimental [17, 21, 22, 39, 44–51] and theoretical studies [50, 52–56]. Most of the cationic, neutral, and anionic gold clusters (with some exceptions as will be discussed below) react with CO. However, the reactivity of cationic clusters is significantly higher than the reactivity of anionic clusters. This charge state dependence is reflected in the reaction rate constants for the adsorption of a first CO molecule [21, 22] as shown in Fig. 6 as well as also in the cluster-CO bond strength [50, 52, 56].

The binding and the capability of the gold clusters to activate the adsorbed CO molecule are determined by the balance between molecule-to-cluster charge donation and cluster-to-molecule charge back-donation as well as electrostatic effects. The intermolecular CO bond is strengthened by charge donation from the CO antibonding $2s\sigma^*$ orbital to the cluster and weakened by charge back-donation from the cluster into the antibonding $2p\pi^*$ orbital of CO. Thus, adsorption of CO on anionic clusters leads to a significant weakening of the CO bond compared to adsorption on cations [46] which is reflected in a red-shift of the CO stretch frequency when going from the positively charged to the neutral and the negatively charged gold clusters [45, 46, 52, 56]. However, theoretical simulations revealed that the intramolecular C–O bond is only slightly elongated (typically by less than 5 % compared to the free molecule [52]) which indicates that small gold clusters cannot considerably activate or even dissociate CO molecules.

Figure 6 shows the cluster size dependent rate constants for the adsorption of a first CO molecule on anionic (Fig. 6a) and cationic (Fig. 6b) gold clusters. Although strong size effects are apparent for both charge states, a clear pattern as for the reactivity with $O_2$ (odd–even alternation) is not recognizable. This indicates that the reactivity toward CO does not correlate with the cluster electron affinity or HOMO/LUMO energies due to the mutual molecule-metal charge donation and back-donation effects. Consequently, the cluster size dependent reactivity toward CO cannot be explained in terms of simple frontier orbital theory. Instead, several different effects have been discussed in literature such as electron shell closings [17, 21, 22, 39, 48], the symmetry of the delocalized valence orbitals and orbital overlap [17, 53], the geometry of individual clusters and the coordination number of the atoms in the cluster [17, 21, 55], as well as the available degrees of freedom in the cluster [39, 47].

**Low temperature CO oxidation** The catalytic activity of free gold clusters has first been predicted theoretically by Häkkinen and Landman in 2001 [57]. In this study the authors showed that an isolated negatively charged gold anion $Au_2^-$ can mediate the CO oxidation reaction. Most importantly, oxygen was found to bind molecularly on $Au_2^-$ and a carbonate complex $Au_2CO_3^-$ had been identified as the most likely reaction intermediate. In 2003 Bernhardt et al. then confirmed experimentally the catalytic activity of the free gold dimer [58].

Reactivity studies in an octopole ion trap revealed that $Au_2^-$ does not react with CO at room temperature but adsorbs one oxygen molecule yielding $Au_2O_2^-$ [38, 47, 58]. Adding a mixture of CO and $O_2$ to the ion trap at room temperature does not change the ion mass distribution and $Au_2O_2^-$ is still detected as the only reaction product (cf. Fig. 7a) [38, 58]. However, the kinetic data (Fig. 7b) show that in the presence of both CO and $O_2$ $Au_2^-$ is no longer completely transformed into $Au_2O_2^-$, but an offset appears in the gold cluster concentration at longer reaction times. This indicates that CO must be involved in the reaction although no reaction products containing CO can be detected at this reaction temperature.

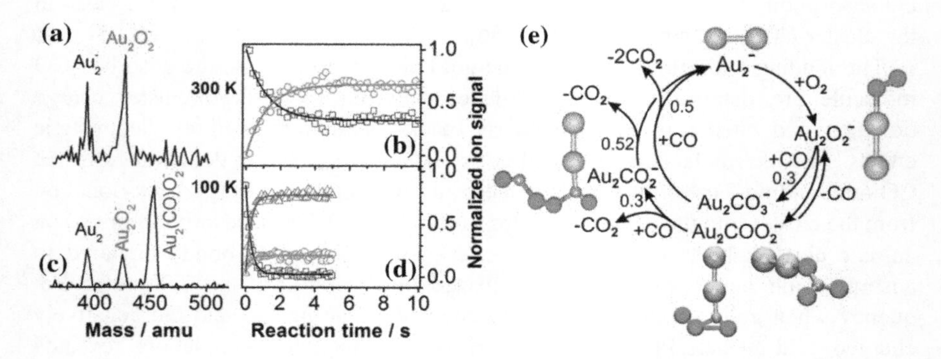


**Fig. 7** Ion mass distributions and kinetic data (*open symbols*) obtained after reaction of $Au_2^-$ with a $CO/O_2$ mixture at **a**, **b** 300 K and **c**, **d** 100 K. The *solid lines* are obtained by fitting the integrated rate equations of the catalytic reaction mechanism to the experimental data. **e** Schematic representation of the reaction mechanism together with calculated structures of the intermediates. Au, O, and C atoms are represented by *yellow*, *red*, and *grey* spheres, respectively [20, 58]. (a–d) Reprinted with permission from J. Am.Chem. Soc. 125, 10437 (2003). Copyright (2003) American Chemical Society. (e) Reprinted from Int. J. Mass Spectrom. 243, 1 (2005), Copyright (2005), with permission from Elsevier

To gain more insight into the reaction mechanism and the involvement of CO the experiments have been repeated at lower temperatures [38, 58]. Figure 7c displays a mass spectrum obtained after reaction of $Au_2^-$ with a mixture of CO and $O_2$ at 100 K. At this low reaction temperature the dimer forms the dioxide $Au_2O_2^-$ (which is also observed at room temperature) as well as an additional product corresponding to the stoichiometry $Au_2(CO)O_2^-$ which clearly shows the co-adsorption of CO and $O_2$ on the cluster. This complex is identical to the key intermediate predicted in the earlier theoretical study for catalytic CO oxidation by Häkkinen and Landman [57]. Furthermore, the kinetic data (cf. Fig. 7d) reveal that at this low temperature the $Au_2^-$ concentration strongly decreases at longer reaction times and that the products $Au_2O_2^-$ and $Au_2(CO)O_2^-$ are in equilibrium with each other.

Based on a series of pressure and temperature dependent experimental data the most simple reaction mechanism which is able to fit all the kinetic data has been determined. This mechanism is shown in Fig. 7e. Since $Au_2^-$ is non-reactive toward CO, the first reaction step must be represented by the adsorption of one oxygen molecule. $Au_2O_2^-$ then reacts with CO to form $Au_2(CO)O_2^-$ which will either re-dissociate to the oxide or further react with a second CO molecule to re-form $Au_2^-$ while liberating two $CO_2$ molecules.

Figure 7e also shows the geometric structures of the reaction intermediates calculated by Landman et al. [57, 58]. $O_2$ binds to $Au_2^-$ in an end-on configuration which leads to a partial charge transfer of 0.4 e to the oxygen molecule transforming $O_2$ into a superoxo-like species. This complex is then able to coadsorb one CO molecule via direct interaction of CO with the pre-adsorbed oxygen molecule yielding either a peroxyformate complex $Au_2COO_2^-$ which contains a reacted O–O–C–O group (*cf.* left structure in the lower middle of Fig. 7e) or a carbonate complex $Au_2CO_3^-$ (*cf.* right structure in the lower middle of Fig. 7e). In the final reaction step this intermediate reacts with a second CO molecule before two $CO_2$ are released. The formation of $CO_2$ on the peroxyformate structure involves two activation barriers of 0.3 and 0.52 eV which are associated with the transition state structure $Au_2CO_3CO^-$ (shown on the left side of Fig. 7e). Finally, the reaction proceeds via formation of $Au_2CO_2^-$ and a neutral $CO_2$ and the elimination of the second $CO_2$, respectively. In contrast, the carbonate complex $Au_2CO_3^-$ is predicted to react with CO under simultaneous elimination of both $CO_2$ molecules which requires an energy of 0.5 eV. It should be noted that the intermediate $Au_2CO_2^-$ has only been theoretically predicted but has not been observed in this experimental study. However, a similar intermediate $Au_6CO_2^-$ has been directly observed in an independent gas phase investigation performed in a fast flow reactor [15].

As discussed above the activity and operation temperature of a catalyst is usually determined by the ability of the catalyst to dissociatively adsorb $O_2$ and the bond strength of CO. Most importantly, the experimental and theoretical investigations of free $Au_2^-$ showed that molecular oxygen does not dissociate on small gold clusters but is instead adsorbed in an activated superoxo-like state. Nevertheless, catalytic CO oxidation is possible and the formation of the super-oxo-like oxygen species opens an alternative reaction mechanism with a peroxy-formate complex $Au_2COO_2^-$ or a carbonate complex $Au_2CO_3^-$ as reaction intermediates. Furthermore, $Au_2^-$ does not react with CO at room temperature and also at low temperatures reactivity is rather low. Consequently, the cluster is not poisoned by CO. On the contrary, the coadsorption of CO is cooperative, i.e. CO adsorption is enhanced after pre-adsorption of oxygen on the cluster. Thus, these gas phase findings can provide detailed molecular level insight into the origin of the *low temperature* catalytic activity of small gold particles.

Based on this knowledge a targeted further improvement of the catalyst in particular with respect to operation temperature and activity (turn-over-frequency) is possible. This can for example be realized by doping the gold particles with other transition metals which are known to dissociatively adsorb $O_2$ or which are more reactive toward CO. However, investigations, in particular experimental investigations, on free binary gold clusters are rather scarce.

### 3.1.2 Activation of Molecular Oxygen by Binary Gold Clusters

So far there are only few studies on the adsorption and activation of molecular oxygen on free binary gold clusters [20, 41, 59–66] which demonstrate that the reaction properties often change in a non-linear way.

**Anionic gold–silver clusters** Fig. 8 displays the composition dependent binding energies of $O_2$ to anionic gold silver dimers $Au_{2-x}Ag_x^-$ and trimers $Au_{3-x}Ag_x^-$. The binding energies have been obtained by kinetic measurements in an octopole ion trap and subsequent RRKM analysis under assumption of "tight" (red circles) and "loose" (black squares) transition states [20, 59, 60]. As shown in Sect. 3.1.1 the gold dimer $Au_2^-$ reacts with $O_2$ to form $Au_2O_2^-$ at room temperature and the experimentally obtained binding energy amounts to $0.60 \pm 0.10$ and $0.93 \pm 0.10$ eV for a "tight" and "loose" TS, respectively. Similarly, $AuAg^-$ and $Ag_2^-$ also adsorb one single oxygen molecule, however, these reactions are

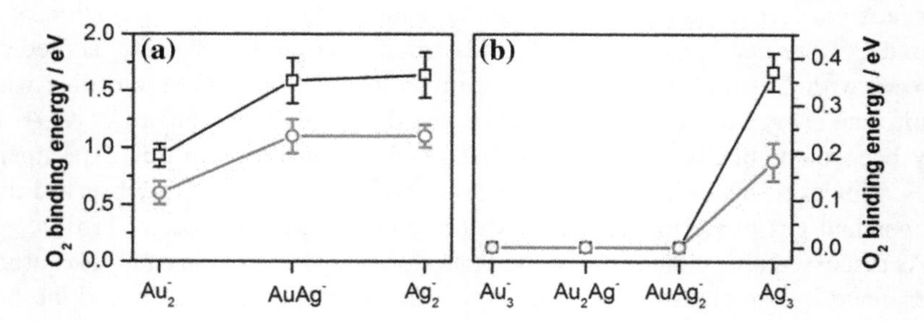


**Fig. 8** Composition dependent binding energies of $O_2$ to **a** gold–silver dimers $Au_{2-x}Ag_x^-$ and **b** gold–silver trimers $Au_{3-x}Ag_x^-$. Binding energies determined by assuming a "tight" and "loose" transition state are indicted by red circles and black squares, respectively. The *lines* are drawn to guide the eye [60]. Reprinted with permission from J. Phys. Chem. A 113, 2724 (2009). Copyright (2009) American Chemical Society

considerably faster and thus the binding energies are about 0.5 eV higher than for $Au_2^-$.

This composition dependence can again be understood by considering the frontier orbital interactions. As discussed above, the VDE of the cluster is a measure for the energetic location of the HOMO and provides insight into the ability for electron donation to a potential ligand molecule. Since molecular oxygen is an electron acceptor, the VDE is inversely related to the reactivity and the binding energy of molecular oxygen. The measured VDEs of the dimers decrease non-linearly with decreasing gold content ($Au_2^-$: 2.01 eV [67, 68]; $AgAu^-$: 1.43 eV [69]; $Ag_2^-$: 1.06 eV [67] and 1.1 eV [70]) which is in perfect agreement with the non-linear increase of the $O_2$ binding energy with decreasing gold content.

Most interestingly, the influence of doping the gold cluster with silver is less pronounced for the trimers. Due to its even number, spin paired valence electron structure, the anionic gold trimer does not react with molecular oxygen. Successive replacement of gold atoms by silver atoms does not change the inertness of the cluster for the binary $Au_2Ag^-$ and $AuAg_2^-$ (resulting in zero binding energy) while the pure silver trimer $Ag_3^-$ readily adsorbs two oxygen molecules at room temperature [20, 59, 60]. This means that the substitution of gold atoms by silver atoms does not increase the reactivity toward $O_2$, while, however, the mere substitution of one silver atom of $Ag_3^-$ by one gold atom immediately inhibits $O_2$ adsorption. Again this composition dependent reactivity is in agreement with the high VDEs of $Au_3^-$ (3.5–3.89 eV [43, 67, 68]), $Au_2Ag^-$ (3.86 eV [69]), and $AuAg_2^-$ (2.97 eV [69]). Even the silver trimer exhibits a comparably high VDE of 2.43 eV [67, 70] leading to a rather low binding energy of only $0.18 \pm 0.04$ and $0.37 \pm 0.05$ eV for a “tight” and a “loose” TS, respectively. Furthermore, this composition dependent reactivity can be attributed to differences in the energetic location of the cluster HOMO. The atomic 6s orbital of gold is 1.36 eV [71] lower in energy than the atomic 5s orbital of silver due to the relativistic contraction of the gold s orbitals, making the charge donation from gold cluster and small mixed gold–silver cluster to the electron acceptor $O_2$ less favorable than from bare silver clusters.

Thus, this example demonstrates that the adsorption and activation of molecular oxygen might be enhanced by doping small gold clusters with silver atoms. However, the cluster size, in addition to the exact elemental composition of the cluster, appears to play an important role for this reaction.

**Cationic gold–palladium clusters** A second example to demonstrate how doping gold clusters with other transition metal atoms can change the reactivity towards $O_2$ is a recent investigation of cationic gold–palladium clusters [61].

Cationic gold clusters $Au_x^+$ are completely non-reactive toward $O_2$ at thermal reaction conditions [33, 34, 40, 41]. Therefore an upper limit for the termolecular rate constant for adsorption of a first $O_2$ molecule onto $Au_x^+$ of $3 \cdot 10^{-31}$ $cm^6$ $s^{-1}$ has been estimated based on the accessible reaction conditions in an octopole ion trap experiment (cf. Fig. 9). On the contrary, the investigated cationic palladium clusters $Pd_2^+$, $Pd_3^+$, $Pd_4^+$, and $Pd_5^+$ react fast with up to two $O_2$ molecules at room temperature. Most interestingly, the reactivity changes in a non-linear way upon successive exchange of the gold by palladium atoms. Similar to $Au_3^+$ the binary

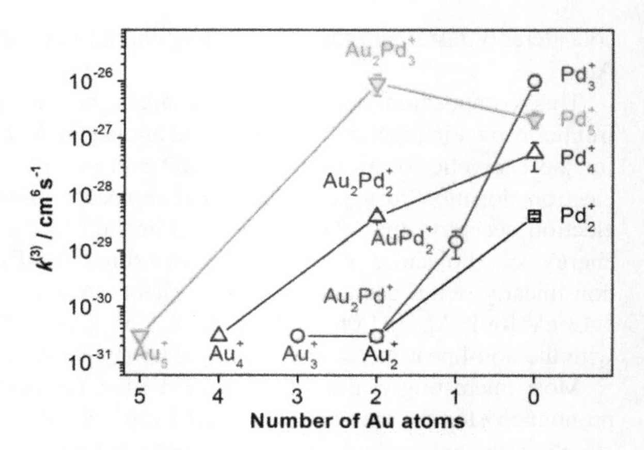


**Fig. 9** Composition and cluster size dependent termolecular rate constants $k^{(3)}$ for the adsorption of a first $O_2$ molecule on pure gold and palladium as well as binary gold–palladium clusters at room temperature. The *solid lines* are drawn to guide the eye and connect the clusters with the same number of atoms [61]. With kind permission of The European Physical Journal (EPJ)

cluster $Au_2Pd^+$ is non-reactive at room temperature and the reactivity of $AuPd_2{}^+$ is also very low, i.e. the reaction product $AuPd_2O_2{}^+$ is only observed at reaction times of at least 2s. The binary tetramer $Au_2Pd_2{}^+$ is considerably more reactive than the bare gold cluster, however, the rate constant is about one order of magnitude lower than for the palladium tetramer. In marked contrast to the binary trimers and the tetramer $Au_2Pd_2{}^+$, the pentamer $Au_2Pd_3{}^+$ is very reactive and even exhibits enhanced room temperature reactivity compared to $Pd_5{}^+$.

More importantly, Fig. 9 shows that all clusters containing an identical number of palladium atoms exhibit similar rate constants. This behavior has further been confirmed by temperature dependent measurements in a temperature range of 300 K to about 100 K which reveal similar product pattern for $Pd_2{}^+$, $AuPd_2{}^+$ and $Au_2Pd_2{}^+$ as well as $Pd_3{}^+$ and $Au_2Pd_3{}^+$. This indicates that the addition of gold atoms to palladium clusters does hardly change their reactivity, i.e. the rate constant and the product pattern and the gold atoms merely act as a spectator in the reaction of the binary clusters with oxygen. However, from the viewpoint of optimizing the reactive and catalytic properties of gold-based materials, the replacement of gold by palladium atoms considerably increases the reactivity towards $O_2$. This finding is extremely important since the fast adsorption of $O_2$ represents one of the crucial reaction steps of the catalytic CO oxidation reaction and thus, small cationic gold–palladium particles can be expected to have enhanced catalytic properties for CO oxidation compared to the cationic gold particles.

### 3.1.3 Adsorption of CO on Binary Gold Clusters

The reaction of binary gold clusters with carbon monoxide has been investigated both experimentally and theoretically. Among the published data, gold–silver clusters are the most intensively and systematically studied systems [20, 44, 50, 51, 59, 60, 72–75] while only few investigations of gold–palladium [63, 64, 74], gold–platinum [65, 66, 74, 76], gold–copper [74, 77], gold–titanium [78, 79], gold–iron [78], and gold–yttrium [80] clusters have been reported so far.

**Anionic and cationic gold–silver clusters** Fig. 10 displays a compilation of experimental and theoretical binding energies between one CO molecule and pure and binary gold–silver clusters of different composition, size, and charge [44, 50–52, 55–57, 60, 72, 81]. Independent of the cluster size and the charge state the CO binding energy decreases with increasing silver content in the clusters. So far mainly two factors have been identified which determine this composition dependent reactivity of binary gold–silver clusters: (1) the energetic location of the metal d-orbitals and (2) the charge transfer between silver and gold atoms in the cluster.

Relativistic effects in the gold atom cause an expansion and destabilization of the atomic d orbital resulting in an energetic location 0.18 eV higher than in the silver atom [71]. This destabilization of the gold d-orbitals is expected to increase the ability for charge back-donation from the cluster to the CO molecule and thus, a stronger binding of the CO molecule to gold and gold-rich clusters compared to silver and silver-rich clusters. The lack of reactivity of the anionic dimer $AuAg^-$ and trimers $Au_2Ag^-$ and $AuAg_2^-$ (which results in an experimental binding energy of zero) indicates that already one silver atom in these small clusters is sufficient to stabilize the metal cluster d-orbitals and to thus make them energetically unavailable for charge back-donation. This effect is clearly reduced for neutrals and cations. While for *anions*, charge back-donation is the dominant process, IR spectroscopic experiments clearly showed that for *cationic* gold clusters the contribution of charge back-donation decreases while charge donation from the

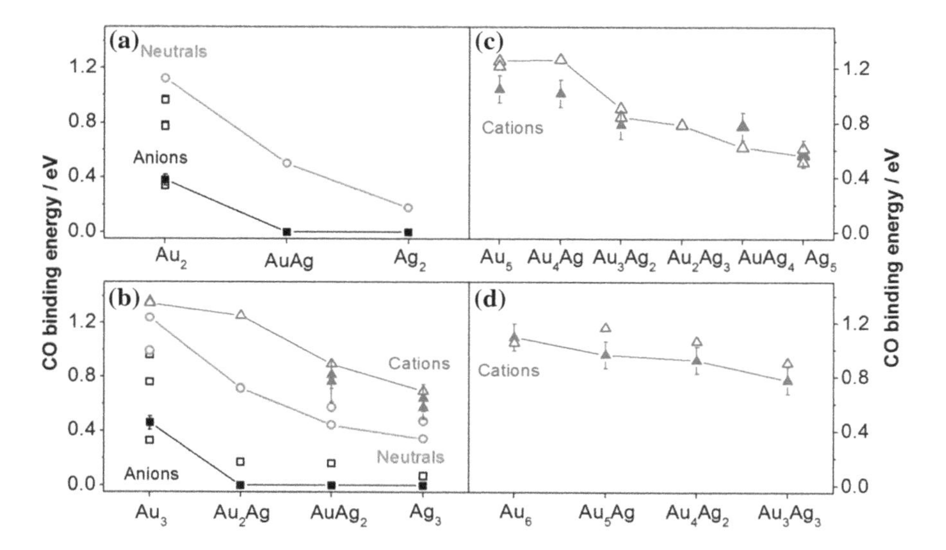


**Fig. 10** Composition dependent binding energies of one CO molecule to binary gold–silver **a** dimers, **b** trimers, **c** pentamers, and **d** hexamers. *Filled symbols* represent experimental data [44, 50, 51, 60, 72] while *open symbols* represent theoretical data [52, 55–57, 81]. The *solid lines* are drawn to guide the eye

σ-orbital of CO to the cluster LUMO becomes considerably more important. This indicates that the consideration of the energetic location of the metal d-orbitals is not sufficient anymore to explain the decreasing bond strength of CO with increasing silver content in the cationic clusters, but instead additional effects must be taken into account.

For example, in a theoretical study Kuang et al. attributed the decrease in CO binding energy to *neutral* $Au_xAg$ compared to neutral $Au_{x+1}$ to a reduced charge transfer from the binary cluster to CO [75]. In contrast, Joshi et al. did not find any correlation between the CO binding energy to neutral $Au_{2-x}Ag_x$ and $Au_{3-x}Ag_3$ and the natural bond orbital charge on the adsorbed CO. This was explained by the complicated interplay between CO-to-cluster charge donation and cluster-to-CO charge back-donation [74].

For binary gold–silver *cations* the charge distribution inside the cluster was proposed to cause the experimentally observed dopant effects. Since gold is more electronegative than silver, the gold atoms are prone to draw electron density which leads to an increased electron density on the gold atom and a stabilization of the cluster LUMOs. This in turn is expected to weaken the bond of CO to the respective gold atom in the mixed gold cluster [44, 72]. Additionally, the exact charge distribution in the cluster is related to the geometry of the cluster which can change by doping the gold clusters with silver and thus also influences the electron distribution [82]. Furthermore, theoretical simulations on neutral as well as cationic gold–silver clusters revealed that for silver-rich clusters such as $AuAg_2$, $Au_2Ag_3^+$ or $AuAg_4^+$ the CO molecule preferably binds to the silver atoms instead of to the gold atoms which must also influence the binding energy [51, 74].

Thus, it can be concluded that the composition dependent CO binding energy is most likely determined by a rather complex interplay of different effects and is still far from being completely understood yet.

**Neutral binary gold clusters** Lievens et al. investigated the influence of transition metal doping on the reactivity of neutral gold clusters toward CO in a wide cluster size range [17, 73, 83]. Since neutral clusters cannot be stored in an ion trap the reaction properties have been studied in a collision cell under few collision conditions and the reactivity is typically quantified in terms of sticking probabilities S which represents the combined probability of complex formation and stability of the complex until detection (lifetime).

Figure 11 displays the cluster size dependent sticking probabilities of a first CO molecule to neutral $Au_x$, $Au_{x-1}Ag$, and $Au_{x-1}V$ clusters obtained at cryogenic temperature [17, 73, 83]. Most interestingly, the size dependence of S for pure gold clusters $Au_x$ shows a pronounced odd–even staggering up to $x = 30$ with even x cluster sizes being more reactive. Such a pronounced odd–even alternation in the reactivity towards CO has neither been observed for cationic nor anionic gold clusters (cf. Fig. 10). For $x > 30$, the sticking probability starts to level off. Furthermore, clear drops in S are found after $x = 14$, 18, 20, 26, and 34 and pronounced maxima at $x = 18$ and 30. Surprisingly, although gold is not an ideal free-electron metal and hybridization of the s- and d-atomic orbitals is known to play an important role in the binding of CO to gold clusters, the size dependent

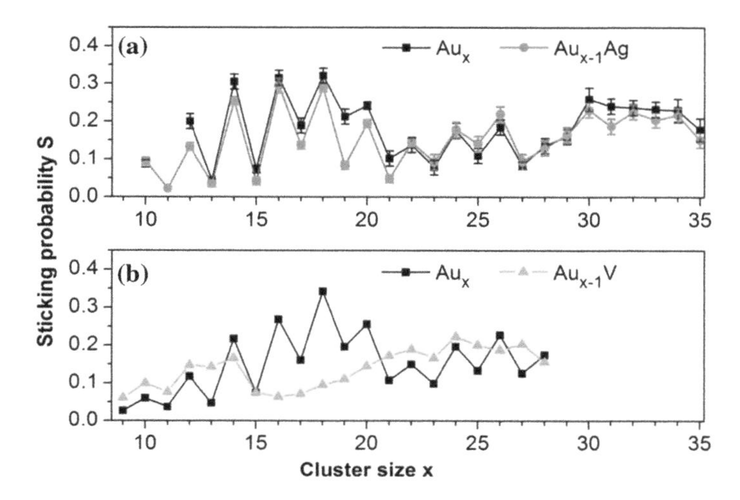


**Fig. 11** Cluster size dependent sticking probabilities of a first adsorbed CO molecule on neutral pure gold $Au_x$ (*black squares*), gold–silver $Au_{x-1}Ag$ (*red circles*), and gold–vanadium $Au_{x-1}V$ clusters (*green triangles*) obtained at **a** 140 K and **b** 124 K, respectively [73, 83]. (a) Reprinted with permission from J. Phys. Chem. A 115, 2103 (2011). Copyright (2011) American Chemical Society. (b) Reproduced from Ref. [83] with permission from the PCCP Owner Societies

reactivity was found to correlate very well with electronic shell closings according to the spherical jellium model [17, 73].

Substituting one gold atom by one silver atom (Fig. 11a) only slightly changes the size dependent sticking probabilities for CO [73]. Since both, gold and silver have a $s^1$ valence electron configuration this composition dependent reactivity further confirms that electronic effects are responsible for CO binding. In marked contrast, exchanging gold by vanadium atoms changes the cluster size dependent sticking probability considerably [83]. Figure 11b shows that for $x < 16$ and $x > 20$ the sticking probabilities of CO to $Au_{x-1}V$ and $Au_x$ are similar, while for $Au_{16-20}V$ the sticking probabilities are considerably decreased compared to $Au_x$. However, more importantly, the odd–even staggering for $x < 16$ is less pronounced for $Au_{x-1}V$ compared to $Au_x$ and completely disappears for $Au_{x-2}V_2$ (not shown here) [83]. Furthermore, a remarkable drop in the sticking probability is observed after $Au_{13}V$ which can also be explained by an electronic shell closing effect, if five electrons (i.e., the atomic $3d^34s^2$ electrons) are counted for vanadium. However, the simple electron-counting rules appear to be not readily applicable anymore for multiple doped $Au_{x-2}V_2$ clusters.

Thus, it can be concluded that although electronic counting rules are able to provide a qualitative understanding of the cluster size dependent reactivity in an extended size range (for gold as well as binary gold clusters), more detailed models which take additional effects (such as the sd hybridization, the symmetry of the valence orbitals, and mutual charge donation and back-donation) into account are required for a deeper analysis and understanding of the cluster-molecule interaction [17], also for the systems presented in this section.

### 3.1.4 CO Oxidation Mediated by Binary Gold Clusters

The examples discussed so far demonstrate that doping anionic and cationic gold clusters with other transition metals can enhance the reactivity towards $O_2$ which represents an important reaction step for CO oxidation (cf. reaction (2a)). In contrast, CO binding appears to decrease with increasing dopant concentration for all of the shown examples. In case of anionic gold clusters which have generally a low reactivity toward CO this can lead to a complete inhibition of the CO oxidation reaction. In case of neutral and cationic gold clusters which are much more reactive toward CO this reduction of reactivity might decrease the probability for CO poisoning. Thus, due to the decrease in CO bond strength in conjunction with an enhanced reactivity toward $O_2$ it can be speculated that neutral and cationic *binary* gold clusters should be more active for the catalytic CO oxidation reaction than the corresponding pure gold clusters.

Bonačić-Koutecký et al. studied the catalytic CO oxidation on anionic $AuAg^-$ by applying density functional theory [84]. The proposed reaction mechanism is very similar to the one found for the gold dimer $Au_2^-$ (cf. Sect. 3.1.1). However, due to the charge distribution in the bimetallic cluster, the silver atom represents the preferred adsorption site of molecular oxygen which is activated to an super-oxide unit. In the next step, $AuAgO_2^-$ can react with CO to form the very stable complex $AuAgCO_3^-$ containing a carbonate-like species bound to the silver atom. The catalytic cycle can then be closed by (1) reaction of $AuAgCO_3^-$ with a second CO molecule and simultaneous loss of two $CO_2$ or (2) unimolecular dissociation of $AuAgCO_3^-$ to yield $AuAgO^-/CO_2$ and subsequent reaction of $AuAgO^-$ with CO and formation and elimination of $CO_2$. Although the first reaction path is energetically more favorable both reaction mechanisms should be possible to proceed at room temperature.

The theoretically obtained binding energy of $O_2$ to $AuAg^-$ is in favorable agreement with the experimentally obtained value. However, in marked contrast to the theoretical study, no indications for the prevalence of a catalytic reaction cycle could be identified experimentally. This has been attributed to the inertness of $AuAg^-$ towards CO at all reaction conditions. Similarly, no coadsorption complex formation or catalytic CO oxidation have been observed for the anionic gold–silver trimers $AuAg_2^-$ and $Au_2Ag^-$ which are completely inert toward $O_2$ and CO in the whole investigated temperature range of 300–100 K [59].

Wang et al. theoretically investigated the properties of a series of $Au_{4-x}Pt_x$ ($x = 0$–4) clusters in the catalytic CO oxidation [85]. It was shown that for all binary gold–platinum clusters $O_2$ and CO bind much stronger on the Pt atoms compared to the Au atom which was attributed to the location of both the HOMO and the LUMO on the Pt atoms. Accordingly, for all Pt containing clusters the CO oxidation has been shown to favorably proceed via a single-center mechanism, i.e. a mechanism in which the adsorption of CO and $O_2$ as well as the subsequent CO oxidation occur on a single Pt site. Consequently, a very similar reaction mechanism with comparable energetics has been found for all investigated clusters. Thus, it has been concluded that Au atoms in the binary gold–platinum clusters are "formally spectators", i.e. they do not directly participate in the reaction. However, the authors have conjectured that despite the similar energetics bimetallic clusters are most likely more efficient than pure gold and platinum clusters. The reason for this is that the introduction of Pt atoms to gold clusters leads to a stronger binding of CO and thus an increased CO coverage and a higher efficiency compared to pure gold particles. On the contrary, the introduction of Au atoms to platinum clusters decreases the CO coverage which leaves room for coadsorption of $O_2$ molecules and thus reduces the catalyst poisoning compared to pure platinum particles.

Similarly to gold–platinum clusters, also for neutral gold–palladium clusters $Au_xPd_y$ ($x + y = 2$–$6$) CO and $O_2$ adsorption has theoretically been found to be preferred on the Pd atom due to the localization of the HOMO and LUMO [64]. Most interestingly, comparison of the reaction pathways for CO oxidation mediated by $Au_3$, $Au_2Pd$, and $Pd_3$ revealed that the activation barriers for the binary cluster $Au_2Pd$ are lower than those for $Au_3$ and $Pd_3$ implying that bimetallic AuPd clusters could have higher activity for CO oxidation. To our knowledge this proposal has not been verified experimentally so far.

To summarize, both experimental and theoretical studies have already been able to provide some basic insight into the reactivity of binary gold clusters towards CO and $O_2$. Unfortunately, so far only few theoretical investigations on the CO oxidation catalyzed by transition metal doped gold clusters have been reported [64, 66, 84, 85]. All these studies predict a similar or even superior catalytic activity of the binary clusters compared to the pure analogues. However, only a very limited amount of experimental studies are available so far which could not confirm the catalytic activity of binary gold clusters yet. This shows that more theoretical and in particular experimental studies on binary gold clusters are mandatory to understand the properties of these clusters on a molecular level and to potentially improve the activity of gold-based catalytic materials.

## 3.2 Activation and Catalytic Conversion of Methane

Methane is nowadays mostly consumed for heating purposes and the generation of electrical power. However, due to its large abundance as natural gas as well as a main component in biogas it also represents a desirable feedstock for the production of liquid fuels and chemicals. Thus, the activation of the methane molecule and its direct catalytic conversion into more valuable products like methanol, formaldehyde, or ethylene is highly desirable. However, the required high reaction temperature and/or the use of highly active reactants to activate the stable C–H bond of methane (bond dissociation energy of 440 kJ $mol^{-1}$ [86]) renders a selective synthesis of the desired products difficult. In order to find suitable catalytic materials that are able to activate the strong C–H bond without leading to complete dehydration, the detailed investigation of elementary reaction steps and the influence of important parameters, like, e.g. the particle size and the material composition which determine the catalytic properties are required. Toward this goal several gas phase investigations employing gold [33, 34, 41, 87–91] as well as binary gold clusters [41, 63, 92, 93] have been performed which are able to provide important insight into these processes.

### 3.2.1 Activation of Methane by gold clusters

Free gold cations were found to be completely non-reactive toward methane at single collision conditions [41]. In contrast, an uptake of several $CH_4$ molecules was observed at multi-collision conditions. Although dehydrogenation of $CH_4$ was not directly observed in this study, it was proposed that methane is dissociatively bound as methyl plus hydride ($CH_3$–$Au_x^+$–H) [33, 34]. Recently, a temperature dependent kinetic study in an octopole ion trap in conjunction with first-principles density functional theory calculations revealed that a first methane molecule is adsorbed molecularly on $Au_{2-6}^+$ [87]. Figure 12 displays the experimentally and theoretically obtained binding energies of the first adsorbed methane molecule on $Au_x^+$ (x = 2–6) together with the calculated minimum energy structures [87]. In all cases the $CH_4$ molecule has been found to bind to a single gold atom of the cluster, with two of its hydrogen atoms lying in the plane of the gold cluster and oriented toward it, while the other two hydrogen atoms are located in the plane normal to the plane of the cluster.

Although the first methane molecule binds molecularly to the gold clusters without experimental indication for C–H bond activation, the dimer $Au_2^+$ has been

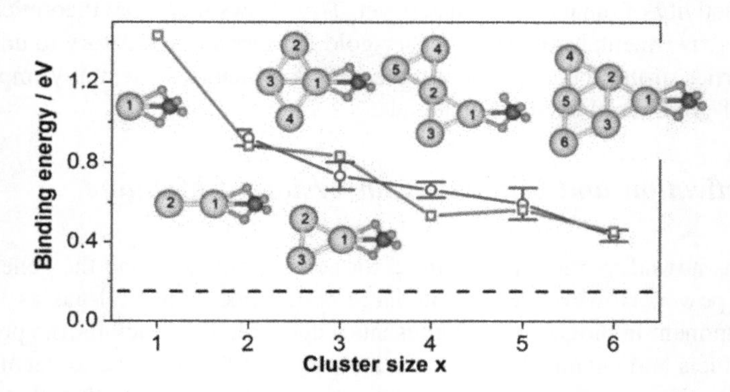


**Fig. 12** Experimental (*blue circles*) and theoretical (*red squares*) cluster size dependent binding energies of a first adsorbed methane molecule to gold clusters $Au_x^+$ together with calculated minimum energy structures. Au, C, and H atoms are indicated by yellow, green, and blue spheres, respectively. The lines are drawn to guide the eye [87]. The *dashed black line* indicates the reported literature value for the binding energy of $CH_4$ on an extended Au(111) surface [94]. Reprinted from Chem. Phys. Chem. 11, 1570 (2010), Copyright (2011), with permission from John Wiley & Sons, Inc.

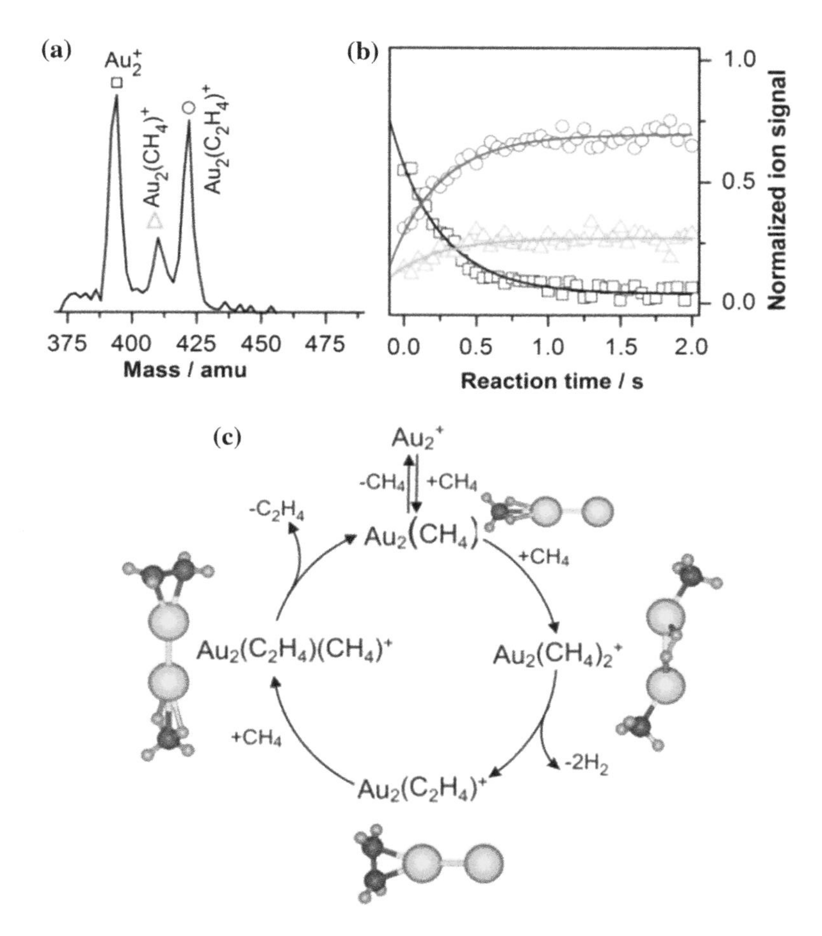


**Fig. 13** **a** Ion mass distribution and **b** kinetic data obtained after reaction of $Au_2^+$ with $CH_4$ in an ion trap experiment. **c** Catalytic reaction cycle proposed on the basis of the experimental kinetic data in conjunction with first-principles simulations together with calculated minimum energy structures of the reaction intermediates. Au, C, and H atoms are indicated by *yellow*, *green*, and *blue* spheres, respectively [88]. Reproduced from Ref. [4] with permission from the PCCP Owner Societies

found to activate and dehydrogenate methane upon adsorption of a second $CH_4$ molecule [88]. Figure 13 displays the detected ion mass distribution and the corresponding kinetics obtained after reaction of $Au_2^+$ with $CH_4$ at room temperature. Also shown is the proposed catalytic reaction mechanism which has been derived on basis of the temperature dependent kinetic measurements in conjunction with theoretical simulations. The ion mass distribution reveals that the cooperative action of two adsorbed $CH_4$ molecules enables C–H bond activation and subsequent dehydrogenation to form ethylene, $Au_2(C_2H_4)^+$. The direct desorption of the formed

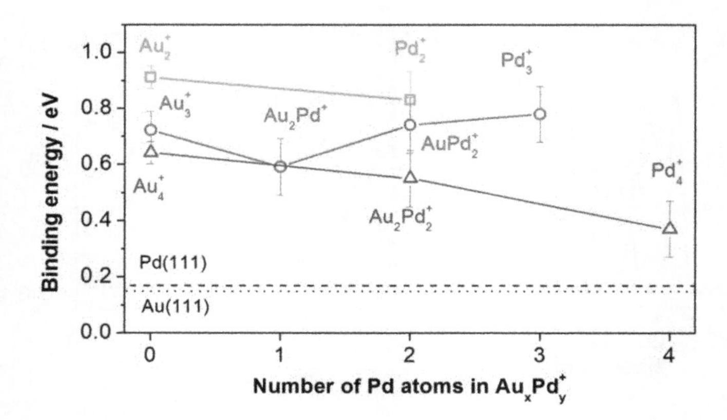


**Fig. 14** Cluster size and composition dependent binding energies of $CD_4$ to $Au_{2-x}Pd_x^+$ (x = 0, 2), $Au_{3-x}Pd_x^+$ (x = 0–3), and $Au_{4-x}Pd_x^+$ (x = 0, 2, 4) [87, 90, 92]. The *solid lines* are drawn to guide the eye and connect clusters with the same total number of atoms. The *dashed* and *dotted black lines* represent literature values for the binding energy of $CH_4$ to extended Pd(111) [96] and Au (111) [94] surfaces, respectively. Reprinted from Int. J. Mass Spectrom. 354–355, 365 (2013), Copyright (2013), with permission from Elsevier

ethylene from $Au_2(C_2H_4)^+$ was calculated to be non-feasible, instead the coadsorption of a third $CH_4$ is necessary yielding the intermediate $Au_2(C_2H_4)(CH_4)^+$ which is experimentally observed at temperatures below 300 K. In this complex the bonding of $C_2H_4$ to $Au_2^+$ is weakened and activated liberation of ethylene is possible to close the catalytic reaction cycle by re-formation of $Au_2(CH_4)^+$.

This example demonstrates that small gold clusters are able to activate the strong C–H bond of methane and mediate the subsequent C–C bond formation due to the cooperative action of two methane molecules. A similar activity has also been observed for the palladium dimer $Pd_2^+$ [91, 95].

### 3.2.2 Activation of Methane by Binary Gold–Palladium Clusters

Investigations of methane activation on binary gold clusters are very scarce. For example, Schwarz et al. showed that the binary gold–platinum clusters $AuPt^+$, $AuPt_2^+$, and $Au_2Pt_2^+$ are able to induce dehydrogenation of a first adsorbed $CH_4$ similar to the pure platinum clusters $Pt_{2-4}^+$. In contrast, $Au_2Pt^+$ as well as the gold cations $Au_2^+$ and $Au_3^+$ were found to be completely inert toward $CH_4$ under these experimental conditions [41].

Recently, a detailed experimental study concerning the activity of gold, palladium, and binary gold–palladium clusters in the activation and dehydrogenation of methane has been reported [87, 88, 90–92, 95]. Temperature dependent kinetic studies in an octople ion trap revealed a strongly cluster size and composition dependent reactivity and ability of the trimers $Au_{3-x}Pd_x^+$ and tetramers $Au_{4-x}Pd_x^+$ to dehydrogenate methane. All investigated tetramers $Au_4^+$, $Au_2Pd_2^+$, and $Pd_4^+$ are non-reactive toward methane at room temperature but adsorb several methane molecules at lower temperatures. The reactivity has been observed to decrease with increasing Pd content in the cluster which is also reflected in the experimentally obtained binding energies of a first adsorbed methane molecule as displayed in Fig. 14 (blue triangles). The mass spectra do not indicate dehydrogenation of methane for any of these clusters [92].

In marked contrast, the trimers $Au_{3-x}Pd_x^+$ react with methane already at room temperature and there is clearly no linear relation between the cluster composition and the reactivity. The binding energies of a first methane molecule to the trimers (cf. Fig. 14, red circles) show that the reactivity increases in the order $Au_2Pd^+ < Au_3^+ \leq AuPd_2^+ < Pd_3^+$, yet, it should be noted that the differences in the binding energies are only small within the error limits [92]. However, the composition dependent reactivity toward $CD_4$ becomes apparent when comparing the mass distributions displayed in Fig. 15.

$Au_3^+$ adsorbs up to two methane molecules yielding $Au_3CD_4^+$ and $Au_3(CD_4)_2^+$ (cf. Fig. 15a) while the binary cluster $Au_2Pd^+$ adsorbs only one molecule to form $Au_2PdCD_4^+$ (Fig. 15c) without indication for methane dehydrogenation. Most interestingly, the product distribution changes completely for $AuPd_2^+$ (Fig. 15e). This cluster reacts with $CD_4$ to form the association product $AuPd_2CD_4^+$ as well as the complexes $AuPd_2C_2D_4^+$ and $AuPd_2C_3D_8^+$ which clearly must be the result of the dehydrogenation of methane. The product distributions measured for $Pd_3^+$ (cf. Fig. 15g) exhibit no simple association products but instead the dehydrogenated complexes $Pd_3CD_2^+$ and $Pd_3C_2D_6^+$ [92].

Based on the ion mass distributions and the kinetic data (also displayed in Fig. 15) a reaction mechanism has been obtained for $Au_2Pd^+$ which is very similar to the catalytic reaction cycle obtained for $Au_2^+$ (cf. Fig. 13c) [92]. This means, that also for $Au_2Pd^+$ the cooperative action of two methane molecules is necessary to enable dehydrogenation. In contrast, the palladium trimer $Pd_3^+$ is able to dehydrogenate a first methane molecule while, however, the kinetics present no clear evidence for the occurrence of a catalytic reaction cycle with hydrocarbon product elimination. Thus, it has been concluded that the ability to dehydrogenate methane increases with increasing palladium content in the cluster following the order $Au_3^+ = Au_2Pd^+ < AuPd_2^+ < Pd_3^+$.

To gain more insight into the role of gold and palladium on the dehydrogenation of methane it is instructive to compare the reactivity of $Au_2Pd_x^+$ (x = 0–2) and $Au_xPd_2^+$ (x = 0–2). As shown in Sect. 3.2.1, the gold dimer $Au_2^+$ dehydrogenates two methane molecules to form ethylene and hydrogen [88]. Adding one or two palladium atoms to $Au_2^+$ ($Au_2Pd^+$ and $Au_2Pd_2^+$, respectively) changes the reactivity completely leading to the adsorption of $CH_4$ without indication of dehydrogenation [92]. Furthermore, Fig. 14 shows that the binding energy of methane decreases in the order $Au_2^+ > Au_2Pd^+ > Au_2Pd_2^+$. This indicates that the Pd atom strongly influences the electronic structure of the gold dimer.

Similar to $Au_2^+$ [88] also $Pd_2^+$ [95] is able to dehydrogenate two methane molecules yielding ethylene and hydrogen. However, the liberation of the formed ethylene from $Pd_2^+$ has been found to be clearly hampered due to the strong binding

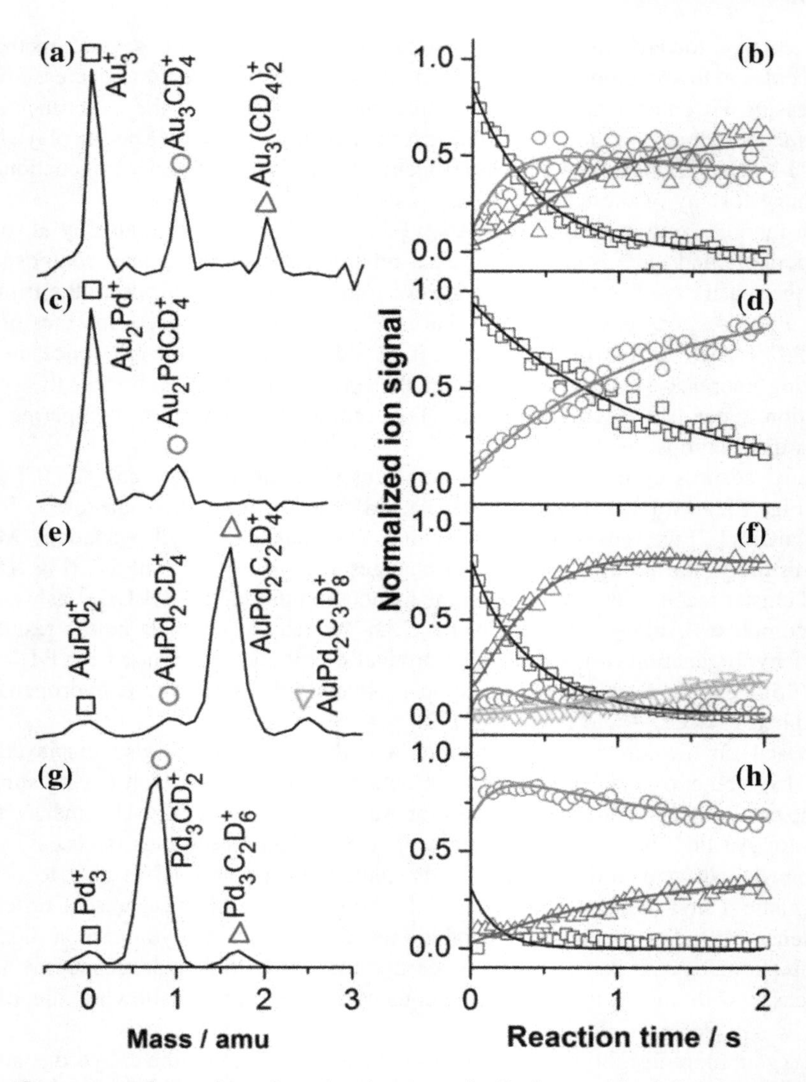


**Fig. 15** **a**, **c**, **e**, **g** Composition dependent ion mass distributions and **b**, **d**, **f**, **h** corresponding kinetic data obtained after reaction of the trimers $Au_{3-x}Pd_x^+$ (x = 0–3) with $CD_4$ at room temperature. The *open symbols* represent the experimental data, the *solid lines* are obtained by fitting the integrated rate equations of proposed reaction mechanisms to the experimental data [92]. Reprinted from Int. J. Mass Spectrom. 354–355, 365 (2013), Copyright (2013), with permission from Elsevier

to $Pd_2^+$. Adding a gold atom to $Pd_2^+$ ($AuPd_2^+$) leads to similar product distributions, though the formed product ethylene can apparently be eliminated from the cluster resulting in the possibility for the closure of a catalytic reaction cycle. Adding a second gold atom ($Au_2Pd_2^+$) changes the product distribution considerably and no dehydrogenation is observed anymore [92]. Additionally, also the binding energy of a first methane molecule decreases with increasing gold content following the order $Pd_2^+ > AuPd_2^+ > Au_2Pd_2^+$. This indicates that also the gold atoms strongly act on the electronic structure of the palladium dimer leading to a reduction of the binding energy and the ability for dehydrogenation. However, it appears that the effect of Pd on $Au_2^+$ is stronger than the influence of Au on $Pd_2^+$.

As already discussed, $Au_2^+$, $Pd_2^+$, and also the binary $Au_2Pd^+$ are able to dehydrogenate methane to form ethylene according to a similar reaction mechanism as shown in Fig. 13c. Recently, a detailed comparison of this reaction mediated by $Au_2^+$ and $Pd_2^+$ has been reported [91]. The lower turn-over-frequency of $0.52 \pm 0.12\ s^{-1}$ for $Pd_2^+$ compared to $26 \pm 7\ s^{-1}$ for $Au_2^+$ has mainly been attributed to a considerably stronger binding of ethylene to $Pd_2^+$ which hampers the final liberation of ethylene and thus the closing of the catalytic reaction cycle [91]. The turn-over-frequency for $Au_2Pd^+$ has been experimentally determined to amount to $0.14 \pm 0.04\ s^{-1}$. However, the different number of constituent atoms of $Au_2Pd^+$ and the dimers renders in this case a direct comparison of the gas phase rate constants difficult. Nevertheless, it can be concluded that $Au_2Pd^+$ represents one among only two experimental examples (see Sect. 3.3) of binary gold clusters in the gas phase which have been observed to mediate a catalytic reaction in a full thermal catalytic cycle. These results might stimulate future investigations of the catalytic properties of such binary gold clusters.

## 3.3 *C–N Coupling*

Hydrogen cyanide (HCN) represents an important feedstock for the production of e.g. the amino acid D,L-methionine, alkali cyanides mainly used in gold mining, as well as polymers such as PMMA (poly(methyl methacrylate)) and nylon [97]. One of the industrially employed processes for HCN synthesis is the BMA (Blausäure–Methan–Ammoniak) or Degussa process. In this process the direct C–N coupling of methane and ammonia to HCN and molecular hydrogen according to

$$CH_4 + NH_3 \rightarrow HCN + 3\,H_2 \tag{3}$$

takes place in alumina tubes which are coated with the catalytic material consisting of 70 % platinum (pure or sometimes together with rhodium, palladium, iridium and/or ruthenium) and as well as aluminum and magnesium in metallic or nitride form [98]. Since the overall reaction is endothermic by 251 kJ $mol^{-1}$ the reaction is most effective at temperatures between 1200 and 1300 °C [99].

Despite the enormous industrial relevance of the Degussa process and its decade long utilization (the development dates back to the 1950ies [99]) mechanistic details, in particular, the nature of key reaction intermediates, the role of possible gas phase reactions steps, the origin of the outstanding activity and selectivity of platinum, and the influence of additives such as Rh, Pd, Ir, Ru is still widely elusive. Furthermore, the catalytic properties (reduction of reaction temperature, increase of product yield) might be improved by the introduction of additional transition metals. Such a binary platinum-based catalyst should be able to (1) activate and dehydrogenate $CH_4$ and (2) mediate the coupling of $CH_x$ ($x < 4$) with $NH_3$ or $NH_x$ ($x < 3$), while at the same time (3) prevent the coupling of $CH_x/CH_x$ and $NH_x/NH_x$ species in order to avoid the formation of soot and $N_2$, respectively.

### 3.3.1 C–N Coupling Mediated by Gold and Platinum Cations

To gain insight into the C–N coupling reaction Schwarz et al. started more than 15 years ago a series of gas phase studies in a Fourier transform ion cyclotron resonance (FTICR) mass spectrometer typically operated at single collision conditions. As already mentioned above (cf. Sect. 3.2.2), gold cations are inert toward $CH_4$ at these experimental conditions [41] and thus cannot mediate the coupling of $CH_4$ and $NH_3$ according to reaction (3). However, the carbene complex $AuCH_2^+$ (generated from other precursors) has been shown to react fast with ammonia yielding cationic $CH_2NH_2^+$ and neutral AuH [100].

In marked contrast, the platinum cation $Pt^+$, which represents the most simple possible model for a platinum catalyst, has been found to be highly reactive toward $CH_4$ and yields $PtCH_2^+$ and molecular hydrogen as reaction products [100–102]. The product $PtCH_2^+$ has also been observed in several other studies performed at different reaction conditions [103, 104] and has theoretically been predicted to have a carbene structure already 15 years ago [100]. However, the carbene structure has only very recently been confirmed experimentally by infrared multiple-photon dissociation spectroscopy using an infrared free electron laser [105]. The interaction of $Pt^+$ with ammonia leads to simple association products $PtNH_3^+$ and $Pt(NH_3)_2^+$ [102] with rate constants about three orders of magnitude smaller than the rate constant for methane adsorption.

Thus, the adsorption and dehydrogenation of $CH_4$ must represent the first reaction step. The formed platinum–carbene has been found to interact with an approaching ammonia molecule leading to C–N bond formation and subsequent generation of HCN via two different possible reaction routes which are shown in

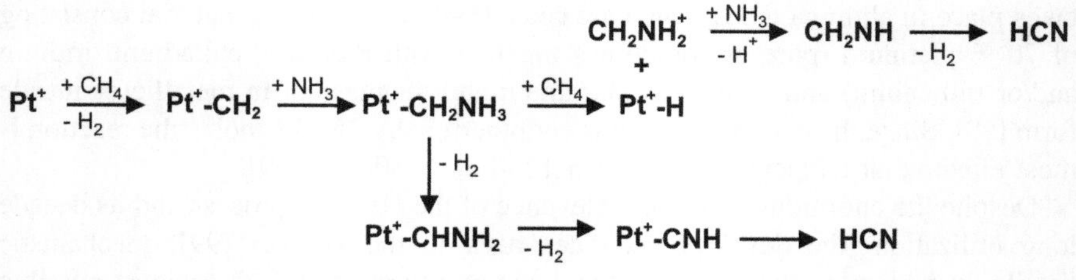

$$Pt^+ \xrightarrow[-H_2]{+CH_4} Pt^+\text{-}CH_2 \xrightarrow{+NH_3} Pt^+\text{-}CH_2NH_3 \xrightarrow{+CH_4} CH_2NH_2^+ + Pt^+\text{-}H$$

$$CH_2NH_2^+ \xrightarrow[-H^+]{+NH_3} CH_2NH \xrightarrow[-H_2]{} HCN$$

$$Pt^+\text{-}CH_2NH_3 \xrightarrow{-H_2} Pt^+\text{-}CHNH_2 \xrightarrow[-H_2]{} Pt^+\text{-}CNH \longrightarrow HCN$$

**Fig. 16** Proposed mechanism for the $Pt^+$-mediated coupling of methane and ammonia to HCN following two different possible reaction routes [100, 102]

Fig. 16. The first reaction path involves the formation of the intermediate $PtCH_2HCH_3^+$ as well as subsequent formation and liberation of an iminium ion $CH_2NH_2^+$. This ion can be deprotonated to neutral formimine and then react to HCN and $H_2$. These latter two reaction steps occur in the gas phase without any metal catalyst involved. In contrast, the second pathway has been proposed to proceed via Pt-bound aminocarbene $PtCHNH_2^+$ which decomposes to yield HCN and the Pt ion [100, 102].

In conclusion, two main factors have been identified which determine the outstanding activity and selectivity of $Pt^+$ for HCN production via C–N coupling, namely, (i) the ability of $Pt^+$ to activate methane but not ammonia and (ii) the preferential attack of the thus formed metal carbene by ammonia rather than methane (avoiding soot formation) [100].

Similarly to the atom, also small platinum cluster cations $Pt_x^+$ have been observed to easily activate and dehydrogenate methane [104, 106], however, the subsequent C–N coupling reaction appeared to be hampered [107].

### 3.3.2 C–N Coupling by Dinuclear Bimetallic Gold–Platinum Cations

In order to study the effect of additives to the platinum catalyst and to potentially improve the catalytic activity of platinum particles by using binary clusters, Schwarz et al. also investigated a series of binary platinum dimers [108]. Motivated by the observation that pre-formed gold–carbene $AuCH_2^+$ undergoes efficient C–N coupling with $NH_3$, in particular the binary gold–platinum dimer $AuPt^+$ has been investigated in detail as the smallest possible gold–platinum cluster.

$AuPt^+$ has been found to adsorb and dehydrogenate methane yielding the carbene $AuPtCH_2^+$ similar to the platinum dimer $Pt_2^+$ but unlike the gold dimer $Au_2^+$ [41, 109]. In contrast, $AuPt^+$ reacts with $NH_3$ and results in cluster degradation (formation of $PtNH_3^+$ + Au) which, however, is considerably less efficient than the reaction with $CH_4$ [108, 109]. Therefore the reaction with $CH_4$ is considered to be the first reaction step. The formed carbene $AuPtCH_2^+$ then reacts with $NH_3$ leading to dehydration and formation of a product of the stoichiometry $AuPtCH_3N^+$. Based on isotopic labeling and collision induced dissociation experiments it has been concluded that this complex corresponds to an aminocarbene $AuPtCHNH_2^+$ structure rather than to a simple adduct $AuPtC^+ \cdot NH_3$ [108, 110]. Thus, it has been argued that $PtAu^+$, unlike $Pt_2^+$ and $Au_2^+$, does mediate C–N coupling between $CH_4$ and $NH_3$ and behaves similar to the platinum cation $Pt^+$. Furthermore, the efficiency for methane dehydrogenation is only slightly smaller for $AuPt^+$ (0.67) compared to $Pt^+$ (0.8) while the efficiencies for C–N coupling are identical (0.3 for both) [41, 102, 109]. This indicates that the gold atom only influences the platinum cation $Pt^+$ to a minor extend, i.e. it acts like a spectator.

Since $AuPt^+$ has been shown to successfully mediate C–N coupling, Schwarz et al. also studied the reactivity and catalytic activity of larger $Au_xPt_y^+$ ($x + y = 3, 4$) clusters. The gold-rich trimer $Au_2Pt^+$and tetramer $Au_3Pt^+$ have been found to be non-reactive toward methane, while the other clusters $AuPt_2^+$, $Au_2Pt_2^+$, $AuPt_3^+$ dehydrogenate methane to form the carbenes $Au_xPt_yCH_2^+$ [41]. Furthermore, all bimetallic clusters adsorb $NH_3$ without degradation of the cluster [109]. Most interestingly, all bimetallic carbene complexes $AuPt_2CH_2^+$, $Au_2Pt_2CH_2^+$, and $AuPt_3CH_2^+$ undergo reaction with $NH_3$ and efficiently lose $H_2$ to form complexes of the stoichiometry $Au_xPt_yCH_3N^+$. In contrast to the binary dimer $AuPt^+$ which was found to mediate C–N coupling, these complexes contain a carbide complex and a simple $NH_3$ adduct, $Au_xPd_yC^+ \cdot NH_3$ due to strong metal–carbene interaction which inhibits C–N coupling. Consequently, $AuPt^+$ appears to represent the only binary cluster which is able to mediate the coupling of methane and ammonia [109] since only this cluster (apart from the atomic $Pt^+$) fulfills the requirements of (1) high activity for $CH_4$ dehydrogenation as a first reaction step and (2) a well-balanced metal–carbene interaction to enable C–N coupling [109].

## *3.4 Other Reactions*

For the sake of completeness, some further investigations on the reactivity of binary gold clusters will be mentioned here. Apart from the above discussed reactions of binary gold clusters with $O_2$, CO, $CH_4$, and $NH_3$ also the interaction between $Au_xM_y$ (M = transition metal) and other molecules such as $H_2$ [63, 111], $H_2S$ [112], $N_2$ [113], $C_2H_4$ [114], $C_3H_6$ [115], and $CO_2$ [63, 78] have been investigated theoretically. Furthermore, a joined experimental and theoretical investigation of the interaction of $Au_{3-x}Ag_x^+$ with an $H_2O$/CO mixture has recently been reported [116].

As displayed in Fig. 10 and discussed above, the binding energy for a first CO molecule decreases with increasing silver content in the cluster: $Au_3^+$ (>1.39 eV) > $AuAg_2^+$ (0.76 ± 0.15 eV) > $Ag_3^+$ (0.57 ± 0.09 eV) [50]. In contrast, the binding energies for a first adsorbed $H_2O$ molecule have been observed to be almost composition independent. From the comparison (cf. Fig. 17a) of the CO and $H_2O$ binding energies it is apparent that a cross-over occurs in the affinity of the trimers towards these molecules. $Au_3^+$ binds CO considerably more strongly than $H_2O$ while the reverse relative bonding behavior is observed for $Ag_3^+$. Consequently, the binding energies of CO and $H_2O$ to the binary cluster $AuAg_2^+$ are very similar. This cross-over has strong implications for a possible CO/$H_2O$ coadsorption.

Figure 17b–d displays the product mass spectra obtained after reaction of $Au_3^+$ (Fig. 17b), $AuAg_2^+$ (Fig. 17c), and $Ag_3^+$ (Fig. 17d) with a mixture of CO and $H_2O$. The bare gold cluster only adsorbs CO as in the experiment without $H_2O$ while the silver cluster only adsorbs $H_2O$ as in the experiment without CO. Most interestingly, the binary $AuAg_2^+$ represents the only cluster which is able to co-adsorb CO

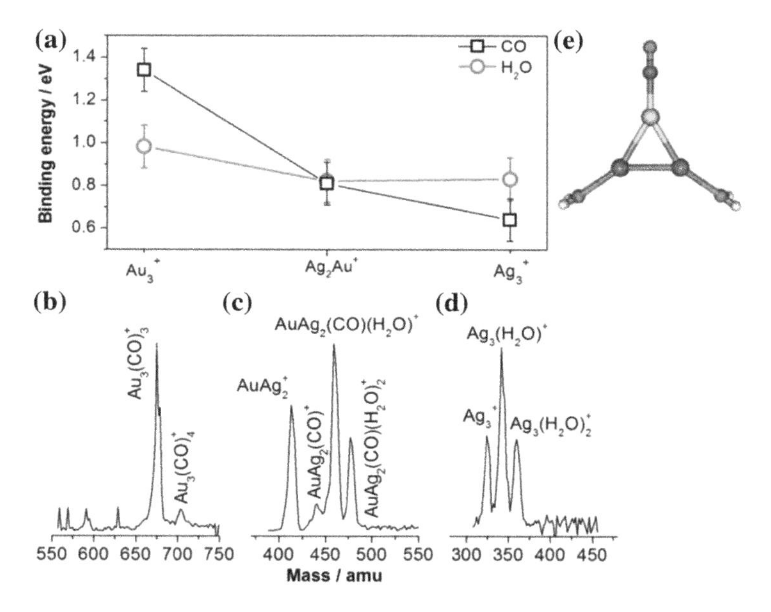


**Fig. 17** **a** Composition dependent binding energy of CO (*black squares*) and $H_2O$ (*red circles*) to binary $Au_{3-x}Ag_x^+$ (x = 0, 2, 3). **b–d** Product distributions obtained after reaction of $Au_{3-x}Ag_x^+$ with a CO/$H_2O$ mixture. **e** Calculated minimum energy structure of the coadsorption complex $AuAg_2(CO)(H_2O)_2^+$ [116]. Reprinted from Chem. Phys. Lett. 565, 74 (2013), Copyright (2013), with permission from Elsevier

and $H_2O$ to form $AuAg_2(H_2O)CO^+$ and $AuAg_2(H_2O)_2CO^+$. This remarkable composition dependent coadsorption behavior has been attributed to the observed cross-over of the CO and $H_2O$ binding energies. Since in the largest co-adsorption product $AuAg_2(H_2O)_2CO^+$ the number of $H_2O$ (CO) molecules corresponds to the number of silver (gold) atoms and it is known that $H_2O$ (CO) binds more strongly to silver (gold) than CO ($H_2O$) it has been speculated that the water and CO ligands are attached to the silver and gold atoms, respectively. Indeed, theoretical calculations predicted this geometry as the minimum energy structure (cf. Fig. 17e). Although no direct catalytic reaction has been observed in this study, it represents another very instructive example how the interplay between two transition metals (in this case gold and silver) can influence the reactivity and thus prepare the ground for potential catalysis.

## 4 Conclusion and Perspective

Experimental and theoretical gas phase studies of small gold clusters demonstrated the surprisingly good catalytic properties of gold for example in the catalytic oxidation of carbon monoxide and the activation of methane. Although the investigation of free binary gold clusters are still limited up today, it could already be shown that doping gold clusters with other transition metal atoms changes the reactivity often in a non-linear and thus rather unpredictable fashion. However, due to the limited amount of studies, the origin of this behavior is often still elusive.

So far only two experimental examples for the catalytic activity of binary gold clusters have been reported: (1) the coupling of methane and ammonia mediated by $AuPt^+$ and (2) the formation of ethylene from methane catalyzed by $AuPd_2^+$. In contrast, theoretical studies have predicted similar or even enhanced properties of binary cluster for the catalytic oxidation of CO. Thus, the investigation of the catalytic properties of free binary gold clusters can still be regarded as being in its infancy and many more surprises can be expected. To achieve a comprehensive conceptual view on the catalytic properties of bimetallic gold clusters on an atomic and molecular level more theoretical and in particular experimental work is mandatory in the future.

## References

1. Lloyd L (2011) Handbook of industrial catalysts. Springer, New York, Dordrecht, Heidelberg, London
2. Sakong S, Mosch C, Groß A (2007) Phys Chem Chem Phys 9:2216; Watanabe M, Motoo S (1975) J Electron Chem 60:267
3. Schwarz H (2011) Angew Chem Int Ed 50:2
4. Lang SM, Bernhardt TM (2012) Phys Chem Chem Phys 14:9255
5. Gross JH (2011) Mass spectrometry, 2nd edn. Springer, Berlin, Heidelberg
6. Haberland H (ed) (1994) Clusters of atoms and molecules: theory, experiment, and clusters of atoms, vol 52. Springer, Berlin; Johnston RL (2002) Atomic and molecular clusters. Taylor & Francis, London
7. Bernhardt TM, Heiz U, Landman U (2007) In: Heiz U, Landman U (eds) Nanocatalysis. Springer, Berlin, Heidelberg, p 1
8. Keller R, Nöhmeier F, Spädtke P, Schönenberg M-H (1984) Vacuum 34:31; Schaffner M-H, Jeanneret JF, Patthey F, Schneider W-D (1998) J Phys D: Appl Phys 31:3177; Leisner T, Vajda S, Wolf S, Wöste L, Berry RS (1999) J Chem Phys 111:1017; Heim HC, Bernhardt TM, Lang SM (2015) Int J Mass Spectrom 387:56
9. Dietz TG, Duncan MA, Powers DE, Smalley RE (1981) J Chem Phys 74:6511
10. Siekmann HR, Lüder C, Faehrmann J, Lutz HO, Meiwes-Broer KH (1991) Z Phys D 20:417
11. Haberland H, Mall M, Moseler M, Qiang Y, Reiners T, Thurner Y (1994) J Vac Sci Technol A 12:2925
12. Bouwen W, Thoen P, Vanhoutte F, Bouckaert S, Despa F, Weidele H, Silverans RE, Lievens P (2000) Rev Sci Instrum 71:54
13. Yasumatsu H, Fuyuki M, Hayakawa T, Kondow T, Phys J (2009) Conf Ser 185:012057

14. Ren X, Hintz PA, Ervin KM (1993) J Chem Phys 99:3575; Whetten RL, Cox DM, Trevor DJ, Kaldor A (1985) J Phys Chem 89:566; Leuchtner RE, Harms AC, Castleman AW Jr (1990) J Chem Phys 92:6527
15. Wallace WT, Whetten RL (2002) J Am Chem Soc 124:7499
16. Andersson M, Persson JL, Rosén A (1996) J Phys Chem 100:12222
17. Veldeman N, Lievens P, Andersson M (2005) J Phys Chem A 109:11793
18. Kemper PR, Weis P, Bowers MT (1997) Int J Mass Spectrom 160:17
19. Fayet P, McClinchey MJ, Wöste LH (1987) J Am Chem Soc 109:1733; Ichihashi M, Hanmura T, Kondow T (2006) J Chem Phys 125:133404; Bell RC, Zemski KA, Justes DR, Castleman AW Jr (2001) J Chem Phys 114
20. Bernhardt TM (2005) Int J Mass Spectrom 243:1
21. Balteanu I, Balaj OP, Fox BS, Beyer MK, Bastl Z, Bondybey VE (2003) Phys Chem Chem Phys 5:1213
22. Neumaier M, Weigend F, Hampe O, Kappes MM (2005) J Chem Phys 122:104702
23. Steinfeld JI, Francisco JS, Hase WL (1999) Chemical kinetics and dynamics, 2nd edn. Prentice Hall, Upper Saddle River
24. Ertl G, Knözinger H, Weitkamp J (eds) (1997) Handbook of heterogeneous catalysis, vol 4. Wiley-VCH, Weinheim
25. Conrad H, Ertl G, Küppers J (1978) Surf Sci 76:323
26. Engel T, Ertl G (1978) J Chem Phys 69:1267
27. Lide DR (ed) (1995) Handbook of chemistry and physics. CRC Press, Inc., Boca Raton
28. Gates BC (1992) Catalytic chemistry. John Wiley & Sons Inc, New York, Singapore
29. Szanyi J, Kuhn WK, Goodman DW (1994) J Phys Chem 98:2978
30. Hammer B, Nørskov JK (1995) Nature 376:238
31. Haruta M, Kobayashi T, Sano H, Yamada N (1987) Chem Lett 16:405
32. Bond GC, Louis C, Thompson DT (2006) In: Catalysis by gold. Imperial College Press, London; Haruta M (2004) Gold Bull 37:27; Haruta M, Daté M (2001) Appl Catal A Gen 222:427; Hashmi ASK, Hutchings GJ (2007) Angew Chem Int Ed 45:7896; Braun I, Asiri AM, Hashmi ASK (2013) ACS Catal 3:1902
33. Cox DM, Brickman R, Creegan K, Kaldor A (1991) Z Phys D 19:353
34. Cox DM, Brickman R, Creegan K, Kaldor A (1991) Mat Res Soc Symp Proc 206:43
35. Huang W, Zhai H-J, Wang L-S (2010) J Am Chem Soc 132:4344; Sun Q, Jena P, Kim YD, Fischer M, Ganteför G (2004) J Chem Phys 120:6510
36. Lian L, Hackett PA, Rayner DM (1993) J Chem Phys 99:2583; Stolcic D, Fischer M, Ganteför G, Kim YD, Sun Q, Jena P (2003) J Am Chem Soc 125:2848; Pal R, Wang LM, Pei Y, Wang L-S, Zeng XC (2012) J Am Chem Soc 134:9438; Woodham AP, Meijer G, Fielicke A (2013) J Am Chem Soc 135:1727; Woodham AP, Fielicke A (2014) Angew Chem Int Ed 53:6554
37. Salisbury BE, Wallace WT, Whetten RL (2000) Chem Phys 262:131
38. Hagen J, Socaciu LD, Elijazyfer M, Heiz U, Bernhardt TM, Wöste L (2002) Phys Chem Chem Phys 4:1707
39. Lee TH, Ervin KM (1994) J Chem Phys 98:10023
40. Lang SM, Bernhardt TM, Barnett RN, Yoon B, Landman U (2009) J Am Chem Soc 131:8939
41. Koszinowski K, Schröder D, Schwarz H (2003) Chem Phys Chem 4:1233
42. Woodham AP, Meijer G, Fielicke A (2012) Angew Chem Int Ed 51:4444
43. Taylor KJ, Pettiette-Hall CL, Cheshnovsky O, Smalley RE (1992) J Chem Phys 96:3319
44. Neumaier M, Weigend F, Hampe O, Kappes MM (2008) Faraday Discuss 138:393
45. Fielicke A, van Helden G, Meijer G, Pedersen DB, Simard B, Rayner DM (2005) J Am Chem Soc 127:8416
46. Fielicke A, van Helden G, Meijer G, Simard B, Rayner DM (2005) J Phys Chem B 109:23935
47. Hagen J, Socaciu LD, Heiz U, Bernhardt TM, Wöste L (2003) Eur Phys J D 24:327

48. Wallace WT, Whetten RL (2000) J Phys Chem B 104:10964; Nygren MA, Siegbahn PEM, Jin C, Guo T, Smalley RE (1991) J Chem Phys 95:6181
49. Wallace WT, Wyrwas RB, Leavitt AJ, Whetten RL (2005) Phys Chem Chem Phys 7:930
50. Popolan DM, Nössler M, Mitrić R, Bernhardt TM, Bonačić-Koutecký V (2011) J Phys Chem A 115:951
51. Popolan DM, Nößler M, Mitrić R, Bernhardt TM, Bonačić-Koutecký V (2010) Phys Chem Chem Phys 12:7865
52. Wu X, Senapati L, Nayak SK, Selloni A, Hajaligol M (2002) J Chem Phys 117:4010
53. Phala NS, Klatt G, van Steen E (2004) Chem Phys Lett 395:33
54. Fernández EM, Ordejón P, Balbás LC (2005) Chem Phys Lett 408:252
55. Yuan DW, Zeng Z (2004) J Chem Phys 120:6574
56. Schwerdtfeger P, Lein M, Krawczyk RP, Jacob CR (2008) J Chem Phys 128:124302
57. Häkkinen H, Landman U (2001) J Am Chem Soc 123:9704
58. Socaciu LD, Hagen J, Bernhardt TM, Wöste L, Heiz U, Häkkinen H, Landman U (2003) J Am Chem Soc 125:10437
59. Bernhardt TM, Socaciu-Siebert LD, Hagen J, Wöste L (2005) Appl Catal A: Gen 291:170
60. Bernhardt TM, Hagen J, Lang SM, Popolan DM, Socaciu-Siebert LD, Wöste L (2009) J Phys Chem A 113:2724
61. Lang SM, Frank A, Fleischer I, Bernhardt TM (2013) Eur Phys J D 67:19, 1
62. Joshi AM, Delgass WN, Thomson KT (2006) J Phys Chem B 110:23373; Torres MB, Fernández EM, Balbás LC (2008) J Phys Chem A 112:6678; Manzoor D, Krishnamurty S, Pal S (2014) J Phys Chem C 118:7501–7507
63. Wells BA, Chaffee AL (2008) J Chem Phys 129:164712/1
64. Peng S-L, Gan L-Y, Tian R-Y, Zhao Y-J (2011) Comp Theor Chem 977:62
65. Tian WQ, Ge M, Gu F, Yamada T, Aoki Y (2006) J Phys Chem A 110:6285
66. Mondal K, Banerjee A, Fortunelli A, Ghanty TK (2015) J Comput Chem 36:2177
67. Ho J, Ervin KM, Lineberger WC (1990) J Chem Phys 93:6987
68. Handschuh H, Ganteför G, Bechthold PS, Eberhardt W (1994) J Chem Phys 100:7093; Häkkinen H, Yoon B, Landman U, Li X, Zhai H-J, Wang L-S (2003) J Phys Chem A 107:6168
69. Negishi Y, Nakamura Y, Nakajima A, Kaya K (2001) J Chem Phys 115:3657
70. Handshuh H, Cha C-Y, Bechthold PS, Ganteför G, Eberhardt W (1995) J Chem Phys 102:6406
71. Lee HM, Ge M, Sahu BR, Tarakeshwar P, Kim KS (2003) J Phys Chem B 107:9994
72. Neumaier M, Weigend F, Hampe O, Kappes MM (2006) J Chem Phys 125:104308
73. Haeck JD, Veldeman N, Claes P, Janssens E, Andersson M, Lievens P (2011) J Phys Chem A 115:2103
74. Joshi AM, Tucker MH, Delgass WN, Thomson KT (2006) J Chem Phys 125:194797
75. Kuang X, Wang X, Liu G (2012) Struct Chem 23:671
76. Song C, Ge Q, Wang L (2005) J Phys Chem B 109:22341; Kuang X, Wang X, Liu G (2012) Eur Phys J App Phys 60:31301; Morrow BH, Resasco DE, Striolo A, Nardelli MB (2011) J Phys Chem C 115:5637
77. Zhao Y, Li Z, Yang J (2009) Phys Chem Chem Phys 11:2329
78. Fernandez EM, Torres MB, Balbas LC (2011) Int J Quantum Chem 111:510
79. Li X-N, Yuan Z, He S-G (2014) J Am Chem Soc 136:3617–3623
80. Lin L, Lievens P, Nguyen MT (2010) Chem Phys Lett 498:296
81. Lüttgens G, Pontius N, Bechthold PS, Neeb M, Eberhardt W (2002) Phys Rev Lett 88:076102
82. Wang L-M, Pal R, Huang W, Zeng XC, Wang L-S (2010) J Chem Phys 132:114306
83. Le HT, Lang SM, Haeck JD, Lievens P, Janssens E (2012) Phys Chem Chem Phys 14:9350
84. Mitrić R, Bürgel C, Burda J, Bonačić-Koutecký V, Fantucci P (2003) Eur Phys J D 24:41
85. Wang F, Zhang D, Ding Y (2010) J Phys Chem C 114:14076
86. Blanksby SJ, Ellison GB (2003) Acc Chem Res 36:255
87. Lang SM, Bernhardt TM, Barnett RN, Landman U (2010) Chem Phys Chem 11:1570

88. Lang SM, Bernhardt TM, Barnett RN, Landman U (2010) Angew Chem Int Ed 49:980
89. Lang SM, Bernhardt TM (2009) Eur Phys J D 52:139
90. Lang SM, Bernhardt TM (2011) Faraday Discuss 152:337
91. Lang SM, Frank A, Bernhardt TM (2013) Catal Sci Technol 3:2926
92. Lang SM, Frank A, Bernhardt TM (2013) Int J Mass Spectrom 354–355:365
93. Xia F, Cao Z (2006) J Phys Chem A 110:10078
94. Wetterer SM, Lavrich DJ, Cummings T, Bernasek SL, Scoles G (1998) J Phys Chem B 102:9266
95. Lang SM, Frank A, Bernhardt TM (2013) J Phys Chem C 117:9791
96. Weaver JF, Hakanoglu C, Hawkins JM, Asthagiri A (2010) J Chem Phys 132:024709; Kao C-L, Madix RJ (2002) J Phys Chem B 106:8248
97. Gail E, Gos S, Kulzer R, Lorösch J, Rubo A, Sauer M, Kellens R, Reddy J, Steier N, Hasenpusch W (2000) In Ullmann's encyclopedia of industrial chemistry. Wiley-VCH Verlag GmbH & Co. KGaA
98. Endter F, Svendsen M, Raum L (1958) Germany
99. Endter F (1958) Chemie-Ing-Techn 5:305
100. Diefenbach M, Brönstrup M, Aschi M, Schröder D, Schwarz H (1999) J Am Chem Soc 121:10614
101. Wesendrup R, Schröder D, Schwarz H (1994) Angew Chem Int Ed 33:1174
102. Aschi M, Brönstrup M, Diefenbach M, Harvey JN, Schröder D, Schwarz H (1998) Angew Chem Int Ed 37:829
103. Irikura KK, Beauchamp JL (1991) J Phys Chem 95:8344; Zhang X-G, Liyanage R, Armentrout PB (2001) J Am Chem Soc 123:5563
104. Kummerlöwe G, Balteanu J, Sun Z, Balaj OP, Bondybey VE, Beyer MK (2006) Int J Mass Spectrom 254:183
105. Lapoutre VJF, Redlich B, van der Meer AFG, Oomens J, Bakker JM, Sweeney A, Mookherjee A, Armentrout PB (2013) J Phys Chem A 117:4115
106. Achatz U, Berg C, Joos S, Fox BS, Beyer MK, Niedner-Schattenburg G, Bondybey VE (2000) Chem Phys Lett 320:53; Kaldor A, Cox DM (1990) Pure Appl Chem 62:79; Koszinowski K, Schröder D, Schwarz H (2003) J Phys Chem A 107:4999; Hanmura T, Ichihashi M, Kondow T (2002) J Phys Chem A 106:11465; Adlhart C, Uggerud E (2006) Chem Commun 2581
107. Koszinowski K, Schröder D, Schwarz H (2003) Organometallics 22:3806
108. Koszinowski K, Schröder D, Schwarz H (2004) Angew Chem Int Ed 43:121
109. Koszinowski K, Schröder D, Schwarz H (2004) Organometallics 23:1132
110. Koszinowski K, Schröder D, Schwarz H (2003) J Am Chem Soc 125:3676
111. Olvera-Neria O, Cruz A, Luna-García H, Anguiano-García A, Poulain E, Castillo S (2005) J Chem Phys 123:164302; Zhao S, Tian XZ, Liu JN, Ren YL, Wang JJ (2015) Comp Theo Chem 1055:1; Zhao S, Tian XZ, Liu JN, Ren YL, Ren YL, and Wang JJ (2015) J Clust Sci 491
112. Aktürk OÜ, Tomak M (2010) Thin Solid Films 518:5195; Pakiari AH, Jamshidi Z (2010) J Phys Chem A 114:9212
113. Zhao S, Tian XZ, Liu JN, Ren YL, Wang JJ (2014) J Mol Model 20:2467
114. Zhao SZ, Li GZ, Liu JN, Ren YL, Lu WW, Wang JJ (2014) Eur Phys J D 68:254
115. Chrétien S, Gordon MS, Metiu H (2004) J Chem Phys 121:9931
116. Fleischer I, Popolan DM, Krstić M, Bonačić-Koutecký V, Bernhardt TM (2013) Chem Phys Lett 565:74

# Index





M.T. Nguyen and B. Kiran (eds.), *Clusters*, Challenges and Advances in Computational Chemistry and Physics 23,
DOI 10.1007/978-3-319-48918-6

CPSIA information can be obtained
at www.ICGtesting.com
Printed in the USA
LVHW01*0331220118
563432LV00001B/103/P

9 783319 489162